Biographical History of Behavioral Neuroendocrinology

Randy J. Nelson • Zachary M. Weil
Editors

Biographical History of Behavioral Neuroendocrinology

Springer

Editors
Randy J. Nelson
Department of Neuroscience
Rockefeller Neuroscience Institute
West Virginia University
Morgantown, WV, USA

Zachary M. Weil
Department of Neuroscience
Rockefeller Neuroscience Institute
West Virginia University
Morgantown, WV, USA

ISBN 978-3-031-12969-8 ISBN 978-3-031-12970-4 (eBook)
https://doi.org/10.1007/978-3-031-12970-4

This Springer imprint is published by the registered company Springer Nature Switzerland AG
The registered company address is: Gewerbestrasse 11, 6330 Cham, Switzerland

Painting by Jay Rosenblatt (Chapter 13). Rosenblatt explored themes in his art that mirrored his interest in the behavioral neuroendocrinology of parenting. Image courtesy of Nina Rose

Dedicated to our families …

Foreword

Behavioral neuroendocrinology is the scientific study of the interaction between hormones and behavior. This interaction is bidirectional: hormones can affect behavior, and behavior can influence hormones. Hormones, chemical messengers released from endocrine glands, travel through the blood system to influence the nervous system to regulate the physiology and behavior of an individual. Hormones change gene expression or the rate of cellular function, and they affect behavior generally by increasing the probability that a given behavior will occur in the presence of a specific stimulus. Hormones achieve this by affecting individuals' sensory systems, integrators, and/or effectors (output systems). Because certain chemicals in the environment can mimic natural hormones, these chemicals can profoundly affect the behavior of humans and other animals. Behavior is generally thought of as involving movement, but nearly any type of output, such as color change, can be considered behavior; for example, color change among chameleons is a behavioral response. A complete description of behavior is required before researchers can address questions of its causation. All behavioral biologists study a specific version of the general question "What causes individual A to emit behavior X?" Behavioral endocrinologists are interested in the interactions between hormones and behaviors.

The study of the interaction between hormones and behavior has been remarkably interdisciplinary since its inception; methods and techniques from other scientific disciplines have been borrowed and refined to shed light on this relationship. Psychologists, endocrinologists, neuroscientists, entomologists, zoologists, geneticists, molecular and cellular biologists, anatomists, physiologists, behavioral ecologists, psychiatrists, and other behavioral biologists have all made contributions to the understanding of hormone-behavior interactions. This exciting commingling of scientific interests and approaches, with its ongoing synthesis of knowledge, has led to the emergence of behavioral neuroendocrinology as a distinct and important field of study. The scientific journal *Hormones and Behavior* began publication in April 1967, and a scientific organization devoted to the study of hormones and behavior, the Society for Behavioral Neuroendocrinology (SBN), was founded in 1996. Both the journal and scientific society are growing with membership in SBN now over 500 members. The number of Behavioral Neuroendocrinology undergraduate courses has grown from about 20 in North America in 1995 to hundreds in 2022.

Hermann Ebbinghaus stated that psychology has a short history but a long past, and the same can be said of behavioral neuroendocrinology. Although the modern era of the discipline is generally recognized to have emerged during the middle of the twentieth century with the publication of the classic book *Hormones and Behavior* by Frank A. Beach in 1948, some of the relationships among the endocrine glands, their hormone products, and behavior have been implicitly recognized for centuries. The goal of this book is to track the development of the field from the first recognized paper in the field by Arnold Berthold in 1849 (although it was mostly ignored for the ensuing 50 years) to the major contributors of the past century.

A useful starting point for understanding research in hormones and behavior is a classic nineteenth-century experiment that is now considered to be the first formal study of endocrinology (Chap. 1). This remarkable experiment conclusively demonstrated that a substance produced by the testes could travel through the bloodstream and eventually affect behavior. Professor Arnold Adolph Berthold, a Swiss-German physician and professor of physiology at the University of Göttingen, demonstrated experimentally that a product of the testes was necessary for a cockerel (an immature male chicken) to develop into a normal adult rooster and display typical rooster behaviors such as crowing and fighting.

One way in which to explore the history and development of this field is by exploring the women and men who conducted the studies that revealed these hormone-behavioral relationships. To that end we will enlist the help of the individuals who knew these pioneers best to describe their backgrounds and discuss the way in which their work has shaped the field.

Now is the perfect time for this book. The field is burgeoning and interest in the development of theoretical perspectives is thriving. Moreover, although this field was dominated by men early on, it has become a field with near sexual parity among its faculty, society membership and leadership, and thus serves as an example of equitable science, training, and advocacy. As noted in our final chapter, the same is not true for individuals from other underrepresented groups. We hope this recognition of so-called hidden gems of scientists promotes more advocacy for equitable representation of scientists within behavioral neuroendocrinology.

Individuals were selected for inclusion based on several criteria. First, we made a list of individuals who have made significant contributions. Second, we limited inclusion to only individuals who had retired or passed away. Third, we circulated the list among the contributors and invited them to add additional names that we may have overlooked. Obviously, this involves a series of judgment calls that are informed by our own biases and experience, and we apologize in advance for any omissions. We hope you enjoy reading this volume as much as we did compiling it.

Morgantown, WV, USA — Randy J. Nelson
15 March 2022 — Zachary M. Weil

Contents

Contributors

Elizabeth Adkins-Regan Department of Psychology Department of Neurobiology and Behavior Cornell University, Ithaca, NY, USA

H. Elliott Albers Neuroscience Institute, Center for Behavioral Neuroscience, Georgia State University, Atlanta, GA, USA

Julie Bakker GIGA Neurosciences, University of Liège, Liège, Belgium

Gregory F. Ball Department of Psychology, University of Maryland, College Park, MD, USA

Jacques Balthazart University of Liège, GIGA Neurosciences, Research Group in Behavioral Neuroendocrinology, Liège, Belgium

Nicole M. Baran Department of Psychology, Emory University, Atlanta, GA, USA

Michael J. Baum Department of Biology, Boston University, Boston, MA, USA

Seema Bhatnagar Department of Anesthesiology and Critical Care, Children's Hospital of Philadelphia, The Perelman School of Medicine at the University of Pennsylvania, Philadelphia, PA, USA

Jacob R. Bumgarner Department of Neuroscience and Rockefeller Neuroscience Institute, West Virginia University, Morgantown, WV, USA

Charlotte A. Cornil University of Liege, GIGA Neurosciences, Liège, Belgium

Jennifer A. Cummings Department of Psychology, University of Michigan, Ann Arbor, MI, USA

A. Courtney DeVries Department of Neuroscience and Medicine, West Virginia University, Morgantown, WV, USA

Juan M. Dominguez Department of Psychology, Institute for Neuroscience, Pharmacology & Toxicology, The University of Texas at Austin, Austin, TX, USA

Christine M. Drea Department of Evolutionary Anthropology, Duke University, Durham, NC, USA

Alison S. Fleming Department of Psychology, University of Toronto at Mississauga, Mississauga, ON, Canada

Andrea Gonzalez Department of Psychiatry & Behavioural Neurosciences, Offord Centre for Child Studies, McMaster University, Hamilton, ON, Canada

Andrea C. Gore Division of Pharmacology & Toxicology, College of Pharmacy, The University of Texas at Austin, Austin, TX, USA

Wolfgang Goymann University of Konstanz, Department of Biology, Konstanz, Germany

Michaela Hau Max Planck Institute for Ornithology, Eberhard-Gwinner-Straße, Seewiesen, Germany and University of Konstanz, Department of Biology, Konstanz, Germany

Barbara Helm Department for Bird Migration, Swiss Ornithological Institute, Sempach, Switzerland

Casey L. Henley Neuroscience Program, Physiology Department, Michigan State University, East Lansing, MI, USA

Gretchen L. Hermes Department of Psychiatry, Yale University, New Haven, CT, USA

Ilia N. Karatsoreos Neuroscience and Behavior Program, Center for Neuroendocrinology, Department of Psychological and Brain Sciences University of Massachusetts, Amherst, MA, USA

Lance J. Kriegsfeld Departments of Psychology and Integrative Biology, Graduate Group in Endocrinology, and The Helen Wills Neuroscience Institute, University of California, Berkeley, CA, USA

Michael N. Lehman Brain Health Research Institute, Kent State University, Kent, OH, USA

Jennifer A. Liu Department of Neuroscience and Rockefeller Neuroscience Institute, West Virginia University, Morgantown, WV, USA

Joseph S. Lonstein Department of Psychology & Neuroscience Program Interdisciplinary Science & Technology Building (ISTB), Michigan State University, East Lansing, MI, USA

Carmel Martin-Fairey Department of Biology, Harris-Stowe State University, St. Louis, MO, USA

Robert L. Meisel Department of Neuroscience, University of Minnesota, Minneapolis, MN, USA

Olga Hecmarie Melendez-Fernandez Department of Neuroscience and Rockefeller Neuroscience Institute, West Virginia University, Morgantown, WV, USA

Paul E. Micevych Department of Neurobiology, David Geffen School of Medicine at University of California, Los Angeles, Los Angeles, CA, USA

Isaac Miller-Crews Department of Integrative Biology, The University of Texas at Austin, Austin, TX, USA

Randy J. Nelson Department of Neuroscience, Rockefeller Neuroscience Institute, West Virginia University, Morgantown, WV, USA

Robin B. Oliverio Department of Neuroscience and Rockefeller Neuroscience Institute, West Virginia University, Morgantown, WV, USA

Brian J. Prendergast Department of Psychology, University of Chicago, Chicago, IL, USA

Marilyn Ramenofsky Department of Neurobiology, Physiology and Behavior, University of California, Davis, CA, USA

Luke Remage-Healey Center for Neuroendocrine Studies, Neuroscience and Behavior Program, University of Massachusetts, Amherst, MA, USA

Heather N. Richardson Psychological and Brain Sciences Department, University of Massachusetts Amherst, MA, USA

Russell D. Romeo Departments of Psychology and Neuroscience and Behavior, Barnard College of Columbia University, New York, NY, USA

L. Michael Romero Department of Biology, Tufts University, Medford, MA, USA

Benjamin D. Sachs Department of Psychological Sciences, University of Connecticut, Storrs, CT, USA

Colin J. Saldanha Departments of Neuroscience and Center for Behavioral Neuroscience, American University, Washington, DC, USA

Barney A. Schlinger Department of Integrative Biology and Physiology and Department of Ecology and Evolutionary Biology, University of California, Los Angeles, Los Angeles, CA, USA

Harold I. Siegel Department of Psychology, Rutgers University, Newark, NJ, USA

Rae Silver Department of Psychology, Columbia University, New York, NY, USA

Department of Neuroscience, Barnard College, New York, NY, USA

Jennifer M. Swann Department of Biological Sciences, Lehigh University, Bethlehem, PA, USA

Zuleyma Tang-Martínez Department of Biology, University of Missouri – St. Louis, St. Louis, MO, USA

Richmond Thompson Departments of Neuroscience and Behavioral Biology, Psychology, Oxford College of Emory University, Oxford, GA, USA

Michelle Tomaszycki Program in Neuroscience, University of Illinois, Urbana-Champaign, Urbana, IL, USA

Claire-Dominique Walker Department of Psychiatry, McGill University, Douglas Institute, Montreal, QC, Canada

Kim Wallen Department of Psychology, Emory University, Atlanta, GA, USA

Zachary M. Weil Department of Neuroscience, Rockefeller Neuroscience Institute, West Virginia University, Morgantown, WV, USA

Brishti A. White Department of Neuroscience and Rockefeller Neuroscience Institute, West Virginia University, Morgantown, WV, USA

John C. Wingfield Department of Neurobiology, Physiology and Behavior, University of California, Davis, CA, USA

Irving Zucker Departments of Psychology and Integrative Biology, University of California, Berkeley, CA, USA

Arnold Adolph Berthold

1

Zachary M. Weil and Randy J. Nelson

Abstract

Arnold Adolph Berthold was a German physician-scientist and textbook author who is most recognized as the author of the first published experiment in endocrinology. This study reports several morphological and behavioral outcomes of an endocrine manipulation that involved castration and reimplantation of rooster testes. His insightful experiment conclusively demonstrated that a substance produced by the testes could travel through the bloodstream and affect morphology and behavior and set the stage for understanding how blood-borne products could affect brain and behavior.

Keywords

Endocrinology · Rooster · Testes · Castration · Male behavior

One goal of this volume is to track the development of the field of Behavioral Neuroendocrinology by highlighting major contributors to the field. We start here with Arnold Adolph Berthold (Fig. 1.1), the recognized author of the first published experiment in endocrinology that reports several behavioral outcomes of an endocrine manipulation (Berthold, 1849, Translation by Quiring, 1944). His remarkable experiment conclusively demonstrated that a substance produced by the testes could travel through the bloodstream and eventually affect behavior.

Berthold was born in 1803 in the small Westphalian town of Soest near Münster. His early life was roiled by the political instability surrounding the Napoleonic

Z. M. Weil (✉) · R. J. Nelson (✉)
Department of Neuroscience, Rockefeller Neuroscience Institute, West Virginia University, Morgantown, WV, USA
e-mail: Zachary.weil@hsc.wvu.edu; Randy.Nelson@hsc.wvu.edu

R. J. Nelson, Z. M. Weil (eds.), *Biographical History of Behavioral Neuroendocrinology*, https://doi.org/10.1007/978-3-031-12970-4_1

Fig. 1.1 Arnold Adolf Berthold. (Courtesy of the National Library of Medicine)

wars, and his birthplace was variously part of an independent kingdom, French client state, and Prussian province. His father was a master carpenter, and his childhood was spare (he later regretted the lack of Christmas presents), but relatively comfortable (Rush, 1929). According to a biographer, one fond memory that he would later recount was the receipt of an ABC book with a rooster and eggs on the cover containing the words "early the rooster crows and early starts to learn" (Rush, 1929). Whether this experience may have influenced his later work is unknown but intriguing to speculate! At age 16, he followed his older brother to medical school at The University of Göttingen, then part of the Kingdom of Hanover. Following Napoleon's abdication in 1814, Hanover was ruled in "personal union" by the British Kings George III and later George the IV. This practice ended, however, when Queen Victoria ascended the throne as Hanoverian law prevented a female from becoming monarch while a male-line successor lived.

In Göttingen, Berthold received his medical degree and spent the first years of his postgraduate training visiting with the outstanding medical and natural scientific men of his era in other German cities and in Paris. He briefly practiced medicine but apparently possessed a restless mind that brought him back to science, and eventually he became a professor and the director of the zoological division of the museum at Göttingen (Loriaux, 2016). Although he is most famous, at least in endocrinological circles, for his testicular transplantation studies, he also published on a wide variety of other topics including myopia (of which he suffered), hair formation, the structure of the head bones in gnawing animals, the actions of mercury on the salivary glands, and the anatomy of the thyroid gland of the parrot; he also published a paper with Robert Bunsen (inventor of the eponymous laboratory tool) describing the use of hydrated iron oxides for the treatment of arsenic poisoning. Berthold also mentored, among others, Carl Bergmann who studied body heat regulation and would go on to coin the terms "poikilotherm" and "homeotherm" (Loriaux, 2016; Medvei, 2012).

But most importantly for the history of behavioral neuroendocrinology, Professor Arnold Adolph Berthold demonstrated experimentally that a then unknown secretory product of the testes was required for the development of a cockerel into a rooster, both morphologically and behaviorally. In sharp contrast to hens or chicks, roosters often behave aggressively, are physically larger than hens, and have characteristic plumage; roosters also direct sexual behavior towards hens and crow. Capons, male chicks that have been castrated early in life, do not exhibit these rooster-typical behaviors or morphological characteristics. The behavioral and physical differences, among roosters, hens, capons, and immature chickens were likely familiar to Berthold (Berthold, 1849) (Fig. 1.1).

Although the precise motivation for Berthold's experiment is not known (i.e., it is not clear that he performed this experiment for the same reason that modern behavioral endocrinologists would have; see below), he employed an approach that would become central to the field of behavioral endocrinology: that is, removal and replacement of the source of a hormone comprised three groups. The first group was caponized (i.e., the testes were removed early in life) and as expected these birds acquired the morphological and behavioral characteristics of capons. The birds in the second group were also castrated, but Berthold reimplanted one testis from each bird back into its abdominal cavity (chicken testes are located in the abdomen) cutting all the vascular and neural connections. Birds in the third group were also castrated, but after the testes were removed, Berthold placed a single testis from each bird into other birds' abdominal cavities. The birds in both groups in which a testis was reimplanted (either an autograft or allograft) developed normally as roosters. Critically, when the birds were dissected, Berthold discovered that the reimplanted testes had developed vascular connections to the viscera, had nearly doubled in size (an early example of compensatory hypertrophy), and contained mature motile sperm.

Based on these results, Berthold drew three major conclusions from this work. First, the testes could be successfully transplanted and reestablish a vascular supply. Second, the implanted testes could produce sperm, and if properly connected to sexual organs, Berthold argued that they could still reproduce. Third, that because vascular, but not neural connections, had been reestablished, that neural inputs were not required for the normal function of the testis. To account for these findings, Berthold proposed that a "secretory blood-borne product" of *the transplanted testes (productive Verhältniss der Hoden)* was responsible for the typical development of the birds in the second and third groups (Forbes, 1949).

As noted, Berthold's experiment has been credited as the genesis of the field of endocrinology (and of behavioral neuroendocrinology). However, Berthold's intriguing demonstration of nonneural control of behavior was apparently not embraced with any enthusiasm by his scientific contemporaries; we find no citations to his paper for nearly 60 years after its publication. In addition to his research activities, Berthold authored a well-known physiology textbook, *Lehrbuch der Physiologie des Menschen und der Thiere* (*Textbook on the Physiology of Humans and Animals*) (Berthold, 1849). His textbook makes it apparent that Berthold was a proponent of the pangenesis theory of inheritance. This theory, endorsed by many biologists prior to the discovery of how chromosomes and genes function, held that all body parts actively discharge bits and pieces of themselves into the blood system, where they are transported to the ovaries or testes and assembled into miniature offspring resembling the parents. Moreover, it was thought that resorption of sperm from the testes was necessary for maintaining the secondary sexual characteristics of males. Because of this theoretical stance, Berthold had two concepts at hand when evaluating the results of his testicular transplantation study: (1) various parts of the body release specific agents into the blood, and (2) these agents travel through the blood to specific target organs.

Although his work, in retrospect, served as the first study to use several modern behavioral endocrinology techniques, including extirpation and replacement and monitoring of both behavioral and anatomical endpoints, at the time, the understanding of the role of the gonads in sexual development and behavior was understood quite differently than in the modern conception. Unfortunately, there is no introduction to his study, which reads much like a slightly elongated abstract, and it took many years and a great deal of inference from Berthold's other published work, to determine the motivation for and underlying premise of the experiment (Forbes, 1949; Quiring, 1944).

The role of castration in the development of animals had been known since antiquity, but the mechanisms for the behavioral and morphological consequences were unknown. In the eighteenth century, however, John Hunter (1718–1783), a Scottish physician, surgeon, medical researcher, and physiologist had performed critical experiments that may have influenced Berthold. Although the results were never published (at least by Hunter), he had transplanted spurs (claws) from hens onto a cock's comb and reported that the comb grew to the larger size typical of males. Spurs transplanted into a hen did not lead to comb hypertrophy. Further, in a study rather entertainingly published in his book *The Natural History of Human Teeth* (he had also implanted a recently removed human tooth into a cock's comb, preserved samples of which are still housed at the Hunterian Museum in London) (Hunter, 1771), Hunter successfully transplanted a testicle into the abdomen of a hen. Hunter was more interested in what was called at the time the "vital principle" which he held to be responsible for the survival of the transplanted tissue than in any conception of an endocrine mode of action (Jørgensen, 1971). Nevertheless, these experiments which Berthold mentioned in 1849 suggest that he was interested, at least in part, in the physiological changes that would occur in the testicle itself.

That the testicular transplant work was not followed up either by Berthold himself (he died 12 years later in 1861) or the broader scientific community is perhaps a bit puzzling. One potential answer, however, is that Rudolph Wagner, an anatomist, and physiologist, who would head the medical faculty at Göttingen while Berthold was on the faculty there, did attempt to replicate the experiments and was unable to do so as many testicular transplants failed to engraft (Loriaux, 2016; Forbes, 1949; Benedum, 1999). These difficulties, moreover, were not unique to the avian model system as similar experimental failures were reported in both amphibians and mammals (Medvei, 2012). It is unknown though tempting to speculate that his department heads' inability to replicate his work made continuing this line untenable and perhaps even led Berthold himself, to discount the work. Given the broad range of topics he addressed, it is also possible that Berthold simply lost interest and moved on to another problem or perhaps his failing health in the early 1850s curtailed his research (Benedum, 1999). What is known, however, is that interest in Berthold's work was not revived until a half century later following Charles Edouard Brown-Sequard's ideas surrounding testicular rejuvenation. Scientific discovery always builds on the work of earlier thinkers. Berthold's ideas were neither completely original nor fully modern but instead serve as a critical early step to the modern conception of endocrine function.

References

Benedum, J. (1999). The early history of endocrine cell transplantation. *Journal of Molecular Medicine, 77*(1), 30–35.

Berthold, A. A. (1849). Transplantation der Hoden. *Archiv für Anatomie, Physiologie und Wissenschaftliche Medicin*, 42–46.

Forbes, T. R. (1949). A.A. Berthold and the first endocrine experiment: Some speculation as to its origin. *Bulletin of the History of Medicine, 23*(3), 263–267.

Hunter, J. 1771. The natural history or the human teeth: Explaining their structure, use, formation, growth, and diseases,. J. Johnson.

Jørgensen, C. B. (1971). *John Hunter, A.A Berthold, and the origins of endocrinology*. Odense University Press.

Loriaux, L. (2016). *A biographical history of endocrinology*. Wiley.

Medvei, V. C. (2012). *A history of endocrinology*. Springer.

Quiring, D. P. (1944). The transplantation of the testes. *Bulletin of the History of Medicine, 16*, 399–401.

Rush, H. P. (1929). A biographic sketch of Arnold Adolf Berthold, an early experimenter with ductless glands. *Annals of Medical History, 1*(2), 208–214.

Frank Lillie

2

Jacob R. Bumgarner

Abstract

Frank Rattray Lillie was an embryologist and endocrinologist of the nineteenth and twentieth centuries. His meticulous experimentation and carefully constructed scientific ideologies led him to become one of the first scientists to identify the mechanisms of sexual differentiation during development. The research he conducted throughout his career at the University of Chicago and the Marine Biological Laboratory covered an array of topics and questions, leaving a memorable and impactful legacy of scientific contributions. Lillie's early career focused on marine embryology, where he provided extensive descriptions of the characteristics and mechanisms of embryonic development in numerous species. Most importantly, Lillie also was the first scientist to exhaustively characterize and provide mechanistic insight into bovine freemartinism. More than just a scientist, Lillie was also a successful leader, administrator, and educator. In his positions at the University of Chicago and the Marine Biological Laboratory, he advanced education and research initiatives in addition to promoting democratic scientific environments. His early contributions to the field of sexual differentiation and his development of sex hormone research initiatives permanently shaped the burgeoning field of neuroendocrinology.

Keywords

Frank Rattray Lillie · Lillie · Freemartin · Sexual differentiation · Sex hormones · Marine biological laboratory

J. R. Bumgarner (✉)
Department of Neuroscience, Rockefeller Neuroscience Institute, West Virginia University, Morgantown, WV, USA
e-mail: jrbumgarner@mix.wvu.edu

R. J. Nelson, Z. M. Weil (eds.), *Biographical History of Behavioral Neuroendocrinology*, https://doi.org/10.1007/978-3-031-12970-4_2

Childhood and Family History

Frank Rattray Lillie was born in Toronto, Ontario, on 27 June 1870, to mother Emily Ann Rattray and father George Waddell Lillie. From Lillie's own account, his mother was a stay-at-home mother who dedicated her time to her family, church, and friends. His father was an accountant and a pharmacist. Although Lillie's parents did not seem to express interest in academics or intellectual pursuits, there was a clear pedigree of academia, science, and theology in Lillie's family history (Willier, 1957).

Lillie's paternal grandfather, Adam Lillie, studied at the University of Glasgow, had a passion for Greek literature and at one point studied Sanskrit at the University of Toronto. Adam's main passion in life was theological research and Christian education under the Congregational church. Adam served as a missionary in Toronto and founded the "Congregational Academy" at the University of Toronto in 1840, where over the next 24 years, he and his tutor would teach 64 students. In honor of his contributions to the Congregational church, the University of Vermont granted him an honorary Doctor of Divinity in 1854. In 1864, the college of the Congregational Academy was moved to the McGill College; Adam Lillie remained principal of the college until his death in 1869 (Eddy, 1976).

Lillie's maternal grandfather, Thomas Rattray, was the son of the famous Scottish pastor and astronomer, Thomas Dick. Thomas Rattray would devote his life to theology, serving as a minister first in Massachusetts and later in Ontario. As with his uncle, Thomas Lillie also had an interest in the natural world and similarly practiced amateur astronomy (Willier, 1957).

In common with many other prominent scientists, Lillie was interested in the natural world from a young age. In a written account, Lillie described his interest in "object lessons" while attending the Provincial School of Education in Toronto as a child. In one account, he remembers being at the age of 10 and learning in a school lesson that water would expand upon freezing (Willier, 1957). Later, on a cold winter night, he would test this phenomenon by filling a bottle with water, corking it, and leaving it outside while he slept. Lillie woke up in the morning to find with delight that the expanded ice had caused the bottle to shatter, confirming the lesson he had been taught in school. This early example of empirical testing and observation would foreshadow the type of scientist that Lillie would eventually become.

Education and Family Life

Lillie's grandparents' influence over his life extended beyond the development of his interest the natural world and academics. This was clear as he began his undergraduate education at the University of Toronto in 1887. Because both of his grandfathers were devoted to ministry and theology, there was some expectation that Lillie would study theology. During his undergraduate studies, Lillie hoped to maintain a personal separation between his scientific and religious views. He recalls discussions with a close friend on the cognitive dissonance that the subject of evolution caused in relation to his religious views learned as a child. Indeed, early after graduation, Lillie would spend some time as a minister in a Presbyterian church. However, he would later abandon the ministry and religion, writing that "science won out" (Willier, 1957).

While at the University of Toronto, Lillie chose to pursue a degree in Natural Science. His early studies included chemistry, geology, and biology. During his senior year of undergraduate studies under the guidance of Professor Ramsay Wright and Dr. Archibald Macallum, Lillie's career-long interest in embryology would begin. The summer after graduating from the University of Toronto in 1891, Lillie moved to the Marine Biological Laboratory (MBL) in Woods Hole, Massachusetts, to study embryology with Charles Otis Whitman (Moors, 1948).

During his first year at the MBL, Lillie studied alongside several prominent scientists of the time. Under the guidance of Whitman, Lillie would earn his doctorate studying cell lineages and the fate of blastomeres in organ development in the mussel *Unio* (Moors, 1948). During his second year of doctoral studies, Whitman took a chair appointment at the newly founded University of Chicago. Lillie would follow Whitman and earn his doctoral degree in Zoology at the University of Chicago in 1894. In 1893, Lillie helped Whitman to develop the MBL's first embryology course. One year later, Whitman would become the director of the course. In 1895, Lillie married Frances Crane; together they would have six children and adopted an additional three children (Wellner, 2009). His wife's brother, Charles R. Crane, would eventually play a prominent role in the financial backing of the development and expansion of the MBL (Lillie, 1944).

A Distinguished Career

Lillie's academic and administrative career alone could warrant an entire chapter to describe. After his doctoral graduation in 1894, Lillie moved to serve as an instructor at the University of Michigan and then in 1899 briefly taught at Vassar College in New York. He returned as an assistant professor to the University of Chicago in 1900, was promoted to associate professor in 1902, and earned tenure in 1906. After Charles Whitman's death in 1910, Lillie would become the Chair of the Department of Zoology at the University of Chicago. He was later named the Andrew MacLeish distinguished Service Professor of Embryology in 1931. He then served as dean of the division of biological sciences from 1931 until his retirement in 1935.

In addition to his career at the University of Chicago, Lillie held a concurrent career at the MBL. He started as an assistant director of the MBL from 1900 to 1908. He was then promoted as the director in 1908, where he remained until 1926. He was the president of the corporation and board of trustees from 1925 to 1942. Following this, he served as *emeritus* president until his death in 1947. Lillie's contributions of the MBL led it to become a democratic and internationally renowned research institution that remains active to this day.

Moreover, from 1935 to 1939, Lillie was the president of the National Academy of Sciences. From 1935 to 1936, he was the chairman of the National Research Council. He was the president of the Oceanography Institute next to the MBL from 1930 to 1939 and served as the managing editor of the Biological Bulletin from 1902 to 1926. Finally, Lillie's contributions earned him honorary degrees from the University of Toronto, Harvard, Yale, and Johns Hopkins University (Willier, 1957). Lillie was clearly an opportune man capable of impressive leadership.

Research Overview

Lillie could be primarily described as an embryologist, as he focused most of his research on the intricacies of embryonic development. Yet much as his administrative career, his embryological research was not confined to a single context. During his tenure at the University of Chicago and the MBL, Lillie studied the development of several organisms. Throughout the 45+ years of his research career, Lillie's expertise as a scientist would impact numerous subdomains of embryology, including fertilization theory, embryonic spatial and cellular organization, and sexual differentiation. Much of his recognized research can be distilled into four main categories: marine embryology, chick development, freemartinism, and sexual differentiation.

Marine Embryology

Inspired by the work that he conducted as a graduate student in Whitman's lab, Lillie's early work focused on the embryonic development of the mussel *Unio*. In several works published between 1893 and 1909, Lillie detailed the development of *Unio* and other species. His work on *Unio* exhaustively described egg polarity, centromere organization, fertilization localization, chromosomal organization, cellular cleavage planes, and cell lineage and fate (Lillie, 1901).

During this early career stage, Lillie also described the development of other species, including *Chaetopterus pergamentaceus*, a segmented worm. In several papers, he describes that the polarization of the egg of this species is dependent on the "ground substance" or the cellular membrane structure of the egg (Lillie, 1909a, b). Another study of this species detailed the ability for varying potassium chloride concentrations to induce cellular differentiation and organ development even in the absence of fertilization (Lillie, 1902). With this paper, he began laying the framework for a theory that he would later develop on the nature of fertilization.

As with his other lines of research, Lillie's scientific creed led him to constantly develop new hypotheses and frameworks to explain embryonic development and fertilization. For example, he was able to determine that egg polarization, nuclear division, cellular cleavage and differentiation, and cellular cleavage are all independent processes, but that typical development is dependent on the correlative timing of these processes (Lillie, 1902). He presented a hypothesis that there was an adaptive function to the organization of embryonic cleavage. In this hypothesis he argued that different features of cleavage produced adaptive structures that were based on the future needs of larvae in their environments (Lillie, 1899).

Finally, one of Lillie's other notable contributions to the early field of embryology was his "Fertilizin Theory," which he formulated in a series of work from 1910 to 1921. In this publication series, Lillie characterized fertilization in numerous species, including the sea urchin *Arbacia*. His developed "Fertilizin Theory" related to his discovery of what he dubbed *fertilizin* and *anti-fertilizin*, the extracellular material that covered eggs and sperm, respectively (Lillie, 1913). This theory encompassed the species-specificity, sperm and egg linkage, and egg activation aspects of fertilization (Lillie, 1914).

The extensive publication record that Lillie produced in this field is a legacy that any scientist would be proud to have. Yet, his focus on marine embryology is only a piece of his entire publication legacy.

Chick Embryology and Sexual Differentiation

In addition to the marine embryological research that Lillie conducted mainly at the MBL, Lillie became interested in sexual differentiation. As a true testament to his logistical and scientific fortitude, the research he conducted on sexual differentiation at the University of Chicago would occur simultaneously to his marine research at the MBL.

Lillie's interest in the mechanisms of sexual differentiation began around the time when much of his work at the University of Chicago was focused on teaching and chick development. After starting his faculty position at the University of Chicago in 1900, Lillie taught courses on embryology for undergraduate, medical, and beginning graduate students. In these courses, he used chicks as a model system to describe development. He felt that studying a single organism's development in detail served as an ideal introduction to the field of embryology and that chicks were the ideal organism to achieve this goal (Wellner, 2009). As a result of his devoted time to the undergraduate courses, Lillie published his introductory textbook to embryology: *The Development of the Chick*, "meant for the use of beginners in embryology" (Lillie, 1908). Additional revisions of this textbook would follow in the decades after its original publication, and it would be widely used as one of the best resources on chick development (Watterson, 1979).

Lillie did not publish extensively on chick development, but his work on this topic furthered his understanding on the fine-tuned sensitivity of the development process. In his experiments, he observed that ablation or perturbation of certain embryonic structures would eliminate the presence of these structures in adulthood or even completely impair overall development (Lillie, 1903). He also described the importance of some embryonic organs for the development of others as a "correlative differentiation" process. It is possible that this understanding of the linear and irreversible aspects of embryonic development led to his early conceptualizations of sex development and sex differentiation. In his first publication on the topic of sexual differentiation, Lillie seemed to have been aware that sexual differentiation was tied to early development, writing that "sexual differentiation is a phenomenon of irritability or response to stimulus, which lasts throughout the life history of the growing organism" (Lillie, 1907). Lillie seemed determined to identify the cause of sexual differentiation.

Freemartinism

One of Lillie's most famous contributions to the emergent field of endocrinology may have happened because of familial and geographical serendipity. Lillie was one of the first scientists to characterize embryonic sexual differentiation. Lillie's family owned a purebred cattle farm in Wheeling, Illinois, and it was here that his fascination with freemartins would begin. This fascination was accelerated when in 1914, his farm manager sent him a pair of calf fetuses twins still encased in their membranes (Watterson, 1979).

Freemartins are female cattle that are born with a male twin. Although these female twins appear to be typical females externally (typical external genitalia and mammary gland development), they are almost always infertile. They often have gonads that more closely resemble testes than ovaries, and the internal genitalia tracts are typically altered. Prior to Lillie's research on the topic, it was understood that freemartins could only be born as twins to a male. Despite freemartins being well-known throughout history, the precise mechanisms responsible for producing this type of animal were unknown.

After receiving the first pair of twins, Lillie began an extensive characterization of freemartins, thanks to the proximity of a cattle slaughterhouse to the University of Chicago. Lillie was able to convince the foreman of this slaughterhouse to notify him when cow uteri were found containing twins. Then, thanks to his relationship with his family farm, Lillie was able to quickly collect these uteri and perform careful dissections. One of his students recounts seeing him "garbed immaculately in a white gown and wearing rubber gloves, examining and dissecting pregnant uteri containing young twins…" (Willier et al., 1948).

Over the next 2–3 years, Lillie's careful dissections and diligent observations would bear two seminal publications. First in 1916, he published "The Theory of the Free-Martin," in *Science* (Lillie, 1916). In the publication, he corrects the previous conception that freemartins were monozygotic male twins that failed to develop external genitalia. Instead, his observation of two corpora lutea indicated that the freemartins were in fact a dizygotic twin. He then used the expected sex ratio of the birthed twins to confirm the observation that the freemartin twins were in fact female. Lastly, he describes that the twins have connected arterial and venous vascular systems. From the observation that the females only become freemartins in the presence of a male twin, he determines that the effect of the male on the female was "unquestionably to be interpreted as a case of hormone action" (Lillie, 1916).

In 1917, Lillie published a follow-up definitive and exhaustive description of freemartin calves (Lillie, 1917). In this publication, he describes freemartins as a form of a natural experiment that allowed for greater insight into the problem of sex-differentiation and sex-determination. His results lead him to conclude that sexual differentiation in mammals is largely determined by humoral sex hormones present in the embryo. Moreover, he also characterizes the effects of the putative sex hormones on duct differentiation. In typical undifferentiated fetuses, there are two sets of reproductive tracts present: the Wolffian and Müllerian ducts. In females, the absence of fetal androgens causes the Wolffian ducts to regress, and the Müllerian ducts will develop into the fallopian tubes, the uterus, and the upper vagina. In males, the presence of circulating androgens leads the Müllerian ducts to regress and cause the Wolffian ducts to form the seminal vesicles and vas deferens.

Lillie concluded that the presence of the putative sex hormone (at the time, no sex hormones had been isolated) that was being secreted by the male twin was leading to atypical development and prevented regression of the Wolffian ducts in the females, leading to a dysfunctional reproductive system and ultimately freemartinism. Lillie observed that the freemartin phenotype was greater when earlier vascular connections were made between the fetuses and that male development was almost always typical (Lillie, 1917).

Lillie's careful and tireless descriptions of freemartins would permanently sway the field of embryology. From his studies, he was able confirm his belief that sexual differentiation occurred during early development. Moreover, his observations on freemartins hold steadfast even by today's scientific standards.

Brown Leghorn Sexual Differentiation

After his discoveries on freemartins, Lillie then moved to organize a research program that focused on the isolation of sex hormones to examine their roles in sexual differentiation (Watterson, 1979). Lillie's own contribution to this program focused on the role of female sex hormones and thyroxin in feather development in the Brown leghorn fowl species. In one publication with Mary Juhn in 1932, Lillie and Juhn described the role that injected estrogen had on female feather pattern development and plumage (Lillie & Juhn, 1932). In this same publication, the role of thyroxin on male plumage was also extensively detailed. Lillie's contributions to the field of endocrinology and sexual differentiation would continue until his retirement.

Involvement in the Eugenic Society

As an unfortunate mark on Lillie's history, it is apparent that he subscribed to eugenic ideologies that were popular at the time. He held membership in the Chicago Eugenics Education Society, was a committee member of the Second International Eugenics Congress, and was on the advisory council of the Eugenics Committee of the United States (Wellner, 2009). According to Wellner (Wellner, 2009), "In the early 1920s Lillie envisioned an Institute of Genetic Biology that would gather data to examine population problems, public health, and social control, but this never came to fruition."

In contrast, it is also reported that Lillie emphasized the recruitment of African American students to the Zoology program at the University of Chicago (Allen, 2008). This effort included the recruitment of Ernest Everett Just, a graduate student and eventual lifelong friend of Lillie; Just would become a well-renowned scientist in the fields of fertilization and development. Nonetheless, it is important to contextualize Lillie's life outside of his contributions to research and remember the historical foundations that modern science stands upon.

Conclusion

Frank Lillie was an esteemed and well-rounded scientist, publishing over 100 communications and research articles throughout the duration of his career. His scientific fortitude, meticulous observations, and natural curiosity helped him to develop numerous veritable theories and make many monumental discoveries, many of which remain credible today. In particular, his contributions to the understanding of freemartins are discussed in undergraduate and graduate courses today (Nelson & Kriegsfeld, 2022). Lillie's view of the environment as a natural source of experimentation will always resonate with future scientists who look to find answers hidden in plain sight.

References

Allen D. (2008). Integrating the life of the mind: African americans at the university of chicago, 1870–1940; Future intellectuals: Ernest Everett Just (PhD 1916). The Hanna Holborn Gray Special Collections Research Center.

Eddy, E. B. (1976). *Lillie, Adam. Dictionary of Canadian Biography* (Vol. 9). University of Toronto/Université Laval.

Lillie F. R. (1899). Adaptation in cleavage. Biological Lectures Delivered at the Marine Biological Laboratory of Wood's Hole, Summers of 1897–1898. Boston: Ginn & Company. pp. 43–67.

Lillie, F. R. (1901). The organization of the egg of Unio, based on a study of its maturation, fertilization, and cleavage. *Journal of Morphology, 17*(2), 227–292.

Lillie, F. R. (1902). Differentiation without cleavage in the egg of the annelid Chaetopterus pergamentaceus. *Archiv für Entwicklungsmechanik der Organismen, 14*, 477–499.

Lillie, F. R. (1903). Experimental studies on the development of the organs in the embryo of the fowl (Gallus domesticus). *The Biological Bulletin, 5*(2), 92–124.

Lillie, F. R. (1907). The biological significance of sexual differentiation – a zoological point of view. *Science, 25*(636), 372–376.

Lillie, F. R. (1908). The development of the chick: An Introduction to Embryology. New York: Holt.

Lillie, F. R. (1909a). Karyokinetic figures of centrifuged eggs; an experimental test of the center of force hypothesis. *The Biological Bulletin, 17*(2), 101–119.

Lillie, F. R. (1909b). Polarity and bilaterality of the annelid egg. Experiments with centrifugal force. *The Biological Bulletin, 16*(2), 54–79.

Lillie, F. R. (1913). The mechanism of fertilization. *Science, 38*(980), 524–528.

Lillie, F. R. (1914). Studies of fertilization. VI. The mechanism of fertilization in Arbacia. *The Journal of Experimental Zoology, 16*(4), 523–590.

Lillie, F. R. (1916). The theory of the free-martin. *Science, 43*(1113), 611–613.

Lillie, F. R. (1917). Sex hormones in the foetal life of cattle. *The Journal of Experimental Zoology, 23*, 371.

Lillie, F. R. (1944). Founding and early history of the marine biological laboratory. In *The woods hole marine biological laboratory*. University of Chicago Press.

Lillie, F. R., & Juhn, M. (1932). The physiology of development of feathers. I. Growth-rate and pattern in the individual feather. *Physiological Zoology, 5*(1), 124–184.

Moors, C. R. (1948). Frank Rattray Lillie 1870–1947. *Science, 107*(2767), 33–35.

Nelson, R. J., & Kriegsfeld, L. J. (2022). *An introduction to behavioral endocrinology*. Oxford University Press.

Watterson, R. L. (1979). The striking influence of the leadership, research, and teaching of Frank R. Lillie (1870–1947) in zoology, embryology and other biological sciences. *American Zoologist, 19*(4), 1275–1287.

Wellner, K. (2009). Frank Rattray Lillie (1870–1947). Embryo Project Encyclopedia (2009-07-22).

Willier, B. H. (1957). *Frank Rattray Lillie. Biographical memoirs* (Vol. 30, pp. 177–236). National Academy of Sciences.

Willier, B. H., Harrison, R. G., Bigelow, H. B., & Conklin, E. G. (1948). Addresses at the Lillie Memorial Meeting Woods Hole, August 11, 1948. *The Biological Bulletin, 95*(2), 151–162.

Calvin Perry Stone

3

Juan M. Dominguez

Abstract

Calvin Perry Stone (1892–1954) was an American psychologist who is most recognized for his work in comparative and experimental psychology, though his research focused primarily on the nervous system and behavior. He is widely recognized for his contributions to our understanding of sexual behaviors in males and the physiological mechanisms that regulate this behavior. His studies using castration and ablations made it abundantly clear that hormonal influences, particular circulatory factors originating in the testes, were of importance to the expression of male sexual behaviors.

Keywords

Castration · Male sexual behavior · Testes

Calvin Alvin Stone was a pioneer in the physiology of reproductive behaviors. Born February 28, 1892, on a farm outside of Portland, Indiana, Stone was the youngest son and seventh of eight children born to Ezekial and Emily Brinkerhoff Stone. His paternal family, the Stones, were North Carolinians of English descent who moved to Indiana by way of Ohio. The Brinkerhoffs, of Dutch descent, settled in New York before gradually moving west, through Pennsylvania and Ohio, reaching Indiana in the late nineteenth century (Pickren, 2006).

By all accounts, Stone grew up in a caring and nurturing environment alongside his mother, sisters, and brothers. His mother played a primary and formative role in

J. M. Dominguez (✉)
Department of Psychology, Institute for Neuroscience, and Pharmacology & Toxicology, The University of Texas at Austin, Austin, TX, USA
e-mail: dominguez@utexas.edu

R. J. Nelson, Z. M. Weil (eds.), *Biographical History of Behavioral Neuroendocrinology*, https://doi.org/10.1007/978-3-031-12970-4_3

this upbringing, after his father died of pneumonia when Stone was only 5 years old. As a young student, Stone was hardworking, dedicated, intelligent, and popular. In 1907, his assiduousness was rewarded when, as a 15 year-old, he qualified to skip high school and enter Valparaiso University, where he went on to earn the degree of Bachelor of Science in 1910 (Hilgard, 1994). He would later receive a second bachelor's degree, this time a Bachelor of Arts in Classics, also at Valparaiso in 1913. From 1910 to 1914, Stone worked as a secondary-school teacher, school principal, and superintendent, before matriculating at Indiana University in 1914 to work towards a master's degree (Hilgard, 1994). While at Indiana, he worked with Ernest H. Lindley and Melvin E. Haggerty in psychology and philosophy. Under Haggerty's direction, Stone completed a master's degree with a thesis titled "Notes on Light Discrimination in Dogs" (Stone, 1921b). In 1915, Haggerty left Indiana and joined the faculty at the University of Minnesota, where he would later be appointed dean of the College of Education in 1920. Stone and Haggerty enjoyed a close working relationship, and so it was fitting that Stone should join Haggerty at Minnesota as a doctoral student in 1916. Before starting graduate studies, however, Stone married Minnie Ruth Kemper of Brook, Indiana, whom he had met at Valparaiso 4 years earlier. The two would later go on to have two sons and a daughter. That same year, Stone deferred his doctoral studies at Minnesota so he might serve 1 year as director of research at a penal institution and 2 years as psychological examiner for the US Army during World War I. After the war, while working on the rehabilitation of soldiers at Walter Reed Medical Center and after attaining the rank of Captain, Stone was discharged from military service in 1919 (Pickren, 2006; Rosvold, 1955).

When Stone resumed his graduate studies, the psychology department at Minnesota had changed. His training would now be primarily provided by A.T. Rasmussen and Karl S. Lashley, in place of Haggerty. Under Rasmussen, a professor of anatomy, and Lashley, a geneticist by training, Stone's graduate training emphasized neurology, anatomy, and psychology. Before returning to Minnesota, Lashley had worked with influential psychologist John B. Watson, studying views on sex and sex education (Watson & Lashley, 1920). While controversial (Benjamin et al., 2007), some credit Watson as one of the earliest pioneers in American sex research. And while Lashley did not focus on sex research, he was interested in the scientific study of sex (Pickren, 2006). Lashley and Watson may have been the first to introduce Stone to the American scientific sex-research community. Under Lashley's guidance, Stone completed his dissertation titled "An Experimental Analysis of the Congenital Sexual Behavior of the Male Albino Rat" (Stone, 1921a), parts of which were later published in The Journal of Comparative Psychology (Stone, 1922). Here Stone provided a detailed and systematic analysis of component actions in the complex sexual behaviors displayed by male albino rats. This classic research has become a seminal piece in our understanding of rodent sexuality, specifically, and a critical impetus to our understanding of sexual physiology, generally. In addition to providing detailed insight into the behavioral physiology of male sexual behaviors, Stone also concluded that sexual behaviors in young male rats occurs independent of previous learning or experience, a topic of much research still to this day. Stone also parted with Freud's ideas on infantile sexuality, noting that sexual behavior in the rat is not initiated until the rat reaches sexual maturity (Stone, 1922). Following his PhD, Stone remained in Minnesota where he taught in the Department of Psychology and the Department of Anatomy. All accounts indicate that he was much respected and admired for his mastery of the course material by both student and colleague. It was to no one's surprise then that Lewis M. Terman, chair of the psychology department at Stanford, recruited Stone to Stanford in 1922. Stone joined the Stanford faculty as a comparative psychologist, where he stayed for the rest of his career.

At Stanford, Stone became the first American Psychologist to develop a sex-research program (Pickren, 1997). From its inception, Stone's program adopted a multidisciplinary approach to research. This cooperative and cross-disciplinary approach was shrewd and in line with the goals of many philanthropic organizations, like the Rockefeller Foundation, which after the War became major sources of support and funding for American scientists. Founded in 1916 against the backdrop of the First World War and made permanent in 1918 by President Wilson (National Academy of Sciences, 2022b), the National Research Council (NRC), similar to philanthropic organizations at the time, also played a pivotal role in guiding our nation's scientific research and advancement through assistance and funding. The goals of the NRC, which was organized by the National Academy of Sciences, align with the Academy's purposes of furthering scientific knowledge and advising the US government on all issues related to science (National Academy of Sciences, 2022b).

However, beyond these primary goals, some at the time believed the role of the NRC to include using science to help ameliorate perceived social problems, which at the time included concerns on sexual behaviors and miscegenation, and often proposed eugenics programs as remedy (Osborn, 1921). Many of these anxieties were propelled by changing sexual mores and customs, along with fears of immigrants from southern and eastern European countries, as well as African Americans migrating to northern states. Concerns by leaders in industry, academia, and philanthropy of perceived sexual problems brought about the Committee for Research in Problems of Sex (CRPS), which was established by the NRC in 1922 with cooperation by the Bureau of Social Hygiene and support from the Rockefeller Foundation (National Academy of Sciences, 2022a). It is worth noting that despite the clear immorality of its founding principles, the CRPS funded seminal research in sexual behavior and physiology. CRPS funding saw the discovery of the first known estrogens, as well as primary research into pituitary hormones (National Academy of Sciences, 2022a). Stone was well funded by the CRPS, supporting his research program from 1922 until 1940 with over $18,000 (Aberle & Corner, 1953).

When other physiologists and emerging endocrinologist were only beginning to think of neural and hormonal interactions to regulate male sexual behaviors (Clarke, 1991), Stone had already written a comprehensive review of this topic (Stone, 1923a). This treatise was one of, if not the, major catalyst to studying brain-hormone interaction in the regulation of male sexual behaviors. In the laboratory, Stone's experiments employed a battery of techniques to elucidate the neuroendocrinological basis of sexual behaviors in males. This included a variety of ablation and castration techniques (Stone, 1923b, 1925a, b, 1926, 1927). His studies using castration made it clear, along with already published findings by others, that hormonal influences, particularly circulatory factors originating in the testes, were essential to the expression of sexual behaviors in males. Further, by using ablations of cortical and subcortical brain regions, he was one of the first to show a prominent role for subcortical structures in the regulation of male sexual behaviors. This is a fact about the neural regulation of sexual behaviors that is still widely studied to this day.

By all accounts, Stone was a nurturing mentor to many graduate students who themselves went on to contribute significantly to our understanding of behavioral neuroendocrinology. This included George T. Avery, who was the first to study sexual behaviors in guinea pigs, both males and females (Avery, 1925), and Clarence Ray Carpenter who, under Stone's guidance, studied the effects of castration on pigeons (Carpenter, 1933). Carpenter would go on to become a prolific researcher working at Yale with Robert M. Yerkes, making major discoveries in primate biology. William Dollard Commins elucidated the effects of castration on sexual behaviors as a function of age (Commins, 1932). Mary Sturman-Hulbe, under Stone's tutelage, gave us a detailed analysis of maternal behaviors in the albino rat (Sturman-Huble & Stone, 1929). Harry Harlow, who credits Stone as his mentor, studied eating behaviors in rats before gaining prominence with his research into cognitive development. Other students who worked with Stone, like Lois Doe-Kuhlmann and Roger Barker, also investigated hormonal influences on different aspects of development particularly in humans.

Stone served on the editorial board or as editor of leading journals including *Annual Review of Psychology*, *Comparative Psychology*, and the *Journal of Comparative & Physiological Psychology*, which would later split into Behavioral Neuroscience and Journal of Comparative Psychology.

Stone's major and formative contributions to our understanding of how hormones act in the brain to regulate sexual behaviors in males were recognized by his election as vice president of the American Association for the Advancement of Science (AAAS) in 1938, as president of the American Psychological Association (APA) in 1941, and as member of the National Academy of Sciences in 1943. Stone served as president of the APA, the leading organization for psychologists, from 1941 to 1942. While wartime forced the cancelation of the APA's annual convention, through his writings and submitted address as President it is evident that Stone advocated for experimental psychology, increased rigor in psychological science, and the application of science to help solve many of the nation's problems (Stone, 1950). Evidence of Stone's prominence and repute were further evidenced when, after World War II, the US Surgeon General invited him to become head of the US Army's new program in clinical psychology. Stone declined as he was reluctant to leave his friends and family (Pickren, 2006).

Before his death in 1954, Stone contributed 100 publications dealing with a wide array of topics in psychological science, but of greatest influence were his studies on the interplay, which we now readily recognize and appreciate, between hormones and the brain to orchestrate sexual behaviors in males. In late December, 1948, while on a class field trip to an asylum, Stone suffered a heart attack. Although

Fig. 3.1 Calvin Perry Stone. (With permission from the National Academies Press)

severe, he recovered sufficiently enough to continue teaching later that year. Stone was active in the classroom and laboratory until his death came suddenly in late December of 1954 (Hilgard, 1994). (Fig. 3.1).

References

Aberle, S. D., & Corner, G. W. (1953). *Twenty-five years of sex research: History of the National Research Council Committee for research in problems of sex, 1922–1947*. Saunders.

Avery, G. T. A. (1925). Notes on reproduction in guinea pigs. *Journal of Comparative Psychology, 5*(5), 373–396.

Benjamin, L. T., Whitaker, J. L., Ramsey, R. M., & Zeve, D. R. (2007). John B. Watson's alleged sex research: An appraisal of the evidence. *The American Psychologist, 62*(2), 131–139.

Carpenter, C. R. (1933). Psychobiological studies of social behavior in Aves II the effect of complete and incomplete gonadectomy on secondary sexual activity with histological studies. *Journal of Comparative Psychology, 16*(1), 59–97.

Clarke, A. E. (1991). Embryology and the rise of American reproductive sciences, circa 1910–1940. In *The expansion of American biology* (pp. 107–132), by Keith Benson (editor), Rutgers University Press, New Brunswick NJ.

Commins, W. D. (1932). The effect of castration at various ages upon the learning ability of male albino rats. *Journal of Comparative Psychology, 14*(1), 29–53.

Hilgard, E. (1994). Calvin perry stone. In *Biographical memoirs: National academy of sciences*. The National Academies Press.

National Academy of Sciences. (2022a). *Committee for research in problems of sex* (pp. 1920–1965). National Academies Press, Washington DC.

National Academy of Sciences. (2022b). *The Organization of the National Research Council.* National Academies Press, Washington DC.
Osborn, H. F. (1921). The second international congress of eugenics address of welcome. *Science, 54*(1397), 311–313.
Pickren, W. (1997). Robert Yerkes, Calvin Stone, and the beginning of programmatic sex research by psychologists, 1921–1930. *The American Journal of Psychology, 110*(4), 603–619.
Pickren, W. E. (2006). Calvin Perry Stone: Solid citizen and scientist. In N. J. Mahwah (Ed.), *Portraits of pioneers in psychology* (Vol. VI). Lawrence Erlbaum Associates Publishers.
Rosvold, H. E. (1955). Calvin Perry Stone: 1892-1954. *The American Journal of Psychology, 68*, 326–329.
Stone, C. P. (1921a). *An experimental analysis of the congenital sexual behavior of the male albino rat.* University of Minnesota.
Stone, C. P. (1921b). Notes on light discrimination in the dog. *Journal of Comparative Psychology, 1*(5), 413–431.
Stone, C. P. (1922). The congenital sexual behavior of the young male albino rat. *Journal of Comparative Psychology, 2*(2), 95–153.
Stone, C. P. (1923a). Experimental studies of two important factors underlying masculine sexual behavior: The nervous system and the internal secretion of the testis. *Journal of Experimental Psychology, 6*(2), 85.
Stone, C. P. (1923b). Further study of sensory functions in the activation of sexual behavior in the young male albino rat. *Journal of Comparative Psychology, 3*(6), 469–473.
Stone, C. P. (1925a). The effects of cerebral destruction on the sexual behavior of rabbits I. The olfactory bulbs. *American Journal of Physiology, 71*(2), 430–435.
Stone, C. P. (1925b). The effects of cerebral destruction on the sexual behavior of rabbits II. The frontal and parietal regions. *American Journal of Physiology, 72*(2), 372–385.
Stone, C. P. (1926). The effects of cerebral destruction on the sexual behavior of male rabbits III the frontal, parietal, and occipital regions. *Journal of Comparative Psychology, 6*(6), 435–448.
Stone, C. P. (1927). The retention of copulatory ability in male rats following castration. *Journal of Comparative Psychology, 7*(5), 369–387.
Stone, C. P. (1950). Presidential address. *American Psychologist, 5*, 57–57.
Sturman-Huble, M., & Stone, C. P. (1929). Maternal behavior in the albino rat. *Journal of Comparative Psychology, 9*(3), 203–237.
Watson, J. B., & Lashley, K. S. (1920). A consensus of medical opinion upon questions relating to sex education and venereal disease campaigns. *Mental Hygiene, 4*, 769–847.

Josephine Ball

4

O. Hecmarie Meléndez-Fernández, Jennifer A. Liu, and Nicole M. Baran

Abstract

Josephine Ball (1898–1977) was an American comparative psychologist, endocrinologist, and clinical psychologist. She was a pioneer in the study of reproductive behavior and neuroendocrinology (1920s–1940s). Her research on the role of steroid hormones and reproductive behavior was contemporaneous with that of the founders of the field of behavioral neuroendocrinology, WC Young and Frank A. Beach. Ball's first paper (Ball, Am J Physiol, 78(3), 533–536, 1926) describes the earliest investigation of the role of steroid hormones on learning. Her meticulous studies on hormones and reproductive behavior in rats and macaques provided foundational contributions to the field. Notably, her work in the early 1930s touched on the organizational and activational roles of steroid hormones, predating, by more than two decades, the influential work led by Young on this topic. Struggling to find steady funding for her research in the 1940s, she transitioned to a career in clinical psychology and retired in 1967.

Keywords

Behavioral neuroendocrinology · Sex hormones · Sexual behavior · Progesterone · Estrogen · Testosterone · Progesterone

O. H. Meléndez-Fernández (✉) · J. A. Liu
Department of Neuroscience, Rockefeller Neuroscience Institute, West Virginia University, Morgantown, WV, USA
e-mail: ohm0001@mix.wvu.edu

N. M. Baran
Department of Psychology, Emory University, Atlanta, GA, USA

R. J. Nelson, Z. M. Weil (eds.), *Biographical History of Behavioral Neuroendocrinology*, https://doi.org/10.1007/978-3-031-12970-4_4

Josephine Ball was born on 28 April 1898 in Chicago, Illinois (Cook County Birth Certificates 1878–1922). Her father, William Dearborn Ball, was born in Turkey to Christian missionaries and worked as an electrical engineer who earned multiple degrees from the University of Michigan. Her mother, Alice May Edwards, was born in Niles, Michigan, to abolitionist parents, Hiram Edwards and Angelina Dickey (EdwardsTract). Indeed, it was reported that her mother's childhood home had been a stop on the Underground Railroad (EdwardsTract). Ball's parents were married at her grandparent's home in Niles, Michigan, in 1896 (Freeman, 2019). Ball was born in Chicago two years later but spent her early infancy in Paris, France, where her father was involved in electrical work for the construction of the Eiffel Tower for the 1899 Exposition Universelle (Freeman, 2019).

According to a diary written by Ball's cousin, Lorna Louise Freeman (née Edwards), the Ball family later returned to Illinois, this time moving to the Chicago suburb of Evanston. When Ball was just 4 years old, her mother died while giving birth to triplets (Freeman, 2019). Ball's extended family suggested that her mother may have avoided medical interventions during childbirth which could have saved her life, because her parents were Christian Scientists who were known to occasionally forgo medical procedures. Ball's siblings all survived, although her sister Dorothy was permanently intellectually disabled. According to Freeman, Ball's father never fully recovered from his grief. Ball's aunt, Fanny Ball, came from Michigan to help raise the children and stayed with them for 6 years. Around 1908, Ball's father married a woman named Emma Redman. Redman "came down on the family with the heavy hand of discipline, which up until then, they had not known. It was hard on all concerned" (Freeman, 2019).

Josephine Ball left home at the age 15 to become a "mother's helper" in Swarthmore, Pennsylvania. After she left, Ball's Aunt Fanny intervened and took Dorothy to raise on a trust set up by another aunt, Louise Ball. Dorothy lived with Fanny until Fanny's death in 1941. Ball's two male siblings, Donald and Douglas, stayed in Illinois with their father and Ms. Redman. Ball worked her way through high school in Swarthmore, employed as a mother's helper. On a Barnard College career development application in 1922, Ball listed her high school as Swarthmore High School and listed as an employment reference the wife of J. Barnard Walton, a prominent member of the Quaker community in Swarthmore.

Ball matriculated at Barnard College in 1918, where she principally studied sociology, psychology, economics, and English. During the summers of 1919 and 1921, she worked in Massachusetts as a governess for the family of Dr. Stanley Cobb, who was a well-known neurologist at Harvard University and who was credited as the founder of biological psychiatry. Ball ultimately earned her AB from Barnard College in 1922 (Ogilvie & Harvey, 2000) (Fig. 4.1).

After college, Ball was employed for a year at the National Research Council (NRC) in Washington DC, where she worked with Dr. Robert M. Yerkes as part of the newly formed Committee for Research in Problems of Sex (Pickren, 1997). Yerkes played an important role in Ball's career trajectory, setting her on the path of studying sexual behavior and serving as a mentor and professional reference throughout her career. They remained in contact via regular letters and a record of

Fig. 4.1 Josephine Ball, Barnard College Yearbook, 1922. (Republished with permission from the Barnard College Archives)

their warm, long-running professional relationship is catalogued in Yerkes' personal correspondence archived at Yale University (Stark, 1985; Yerkes).

After completing summer school at Harvard University in statistics in 1923, Yerkes helped Ball obtain a fellowship and a place in the research lab of Karl Lashley, who was then at the University of Minnesota and was also a member of the NRC Committee for Research in Problems of Sex (Elliott, 1919–1939, R.M. Yerkes to R. E. Elliott, July 4, 1923) (Yerkes). She worked as an assistant in psychology for Lashley from 1923 to 1926. In a letter to Yerkes in February 1924, Ball conveyed her excitement both to be working with Lashley and for her coursework (Yerkes, 1923–1945, J. Ball to R.M. Yerkes, February 18, 1924) (Yerkes). "I have little doubt that my interests will be mainly in physiological and comparative work," she wrote, indicating her intention to pursue psychobiological research.

In the summer of 1924, Yerkes invited Ball to join him and several colleagues on a trip to Cuba to visit Madame Rosalía Abreu's primate colony. Abreu, the daughter of a wealthy Cuban plantation owner and the world's first animal keeper, kept a captive breeding colony of chimpanzees. Yerkes' primary aims for the expedition were to "...get the lady's experiences and observations on record" and to study the behavior of the chimpanzees (Yerkes, 1923–1945, R.M. Yerkes to J. Ball, May 21, 1924) (Yerkes), with the ultimate goal of establishing a long-term colony back in the United States to observe the development and behavior of apes. The establishment of this institution, the Yale Anthropoid Experiment Station (eventually renamed Yerkes Laboratory of Primate Biology) came to fruition in 1930. In a flurry of letters, Ball indicated that she was eager to join the excursion, but only if her expenses could be guaranteed, as she did not have the funds to support herself on the trip. Yerkes was able to arrange funds to cover her travel and lodging while in Cuba, and Ball happily made the trek, hoping to assist both Yerkes and his research associate at Yale University, Harold C. Bingham, in their experimental studies of the chimpanzees. Based on later letters between Yerkes and Ball, it seems that the experimental work did not proceed as hoped, but Ball enjoyed the experience, nonetheless. "The summer, thanks to the opportunity you offered, has been a most stimulating one and I find myself eager for work and full of things I want to do," Ball wrote to Yerkes later that fall (Yerkes, 1923–1945, J. Ball to R.M. Yerkes, October 1, 1924 (Yerkes)).

In 1926, Ball published her first paper, "The Female Sex Cycle as a Factor in Learning in the Rat," which was notably one of the first experiments examining the role of sex hormones in learning and memory (Ball, 1926). In this experiment, Ball tested whether the estrous cycle in female rats influenced the speed with which they learned to navigate a maze. Ball used a slightly modified protocol for collecting and analyzing vaginal smears in rats which had been developed by Long and Evans (1922) 4 years prior, to determine the estrous cycle phase in rats. Ultimately, she did not observe a significant difference between the groups, instead reporting a great deal of variability in maze learning within both groups. However, this work presaged the now extensive evidence that steroid hormones and the estrous cycle influence learning processes (Inoue, 2021). She later published a study with Lashley entitled "Spinal Conduction and Kinesthetic Sensitivity in the Maze Habit," which demonstrated that rats trained to run a maze can still run the maze without afferent sensory input via the spinal cord (Lashley & Ball, 1929). This work provided early support for the idea that the ability to learn a maze occurs centrally and does not require proprioceptive feedback (Adams, 1971).

In 1927, Ball moved to the University of California, Berkeley where she worked as a teaching fellow in psychology and as a research assistant in the lab of anatomist, embryologist, and endocrinologist, Herbert McLean Evans. In 1929, she earned her Ph.D. from the University of California, as well as a diploma from the American Board of Examiners of Professional Psychologists. Her thesis, "Measurement of Sexual Behavior in Male Rats" (Ball, 1929) was an 18 month investigation of the mating behaviors of 61 male rats under repeated and standardized conditions (Einstein, 2007).[1]

After graduation, the newly minted Dr. Ball accepted an assistant psychobiologist position at the Henry Phipps Psychiatric Clinic at Johns Hopkins University Hospital in Baltimore, Maryland, where she collaborated with Carl Gottfried Hartman, the Director of the Carnegie Institution of Washington and head of the Department of Embryology (Fig. 4.2). The behavioral insights gleaned from Ball's careful investigation of mating behaviors in rats during her doctoral work facilitated her initial research at Johns Hopkins and Carnegie Institution. One paper stated that

[1]The authors sought to obtain a copy of Dr. Ball's dissertation. However, the copy held by the University of California Berkeley was found to be missing from the shelves. We sincerely hope that a copy is found before it is lost to history.

Fig. 4.2 Photograph at the Carnegie Institution of Washington department of embryology, 1931. Left to right: George Streeter, Robert Enders, Chester Heuser, Josephine Ball, Carl Gottfried Hartman, P. Mihalic, Warren Lewis, and Sam Reynolds. (Reprinted from Birney & Choate (1994) with permission from the American Society of Mammalogists)

Ball had "learned to interpret the behavior of the male [rat] so as to almost infallibly to diagnose the copulation that resulted in ejaculation" (Hartman & Ball, 1930). This turned out to be the critical insight that enabled her and Hartman to demonstrate that sperm were transported into the uterus within seconds of ejaculation, presumably via uterine contractions. This effectively disproved earlier theories that the male's sperm reached the uterus on their own power by swimming with their tails. Her careful studies of sexual behavior in rats also allowed her to examine the stimuli required to induce pseudopregnancy (an increase in the length of the estrous cycle following sterile mating or cervical stimulation) (Ball, 1934a). She also demonstrated that the surgical removal of the reproductive organs (i.e., the uterus and vagina), but not the ovaries, failed to result in changes to mating behaviors in female rats (Ball, 1934b).

Several years later, Ball's research focused on female sexual excitability (sexual arousal) and sexual receptivity in Rhesus macaques (*Macaca mulatta*) during the menstrual cycle (Ball & Hartman, 1935) and pregnancy (Ball, 1937b) and how these were altered by hormone treatment (Ball, 1936; Ball & Hartman, 1939). Her work demonstrated that although macaques do not limit their sexual activity to their fertile window as some rodents do, they do exhibit a sharp increase in sexual "excitability" (defined as presenting, attempting to attract a partner's attention, or going towards the male) immediately before ovulation, and a decrease after ovulation. Ball later collaborated with Hartman's successor, George Corner, during her position at Johns Hopkins. Corner was the co-discoverer of progesterone (Corner & Allen, 1929), a hormone which Ball also used in her later studies on sexual behavior.

In her article entitled, "Further Evidence on Hormonal Basis of 'Heat' Behavior" (Ball & Hartman, 1936), Ball examined the endocrine contributions to "heat" (induction of estrous), by using either castrated or hypophysectomized female rats and various combinations of gonadotropic hormone, luteinizing hormone (LH), estradiol benzoate (EB, a synthetic benzoate ester of estradiol with similar downstream effects), and/or progestin. She determined that in hypophysectomized rats, EB had no effect on behavior, suggesting that the pituitary is necessary for the production of these behaviors. However, when castrated rats received an EB and LH treatment, she could elicit "heat" behaviors in the rats. She also determined that treatment with progestin in addition to EB had no observable behavioral effect.

Ball complemented her work in females by examining whether sexual behavior in castrated male rats was altered by administration of EB (Ball, 1937a). This experiment was designed with the assumption that EB would stimulate the pituitary to release the "same hormone" the testes released and thereby increase or restore normal sex activity in castrated males. Indeed, EB restored the interest in and ability to copulate as well as the ejaculation behavior, but not the activity of accessory sex glands (i.e., the gonads). These rats did not ejaculate but behaved as if typical of ejaculation in the uncastrated male rats. Ball concluded that this "gonadal hormone does not organize the mating behavior pattern in the adult but merely activates a pattern already present." This work, and Ball's direct reference to both organizational and activational roles for steroid hormones, predates by more than two decades the classic Phoenix et al. (1959) which conclusively demonstrated the existence of organizational roles for steroid hormones during development.

Ball further explored the role of sex hormones and treatments on sexual excitability in both female (Ball, 1938, 1939) and male rats (Ball, 1937a, 1939). In her paper published in 1939, male rats were castrated prior to puberty and treated with estrogens until adulthood. Ball demonstrated that both male and female prepubertally castrated rats treated with estrogens displayed lordosis behavior when mounted but that males required significantly more estrogens compared to females, in order to elicit female-typical behavior. A subsequent experiment in this study examined the role of progesterone administration in conjunction with estrogen and determined that progesterone enhances the effect of estrogen in males. She concluded in this study, that the "neuromuscular pattern underlying male [and female] copulation is organized very early, possibly before birth" and suggested that pubertal hormones alone could not explain mating behaviors in both sexes (Ball, 1939).

Further notable research by Ball at this time investigated the role of testosterone on sexual behavior in female rats, to test her prediction that male hormone (testosterone) inhibited female sexual excitability. She observed disturbances in female estrous cycles with prolonged cornification, one attempted mounting from another female rat and increased aggressive behavior towards the testosterone-treated females. After discontinuing testosterone, she noted that cycles and sexual behavior normalized; however, even after 1 year post-testosterone treatment females had hypertrophied clitorises (Ball, 1940).

Ball's contributions to the field of behavioral endocrinology appear to have largely stopped after 1940. In 1942, she left Baltimore and held a series of short-term positions. The precise reasons why she ultimately left her long-term position at the Henry Phipps Psychiatric Clinic and the Carnegie Institution remain unclear. Based on communications between Drs. Ball and Yerkes, it appears that this series of career moves were related to a lack of availability of financial resources for her research and opportunities for her clinical practice (Yerkes, 1923–1945, J. Ball to R.M. Yerkes, July 5, 1945) (Yerkes). This was also a time of major social upheaval as World War II raged on. Many men took leaves of absence to serve in the military, which meant that many faculty, teaching, and research positions became available to women. Simultaneously, however, funding priorities shifted towards the war effort. In his letters, Yerkes encouraged Ball to apply broadly and reach out to mutual colleagues for potential placements (Yerkes, 1923–1945, R.M. Yerkes to J. Ball, February 7, 1942).

From 1942 to 1943, Ball worked as a research associate at Cornell University's College of Home Economics. While there, she described encountering obstacles regarding her salary and research funds (Yerkes, 1923–1945, J. Ball to R.M. Yerkes, September 9, 1943), and thus, she quickly transitioned to a position as an assistant professor in the psychology department at Vassar College between 1943 and 1945. At the time she was moving to Vassar, she had a proposal under consideration with the Committee on the Problems of Sex to study dysmenorrhea (pain associated with menstruation) in women (Yerkes, 1923–1945, J. Ball to R.M. Yerkes, September 9, 1943). In April 1945, she wrote Yerkes to ask for funding to support a "psychobiological study of the menstrual cycle in women" in relation to various measures of physiological arousal and mood (Yerkes, 1923–1945, J. Ball to R.M. Yerkes, April 30, 1945). It is unclear if this proposed work in women ever occurred and shortly thereafter, she moved institutions yet again. From 1945 to 1947, she held an assistant professorship at Connecticut's Hartford Junior College (Fig. 4.3) teaching nurses and psychiatric aids and was a clinical psychologist at the University of Connecticut's Institute of Living. Although she initially hoped to have time for research, this transition appears to have marked the beginning of Ball's career in the field of clinical psychology.

Starting in 1948, Ball worked as a clinical psychologist for the New York State health system. From 1948 to 1950, she worked as a senior psychologist at the Rockland State Hospital. From 1950 to 1955, she served as a field supervisor for the New York State Psychological Intern Training Program. During this time, she published a short review on the textbook, "Psychology for Nurses," which commented on the poor interpretation, scientific content, and careless statements which were difficult to interpret by nursing students (Ball, 1948). She also wrote another (largely positive) review of the book "Sex and the Social Order" written by Georgene H. Seward, an early feminist psychologist and contemporary of Ball's, which summarized much of the research on sex differences (Ball, 1949). In addition, Ball served as the secretary of the New York State Psychological Association from 1951 to 1952. From 1954 to 1955, she served as the assistant director of psychological services for the New York State Department of Mental Hygiene.

Fig. 4.3 Hartford junior college faculty group photo, 1945. Ball is in the top row, second from left. (Reprinted with permission from the University of Hartford archives and special collections)

In 1955, Ball returned to Maryland as a research psychologist associated with the, now controversial, lobotomy research project (Phillips, 2013) at the Veterans Administration Hospital in Perry Point, Maryland. Given that most lobotomies were conducted between 1947 and 1950 and the procedure fell out of favor as tranquilizer drugs became available in the mid-1950s, it is unlikely that she was involved in performing lobotomies. She did, however, research their effects in a large-scale study. She was lead author on the paper, "The Veterans Administration study of prefrontal lobotomy," published in 1959 (Ball et al., 1959). In 1959, she left lobotomy research to work as a clinical psychologist at the Veterans Administration Hospital, focusing on gerontology. She remained in this position until her retirement in 1967.

An alumni survey conducted by Barnard College between 1956 and 1957 (Barnard College, 1956), Ball indicated that she never married, owned her own house in Maryland, enjoyed both reading and gardening, and was a member of the League of Women Voters. In this questionnaire, she also expressed some ambivalence about her education and career path. According to Ball, "I needed badly to know a lot more than I did about ways of earning one's living—what there is in the way of jobs, what they're like, what they take in the way of training, & personal qualifications—and a lot more than I did about myself in relation to this problem. I don't care much for the profession I've slipped into by chance. I doubt if psychology is ready to do all I hope someday it can for college students in this respect." She also shared her thoughts on what she thought should be the goal of higher education: "...colleges should, I think, be very active in preparing students to vote intelligently—I mean, of course, in a very broad as well as practical sense, — courses in history, economics, sociology, & social psychology should be not only background but should give the student the habit of thinking on present community & world problems." Ball retired to Los Gatos, California, presumably to be near her brothers, where she died on August 1, 1977 (California Death Index).

Although she was conducting research on the role of steroid hormones and behavior that preceded or was contemporaneous with that of scientists such as W.C. Young and Frank Beach, Josephine Ball has not been widely recognized as having played an influential role in the founding of the field of behavioral neuroendocrinology. Indeed, during a speech at a meeting, Beach commented that if a conference on reproductive behavior had been held in the late 1930s, it would have had only three participants, W.C. Young, Josephine Ball, and himself (Brush & Levine, 1989), placing Dr. Ball side by side with the field's founding fathers. Unlike these men, however, she never found steady work as a research professor and struggled to obtain funding to pursue her independent ideas, many of which were notably ahead of their time. Her interest in studying women and female nonhuman animals was evident throughout her career, as well. Ball did not have the opportunity to mentor graduate students the way her male colleagues did, despite support and encouragement from several highly influential scientists. Nevertheless, she persisted in the field for a remarkably long time as a woman in a male-dominated field, producing an impressive body of work that was undoubtedly foundational to the field of behavioral neuroendocrinology.

References

Adams, J. A. (1971). A closed-loop theory of motor learning. *Journal of Motor Behavior, 3*(2), 111–150.

Ball, J. (1926). The female sex cycle as a factor in learning in the rat. *American Journal of Physiology, 78*(3), 533–536.

Ball, J. (1929). *Measurement of sex behavior in male rats.* University of California.

Ball, J. (1934a). Demonstration of a quantitative relation between stimulus and response in pseudopregnancy in the rat. *American Journal of Physiology-Legacy Content, 107*(3), 698–703.

Ball, J. (1934b). Sex behavior of the rat after removal of the uterus and vagina. *Journal of Comparative Psychology, 18*(3), 419.

Ball, J. (1936). Sexual responsiveness in female monkeys after castration and subsequent estrin administration. *Psychological Bulletin, 33*(811), 117–119.

Ball, J. (1937a). Sex activity of castrated male rats increased by estrin administration. *Journal of Comparative Psychology, 24*(1), 135.

Ball, J. (1937b). Sexual responsiveness and temporally related physiological events during pregnancy in the rhesus monkey. *The Anatomical Record, 67*(4), 507–512.

Ball, J. (1938). Partial inhibition of sex activity in the intact female rat by injected estrin. *Endocrinology, 23*(2), 197–199.

Ball, J. (1939). Male and female mating behavior in prepubertally castrated male rats receiving estrogens. *Journal of Comparative Psychology, 28*(2), 273.

Ball, J. (1940). Frequent failure of a single insemination to activate the corpora lutea of the rat sufficiently for implantation of fertilized ova. *American Journal of Physiology-Legacy Content, 130*(3), 471–474.
Ball, J. (1948). Review of psychology for nurses.
Ball, J. (1949). Review of sex and the social order.
Ball, J., & Hartman, C. G. (1935). Sexual excitability as related to the menstrual cycle in the monkey. *American Journal of Obstetrics and Gynecology, 29*(1), 117–119.
Ball, J., & Hartman, C. G. (1936). Further evidence on hormonal basis of "Heat" Behavior. *Proceedings of the Society for Experimental Biology and Medicine, 35*(3), 416–418.
Ball, J., & Hartman, C. G. (1939). A case of delayed ovulation after estrin administration in the intact monkey. *Proceedings of the Society for Experimental Biology and Medicine, 40*(4), 629–631.
Ball, J., Klett, C. J., & Gresock, C. J. (1959). The veterans administration study of prefrontal lobotomy. *Journal of Clinical and Experimental Psychopathology & Quarterly Review of Psychiatry and Neurology, 20*, 205–217.
Barnard College. (1956). *Alumnae affairs biographical questionnaire*. Barnard Archives and Special Collections: Barnard College.
Birney, E. C., & Choate, J. R. (1994). Seventy-five years of mammalogy, 1919–1994 (First Edition). *American Society of Mammalogists*. https://archive.org/details/seventyfiveyears00birn/page/276/mode/2up.
Brush, F. R., & Levine, S. (1989). *Psychoendocrinology*. Academic.
California Death Index. In: State of California, D. O. H. S., Center for Health Statistics, (Ed.). Sacramento, CA.
Cook County Birth Certificates. (1878–1922). In: Illinois Department of Public Health. Division of Vital Records (Ed.). Springfield.
Corner, G. W., & Allen, W. M. (1929). Physiology of the corpus luteum: II. Production of a special uterine reaction (progestational proliferation) by extracts of the corpus luteum. *American Journal of Physiology-Legacy Content, 88*(2), 326–339.
EdwardsTract. Available: https://web.archive.org/web/20151109232002/http://ehutchison.homestead.com/edwardstract.html. Accessed 22 July 2021.
Einstein, G. (2007). *Sex and the brain*. MIT Press.
Freeman, M. (2019). *Freeman/Hoag Family Tree* [Online]. Available: https://wc.rootsweb.com/trees/250392/I1072/-/individual. Accessed 22 July 2021.
Hartman, C. G., & Ball, J. (1930). On the almost instantaneous transport of spermatozoa through the cervix and the uterus in the rat. *Proceedings of the Society for Experimental Biology and Medicine, 28*(3), 312–314.
Inoue, S. (2021). Neural basis for estrous cycle-dependent control of female behaviors. *Neuroscience Research, 176*, 1–8.
Lashley, K. S., & Ball, J. (1929). Spinal conduction and kinesthetic sensitivity in the maze habit. *Journal of Comparative Psychology, 9*(1), 71.
Long, J. A., & Evans, H. M. (1922). *The oestrous cycle in the rat and its associated phenomena*. University of California Press.
Ogilvie, M., & Harvey, J. (2000). *The biographical dictionary of women in science: Pioneering lives from ancient times to the mid-20th century*. Routledge.
Phillips, M. M. (2013, December 12). The lobotomy files. *The Wall Street Journal*.
Phoenix, C. H., Goy, R. W., Gerall, A. A., & Young, W. C. (1959). Organizing action of prenatally administered testosterone propionate on the tissues mediating mating behavior in the female guinea pig. *Endocrinology, 65*(3), 369–382.
Pickren, W. (1997). Robert Yerkes, Calvin Stone, and the beginning of programmatic sex research by psychologists, 1921–1930. *The American Journal of Psychology, 110*(4), 603–619.
Stark, B. P. (1985). The Robert Mearns Yerkes papers. *The Yale University Library Gazette, 59*, 162–167.
Yerkes, R. M. *Robert Mearns Yerkes Papers (MS 569)* [Online]. Yale University Library. Available: https://archives.yale.edu/repositories/12/resources/4430. Accessed 2022.

William Caldwell Young

5

Irving Zucker

Abstract

William C. Young was a major founder of the discipline of behavioral endocrinology. His impact derived from theoretical writing, voluminous important empirical research, and the many scientists he trained and inspired. He was instrumental in establishing that effects of androgens secreted pre-– and/or postnatally organize structures that mediate mating behavior in several mammalian species. His academic descendants are major contributors to modern behavioral neuroendocrinology.

Keywords

Guinea pig · Androgens · Organizational hypothesis · Sex behavior

William C. Young was one of the most important figures in the history of behavioral endocrinology. As enumerated by Beach (1981), Young's impact came from his "many significant research contributions, the breadth and integrative value of his theoretical writings, and the many important contributions to the discipline which have been and continue to be made by scientists he trained and inspired." Several decades later this assessment remains apropos: Young's lineage includes some of the twenty-first century's most prominent behavioral neuroendocrinologists.

Young was born in 1899 and died in 1965. He completed undergraduate studies at Amherst College and obtained a doctorate in Anatomy from the University of Chicago, where he was mentored by the prominent endocrinologist Carl Moore. His

I. Zucker (✉)
Departments of Psychology and Integrative Biology, University of California, Berkeley, CA, USA
e-mail: irvzuck@berkeley.edu

R. J. Nelson, Z. M. Weil (eds.), *Biographical History of Behavioral Neuroendocrinology*, https://doi.org/10.1007/978-3-031-12970-4_5

dissertation consisted of a series of studies on the epididymis of guinea pigs. His first faculty position was at Brown University in 1928, where with several graduate students, who subsequently enjoyed notable scientific careers, he initiated landmark studies of ovarian control of female sexual behavior, relying on the manually elicited lordosis reflex of guinea pigs as the favored endpoint. Their work was foundational in establishing that preovulatory progesterone secretion triggered behavioral estrus; they went on to characterize the sequential interplay of estradiol and progesterone in the control of female sexual behavior. Young's emphasis on behavior received no encouragement from several senior endocrinologists, including his former mentor Moore, who cautioned him that "the behavior of animals was utterly capricious, unordered by hormonal events and unrelated to variables of significance to reproductive biology" (Goy, 1967). Nevertheless, Young persisted. He was not flamboyant; instead, he was described by one of his former associates as "a highly reserved person with minimal extroverted characteristics. He spoke in a soft voice and often, haltingly. He refrained from writing controversial articles and engaging in *ad hominem* criticisms. But his quiet demeanor sheltered intellectual strengths and persistence that enabled him to overcome obstacles and achieve his self-defined goals" (Gerall, 2009). In 1939 Young moved to the Yale Laboratories of Primate Biology in Florida where he embraced the opportunity to work with nonhuman primates. He left in 1943, missing the advantages of an academic setting. He spent several years teaching undergraduates at Cedar Crest College before relocating in 1946 to the University of Kansas where he established an endocrine laboratory. With a multitalented group of graduate students including Elliot Valenstein, Milton Diamond, Jerome Grunt, Walter Riss, Ken Grady, and Harvey Feder and postdoctoral associates Robert Goy, Charles Phoenix, and Arnold Gerall, these investigators contributed to the foundation of a mature science of behavioral endocrinology. Studies by Valenstein, Riss, and Young [4] showed that male guinea pigs' sexual behavior is formed by an interaction of experiential and genetic factors; contact with other animals is critical for acquisition of the mature copulatory pattern. Grunt and Young (Grunt & Young, 1953) reported that differences in sexual behavior of male guinea pigs are attributable in part to differences in reactivity of tissue substrates that mediate responses to gonadal hormones. They categorized guinea pigs into low, medium, and high drive sex groups; after castration mating scores of all three groups decreased to the same low baseline; even after supraphysiological replacement with testosterone the groups recovered to their characteristic pre castration mating levels, strong evidence that substrates, presumably neural in origin, profoundly influence mating behavior. This finding questioned whether differences in androgen secretion are the main cause of individual differences in male sexual behavior.

The 1959 study by Phoenix, Goy, Gerall, and Young (Phoenix et al., 1959) is arguably the single most important report in the history of behavioral endocrinology. The sexual behavior of male and female guinea pig offspring from mothers treated with testosterone during most of pregnancy was studied in adulthood. The larger quantities of gestational testosterone produced hermaphrodites whose external genitalia were indistinguishable macroscopically from those of newborn males.

Lordosis behavior of these females was greatly reduced after gonadectomy and treatment with estradiol and progesterone. Male-like mounting behavior was displayed by many of these animals, approaching levels of castrated males treated with the same amount of testosterone. Suppression of the capacity for displaying lordosis was achieved with testosterone doses lower than that those that masculinized the external genitalia. Phoenix et al. concluded that testosterone administered prenatally organizes the tissues mediating mating behavior, producing a responsiveness to exogenous hormones different from that of normal adult females. Structural and behavioral characteristics of male siblings were essentially normal. The results justified the conclusion that the prenatal period in this long gestation rodent is a time when fetal morphogenic substances have an organizing or "differentiating" action on the neural tissues mediating mating behavior. The organizational action of hormones was contrasted with activational hormonal effects that influence behavior postnatally. Valuable perspectives on this study are related by Gerall (2009), Phoenix (2009), and Wallen (2009). Much earlier, Wilson et al. (1940) anticipated the 1959 report on guinea pigs, showing that development of reproductive function in rats is markedly impaired by postnatally and/or prenatally administered androgen; rats that received only postnatal androgen treatment failed to respond to a regimen of estrogen followed by progesterone, whereas control animals exhibited estrous behavior.

In 1963, desiring to extend the rodent work to primates, Young, Phoenix and Goy relocated to the Oregon Regional Primate Research Center in Beaverton where Young was Chairman of the Department of Reproductive Physiology and Behavior. The goal was to assess whether organizational actions of androgens in rodents would generalize to primates. Findings in the next decades confirmed similar organizational actions on behavior of rhesus monkeys (Eaton et al., 1973). Eaton and colleagues reported that testosterone injected into pregnant rhesus monkeys can modify female fetuses so that they are predisposed to acquire male-like patterns of behavior. A comprehensive review by Wallen (2009b) concluded that prenatal exposure to exogenous androgen either during the second-third or last-third of gestation masculinizes the juvenile behavior of genetic females. The behavioral masculinization is not the result of genital masculinization but an independent effect of prenatal androgen exposure. Blocking endogenous androgen in genetic males reduced genital masculinization, without disrupting masculinization of behavior, establishing the independence of genital and behavioral masculinization.

Young was a scholar of the very first rank. His chapter on hormones and mating behavior in the third edition of *Sex and Internal Secretions*, a landmark publication which he edited, is a masterful exposition that summarizes all that one needs to know on the subject through 1960 (Young, 1961). His treatise on the psychobiology of sexual behavior of the guinea pig is another stellar work (Young, 1969).

I began as Young's postdoc in August 1964 with no background in behavioral endocrinology. The first 3 months were spent browsing the collected works of William C. Young. It was a wonderful education. I was struck by 1930s studies of the role of progesterone in controlling female sex behavior of guinea pigs and subsequently pursued this topic in several studies of this species and rats; the first report was dedicated to Young. His example spurred a career-long interest in female reproductive behavior and physiology. Young and his wife Ruth were kind to me and my wife Ellen. I enjoyed noon time walks with Will (whom I addressed as Dr. Young) around the perimeter of the primate center grounds, where the conversation ranged over nonscientific and scientific topics. It was a severe blow when he died a year after my arrival in Beaverton. He has remained an important influence on me as well as many others. I cherish my time with him and am pleased that his contributions are acknowledged and celebrated by the Young award given to a recently graduated Ph.D. student by the *Society for Behavioral Neuroendocrinology*.

References

Beach, F. A. (1981). Historical origins of modern research on hormones and behavior. *Hormones and Behavior, 15*(4), 325–376.

Eaton, G. G., Goy, R. W., & Phoenix, C. H. (1973). Effects of testosterone treatment in adulthood on sexual behaviour of female pseudohermaphrodite rhesus monkeys. *Nature New Biology, 242*(117), 119–120.

Gerall, A. A. (2009). Recollections of the origins of and reactions to the organizational concept. *Hormones and Behavior, 55*, 567–569.

Goy, R. W. (1967). William Caldwell Young. *The Anatomical Record, 157*(1), 3–11.

Grunt, J. A., & Young, W. C. (1953). Consistency of sexual behavior patterns in individual male guinea pigs following castration and androgen therapy. *Journal of Comparative and Physiological Psychology, 46*(2), 138–144.

Phoenix, C. H. (2009). Organizing action of prenatally administered testosterone propionate on the tissues mediating mating behavior in the female guinea pig. *Hormones and Behavior, 55*(5), 566.

Phoenix, C. H., Goy, R. W., Gerall, A. A., & Young, W. C. (1959). Organizing action of prenatally administered testosterone propionate on the tissues mediating mating behavior in the female guinea pig. *Endocrinology, 65*, 369–382.

Wallen, K. (2009). The organizational hypothesis: Reflections on the 50th anniversary of the publication of Phoenix, Goy, Gerall, and Young (1959). *Hormones and Behavior, 55*(5), 561–565.

Wallen, K. (2009b). Sexual differentiation of behavior in nonhuman primates. In *Hormones, brain and behavior* (Vol. 3, 2nd ed.). Elsevier Academic Press.

Wilson, J. G., Young, W. C., & Hamilton, J. B. (1940). A technic suppressing development of reproductive function and sensitivity to estrogen in the female rat. *The Yale Journal of Biology and Medicine, 13*(2), 189–202. 3.

Young, W. (1961). The hormones and mating behavior. In W. C. Young (Ed.), *Sex and internal secretions* (Vol. 2).

Young, W. C. (1969). Psychobiology of sexual behavior in the guinea pig. In *Advances in the study of behavior*. Elsevier.

6 Dorothy Price

Lance J. Kriegsfeld

Abstract

Dorothy Price was an American zoologist responsible for identifying, characterizing, and formulating the principles of hormone negative feedback. During her discovery, Price worked with Carl R. Moore in the Department of Zoology at the University of Chicago. Through a series of early studies, Price and Moore established that sex hormone administration reduced gonadal and accessory sex gland size, but the cause was enigmatic. The key to determining the mechanism leading to this sex-hormone-induced reduction came when Price speculated that there was "reciprocal influence" between the gonads and the anterior pituitary. Studies designed and performed by Price in which she showed that daily implants of pituitary fragments could reverse the effects of gonadal steroid administration provided strong evidence for this proposition and formed the foundation for hormone negative feedback, a central principle in endocrinology. Following these seminal findings, Price went onto receive her PhD at the University of Chicago, joined the faculty there, and continued her major contributions to the field throughout her career.

Keywords

Negative feedback · Sex steroid antagonism · Moore-Price theory · Reciprocal influence · Neuroendocrinology

L. J. Kriegsfeld (✉)
Departments of Psychology and Integrative Biology, Graduate Group in Endocrinology, and The Helen Wills Neuroscience Institute, University of California, Berkeley, CA, USA
e-mail: kriegsfeld@berkeley.edu

R. J. Nelson, Z. M. Weil (eds.), *Biographical History of Behavioral Neuroendocrinology*, https://doi.org/10.1007/978-3-031-12970-4_6

Fig. 6.1 Dorothy Price at the University of Chicago where she joined the Zoology department in 1947 and retired from full-time teaching in 1965. (Credit: University of Chicago Photographic Archive (apf7–00475), Hanna Holborn Gray Special Collections Research Center, University of Chicago Library)

Dorothy Price (1899–1980), recognized for identifying and formulating the principles of what we now call hormone negative feedback, was born in Aurora, Illinois, in 1899 (Fig. 6.1). Price had a long and distinguished career beginning as an undergraduate at the University of Chicago where she received her BS in 1922. Lamentably, this was a time when women were not readily accepted into scientific society and the field was dominated by men. Despite these challenges, Price worked at the forefront of endocrinology, discovering key findings throughout her career that have driven the field until present day.

As an undergraduate at the University of Chicago in the Department of Zoology, Price was trained in embryology, histology, and vertebrate and invertebrate physiology. At the time, Frank Lillie was chair of the department and was soon to publish his groundbreaking findings on the bovine "freemartin," a sterile female co-twin to a male, in 1923. Based on his findings, Lillie proposed that fetal testicular hormones participated in sexual differentiation and that these secretions led to atypical sexual development of the female twin. As an undergraduate, Price was heavily influenced by Lillie, later describing him as "a most impressive and awe-inspiring man for undergraduate and graduate students alike" (Price, 1975). Despite her interest in Lillie's work, Price joined Benjamin Willier's lab at The University of Chicago as a graduate student in 1922 and studied intracellular digestion in *Hydra viridis*. Her work on *Hydra* progressed slowly, and, while on summer vacation, she wrote the department to indicate that she had decided to leave the program. When Lillie heard of her decision, he asked that she return to Chicago to work as a research assistant to support studies in the department funded by his grant from the National Research Council Committee for Research in Problems of Sex. Price agreed, returned to the department, and was initially tasked with preparing histological samples for several members in the department. Soon thereafter, Carl Moore, a faculty member and former student of Lillie's, sought out Price's work more exclusively. Price described Moore as "unfailingly cheerful and amiable" and someone with whom she got along "splendidly" (Price, 1975).

When Price began working with Moore, she initially continued to prepare histological samples and soon began expanding her role to hormone treatments and supervision of the animal colony. It did not take long for her to become invaluable to the lab, a place where she eventually felt like a research partner, discussing research questions/results with Moore and reading/commenting on manuscripts that he was preparing for publication. Like Price, Moore was heavily influenced by the work of, and discussions with, Lillie. As recounted by Price, soon after Moore received his PhD, Lillie suggested that he attempt to reproduce the freemartin phenomenon by "some method" (Price, 1975). At the time, Lillie believed that testicular hormones passed through the connected bloodstream of the twins, inhibiting ovarian function (a phenomenon referred to at the time as sex hormone antagonism), and masculinizing the reproductive system of the female. There were no purified testicular "extracts" at this time, and the only way such a study could be conducted was through testicular transplantation into pregnant animals. Unfortunately, in Moore's work, these experiments resulted in loss of the pregnancy precluding the ability to answer the question of interest.

Fortunately, the 1920s and 1930s were a time of rapid scientific advancements in endocrinology that resulted in the isolation, purification, and eventual synthesis of estrogens, androgens, and progestins (e.g., Allen & Doisy, 1983). Additionally, the work of Smith and Engle in 1927 showing that hypophysectomy (removal of the pituitary gland) resulted in gonadal regression and that implants of pituitary fragments could restore gonadal function established pituitary control of the gonads (Smith & Engle, 1927; Corner & Allen, 1929). However, how the pituitary was controlled to prevent overstimulation of the gonads was not known and was of great interest to Moore and Price, leading to them to focus on this question in 1929.

Through a series of studies in which gonadally intact and gonadectomized male and female rats were provided bull testis extract or the estrogenic hormone, estrin, alone or in combination, they uncovered puzzling findings that were at odds with the prevailing concept of sex hormone antagonism. For example, in castrated males, a mixture of the two hormones did not negatively affects male sex accessory glands (i.e., estrin did not negate the effects of the bull testis extract in promoting male accessory glands). In intact males, treatment with bull testis extract led to gonadal regression. During this time period, Moore and Price were familiar with the work of Steinach and Kun showing that estrin administration to intact males also led to gonadal regression, and they replicated this finding in their own work (Steinach & Kun, 1926). Their observations that estrin given to intact males led to gonadal regression was consistent with the concept of sex hormone antagonism, but their finding that testicular extracts resulted in the same outcome was not.

Moore pored over all of their findings and those of others to come up with a rationale for their results and asked Price to do the same. After dinner one night, an explanation came to Price while she considered all that she had discovered. Price theorized that, if gonadal hormones are controlled by the pituitary, and the presence of estrin or testicular hormones in the bloodstream negatively impacts the gonads, then the pituitary must be the common link. She speculated that gonadal steroids must affect the pituitary to control gonadal stimulation and levels of hormone in the blood (Price, 1975). Such an arrangement could account for the ability of both estrins and testicular extracts to suppress gonadal function.

The next morning, Price excitedly went into Moore's office to describe her theory and begin new experiments to test the idea. In Price's words, "When morning came I hurried to the Zoology department and almost burst into Moore's office to tell him that I thought I had solved the riddle. When I went over my reasoning in detail, he thought for several minutes and then began to warm up; soon he began to consider it a brilliant idea" (Price, 1975). Price repeated her studies in males given estrins and testicular extracts with the addition of daily implants of pituitary tissue or injections of hebin (a gonad stimulating substance from pregnant urine). As Price postulated, pituitary implants or hebin rescued the gonadal degradation seen with hormone injections providing strong evidence that sex hormones acted to inhibit gonadal stimulation upstream at the level of the pituitary. This concept came be known as reciprocal influence or the Moore-Price theory (of course, later called negative feedback). These findings formed the foundation for understanding the regulation of hormones of the hypothalamo-pituitary-axes more generally and were published in 1930 in the paper, *Reciprocal Influence between the Gonads and Hypophysis* (Moore & Price, 1930). They later wrote a more extensive overview of this work and its significance in 1932 (Moore & Price, 1932).

One might ask why, if Price solved the riddle of negative feedback, was the principle called the Moore-Price theory. As indicated previously, unfortunately, it was a sign of the times and Price came up with the name herself given the *zeitgeist*. Although by Price's own account, she and Moore got along exceptionally well and she felt like a partner, she also indicated that "Moore was a male chauvinist, and women (with the possible exception of a few including me on some occasions) were not really to be considered scientifically equal to men. I think he did not realize the depth of his prejudice. My reaction to this can be imagined but this attitude was (and is) a common characteristic of many men. I chose to disregard it then as much as I could" (Price, 1975).

Following this seminal discovery, Price worked to complete her dissertation work under Moore and was awarded her PhD in 1935. Her PhD work was the first to meticulously detail to the cytology, and hormonal actions on, the prostate, and seminal vesicles. Following her PhD, Price continued to work with Moore's team as an associate and close collaborator. As perhaps another sign of the times, it was not until 1947 that Price obtained a faculty position in the Department of Zoology at The University of Chicago. She was soon promoted to Associate Professor and was made Full Professor in 1958. Much of her work as an independent investigator focused on the role of fetal hormones in sexual differentiation and early postnatal development. Price retired from The University of Chicago in 1965 and moved to the Laboratory of Cell Biology and Histology at The University of Leiden (Ortiz, 1981). At Leiden, Price was appointed Boerhaave Professor where she collaborated with Johanna Zaaijer and Evelina Ortiz, a professor at University of Puerto Rico, who spent her summers and leaves in Leiden. She was awarded the Silver Medal of the University of Leiden and, in 1971, she received the University of Chicago Professional Achievement Award for alumni. In addition to being a creative and dedicated scientist, Dorothy Price was described by her friend and collaborator, Evelina Ortiz, as a "remarkable person…inspiring others with her bright inquiring mind, and her dynamic personality combined with a quick sense of humor made her excellent company both socially and at work" (Ortiz, 1981).

References

Allen, E., & Doisy, E. A. (1983). Landmark article Sept 8, 1923. An ovarian hormone. Preliminary report on its localization, extraction and partial purification, and action in test animals. By Edgar Allen and Edward A. Doisy. *JAMA, 250*, 2681–2683.

Corner, G. W., & Allen, W. M. (1929). Physiology of the corpus luteum. *American Journal of Physiology-Legacy Content, 88*, 326–339.

Moore, C. R., & Price, D. (1930). The question of sex hormone antagonism. *Proceedings of the Society For Experimental Biology and Medicine, 28*, 38–40.

Moore, C. R., & Price, D. (1932). Gonad hormone functions, and the reciprocal influence between gonads and hypophysis with its bearing on the problem of sex hormone antagonism. *American Journal of Anatomy, 50*, 13–71.

Ortiz, E. (1981). In memory of Dorothy Price (1899–1980). *Biology of Reproduction, 25*, 450–451.

Price, D. 1975. Feedback control of gonadal and hypophyseal hormones: Evolution of the concept. In *Pioneers in neuroendocrinology*. Springer.

Smith, P. E., & Engle, E. T. (1927). Experimental evidence regarding the role of the anterior pituitary in the development and regulation of the genital system. *American Journal of Anatomy, 40*, 159–217.

Steinach, E., & Kun, H. (1926). Antagonistische Wirkungen der Keimdrüsen-Hormone. *Biologia generalis, 2*, 815–834.

Frank A. Beach

7

Randy J. Nelson

"I detect very little evidence that individuals conducting research on hormones and behavior are overly concerned with what behavioral endocrinology is all about." (F.A. Beach, 1985, p.9).

Abstract

Frank Ambrose Beach is generally considered to be one of the founders of behavioral neuroendocrinology with the publication of his influential book, *Hormones and Behavior*, in 1948; he continued to provide intellectual leadership in shaping the field for the next 40 years. He received his PhD from University of Chicago, working with Karl Lashley and Harvey Carr. He became focused on hormone-behavior interactions, especially in the realm of male sexual behaviors. He published several important papers promoting the value of comparative behavioral analyses, as well as the value of studies of so-called motivated behaviors when American comparative psychology was focused on studies of learning and memory. With two former trainees, he founded the journal, *Hormones and Behavior*, the flagship journal for behavioral neuroendocrinology. Beach was instrumental in training many of the future leaders of the field, and his legacy in behavioral neuroendocrinology continues through his third, fourth, and fifth generation of academic descendants.

R. J. Nelson (✉)
Department of Neuroscience, Rockefeller Neuroscience Institute, West Virginia University, Morgantown, WV, USA
e-mail: Randy.Nelson@hsc.wvu.edu

R. J. Nelson, Z. M. Weil (eds.), *Biographical History of Behavioral Neuroendocrinology*, https://doi.org/10.1007/978-3-031-12970-4_7

Keywords

Behavioral endocrinology · Male mating behavior · Female mating behavior · Maternal behavior · Hormones and behavior

> *"My conviction is that behavioral endocrinology represents an attempt to discover ways in which the endocrine system helps all kinds of animals, including our own species, to "make a living," to survive, and to perpetuate their own kind."*
> *(F.A. Beach, 1985, p.9).*

Frank Ambrose Beach is credited as one of the founders of behavioral neuroendocrinology with the publication of his influential book, *Hormones and Behavior*, in 1948, and he continued to provide intellectual leadership in shaping the field for the next 40 years. Beach was born in Emporia, Kansas, on 13 April 1911, and he died on 15 June 1988 in Berkeley, California. Bertha Robinson Beach was his mother. His father, Frank A. Beach, Sr., was a prominent music professor at the Emporia State University in Kansas. Beach Music Hall is named in honor of the senior Beach, who was one of the leading architects of the American public school music movement and served as the Music Department chair at Emporia State from 1908 to 1935.

Beach's academic trajectory was somewhat ambagious. He attended Emporia State University with the goal of becoming an English teacher, but his first-year grades were poor and his parents sent him to Antioch College in Ohio for a year to improve his grades (Dewsbury, 1998). Upon returning to Emporia State, he enrolled in his first psychology course that was taught by James B. Stroud. This class likely altered the trajectory of his career. Stroud had been a doctoral student with Harvey A. Carr at the University of Chicago. Beach graduated from college at the peak of the Great Depression and was unable to find a high school teaching job as planned. Stroud offered him a research fellowship that allowed Beach to pursue an MS in psychology that was awarded in 1933 from Emporia State. His thesis project assessed the possibility of color vision in rats. Stroud introduced Beach to Carr who was sufficiently impressed by Beach that he provided him a fellowship that allowed Beach to pursue graduate school for a year at Chicago. Teaching jobs and funds remained scarce, but after a year in graduate school, Beach secured an English teacher job in Kansas. While at Chicago, Beach came to admire the brain research work of Karl Lashley, and after a year teaching English, Beach returned to Chicago to continue his PhD work with Lashley; however, Lashley had recently left for Harvard so Beach's dissertation research on the effects of brain cortical lesions on maternal behavior (Beach, 1937) was largely self-directed. Nonetheless, Beach was able to complete his PhD dissertation research project after only 2 years total at Chicago and was in the process of applying for positions as he worked on his dissertation and completed his foreign language requirements (Dewsbury, 1998). He was offered a research position in Lashley's group at Harvard where he continued his brain lesioning approach, which he referred to later as "slash and burn neurobiology," on mating behavior in male rats. During the next 5 years, Beach married, and his son and daughter were born; it was during this time that Beach also provided the intellectual foundation for behavioral neuroendocrinology.

Beach was an intuitive scientist. I recall sitting in a biopsychology proseminar with Beach, Steve Glickman, Irv Zucker, and Paul Sherman as faculty leads. Zucker was regaling the students about the importance of "strong inference" in conducting clean, efficient experiments, and the four steps that Platt (Platt, 1964) suggested following when conducting research when Beach guffawed, "Well I guess I must do weak inference because I've never followed any of those steps in my life." As he noted in an autobiographical review, after taking the advice of an endocrinologist, he injected testosterone into some of the brain-lesioned male rats that had stopped mating and this restored the behavior. According to Beach, "parenthetically, I might as well confess that is about as close as I ever came to using strong inference in the design of an experiment" (Beach, 1985). Despite his "weak inference" approach to behavioral neuroendocrinology research, Beach's research was best characterized as "consistently significant" (Gandelman, 1985 p.2). During the ensuing 5–10 years after he left Harvard, he conducted fewer studies on brain function and became increasingly focused on hormones and behavior (Beach, 1985).

After a year with Lashley, Beach moved to New York City where he worked as an assistant curator in the Department of Experimental Biology housed in the American Museum of Natural History (Beach, 1974). This position allowed him to conduct research full-time, and he was exposed to a broad range of animal studies across several taxa. At this time, he finally completed the language requirements for his PhD, which was awarded in 1940. The chair of the department at the museum died shortly thereafter, and it appeared that the Experimental Biology Department might expire with him. Beach, however, lobbied forcefully to keep the department open. He was successful in this endeavor and was appointed chair—Beach renamed the unit, the Department of Animal Behavior.

While at the museum, Beach audited a course at New York University to learn more about endocrinology (Dewsbury, 1998). Provoked by the lack of behavioral interactions with hormones in the lectures, he wrote a term paper to review what was known. Later, he expanded that term paper into his classic book, *Hormones and Behavior* (Beach, 1948), in which he synthesized what was known about the topic. For many, the publication date of *Hormones and Behavior* serves as the official birth of the field of behavioral neuroendocrinology.

Beach left the museum for a psychology faculty position at Yale in 1946; despite having no experience in teaching undergraduates or graduate students, or working in an academic setting, he was named a Sterling Professor in 1952. He joined several professional societies during this time and was elected president of the Division of Experimental Psychology of the American Psychological Association. His 1949 inaugural address, entitled "The Snark was a Boojum," openly criticized the narrow approach that was characterizing US comparative psychology at that time. For this talk, Beach reviewed all articles published between 1911 and 1948 in the APA flagship biological psychology journal, *Journal of Comparative and Physiological*

Psychology, and argued that the lack of experimental species used in research and the limited topical focus (primarily on learning and memory) was stunting comparative behavioral research and slowing progress (Beach, 1950). Between 1911 and 1948, Beach reported that ~70% of all papers published reflected *Rattus norvegicus* studies and that ~80% of papers were devoted to learning and conditioning, reflexes, and sensory capacity, whereas virtually no studies on reproduction or social behaviors were reported (Beach, 1950). This paper had a sobering effect on the field of biopsychology generally and on his chair who researched learning and memory in rats.

While at Yale, he became more interested in the endocrine and inter-individual aspects of human sexual behavior and teamed up with a prominent anthropologist, Clelan Ford, to write *Patterns of Sexual Behavior* which broadly reviewed human sexuality (Ford & Beach, 1951). He mentored several PhD students and postdocs at Yale including Harry Fowler, Allan Goldstein, Carolyn Hoffberg, Julian Jaynes, Jerome Kagan, Charles Rogers, Burton Rosner, Richard Whalen, and Marvin Schwartz (McGill et al., 1978).

During a sabbatical at the Center for Advanced Study in the Behavioral Sciences at Stanford (1957–1958), he was approached to join the psychology faculty at Berkeley. In 1958 he accepted the offer with several caveats including that he would be able to determine his own teaching assignments, that he would be assigned a full-time secretary and that he would never be asked to be departmental chair (Dewsbury, 1998). The Psychology Department at Yale was dominated by Clark Hull, also a Sterling professor and department chair, who promoted drive theory and who attempted to uncover the general laws of behavior. Hull debated his perspectives most visibly with Edward Tolman, a professor at Berkeley who developed "purposive behaviorism." Tolman also promoted the concept known as latent learning. Beach did not believe in universal laws of behavior, but rather expressed that individuals of different species would develop their own behavioral repertoires in the context of their specific niches and other environmental factors (Beach, 1947, 1955).

At Berkeley, Beach trained many outstanding PhD and postdoctoral trainees in his lab, many of whom went on to be important contributors to behavioral neuroendocrinology including Norman Adler, Joseph Anisko, Gordon Bermant, Lynwood Clemens, Julian Davidson, Donald Dewsbury, Richard Doty, Ian Dunbar, Joyce Fleming, Stephen Glickman, Lawrence Harper, Ann Johnson, Knut Larsson, Fred Leavitt, Burney Le Boeuf, Thomas McGill, Ralph Noble, Timothy Ransom, Benjamin Sachs, Jeffrey Stern, Leonore Tiefer, William Westbrook, and Dale Wise.

During his Berkeley years, Beach published several influential, and often cleverly titled, papers. In 1969, for example, he published "Locks and Beagles" describing canine sexual behavior (Beach, 1969). Two years later, he followed up on some of the themes of the "Snark was a Boojum" in a paper entitled, "Hormonal factors controlling the differentiation, development, and display of copulatory behavior in the ramstergig and related species" (Beach, 1971). In 1975, he formally conceptualized the field with the publication of "Behavioral endocrinology: An emerging discipline" (Beach, 1975). The following year, his somewhat controversial paper, "Sexual attractivity, proceptivity, and receptivity in female mammals," describing female mammals actively seeking mating opportunities was published (Beach, 1976). In 1981, Beach published "Historical origins of modern research on hormones and behavior" (Beach, 1981). One of my personal favorites, but generally not cited, Beach publications was his chapter defending comparative psychology from sociobiology (Beach, 1978). E.O. Wilson had asserted in his book, *Sociobiology*, that studies of animal behavior would soon be swallowed up by neurobiology at one end and sociobiology at the other (Wilson, 1975). Beach lamented in the introduction that he may be perceived as a ghost standing in an academic grave, then systematically dismantled Wilson's arguments in a witty and compelling manner.

Although Beach was considered academically progressive in his writing about female sex partner choice and proceptivity (e.g., Beach & LeBoeuf, 1967; Beach, 1971), he had a reputation for being an academic and intellectual misogynist (Sachs, 1988). Recently, his perspectives on female graduate students have re-emerged, and Beach has become a controversial figure in the history of behavioral neuroendocrinology. His dated and sexist opinions about women trainees may have reflected in part the early and mid-twentieth century norms that were unfortunately widespread in much of academia, but his outdated perspectives increasingly stood out at Berkeley as women interested in studying hormones and behavior were welcomed into the labs of his colleagues, such as Irv Zucker and Stephen Glickman. While the direct impact of Beach's skewed training record on the progression of the field is difficult to ascertain, several of Beach's trainees at Berkeley, particularly Norm Adler, Lyn Clemens, Julian Davidson, Don Dewsbury, Dick Doty, Steve Glickman, and Burney Le Boeuf were especially dedicated to training women. Notably, and unlike other fields in biology, women have come to dominate this scientific discipline (Baran, 2018), likely reflecting the early mentoring of Danny Lehrman and Jay Rosenblatt, among others, and the progressive increase in the number of female mentors. Currently nearly 60% of US faculty in the field of behavioral endocrinology are women (Baran, 2018).

Beach's perspectives on science and scientific training were "old-school." No trainees were allowed to address him as Frank. He was a highly conservative Midwesterner and was famous for not fitting into the Berkeley culture of the 1960s and 1970s. There is no way to estimate the damage, both individually and to the field, caused by his unwillingness to train women; however, his views were nuanced—for example, he deeply admired Josephine Ball, an early researcher in male sex behavior. By the time I interacted with Beach, his views had evolved substantially, and he had trained several women by the early 1980s.

To his credit, Beach became an advocate for many women in behavioral neuroendocrinology (K. Olsen, A. Etgen personal communication). Moreover, he read and cited the work of women, and I know of no evidence that he prevented the advancement of women in science. Certainly, during the later years of his life, he became more gracious in his comments about women in part reflecting the influence of his second wife, Noel Beach.

In addition to his national and international influence, Beach's influences were also felt locally at Berkeley. For example, he was instrumental at establishing and serving as the founding director of the Berkeley Field Station for Behavioral

Research (Glickman & Zucker, 1989). His colony of beagles was studied there; later, the famous behavioral neuroendocrinology studies of hyenas were conducted at the field station.

Beach's scientific contributions were widely recognized. He was elected to the National Academy of Science in 1949 at the age of 38. He was also elected to the American Academy of Arts and Sciences and the American Philosophical Society. He also received the Distinguished Scientific Contribution of the American Psychological Association (APA), the Howard Crosby Warren Medal of the Society of Experimental Psychologists, and the Distinguished Teaching in Biopsychology award from the American Psychological Foundation.

Beach's focus on behavioral analyses, and his ability to synthesize information from diverse fields kept his theoretical perspectives at the forefront of the field. His research was cutting-edge, well-conceived, and meticulously conducted. His book, *Hormones and Behavior*, summarized the information about the topic in the mid-twentieth century and served as a marker from which progress in the field could be measured. He trained many of the future leaders of the field and as a founding co-editor, he helped to establish the primary disciplinary journal, *Hormones and Behavior*. Beach's frequent writing about the limitations of animal models and his interactions with Kinsey, Masters and Johnson provoked him to write about sex behavior in humans (e.g., Beach, 1977). His contributions to the field remain important, and his foundational conceptualizations remain central in current behavioral neuroendocrinology research.

References

Baran, N.M. (2018). How women came to dominate neuroendocrinology. In *Nautilus* (Vol. 63, p. 3).

Beach, F. A. (1937). The neural basis of innate behavior. I. Effects of cortical lesions upon maternal behavior pattern in the rat. *Journal of Comparative Psychology, 24*(3), 393–436.

Beach, F. A. (1947). Evolutionary changes in the physiological control of mating behavior in mammals. *Psychological Reviews, 54*(6), 297–315.

Beach, F. A. (1948). *Hormones and behavior*. Hoeber.

Beach, F. A. (1950). The snark was a boojum. *American Psychologist, 5*(4), 115–124.

Beach, F. A. (1955). The descent of instinct. *Psychological Reviews, 62*(6), 401–410.

Beach, F. A. (1969). Locks and beagles. *American Psychologist, 24*(11), 971–989.

Beach, F. A. (1971). Hormonal factors controlling the differentiation, development, and display of copulatory behavior in the ramstergig and related species. In L. Aronson & E. Tobach (Eds.), *Biopsychology of development* (pp. 249–296). Academic Press.

Beach, F. A. (1974). Frank A. Beach. In G. Lindzey (Ed.), *History of psychology in autobiography* (pp. 31–58). Prentice-Hall.

Beach, F. A. (1975). Behavioral endocrinology: An emerging discipline. *American Scientist, 63*(2), 178–187.

Beach, F. A. (1976). Sexual attractivity, proceptivity, and receptivity in female mammals. *Hormones and Behavior, 7*(1), 105–138.

Beach, F. A. (Ed.). (1977). *Human sexuality in four perspectives*. Johns Hopkins University Press.

Beach, F. A. (1978). Sociobiology and interspecific comparisons of behavior. In M. S. Gregory et al. (Eds.), *Sociobiology and human nature* (pp. 116–135). Jossey-Bass.

Beach, F. A. (1981). Historical origins of modern research on hormones and behavior. *Archives of Sexual Behavior, 15*(4), 325–376.
Beach, F. A. (1985). Conceptual issues in behavioral endocrinology. In R. Gandelman (Ed.), *Autobiographies in experimental psychology* (pp. 5–18). Erlbaum Associates.
Beach, F. A., & LeBoeuf, B. J. (1967). Coital behavior in dogs. I. Preferential mating in the bitch. *Animal Behaviour, 15*(4), 546–558.
Dewsbury, D. A. (1998). Frank Ambrose Beach. *Biographical Memoirs, 73*, 3–22.
Ford, C. S., & Beach, F. A. (1951). *Patterns of sexual behavior*. Harper.
Gandelman, R. (1985). Introduction. In R. Gandelman (Ed.), *Autobiographies in experimental psychology* (pp. 2–4). Erlbaum Associates.
Glickman, S. E., & Zucker, I. (1989). Frank A. Beach. *American Psychologist, 44*, 1234–1235.
McGill, T. E., Dewsbury, D. A., & Sachs, B. D. (Eds.). (1978). *Sex and behavior: Status and prospectus*. Plenum Press.
Platt, J. R. (1964). Strong inference. *Science, 146*, 347–353.
Sachs, B. D. (1988). In memoriam: Frank Ambrose Beach. *Psychobiology, 16*, 312–314.
Wilson, E. O. (1975). *Sociobiology: The new synthesis*. Harvard University Press.

Arnold A. Gerall

8

H. Elliott Albers

Abstract

Arnold A. Gerall (1927–2013) was an American scientist who was trained as a learning theorist in the 1950s. He promoted the idea that behavior could be quantified as a scientific endpoint, a view not widely held in American Psychology at the time. Gerall transitioned from learning theory to physiological psychology and became part of a team in William Young's lab that conducted landmark research on sexual differentiation. This work culminated in the classic paper (Phoenix et al., 1959). He was among the strongest proponents of the concept that perinatal androgen exposure organized the brain by inducing sexual dimorphisms. As a pioneer in the study of hormones and behavior he also contributed to the establishment of the field of behavioral neuroendocrinology. His subsequent research at Tulane University made major contributions to our understanding of how hormones and the social environment influence brain and behavior. He was also a prolific mentor of trainees, many of whom have also contributed significantly to the field.

Keywords

Sexual differentiation · Steroid hormones · Behavioral neuroendocrinology · Sex behavior

H. E. Albers (✉)
Neuroscience Institute, Center for Behavioral Neuroscience, Georgia State University, Atlanta, GA, USA
e-mail: biohea@gsu.edu

R. J. Nelson, Z. M. Weil (eds.), *Biographical History of Behavioral Neuroendocrinology*, https://doi.org/10.1007/978-3-031-12970-4_8

Arnold A. Gerall

Arnold A. Gerall was born in New Haven Connecticut to Barney and Minnie Gerall on March 14, 1927. During high school he had the unusual opportunity to work part-time as a technician at the John P. Pierce Foundation which is associated with Yale University. It was during these 2 years at the Foundation that he was first introduced to physiology. After service in the US Navy where he served as an electrical technician from 1945 to 1946, he attended the University of Michigan where he received a BS in 1949. He went on to graduate school where he earned a MA from the University of Connecticut in 1950 and his PhD from the University of Iowa in 1951. He was attracted to the University of Iowa by the presence of Kenneth Spence who was a prominent interpreter of Hullian learning theory at the time. He was also drawn to the Psychology Department at Iowa because of their philosophy that the scientific approach to studying behavior should be fundamental to the discipline of

Psychology, a view that was not widespread among psychology departments in the United States. Unlike the vast majority of those in Psychology, Gerall continued to be interested in the physiological basis of behavior. Spence once told him that he was the only graduate student that he had worked with who was interested in the physiological basis of behavior. Although heavily influenced by Spence, Gerall's dissertation entitled "A test of the Mowrer two-factor theory of learning" was supervised by Dr. Judson S. Brown.

After completion of his PhD in 2 years, he accepted a faculty position at the University of Rochester. At Rochester he continued his research on learning theory and studied human tracking performance and classical conditioning of pupillary dilation (Gerall et al., 1957). It was at Rochester that he began to work on neural systems. In collaboration with a local physician, he recorded neural responses in the auditory system. He later said that this was the work that "got me into the brain".[1] Because of his training in the Navy, he was comfortable with electronics and built his own equipment (e.g., amplifiers) to record neural responses. In 1956, after 5 years at Rochester, he left to take a position with Dr. William C. Young at the University of Kansas Medical School to pursue a dramatically different research program from his studies on learning theory. There he joined Charles Phoenix and Bob Goy as trainees in the Young lab. He went with the intention of doing electrophysiological recordings of behavior in guinea pigs but also became involved in ongoing research examining whether developmental manipulations such as social isolation or gonadal hormone treatment could modify adult behavior.

Gerall was an important participant in the development of the field that we now call behavioral neuroendocrinology. Its formation was heavily influenced by Young as well as Frank Beach who was then in the psychology department at the University of California – Berkeley. Young and Beach's contrasting personalities, and their starkly different theoretical perspectives energized the development of the field. While Young was a brilliant scientist, he was not a loud or forceful proponent of his own theoretical views, an approach that Gerall emulated throughout his scientific career. Gerall considered the ability to forge ahead despite distractions and even inconsistencies as an ability that most great scientists have. This was certainly the case with Young who persisted despite having had had cancer and with the knowledge that it would reoccur and leave him with little time to complete his objectives. Before his death Young intended to (1) publish 100 articles, (2) finish his edited two-volume book *Sex and Internal Secretions*, and (3) publish one article that will make a huge impression. All of which he accomplished. Gerall considered him to be a truly historical figure.

One of Gerall's research projects in the Young lab examined whether androgens administered developmentally could induce precocious puberty in guinea pigs. These studies were stimulated by Beach's work indicating that androgens could advance puberty in rats by around 20 days and the inability of other investigators to repeat this effect in guinea pigs. Gerall was not content to write these differences off by simply saying they were the result of species differences. Indeed, one of Gerall's

[1] All quotes of Gerall are taken from the taped interview of Gerall by Dr. Kim Wallen in 2004.

lifelong tenets was "when you see species differences you don't end the discussion there," you determine the variables that are causing them. So Gerall tried prenatal injections of androgen to pregnant females to circumvent the species differences in the developmental timeframe between rats and guinea pigs. The use of prenatal injections of androgens became a critical approach to testing the organizational hypothesis.

According to Gerall, the driving force behind the organizational hypothesis was Young, he was the head of the lab and he put it together. "It was Young!" he said. Although it is hard to believe now, biological scientists of the time thought that the quantification of behavior would never reach the level of a scientific endeavor. In Gerall's view the major contributions of the 1959 paper was methodology (Phoenix et al., 1959). He shared with Young the importance of quantifying behavior "It is about behavior – behavior is the endpoint."

Interestingly, although all the authors made important contributions to the paper, it was rare for the lab group to meet and discuss the research. Whereas the group agreed that neonatal exposure to androgens produced changes to the tissues mediating behavior, there were some differences in opinion as to what those tissues might be. It was Gerall's view that if it is a functional change, then it has to have a structural basis and that those structural changes took place in the nervous system. Even at this early stage, Gerall was the strongest proponent of the concept that if the behavior is sexually differentiated then there must be sexual dimorphisms in the brain mediating the behavioral changes.

Gerall interacted with Frank Beach during the origins of the field and felt that Beach brought a critical perspective by emphasizing more genetic and ethological approaches, approaches that had been largely missing in American Psychology. Beach was an articulate and enthusiastic promoter of his own theoretical views and at times a vehement detractor of other theoretical perspectives. Indeed, Beach's disdain for the concept that gonadal hormones have organizational effects was expressed in a 1971 article (Beach, 1971) stating "… the area of speculation or fantasy wherein flourishes the organizational theory of hormonal action is long overdue for slash and burn treatment." Beach indirectly let Gerall know of his displeasure of several of Gerall's studies in the Young lab. Gerall's response to Beach was "the argument will be settled by the data." Despite the intensity of their early interactions, Gerall and Beach became close friends in later years.

Gerall left Kansas and joined the faculty at Tulane University in 1961. His lab was housed in wooden army barracks that had seen service in World War II. From these humble beginnings, he not only established a world class research program he trained an army of around 40 PhDs (affectionately called Arnie's Army). His first grant was funded to look for sexually dimorphic brain areas that he proposed to underlie sexually differentiated behavior. Although he was not successful in being the first to identify sexual dimorphisms in the brain, he developed an influential research program focused on the activational and organizational effects of gonadal hormones. One early focus of his work was to define the prenatal and neonatal time intervals where the organizational effects of gonadal hormones occurred. He along with his graduate student Ingeborg Ward demonstrated that when timed appropriately testosterone gave female rats the ability to display male copulatory behaviors including ejaculatory behaviors (Ward, 1969; Gerall & Ward, 1966). By manipulating the timing of neonatal castration, he and his students were also able to define the timing of the neonatal surge of endogenous testosterone that produced the organizational effects on male sex behavior (Thomas & Gerall, 1969; Gerall et al., 1967a). In other papers of this era, his work defined the relationship between neonatal androgenization and changes in the behavioral sensitivity to estrogen and progesterone in adults and found that the ovaries of rats are active prenatally and have the potential to contribute to female sex differentiation (Hendricks & Gerall, 1970).

Gerall also returned to a topic he had been interested in since his time in the Young lab, the effects of social isolation on adult behavior. Studies in the Young lab had found that social isolation of neonatal male guinea pigs disrupted their adult mating behavior (Valenstein et al., 1955), while earlier studies by Beach in rats had found that social isolation did not alter mating performance (Beach, 1942; Beach, 1958). As noted in Gerall's 1967 paper (Gerall et al., 1967b), Beach (1958) concluded that the discrepancy in the results obtained with the guinea pigs and rats is due to species differences. As noted above, Gerall's view was that when species differences were identified you do not end the discussion there. So, after beginning his lab at Tulane, Gerall in collaboration with his wife Helene revisited the role of social isolation in rats to see if there was a commonality in the effects of isolation across species (Gerall, Ward, & Gerall, 1967b). Indeed, their study did identify substantial deficits in male sex behavior in rats and suggested that the differences among studies might relate to the amount of general play activity engendered by the environment. In related subsequent work, Gerall discovered that prepubertal social experience could overcome the functional consequences of specific forms of brain damage. It was well known that destruction of the medial preoptic area (MPOA) permanently eliminated adult sexual behavior in males in all species studied. In a fascinating paper published in Science, Gerall and his student Dennis Twiggs (Twiggs et al., 1978) reported that male rats given prepuberal lesions of the MPOA display normal sexual behavior as adults only if they receive specific forms of social experience during the prepubertal period.

Likely as an outgrowth of his early studies looking for sexual dimorphisms in the brain Gerall embraced immunohistochemistry (ICC) and was one of the first investigators along with his postdoc and former student Joan King to employ it. They used this technique to visualize neurons containing gonadotropin releasing factor in male and female rats (King & Gerall, 1976; Elkind-Hirsch et al., 1984). During these early years, ICC was "finicky." The technique would work for some runs and inexplicably not for others. Despite his many administrative duties as department chair and a heavy teaching load Gerall could be found in the ICC lab almost daily trying to diagnose and fix the most recent problem. Another example of his persistence. He loved "hands on" lab work and he was a full participant in the lab's research. Even after retirement in 1997 Gerall joined the lab of his former student Jim Zadina and continued his neuroanatomical studies focusing on the endomorphins which Jim had discovered.

One hallmark of Gerall's work was that he was exceptionally careful, and he was unwilling to publish data without fully believing it and fully understanding it. This characteristic is illustrated by some elegant work he conducted using a model for the premature aging of the reproductive system. He showed that neonatally administered testosterone had an aging-like effect on the length of time a female rat displayed estrous cycle and that the duration of cycle was reduced as the dose of testosterone was increased. He went on to examine how the reproductive system measured the interval of time before it shut down by placing the rats in a variety of different light-dark cycles producing different day lengths. Remarkably he found that reproductive system was not counting 24 h days or circadian cycles. Rather the number of light-dark transitions determined when the reproductive system shutdown. Although these data were solid, he never published this intriguing finding because he could not come up with a putative mechanism that might explain it.

Perhaps Gerall's most significant contribution to the field was his mentoring of undergraduates, graduate students, and postdoctoral fellows. Throughout his career Gerall trained a large number of students and many went on to very successful careers of their own ranging from a university president to NIH program officers. Most of his students, however, ended up starting their own research labs. Despite the large number of students in his lab he always found a way to engage each one individually in such a way that a personal link was forged. He was very supportive of his students and encouraged them to follow their interests even when those interests were outside the major focus of his lab (e.g., circadian rhythms (Albers et al., 1981)). Despite these personal connections, Gerall was a formal man, and he did not encourage his students to call him Arnie but when some of them eventually did, he accepted it. He was also a pioneer in promoting the participation of women in behavioral neuroendocrinology, successfully graduating some of the top scientists in the field who happened to be women.

Gerall also impacted the field of behavioral neuroendocrinology in other very significant ways. He was one of the small group of investigators primarily from the Beach and Young labs that formed the nucleus of the field that eventually became the Society for Behavioral Neuroendocrinology (SBN). The development of the field has been detailed elsewhere (Dewsbury, 2003); however, it is important to note Gerall's contributions. Gerall was one of the speakers at the first meeting of the precursor group known as the Eastern Conference on Reproductive Behavior and he hosted the meeting twice at Tulane during its infancy in the 1970s. Gerall was also part of a roundtable held 1976 on the status of the concepts coming out of the research on the organizational effects of perinatal hormone exposure. Although Gerall felt that he was not as articulate as Beach or Goy, he was perhaps the strongest proponent of the idea that perinatal androgen exposure organized the brain by inducing sexual dimorphisms. It was at the roundtable that Beach first acknowledged that these perinatal hormone effects influence the structure of the brain.

Gerall also contributed to the field in numerous other ways. He was a longstanding member of that Biopsychology Study Section that was influential in determining the focus and direction of the field for many years. He was considered one of the best grant reviewers on the panel. He was elected a Fellow of the American

Psychological Association and served the Association in a number of important roles. He was on the Executive Committee of Division 6 – Behavioral Neuroscience and Comparative Psychology from 1989 to 1990 and served as President in 1997. He also served as President of the International Society of Developmental Psychobiology, and he co-edited an influential volume of the *Handbook of Behavioral Neurobiology* (Gerall & Givon, 1992). In 2008, Gerall received the highest honor of the Society for Behavioral Neuroendocrinology, the Daniel S. Lehrman Lifetime Achievement Award.

His personal belief in persistence in the face of adversity was also evident in his personal life. The love of his life, Helene, who was also a talented colleague in the Psychology Department at Tulane (some say exceeding Arnie) suffered a massive stroke in 1968. Despite carrying on with all his professional duties, he was her devoted caregiver for the remaining 44 years of her life. Gerall had many traits that contributed to his success as a scientist and mentor. He was demanding, fair, kind, and tough. He was also defined by a contradiction. He was bold and yet cautious.

Acknowledgments The sources of information for this biography were many. These include my personal memories from when I was a graduate student in Gerall's lab from 1974 to 1979. I also relied on numerous discussions with the Gerall family of trainees over the years. There are also several excellent published accounts of Gerall's life and accomplishments from his students and colleagues (Ward & Ward, 2009; Sachs, 2014). A very valuable source was a Commentary by Gerall entitled "Recollections of the origins of and reactions to the organization concept" published in *Hormones and Behavior* (Gerall, 2009). Finally, I was very lucky to have access to a taped interview of Gerall discussing his time in the Young lab conducted by Dr. Kim Wallen in 2004.

References

Albers, H. E., Gerall, A. A., & Axelson, J. F. (1981). Effect of reproductive state on circadian periodicity in the rat. *Physiology & Behavior, 26*, 21–25.

Beach, F. A. (1942). Comparison of copulatory behavior in male rats raised in isolation, cohabitation, and segration. *The Journal of Genetic Psychology, 60*, 121–136.

Beach, F. A. (1958). Normal sexual behavior in male rats isolated at fourteen days of age. *Journal of Comparative and Physiological Psychology, 51*, 37–38.

Beach, F. A. (1971). Hormonal factors controlling the differentiation, development, and display of copulatory behavior in the ramstergig and related species. In E. Tobach, L. R. Aaronson, & E. Shaw (Eds.), *The biopsychology of development*. Academic Press.

Dewsbury, D. A. (2003). The conference on reproductive behavior: A history. *Hormones and Behavior, 43*, 465–473.

Elkind-Hirsch, K., King, J. C., Gerall, A. A., & Leeman, S. E. (1984). Neonatal estrogen affects preoptic/anterior hypothalamic LHRH differently in adult male and female rats. *Neuroendocrinology, 38*, 68–74.

Gerall, A. A. (2009). Recollections of the origins of and reactions to the organizational concept. *Hormones and Behavior, 55*, 567–569.

Gerall, A. A., & Givon, L. (1992). Early androgen and age-related modifications in female rat reproduction. In A. A. Gerall, H. Moltz, & I. L. Ward (Eds.), *Sexual differentiation*. Plenum Press.

Gerall, A. A., & Ward, I. L. (1966). Effects of prenatal exogenous androgen on the sexual behavior of the female albino rat. *Journal of Comparative and Physiological Psychology, 62*, 370–375.

Gerall, A. A., Sampson, P. B., & Boslov, G. L. (1957). Classical conditioning of human pupillary dilation. *Journal of Experimental Psychology, 54*, 467–474.
Gerall, A. A., Hendricks, S. E., Johnson, L. L., & Bounds, T. W. (1967a). Effects of early castration in male rats adult sexual behavior. *Journal of Comparative and Physiological Psychology, 64*, 206–212.
Gerall, H. D., Ward, I. L., & Gerall, A. A. (1967b). Disruption of the male rat's sexual behaviour induced by social isolation. *Animal Behaviour, 15*, 54–58.
Hendricks, S. E., & Gerall, A. A. (1970). Effect of neonatally administered estrogen on development of male and female rats. *Endocrinology, 87*, 435–439.
King, J. C., & Gerall, A. A. (1976). Localization of luteinizing hormone-releasing hormone. *The Journal of Histochemistry and Cytochemistry, 24*, 829–845.
Phoenix, C. H., Goy, R. W., Gerall, A. A., & Young, W. C. (1959). Organizing action of prenatally administered testosterone propionate on the tissues mediating mating behavior in the female guinea pig. *Endocrinology, 65*, 369–382.
Sachs, B. D. (2014). In memoriam Arnold A. Gerall, Ph.D. *Hormones and Behavior, 66*, 401–408.
Thomas, T. R., & Gerall, A. A. (1969). Dissociation of reproductive physiology and behavior induced by neonatal treatment with steroids. *Endocrinology, 85*, 781–784.
Twiggs, D. G., Popolow, H. B., & Gerall, A. A. (1978). Medial preoptic lesions and male sexual behavior: Age and environmental interactions. *Science, 200*, 1414–1415.
Valenstein, E. S., Riss, W., & Young, W. C. (1955). Experiential and genetic factors in the organization of sexual behavior in male guinea pigs. *Journal of Comparative and Physiological Psychology, 48*, 397–403.
Ward, I. L. (1969). Differential effect of pre- and postnatal androgen on the sexual behavior of intact and spayed female rats. *Hormones and Behavior, 1*, 25–36.
Ward, I. L., & Ward, O. B. (2009). Conceptual and technical contributions from a pioneer in behavioral endocrinology. *Hormones and Behavior, 55*, 480–481.

Roger A. Gorski

9

Zachary M. Weil and Paul E. Micevych

Abstract

Roger A. Gorski was a neuroendocrinologist who helped shape the understanding of the role of sex steroid hormones in the regulation of neural circuitry and behavior. He was trained at the University of Illinois and the University of California at Los Angeles (UCLA) where he spent the remainder of his long career. Early in his career, he helped reveal the role of sex steroids and the relevant neuroanatomy that organizes the preovulatory LH surge in rats. Later, Gorski identified the sexually dimorphic nuclei of the rat anterior hypothalamus and contributed to the discovery of roughly homologous nuclei in human brains. His work during his long career at UCLA contributed fundamental elements to the understanding of sexual differentiation of brain and behavior, as well as the role of sex hormones in this process.

Keywords

Sex differentiation · Sexual dimorphism · Neuroanatomy · Sexual orientation

Z. M. Weil (✉) Department of Neuroscience, Rockefeller Neuroscience Institute, West Virginia University, Morgantown, WV, USA
e-mail: Zachary.weil@hsc.wvu.edu

P. E. Micevych
Department of Neurobiology, David Geffen School of Medicine at University of California, Los Angeles, Los Angeles, CA, USA

R. J. Nelson, Z. M. Weil (eds.), *Biographical History of Behavioral Neuroendocrinology*, https://doi.org/10.1007/978-3-031-12970-4_9

Roger A. Gorski was born on 30 December 1935 in Chicago, Illinois. Born into a family of Polish origin, that he described as "modest but not quite modest," he developed an interest in meteorological phenomena and wanted to attend the University of Chicago to pursue the weather as a career (Stein, 1990). However, when he did not receive a scholarship to attend the expensive private school, he shifted gears and enrolled in the University of Illinois instead where he received BS (1957) and MS (1959) degrees in Physiology. At Illinois, he worked as a research assistant in an animal science lab and was investigating electrophysiological correlates of ovulation in sheep and found it fascinating. As he was looking to apply to PhD programs, an academic advisor recommended that he apply to the University of California at Los Angeles (UCLA).

Gorski, "got into his 1947 DeSoto, drove straight to Los Angeles and the Anatomy Department and was assigned to Charlie Barraclough" (Stein, 1990). This fortuitous decision would set the stage for a long and fruitful career in neuroendocrinology at UCLA. Barraclough was a freewheeling and creative mentor and often dazzled his students with jazz riffs (Stein, 1990). Barraclough had previously shown that postnatal injections of androgens could permanently block ovulation and in many cases produced persistent vaginal estrus but the neural site of this response wasn't then known (Barraclough & Leathem, 1954). In a series of studies, Gorski and Barraclough identified the preoptic area of the hypothalamus as the key locus of the androgen effect. The ovaries and the pituitary largely retained their ability to respond to trophic hormones. Moreover, electrical stimulation of the preoptic area (POA) could induce ovulation in both androgenized and vehicle-treated rats, though the androgenized animals required pretreatment with progesterone. The effect of early sex steroids on the ability of electrical stimulation of the POA to induce ovulation was both steroid – and dose-dependent. Where androgen-treated female rats would continue to ovulate, those treated with estrogens almost never did. The converse was also true, early castration could produce a brain capable of inducing ovulation in transplanted ovaries. Together these data provided strong evidence that sex steroids were responsible for masculinizing the brain, but in the absence of those steroids, a female-typical brain developed. In 1962, on the strength of these studies, Gorski earned a PhD in Anatomy.

After only 2 years together Barraclough would depart on sabbatical and never return. Gorski would thus inherit his grant, postdoc, and technician. The same year as he earned his degree, he was appointed assistant professor of anatomy at UCLA. His early work would follow along similar lines, continuing to isolate the neurochemical and neuroanatomical substrates that mediated hormone responses in the hypothalamus. Data from Gorski's thesis studies provided support for the idea that the control of gonadotrophin secretion was under the dual anatomical control of the POA, which appeared to drive cyclical secretion (the "surge" release and ovulation) while the ventromedial/arcuate region provided tonic inputs. This idea has stood the test of time and numerous experiments. In the mid-1960s, Hungarian neuroanatomist Béla Halász spent a year at UCLA working with Gorski and Charles Sawyer as a Ford Foundation fellow. Halász invented a knife which made precise, stereotactically controlled cuts would precisely deafferent the mediobasal hypothalamus but leave the portal blood vessels intact. Using this technique, Halász and Gorski were able to confirm that inputs from the POA were necessary for ovulation and that animals with deafferentation of the posterior of the MBH (thus sparing inputs from the POA) could still ovulate (Halasz & Gorski, 1967).

Gorski would recall years later that it was a visit from Seymour "Gig" Levine, that prompted him to think about sexual differentiation of the brain beyond the neural control of ovulation. "As we talked, it suddenly dawned on me that the neural control of ovulation was just the tip of the iceberg. We were actually dealing with a much more fundamental process: How does the brain differ between the sexes, and how do sex hormones bring about these differences? From that day I've devoted my career to these questions" (Stein, 1990).

Gorski's most famous discovery was not actually made by him and in fact was contrary to his thinking at the time about sexual differentiation of the brain. Gorski believed until the early 1970s that structural sex differences in the brain were unlikely. This belief stemmed from his observations and numerous findings from the literature that behavioral sex differences were quantitative but not qualitative. Female rats are capable of mounting, and various surgical procedures could disinhibit the lordosis reflex even in normal male rats (Yamanouchi & Arai, 1985; Stone, 1924; Yamanouchi & Arai, 1975). Thus, it seemed to Gorski that sex differences existed in sensitivity of these circuits to the activational effects of sex steroid hormones but that similar structures must exist in both sexes. This belief began to change after the 1971 publication by Raisman and Field that demonstrated sex differences in the location of synapses in the preoptic area (males had nearly all their synapses on dendritic shafts while females had far more on dendritic spines) (Raisman & Field, 1971). Moreover, Gorski's UCLA colleague Art Arnold had recently published with Fernando Nottebohm, the description of marked anatomical sex differences in the songbird brain (Nottebohm & Arnold, 1976). Finally, given his longstanding interest in the anatomy of the POA, Gorski thought it was highly unlikely that his postdoc Larry Christensen, had identified previously unknown sex differences there. However, when Christensen projected slides of the male and female rat brain on Gorski's wall, he was astonished to see that in fact the sexual dimorphism was so dramatic that in the future they would be able to identify the sex of brains without magnification! Gorski later marveled that this relatively obvious dimorphism had been missed even by those looking for anatomical sex differences. He reported that this was one of the only times in his career that he rushed to get data published because he was sure that a sex difference this obvious would inevitably be discovered by others (Stein, 1990).

Gorski named the nuclei the sexually dimorphic nuclei (SDN) and reported the findings in a series of papers thereafter (Gorski et al., 1978; Gorski et al., 1980). The SDN was larger in males, but the cell size and density were similar indicating that sex differences were derived from differences in cell number. The larger number of cells appeared to reflect the increased propensity for survival of cells in the male SDN relative to the female, an effect apparently mediated by inhibition of apoptosis (Davis et al., 1996). Critically, steroid hormones, when administered early in life, could significantly and persistently alter the size of the SDN. Neonatal castration reduced the size of the male SDN by around 50%, and this reduction could be prevented by treatment with testosterone propionate (TP) (Jacobson et al., 1981). Additionally, treatment of neonatal females with TP could greatly increase the size of the SDN but did not completely eliminate sex differences (Jacobson et al., 1981). For female rats, only exposure to testosterone both in utero (from e16) and directly until 10 days postnatal was sufficient to fully masculinize the SDN of female rats (Dohler et al., 1984). Intriguingly, this massive and nonphysiological exposure to androgens did not further increase the size of the SDN in the males though treatment over a similar period with tamoxifen could eliminate the anatomical sex difference. These both strongly implicated gonadal steroids in the development of neuroanatomical sex differences and raised the possibility that they were not the entire story and that genetically encoded differences may also contribute (Gorski, 2002).

Remarkably however, after trying to find differences in the brain that mediated sex differences in behavior, Gorski and his team were stuck with the inverse problem; that is, to try to determine what function this sexually dimorphic structure played in behavior. Unexpectedly, rats that Gorski's lab castrated neonatally and were thus feminized (he called them *fales*) could exhibit estrogen positive feedback (surge release of luteinizing hormone) and lordosis only exhibited partial reduction in SDN sex differences (Gorski, 1967). Moreover, both sexually experienced and naïve rats with lesions in the SDN exhibited normal mating behavior or at most slight disruptions (Arendash & Gorski, 1983; De Jonge et al., 1989). One study that produced truly surprising results was that neonatal transplantation of brain tissue punches containing the SDN could facilitate male sexual behavior in female rats that were primed with testosterone as adults (Arendash & Gorski, 1982).

Perhaps not surprising the identification of the rat SDN set off a search for similar sex differences in the human brain. By the mid-1980s there had been reports of subtle sex differences in corpus callosum anatomy (Delacoste-Utamsing & Holloway, 1982), but in 1985 Dick Swaab reported the existence of a set of nuclei in the human hypothalamus that appeared to both be analogous to the SDN and exhibited similar patterns of sexual dimorphism (Swaab & Fliers, 1985). With his usual attention to anatomical detail Gorski's lab contributed a series of studies that identified four nuclei in the human hypothalamus, that he termed the interstitial nuclei of the anterior hypothalamus (INAH1-4) (Allen et al., 1989). Further, Gorski's group reported that there were sex differences in INAH-2 and -3 (Allen et al., 1989).

Public interest into the INAH exploded in 1991 when Simon Levay reported that while INAH-3 was larger in individuals that were sexually oriented towards women and smaller in individuals sexually oriented toward men. There are numerous methodological and interpretation issues relating to this study that are beyond the scope of the current chapter. However, what is not disputable is that the story exploded in the popular imagination. It has been postulated that interest in the work was magnified because of the political interest in the origins of homosexuality, the relative simplicity of the findings, and because it neatly supported a theory from Gunter Dorner's lab that homosexuality reflected atypical brain development and predicted that homosexual males would have "female" brains. Gorski's group had previously shown that the anterior commissure was sexually dimorphic in the human hypothalamus (larger in females) (Allen & Gorski, 1991), and in 1992, they followed up to report that the anterior commissure was 18% larger in homosexual men compared to heterosexual men and 34% larger than heterosexual women (Allen & Gorski, 1992). Gorski was the subject of much media attention and was often quoted in media reports about the story. He related that when he spoke about the biology of homosexuality to lay audiences, he got one of three types of responses. One: "My son or daughter had 'an accident' and is gay; what can, or should I do?" Two: "I'm gay and thank you for showing it's not my fault." Three: "I'm gay, and how dare you try to tell me it wasn't my choice?" Can't we accept that many people are perfectly happy being gay? (Stein, 1990).

Gorski proceeded through the ranks, earning the title of Full Professor in 1970. In 1980, he was appointed Chair of the Anatomy department, a position in which he served until 1992. He has been recognized by the scientific community on many occasions for his excellence in research. A partial list includes the Oppenheimer Award from the Endocrine Society (1976), membership on the NIH Reproductive Biology Study Section (1974–1978), the Society for the Study of Reproduction Research Award (1983), Honorary Member of The Japan Endocrine Society (1986), Fellow of the American Academy of Arts and Sciences (1990), Carl G. Hartman Award of the Society for the Study of Reproduction (2002), and the Professional Achievement Award from the UCLA Alumni Association (1982). Gorski was an outstanding mentor of graduate students and postdoctoral fellows. Gorski was an exemplary professor of gross and microscopic anatomy to medical students, winning numerous awards including the UCLA School of Medicine Golden Apple Award and the Distinguished Teaching Award from the UCLA Alumni Association. Gorski retired from his beloved UCLA in 2005 and passed away in 2021 after a long battle with dementia.

The Gorski Lab was a large active group that attracted investigators and trainees from around the world. They investigated the long-term organizational influences of gonadal hormones early in life and the short-term activational effects in the adult, on such processes as circadian rhythms, ovulation, reproductive behavior, and synaptic transmission. In this stimulating environment, Gorski trained generations of young scientists who themselves went on to productive scientific careers. His students appreciated Gorski's warmth, personal attention, and scientific acumen. At UCLA, Gorski founded the Laboratory of Neuroendocrinology (LNE), an interdepartmental federation of scientists with common interests in neuroendocrinology and sexual differentiation. Gorski's spirit of collegiality and a transdisciplinary approach to neuroendocrinology continues to spark the LNE.

Personally, Gorski was a gregarious man, who made strong friends in the international neuroendocrine community. He loved to cook and to give a party. Gorski often welcomed UCLA colleagues and international visitors to his home for an evening of food, drink, and lively conversation. Each year, when he was Chair, Gorski gave a New Year's Eve party for his department, for which he had spent days cooking all the food. At these parties, Gorski had a special impact on students and postdocs, often giving the scientific process a human face. His career spanned a time when many significant discoveries were made in the fields of neuroendocrinology and behavior, the results of which were often first made public at national or international venues. Gorski would regale the students with stories about these conferences and the scientists involved. He described their personalities and perspectives, their clever experimental designs, and their noteworthy impact on concepts at the time. More often than not, he included funny anecdotes to which he often chuckled the loudest.

References

Allen, L. S., & Gorski, R. A. (1991). Sexual dimorphism of the anterior commissure and massa intermedia of the human brain. *The Journal of Comparative Neurology, 312*, 97–104.

Allen, L. S., & Gorski, R. A. (1992). Sexual orientation and the size of the anterior commissure in the human brain. *Proceedings of the National Academy of Sciences of the United States of America, 89*, 7199–7202.

Allen, L. S., Hines, M., Shryne, J. E., & Gorski, R. A. (1989). Two sexually dimorphic cell groups in the human brain. *The Journal of Neuroscience, 9*, 497–506.

Arendash, G. W., & Gorski, R. A. (1982). Enhancement of sexual behavior in female rats by neonatal transplantation of brain tissue from males. *Science, 217*, 1276–1278.

Arendash, G. W., & Gorski, R. A. (1983). Effects of discrete lesions of the sexually dimorphic nucleus of the preoptic area or other medial preoptic regions on the sexual behavior of male rats. *Brain Research Bulletin, 10*, 147–154.

Barraclough, C. A., & Leathem, J. H. (1954). Infertility induced in mice by a single injection of testosterone propionate. *Proceedings of the Society for Experimental Biology and Medicine, 85*, 673–674.

Davis, E. C., Popper, P., & Gorski, R. A. (1996). The role of apoptosis in sexual differentiation of the rat sexually dimorphic nucleus of the preoptic area. *Brain Research, 734*, 10–18.

De Jonge, F. H., Louwerse, A. L., Ooms, M. P., Evers, P., Endert, E., & Van De Poll, N. E. (1989). Lesions of the SDN-POA inhibit sexual behavior of male Wistar rats. *Brain Research Bulletin, 23*, 483–492.

Delacoste-Utamsing, C., & Holloway, R. L. (1982). Sexual dimorphism in the human corpus callosum. *Science, 216*, 1431–1432.

Dohler, K. D., Coquelin, A., Davis, F., Hines, M., Shryne, J. E., & Gorski, R. A. (1984). Pre- and postnatal influence of testosterone propionate and diethylstilbestrol on differentiation of the sexually dimorphic nucleus of the preoptic area in male and female rats. *Brain Research, 302*, 291–295.

Gorski, R. A. (1967). Localization of the neural control of luteinization in the feminine male rat (FALE). *The Anatomical Record, 157*, 63–69.

Gorski, R. A. (2002). Hypothalamic imprinting by gonadal steroid hormones. In S. A. Zderic, D. A. Canning, M. C. Carr, & H. M. Snyder (Eds.), *Pediatric gender assignment: A critical reappraisal*. Springer.

Gorski, R. A., Gordon, J. H., Shryne, J. E., & Southam, A. M. (1978). Evidence for a morphological sex difference within the medial preoptic area of the rat brain. *Brain Research, 148*, 333–346.

Gorski, R. A., Harlan, R. E., Jacobson, C. D., Shryne, J. E., & Southam, A. M. (1980). Evidence for the existence of a sexually dimorphic nucleus in the preoptic area of the rat. *The Journal of Comparative Neurology, 193*, 529–539.

Halasz, B., & Gorski, R. A. (1967). Gonadotrophic hormone secretion in female rats after partial or total interruption of neural afferents to the medial basal hypothalamus. *Endocrinology, 80*, 608–622.

Jacobson, C. D., Csernus, V. J., Shryne, J. E., & Gorski, R. A. (1981). The influence of gonadectomy, androgen exposure, or a gonadal graft in the neonatal rat on the volume of the sexually dimorphic nucleus of the preoptic area. *The Journal of Neuroscience, 1*, 1142–1147.

Nottebohm, F., & Arnold, A. P. (1976). Sexual dimorphism in vocal control areas of the songbird brain. *Science, 194*, 211–213.

Raisman, G., & Field, P. M. (1971). Sexual dimorphism in the preoptic area of the rat. *Science, 173*, 731–733.

Stein, D. 1990. Interview: Roger Gorski. *Omni.*

Stone, C. P. (1924). A note on "feminine" behavior in adult male rats. *American Journal of Physiology, 68*, 39–41.

Swaab, D. F., & Fliers, E. (1985). A sexually dimorphic nucleus in the human brain. *Science, 228*, 1112–1115.

Yamanouchi, K., & Arai, Y. (1975). Female lordosis pattern in the male rat induced by estrogen and progesterone: Effect of interruption of the dorsal inputs to the preoptic area and hypothalamus. *Endocrinologia Japonica, 22*, 243–246.

Yamanouchi, K., & Arai, Y. (1985). Presence of a neural mechanism for the expression of female sexual behaviors in the male rat brain. *Neuroendocrinology, 40*, 393–397.

Robert W. Goy

10

Kim Wallen

Abstract

Phoenix, Goy, Gerall, and Young's 1959 publication radically changed the field of hormones and behavior which focused on investigating short-term hormonal activation of sexual behavior. The paper's demonstration that fetal testosterone exposure produced long-term behavioral change led to the Organizational Hypothesis that exposure to androgens during pregnancy permanently altered adult behavior. Robert W. Goy, who came to the WC Young lab with a history of studying conditioning, seemed an unlikely contributor to this revolutionary hypothesis. After joining the Young lab, Goy quickly mastered hormonal research, becoming one of the founders of the Organizational Hypothesis. The hypothesis was controversial and Frank Beach, in particular, publicly argued that hormones did not permanently alter brain development. Goy defended organization, as evidenced in an extensive private correspondence with Beach. In 1976 Beach publicly conceded that the organizational hypothesis was correct. Young, Goy, and Phoenix moved from Kansas to the Oregon Regional Primate Research Center (ORPRC) to develop studies with nonhuman primates and to investigate the development of sex differences in social behavior. With Young's untimely death in 1966, Goy became the head of the ORPRC lab and director of the Division of Reproductive Physiology and Behavior. In 1971 he became director of the Wisconsin Regional Primate Research Center (WRPRC), where he continued developmental studies of monkeys. These studies demonstrated that administering androgens prenatally, depending on timing and dosage, could masculinize reproductive anatomy without also masculinizing behavior and vice versa. Goy was an important founder of behavioral neuroendocrinology and promotor of the role that hormones played in development.

K. Wallen (✉)
Department of Psychology, Emory University, Atlanta, GA, USA
e-mail: kim@emory.edu

R. J. Nelson, Z. M. Weil (eds.), *Biographical History of Behavioral Neuroendocrinology*, https://doi.org/10.1007/978-3-031-12970-4_10

Keywords
Sex difference · Androgens · Testosterone · Development · Organization · Activation · Sexual differentiation

Robert W. Goy, Ph.D. in 1971 on the announcement of his leaving the Oregon Regional Primate Research Center to become the Director of the Wisconsin Regional Primate Research Center. (Photo credit: Oregon National Primate Research Center)

Most behavioral neuroendocrinologists know of Robert W. Goy (Bob) as one of four authors on the 1959 paper from W.C. Young's Kansas laboratory that permanently altered the study of what became behavioral neuroendocrinology (Phoenix et al., 1959). This single study argued that hormones not only had short-term effects that activated behavior but were also involved in organizing the development of the substrate of behavior biasing the individual's development and adult behavior. This completely new type of hormonal action, organizational, significantly expanded the endpoints and manner that hormones were investigated and how they might affect behavior. After this study hormones and behavior went from a discipline that investigated how steroids activated behavior, primarily reproductive behavior, to one in which activational effects of hormones worked in concert with organizational effects of hormones in affecting a wide range of behavior. This view was not initially accepted by some investigators, but Goy was a powerful champion of the organizational hypothesis, helping it to become an essential principle of behavioral neuroendocrinology.

Goy's Early Career

Surprisingly, it was almost accidental that Goy joined the W.C. Young lab and became an author on the 1959 paper. Some historical background illustrates the serendipity in Goy's research career.

Born in Detroit, Michigan on 25 January 1924, Goy's father was a dentist and his mother a homemaker. Little record is available about Goy's childhood and early experience. In 1948 Goy received a BA in psychology from the Michigan State College of Agriculture and Applied Science (later to become Michigan State University) with a senior thesis entitled *"Learning of a Differential Response as a Function of Stimulus-response Asynchronism (!),"* which reflected his interest in behaviorism and its dominant influence on psychology. After graduation Goy left Michigan and pursued graduate study in psychology in the laboratory of Howard F. Hunt at Chicago University, receiving his PhD in 1953. Goy's dissertation, entitled *The effect of electro-convulsive shock on a conditioned emotional response: the relationship between amount of attenuation and strength of the conditioned emotional response* reflected the primary focus of Hunt's laboratory on conditioning (Hunt et al., 1953). While the dissertation was never published, it clearly reflected Goy's interest in conditioning, as well as the interaction between conditioning and physiological events. This was Goy's introduction to what later would become behavioral neuroscience, and it is a short distance from behavioral neuroscience to behavioral neuroendocrinology. None of these terms were in vogue at that time and there is no evidence that Goy was aware of what soon became hormones and behavior.

After receiving his PhD in 1953, Goy, along with his wife Barbara, moved to Hot Springs, Arkansas, to start postdoctoral work in the laboratory of Keller and Marian Breland, two behaviorists who had studied with B.F. Skinner and are credited with creating the first commercial application of behaviorism (Breland & Breland, 1951). They set out to revolutionize animal training, which they did using operant techniques to train animals to perform tasks not native to the animals, such as a guitar-playing duck, a baseball-playing chicken, and many others. Each activity had an operant box where the animal would perform the task each time a cue was given. Initially, the Brelands developed these show boxes for the Larro-feed division of General Mills, and the boxes were placed in feed stores and used in television commercials. Some boxes were designed to take coins and were placed in penny arcades and fair exhibits where a coin dropped into the apparatus became the cue to perform whatever the animal had been trained to do. They created Animal Behavior

Enterprise (ABE) and left Minnesota and moved to Arkansas. Although they published some research, *The Misbehavior of Organisms* (Breland & Breland, 1961), being one that introduced the concept of instinctual drift, which argued that over time animals' behavior would drift towards instinctual behavior to the detriment of conditioned behavior. This contradicted behaviorist dogma, which argued that animals build behavior solely by responding to reinforcement contingencies, whereas the Brelands showed that the same contingencies produced different behavioral responses depending upon the species studied. Goy had an enduring interest in behaviorism, as was common in psychology of the 1950s, and it seems likely that it was the Breland's focus on behaviorism and the availability of a position that attracted Goy to Arkansas. There is no evidence that Goy knew anything about the W.C. Young Lab about 450 miles North in Lawrence, Kansas, or anything about hormones and behavior.

Elliot Valenstein, a graduate student in W.C. Young's laboratory, was, like Goy, a Michigan native, who graduated from the University of Michigan. While Goy was in Chicago, Valenstein's wife, Theresa, met Goy and his wife at a meeting, and she reported to Elliot that she had met a very nice couple in Chicago (Baum et al., 1999). Some months later Valenstein and his wife were driving from Lawrence, Kansas, to a scientific meeting in Galveston, Texas. As described by Valenstein (Baum et al., 1999), their route took them through Hot Springs, Arkansas, which they remembered was where the Goys lived. They called the Goys and were invited to visit them. They were surprised when they arrived at the Goy's house that it appeared that most of the Goy's possessions were on the porch and that Bob and Barbara were preparing to leave (Baum et al., 1999). Goy could not tolerate doing commercial animal training for the Brelands instead of research. Even though the operant conditioning was automated, the ABE was so successful that little or no time was left for research. Typical of Goy when he was fed up with some activity, he would make a snap decision, even if it meant an uncertain future. Thus, the Goys were leaving Hot Springs and heading to Chicago to seek Hunt's help in finding a position. According to Valenstein in meeting with the Goys in Hot Springs, he wondered out loud whether Young would hire Bob. A few weeks after moving to Chicago, Goy asked Valenstein whether Young would hire him. It turned out that Young, who was in an anatomy department but studied behavior, had been thinking of adding another psychologist in a postdoctoral position (Baum et al., 1999). Thus, Goy joined the lab in 1954, followed soon by Charles H Phoenix in 1954 and Arnold A. Gerall in 1956.

Goy entered the Young lab supported as a Public Health Service Research Fellow of the National Institute of Mental Health. His entry must have been daunting; Goy at that time had never published a scientific paper and had no experience with hormones, behavior, or anatomy, which was a problem for a position in a department of anatomy. The field Goy was entering was then dominated by Frank A. Beach and especially by W.C. Young. In the 1930s Young and collaborators had shown that the estrous behavior of female guinea pigs varied with the state of her ovaries. Steroid assays were not to come about until 1967; thus, the anatomy of the ovary, which indicated follicular development, ovulation, and corpus luteum formation served as a proxy for the underlying hormonal changes. Young's lab in the mid-1930s developed the first hormonal replacement therapy for ovariectomized guinea pigs (Dempsey et al., 1936) demonstrating that the female had to be exposed to at least 24 h of exogenous estradiol followed by a single injection of progesterone which activated the female's expression of lordosis indicating her sexual receptivity. This hormonal regimen has been found effective in multiple rodent species, but does not work in many nonrodent mammals, such as nonhuman primates and humans.

Goy was a quick study and rapidly integrated himself into the behavioral work as well as learning gross anatomy so that he could teach in the department. By 1957 Goy published his first paper and published five papers in total that year on a range of topics. Appropriately, Goy's first publication was as a co-author with Valenstein entitled *"Further studies of the organization and display of sexual behavior in male guinea pigs"* (Valenstein & Goy, 1957). The use of "organization" in this article might be seen as foreshadowing what was to come, but "organization" was not used in a manner having anything to do with hormones and simply meant how a behavior was put together, whether that organization resulted from experiential and/or physiological factors. This paper, published in 1957, was submitted for publication in October of 1955, a year after Goy arrived in the Young lab illustrating how rapidly Goy developed research. In the case of this article, he may have been invited to work on a topic already developed by Valenstein as part of his ongoing dissertation.

Goy's publications reflected an eclectic range of research interests ranging from the length of gestation (Goy, Hoar, & Young, 1957) in guinea pigs to the role of soma in sexual behavior (Goy & Young, 1957). Soma, a term that has fallen out of favor, refers to what we would now call the body but was used in more limited fashion in the Young lab becoming a synonym for neural systems. Goy's initial first-authored paper, with Young, "Somatic basis of sexual behavior patterns in guinea pigs: Factors involved in the determination of the character of the soma in the female," addressed what was meant by 'soma' and what factors might be considered relevant to understanding the role that soma played in behavioral effects of hormones. The paper starts with the following:

> … once the threshold necessary for hormonal stimulation has been reached, the character of sexual behavior displayed in response to that stimulation is determined by the nature of the soma or substrate on which the hormones act rather than by qualitative or quantitative variations in endogenous hormones. (Goy & Young, 1957, p.144)

This focus on the nature of the underlying substrate that is responsive to hormone action was only a hypothesis at this point and presented a relatively radical departure from the stimulus-response explanations used by behaviorists. Little was actually known at this time about the nature of soma and the central problem facing Young's lab was how to identify factors that determine soma and elucidate how hormones contributed to the nature of soma. This focus, however, asserted that there was a physical substrate underlying behavior and that substrate was likely neural.

Key evidence suggesting the consistent nature of the substrate comes from studies in the Young lab of two inbred strains of guinea pig. Strains 2 and 13 were the last remaining strains from Sewell Wright's original inbreeding study (Wright, 1923). Goy and Valenstein led studies of these strains assessing their sensitivity to steroids for activating male and female sexual behavior (Goy & Young, 1956; Goy & Jakway, 1959; Valenstein et al., 1955). These studies showed strain differences in male and female sexual behavior and that these differences had high heritability, supporting the notion that the substrate underlying sexual behavior differed consistently between the strains. This provided the basis for identifying hormonal factors that could permanently modify the soma.

Phoenix, Goy, Gerall, and Young, 1959 and Beyond

At some point during the period from 1952 to 1958, Young focused his work on three things he wanted to accomplish in the coming years. Young had experienced a bout of cancer that was in remission, but according to Gerall (Gerall, personal communication, 2004), Young was convinced that the cancer would reoccur, which it did in 1965 (Goy, 1967). Young felt his time was severely limited (Wallen, 2004) and thus Young created the three goals. The first was to complete the third edition of *Sex and Internal Secretions*, which was the "bible" of hormones and behavior and which Young had taken over from Edgar Allen (Allen, 1932). Progress on the new edition was slow, but Young did complete it in 1961 (Young, 1961a). The second goal was to publish 100 research papers, which Young did. Third was to complete a study that would have a significant impact on the field of hormones and behavior. Bringing together a number of threads from previous studies, Young proposed investigating whether exposing genetic females to androgens prenatally would alter their behavioral development. It is not clear whether Young proposed that the prenatal effects of exposure to androgens would be permanent, but it likely was his hypothesis that the effects of prenatal androgens would differ from adult activational effects which are transitory. Young was aware of Vera Danchakoff's work in the middle 30s in which she injected testosterone directly into fetal guinea pigs and investigated their adult sexual behavior (Dantchakoff, 1938a, b). She reported that the injected guinea pigs showed genital masculinization as well as behavioral masculinization. Unfortunately, there was no control group and Danchakoff was apparently unaware that female guinea pigs show significant mounting behavior, which is increased by injections of androgens. Danchakoff's work foreshadowed the organizational hypothesis but was poorly controlled leading to little adoption of her views on hormones and development. Young, on the other hand, interpreted Danchakoff's work as demonstrating the possibility that prenatal androgens could permanently alter the development of the nervous system. This seemed to be an issue worth pursuing that might have a substantial impact on the field.

Exactly how the study was developed and how it was decided who would work on it is unclear. Gerall reports (Gerall, personal communication, 2004) that the investigators working on the project worked relatively independently; there were no lab meetings to develop the project or to discuss how the data would be analyzed. Even the final write up was done relatively independently with pages passed around but reflecting essentially independent work. When the study produced evidence that suggested that prenatal exposure to androgens resulted in permanent behavioral masculinization of genetic females, there was disagreement about how to interpret this. One faction argued that the nervous system had been permanently modified, while the other faction argued that this was not the case and that androgens simply modified function and not anatomy. Young, in particular, argued that as an anatomist that there was no anatomical evidence to support the argument that prenatal androgens masculinized the nervous system as they did genitalia. This issue was unresolved when the paper was submitted to Endocrinology for publication. The legacy of not resolving this issue resulted in one of the striking aspects of the 1959 "organization" paper (Phoenix et al., 1959) in that it has two ending paragraphs that disagree with each other. The first, written by Goy (Gerall, personal communication, 2004) is as follows:

> The nature of the modifications produced by prenatally administered testosterone propionate on the tissues mediating mating behavior and on the genital tract is challenging. Embryologists interested in the latter have looked for a structural retardation of the Mullerian duct derivatives culminating in their absence, except perhaps for vestigial structures found in any normal male. Neurologists or psychologists interested in the effects of the androgen on neural tissues would hardly think of alterations so drastic. Instead, a more subtle change reflected in function rather than in visible structure would be presumed (Phoenix et al., 1959, Page 381)

Goy argued that the influence of prenatal androgen is on function instead of physical structure. In other words, androgen-influenced structural modification of the central or peripheral nervous system was rejected. Young proposed using the phrase "tissues mediating mating behavior" never letting the reader know what comprised those tissues. The penultimate sentence rules out that neural tissues are under discussion.

Gerall (personal communication, 2004) contributed the last paragraph which stated:

> Involved in this suggestion is the view that behavior may be treated as a dependent variable and therefore that we may speak of shaping the behavior by hormone administration just as the psychologist speaks of shaping behavior by manipulating the external environment. An assumption seldom made explicit is that modification of behavior follows an alteration in the structure or function of the neural correlates of the behavior. We are assuming that testosterone or some metabolite acts on those central nervous tissues in which patterns of sexual behavior are organized. We are not prepared to suggest whether the site of action is general or localized. (Phoenix et al., 1959, Page 381)

This paragraph leaves little doubt as to that the "tissues mediating mating behavior" are neural tissues. Young expressed little concern about the contrast between the two views arguing that history would decide which was correct (Gerall, personal communication, 2004). It is ironic that Goy championed the functional argument over the anatomical argument as he became known for his view that prenatal androgens modified the nervous system, having abandoned the functional argument by 1964 when Young et al. (1964) published "Hormones and Sexual Behavior" in *Science*. In this article the authors argued that prenatal hormones modified the substrate that hormones acted on (soma) to activate sexual behavior. Soma was presumed to be neural tissue. Aside from Phoenix et al. (1959), the notion of altering function without altering neural anatomy was not argued by the Young lab. In addition to supporting the idea that hormones could alter neural anatomy, Young et al. (1964) argued for the more radical notion that these findings, obtained from nonhuman mammals, applied to humans as well, an idea that remains controversial to the present but identifies a primary driving force of Young's research program.

After the publication of the 1959 paper, Goy followed up with a study that expanded our understanding of the parameters of organizational effects of androgens. The primary concern, which would be addressed several times in Goy's career, was the timing of androgen exposure on masculinization and defeminization. Goy, Bridson, and Young et al. (1964) administered testosterone propionate (TP) starting at gestation day (GD) 15, 20, 25, 30, 35, 40, or 50 of the 70 day guinea pig pregnancy. TP was administered in different amounts and for different durations, from 15 to 30 days. Since TP was injected daily (5 mg/day for days 1–6 and 1 mg/day for the rest of the treatment), total androgen exposure varied between groups, varying from 40 mg to 75 mg. Androgen-exposed females and control males and females were gonadectomized as adults and tested for lordosis response to a sequential estradiol (E_2) and progesterone (P) regimen that activates female sexual receptivity in untreated females. It was apparent from the findings that one of the most critical variables affecting masculinization and defeminization of genetic females was the timing of the treatment. One hundred percent of females exposed to 15 days of TP treatment, starting on GD15, but only 44% of females whose TP treatment started on GD 30 became sexually receptive after the sequential E_2 and P treatment (e.g., the GD30 females had become defeminized). Duration of treatment (which also affected total TP exposure) also had an effect in that extending the duration of treatment to from 15 to 25 days resulted in 88% of the females started on GD15 becoming sexually receptive as adults, but only 8% of the females whose 25 day treatment started on GD 30 became sexually receptive. This was an important finding as it not only provided a replication of the 1959 paper's findings but also suggested that the developing nervous system had very specific periods of sensitivity to prenatal androgen (Young et al., 1964).

Monkey Studies and the Move to the Oregon Regional Primate Research Center

Soon after publication of the 1959 paper, the Young group in Kansas, now missing Valenstein and Gerall, who had both followed their independent lines of research, thought it important to investigate organization in a non-rodent species and settled on studies of rhesus monkeys. At the time there was no national primate research center program, so the group took advantage of other facilities to start monkey research. Phoenix moved to Cincinnati, OH where the Christ Hospital Laboratory had a small monkey colony that could be used to create timed pregnancies that they thought would allow accurate timing of testosterone administration to the pregnant females (Baum et al., 1999). Phoenix was attempting to do something that had not been previously done and where there was little background information. His goal in Cincinnati was to create monkey pseudohermaphrodites by administering prenatal testosterone as had been done in the guinea pig.

Goy went to Madison, WI, and the laboratory of Harry Harlow who headed the Primate Lab of the Department of Psychology of the University of Wisconsin-Madison to learn how to observe juvenile behavior in monkeys. Leonard Rosenblum, a postdoctoral fellow in Harlow's lab, had collected the first data showing that juvenile males, long before puberty engaged in quite different behavior than did juvenile females, particularly, play and mounting behavior. Goy was to learn how to observe juvenile behavior to be used in evaluating Phoenix's pseudohermaphrodites to see if the females' juvenile behavior had been masculinized by prenatal TP. From a theoretical standpoint, this was a very important investigation as the juvenile behaviors that showed sex differences occurred during a time when the monkey's gonads are quiescent. Thus, these sex differences were not in hormonally activated behavior. If the females prenatally exposed to androgens showed a masculine pattern of juvenile behavior, then it would be irrefutable evidence that the difference in the treated females' behavior was not the effect of hormonal activation but reflected that the function of the nervous system had likely been modified by prenatal androgen exposure. Once modified hormones were not necessary for exhibiting behavior. If such effects were seen, then it could be interpreted as supporting Goys' original interpretation of the 1959 study but would integrate function with the actions of prenatal hormones.

The plan was to transport the treated pregnant females created by Phoenix by van from Cincinnati to Madison and place them under the care of Goy. The young were born in Madison and Goy observed and recorded their behavior in what became a long-term systematic study of their behavior (Baum et al., 1999).

This was a very risky project as little was known about monkey social behavior. In addition, steroid assays had not yet been invented so timing pregnancies had to be done using a calendar method that started counting with the onset of menstruation. It wasn't discovered until after the advent of steroid assays that the relationship between menstrual onset and ovulation was highly variable across females, but at the time there was no alternative. Phoenix was successful in creating the first pseudohermaphrodite monkeys in Cincinnati. The first images of these masculinized females and evidence that the juvenile behavior of pseudohermaphrodite females was masculinized appeared in Young et al. (1964). The process of creating these monkeys had been difficult, but it had been successful.

Fortunately for the Young lab the US government created a national primate center program to greatly increase laboratory primate research. In 1962 the Oregon Regional Primate Research Center (ORPRC) opened and recruited Young to create and direct a Division of Reproductive Physiology and Behavior. Young accepted and moved to Oregon in 1963. Part of the agreement with Young was positions for Goy and Phoenix who were also hired as Associate Professors. ORPRC became the only primate center with a division focusing on hormones and behavior and was the only place in the world where the effects of prenatal androgen exposure on sexually differentiated behavior was being studied in both guinea pigs and rhesus monkeys.

The Young lab continued work in both guinea pigs and rhesus monkeys. As the only place where studies in both species were possible the lab recruited a number of graduate students and post docs. Several of the postdocs were Frank Beach PhD's (Lynwood Clemens, Norman Adler, Gray Eaton) reflecting the close association of the Beach and Young labs.

As Young had predicted his cancer did return leading to his death in April 1966, less than 3 years after he moved to Oregon. On Young's death, Goy became the director of the Division of Reproductive Physiology and Behavior and the principal investigator on the NIMH grant that funded the laboratory's work.

The Organizational Hypothesis and the Ramstergig

Publication of the Organizational Hypothesis in Phoenix et al. (1959) did not immediately have a noticeable influence on the field. During the first decade (1959–1969) after publication, the paper was cited approximately 50 times (Wallen, 2009). This likely reflects that until the publication of the hypothesis no laboratories were working on permanent effects of hormones on sex-specific behavior. Anatomical effects of prenatal steroids had been actively pursued by Alfred Jost (Jost, 1953) and Dorothy Price (Price et al., 1967), but behavior was not an endpoint in their studies. As the idea that hormones could permanently alter neural anatomy and function permeated the field of hormones and behavior, the impact of the 1959 paper increased dramatically. The field went from one that studied factors that affected hormonal activation of hormone-sensitive behavior to one that focused on two different but related processes, activation, and organization of behavior. This firmly established behavior as an important window into neural function.

Not everyone accepted the validity of the organizational hypothesis. Frank Beach argued strongly that steroid hormones during pregnancy or early development did not organize behavior. His objection to the organizational hypothesis culminated in the publication of a paper entitled "Hormonal factors controlling the differentiation, development, and display of copulatory behavior in the ramstergig and related species" (Beach, 1971), which was shockingly critical of the hypothesis. Interestingly, the only member of the Young laboratory who was criticized by name, was Gerall, which Gerall noted (Gerall personal communication, 2004). This may have reflected that Beach was concerned about the possible negative response the Young lab (now the Goy lab) might have. By the time the ramstergig paper was published Gerall had left the Young lab and started a faculty position in Tulane and thus his response to Beach's paper was of less concern than the rest of the Young Lab, Goy in particular. Beach and Goy extensively corresponded and Beach's concern about Goy's reaction is apparent in an undated letter to Goy that begins with "I send you the enclosed manuscript with some trepidation." The manuscript was a not-yet-published version of the Ramstergig paper that Beach was sharing with Goy (personal communication, Beach-Goy correspondence). The rest of the letter continues to elaborate on

Beach's trepidation, saying, for example, "What I hope does not need saying is that I have always held Bill Young in very high esteem and as a close personal friend." Later Beach writes "I sincerely hope that what I here intend to be a totally impersonal, objective and scientific critique will not be interpreted by you as a devaluation of your research or a personal assault." Beach was clearly aware that he might be stepping over a line with Goy, his good friend. The extent to which he had stepped out was apparent in a letter dated June 22, 1970, from Goy to Beach in which Goy, in his somewhat opaque style, described his reaction to the Ramstergig paper as follows (Beach-Goy correspondence): "The ramstergig was delightful, clarifying, and in some parts, hilarious. I especially liked the Sperry-type explanation for organization." [Beach had presented images of an organized and disorganized brain, that were actually of a frog tectum like those studied by Sperry.] Goy continued: "I have to admit that I dismissed some parts (with pique) as too banal, exaggerated, and misrepresentative, but then why shouldn't I? Keep up the good work. Charles and I hope next year to do a devastating rebuttal." Was Goy offended and if so, how much? I suspect Goy's graduate students would recognize his construction of very positive comments intertwined with negative comments leaving one to guess the depth and direction of Goy's true feelings. Personally, it took me several years in Goy's lab to realize that "I wouldn't do it that way" strictly meant "don't do it!" On 25 June 1970, Beach replied to Goy's letter with "I am …pleased to learn that you find you can at least tolerate the Ramstergig paper. I would have been greatly surprised if you failed to react quite negatively to some of the sections but am glad that this wasn't your overall response."

The Ramstergig paper again appears later in the Goy correspondence, brought up by Beach. Three years later, on 17 October 1973, Beach sent a mimeographed letter entitled "Verbatim Quotation" to both Goy and Phoenix. Whether it was sent to anyone else is not known. The letter had a quotation from 1948 and asked recipients to identify who made the statement, which addressed what type of actions hormones during "embryonic differentiation" exerted and contrasted activation effects on pre-existing "arcs," presumably neural, or as "… organizers inducing certain connections amongst special nervous centers." It appeared that Beach's intent was to show that the issues raised in Phoenix et al. (1959) moving the field to consider hormones as organizers or directors of development had already been raised in 1948. Goy and Phoenix both recognized that the quote was from work by Martins and Valle (1948) in a paper on micturition patterns in the dog, a behavior that ultimately convinced Beach that the hormonal organization concept was real. On 25 October 1973, Goy wrote Beach a defensive letter where he identified the source of the quotation and delineated all of the places, he had presented Martins and Valles findings and considerations, ending the letter with "I agree with your implied opinion that these workers should not endure further neglect by the scientific community, but they really haven't done much since then have they?" Interestingly, Martins and Valle (1948) paper published 11 years before the 1959 paper was not cited in Phoenix, Goy, Gerall, and Young (1959).

Beach replied on 5 November 1973 with "Whoa! Down boy, Down! I was just having fun when I sent out this quote from M&V but I can see from replies I received from you and Charlie that I touched a nerve. … Guess my Ramstergig paper must have left a scar." Beach went on to describe his work on dog urination posture which is sexually dimorphic with females squatting and males lifting a leg allowing them to urinate to the side. Beach was finding that urination posture was affected by hormones during pregnancy, but did not require any activation by hormones, testosterone in particular. Beach added "To some extent it resembles your own evidence concerning sex differences in play in the pseudohermaphroditic female" (monkey). "The effects of early treatment are there, Brother, without any necessity for concurrent stimulation by exogenous hormones. Who was the idiot who claimed that all that prenatal treatment does is to change thresholds to concurrent hormonal stimulation???" In this letter Beach capitulates to the new paradigm that Goy has been championing since Young et al. (1964) that hormones do not simply have activational effects but also organizational effects that direct development. Goy's response 10 days later, while defensive is also typical Goy as after one reads his statement one is not completely certain of his argument. Goy writes "I have no scars from the Ramstergig paper, which I thought did a much-needed job of exorcising. While it did not put the devil in hell with as much artistry as Boccaccio, it at least removed God from the Heavens. I really thought you were worried about where the quotation came from…" Goy proceeds with a long description of all the cases where hormones have organizational, but not activational effects, including cases of dimorphic characteristics, such a canine size, where hormones appear to have no effects on the difference between males and females. Goy ends the letter with "… Please don't ever stop sending your little epistles, whether in anger or despair, or both."

Phoenix's response to Beach's query gives some idea of why Beach took the approach with the mystery quotation. Phoenix wrote to Beach that the Martins and Valle had not gone unnoticed "… nor did the following statement by an eminent scientist published 4 years later (1952): 'It is conceivable that prenatally secreted gonadal hormones might act as 'organizers', influencing the laying down of nervous connections which later are involved in the mediation of sexual behavior,' Beach (1952) Page 214. The author rejected the possibility." The 1952 paper was one of several points where Beach had data consistent with the organizational hypothesis but rejected that explanation. Instead, Beach saw the data as reflecting the effect of prenatal hormones masculinizing female genital anatomy, but not the neural structures underlying masculine sexual behavior (Baum 1990). Beach's dog urination studies led him to the realization, as he confessed to Goy (above), that the dog and monkey studies showed behavioral masculinization without the need of hormonal activation and were compatible with the organizational hypothesis. One can only imagine how challenging it must have been for Beach to realize that the construct that radically changed the field of hormones and behavior was once within his grasp, but he had rejected it.

Many date the resolution of the debate about the organizational hypothesis to 1976 at the Eastern Conference on Reproductive Behavior (ECRB) meeting in Saratoga Springs, New York. Goy and Beach agreed to participate in a roundtable on sexual differentiation where many anticipated verbal fireworks when each would argue the position on organization for which they were known (Dewsbury, 2003). To the surprise of many, including myself, Beach announced that, primarily because of his dog work, he now agreed with Goy and Phoenix that hormones early in development organized neural structures (Dewsbury, 2003). The correspondence between Beach and Goy discussed previously suggests that Goy and Phoenix were not likely surprised by Beach's change of heart, though I never remember Goy ever suggesting it was a possibility that Beach would accept the organizational hypothesis.

After that meeting, the organizational hypothesis was widely accepted. Researchers clarified aspects of the hypothesis and sought organizational effects during other times than the fetal and perinatal period. The most promising time is pubertal organization when many species undergo substantial reorganization (Sisk & Zehr, 2005). These clarifications further defined the parameters of organization but didn't challenge the basic concept (Wallen, 2009). The field of hormones and behavior had been transformed from a field focused on hormonal switches that activated preexisting neural structures to one in which there were activational switches, but also hormonally directed permanent alterations to, and creation of, neural structures. Beyond championing the organization hypothesis, Goy had an impact on a variety of activities that influenced behavioral neuroendocrinology.

Director Wisconsin Regional Primate Research Center

In 1971 Goy was named Director of the Wisconsin Regional Primate Research Center (WRPRC), a position he held for 18 years. He moved his laboratory from Oregon to Wisconsin and succeeded Harry Harlow as director of the WRPRC. Goy brought the first behavioral neuroendocrine lab to the WRPRC. All was not smooth sailing though. In addition to being director of the WRPRC before Goy, Harlow was also director of the Primate Lab, which was part of the University of Wisconsin – Madison Psychology Department. Because Harlow was in charge of both facilities, it apparently was unclear what expenditures of the WRPRC, funded by an NIH base grant, were for WRPRC researchers and which served the Primate Lab researchers. NIH tasked Goy with clearly separating the two facilities (Goy, personal communication). This meant that services paid from, the WRPRC base grant, such as a nursery for infants, had to be put on a charge-back basis for Primate Lab researchers, leading to significant enmity between the Primate Center and the Primate Lab. The conflict between the Primate Center and the Primate Lab never really disappeared during Goy's 18 year tenure as director (Phoenix, 1999),

At Wisconsin, Goy initiated studies on how early experience affects the development of adult reproductive behavior in rhesus monkeys. He was the first to recognize that the standard laboratory rearing paradigm, invented at the University of Wisconsin by Harlow produced seemingly appropriate juvenile social behavior but deficient adult sexual behavior, particularly for males (Goy & Wallen, 1979). Goy developed a unique laboratory rearing environment using carefully selected 4–5 member groups of mothers and infants. The environment preserved important aspects of the social environment a rhesus monkey would normally encounter in its natural habitat. With colleagues David Goldfoot and Kim Wallen, Goy demonstrated the important role that early experience plays in the expression of juvenile and adult sex differences in behavior. This research, in addition to continuing studies of the prenatal hormone role in behavioral development, advanced the notion that the prenatal hormonal environment produces behavioral predispositions which are then shaped and molded by early social context. In Goy's view, both biological and social influences were crucial to the development of masculine and feminine patterns of behavior.

Goy continued studies of monkey development and the role prenatal hormones had in the development of sex differences in behavior, advancing our understanding of the scope of organizational effects of hormones on behavioral development. What was met with skepticism in 1959 is now a central part of behavioral neuroendocrinology. Goy was present at the beginning. After serving 18 years as Director of the WRPRC, Goy retired in 1989 and died 14 January 1999, in Madison, Wisconsin.

Other Contributions to Behavioral Neuroendocrinology

Goy published over 110 refereed papers, each one making an important contribution to behavioral neuroendocrinology. He also was active in serving on NIMH study sections to assist in evaluating research, proposals I have selected some of his activities and several of his papers that were of particular importance to the field.

Editor of *Hormones and Behavior*

In 1969 Beach created the journal, *Hormones and Behavior*, edited by Beach, Richard Whalen, and Julian Davidson, members of Beach's lab (Wallen, 2020). *Hormones and Behavior* was the first journal dedicated to the emerging field of behavioral neuroendocrinology and developed into the primary outlet for behavioral neuroendocrinologists. From 1969 to 1986, the journal changed its structure, adding associate editors from outside of Beach's laboratory in 1973, the year that Goy was named an Associate Editor. In 1977 Goy became one of the Editors in Chief (EIC) along with Beach and Whalen. One characteristic that was constant until 1986 was that all EICs and associate ditors were men. Starting with Issue 3 of Volume 20, Beach became an emeritus editor and Goy and Whalen became co-EICs. With this issue four women were named associate editors signaling the start of a gender balance in the journal that continues today. Goy and Whalen remained as Co-EICs until 1997 when *Hormones and Behavior* became the official journal of the new Society for Behavioral Neuroendocrinology, a society that Goy had championed, and Michael Baum became the EIC (Wallen, 2020).

The Aromatization Hypothesis

In the penultimate sentence of the discussion in the 1959 paper, the authors raise the possibility that although they had demonstrated organization via prenatal testosterone exposure they presciently hedged their bets stating "We are assuming that testosterone or some metabolite acts on those central nervous tissues in which patterns of sexual behavior are organized" (Phoenix et al., 1959 page 381). Years later Reddy et al. (1974) demonstrated that fetal rat limbic and hypothalamic homogenates were capable of metabolizing testosterone into estrone through aromatization. Additionally, the level of aromatization was much higher in male than in female fetuses raising the possibility that testosterone's effects on sexual differentiation might be mediated by conversion to an estrogen (Reddy et al., 1974). McDonald et al. (1970) reported that treating castrated male rats with testosterone propionate (TP) reinstated sexual behavior, but treatment with 5α-dihydrotestosterone propionate (DHTP), a nonaromatizable androgenic metabolite of testosterone, did not reinstate male sexual behavior. This finding was later replicated by Whalen and Luttge (1971). These and other studies provided evidence that estrogenic metabolites, both during fetal development and in adulthood, were necessary for masculinization of behavior in rats, mice, and hamsters. The necessity of the aromatization of testosterone to an estrogen became known as the Aromatization Hypothesis and has been a central dogma of behavioral, neuroendocrinology since the late 1970s. Goy was interested in whether estrogenic metabolites of testosterone were necessary for reinstating male sexual behavior in guinea pigs as was the case in other rodents. With his graduate student, Pamela Alsum, they found that DHTP was as effective as was TP in reinstating male sexual behavior in castrated males. Furthermore, estradiol had no effect on reinstating masculine behavior (Alsum & Goy, 1974). At the same time as the guinea pig work was published so was a study by Phoenix (Phoenix, 1974) showing that long-term castrated male rhesus monkeys' sexual behavior was reinstated by DHTP or TP as was the case in guinea pigs. Additional evidence that aromatization of testosterone was not required was presented by Goy et al. (1988) who found that blocking aromatization of T by administering 1,4,6-androstatriene-3,17-dione (ATD), an aromatase inhibitor, concurrently with testosterone to castrated males did not prevent testosterone reinstating male sexual behavior. Further evidence that testosterone's effects on sexual behavior did not rely on aromatization was found in studies of prenatally androgenized females who had received either TP or DHTP during gestation. Goy along with his postdoc, Steven Pomerantz, and graduate students Marc Roy and Janice Thornton tested androgen-exposed females for evidence of behavioral masculinization and defeminization (Pomerantz et al., 1985; Thornton & Goy, 1986). Unlike all rodent species studied, primate females do not exhibit lordosis or have a behavior comparable to lordosis. Instead the primary female sexual behavior is sexual initiation or solicitation (Wallen, 1990). When genetic females were treated prenatally with either TP or DHTP, they displayed masculine sexual behavior as adults (Thornton et al., 2009). Strikingly when these females were treated as adults with estradiol and tested for sexual initiation and solicitation with adult males, both the TP- and DHTP-treated females did not show solicitation and sexual initiation. Thus, unlike rodents where only aromatizable androgens would defeminize females, in monkeys either aromatizable or nonaromatizable androgens administered prenatally defeminized the female's behavior.

The Aromatization Hypothesis, while widely supported in short gestation (altricial) mammals, where much of sexual differentiation occurs neonatally, did not appear to apply to long gestation (precocial) mammals where sexual differentiation occurs primarily during gestation (Wallen & Baum, 2002). Of particular interest is that rhesus monkeys do not appear to rely on aromatization for either masculinization or defeminization of behavior. This makes it likely that estrogenic metabolites are not necessary for masculinization and possibly defeminization in humans.

Masculinization, Defeminization, and Bisexuality

Studies of sexual differentiation of reproductive anatomy have identified that two types of gonadal hormonal action are necessary, masculinization and defeminization, to produce male anatomy. Sexual differentiation of female reproductive anatomy doesn't appear to involve gonadal hormones reflecting that the sexually dimorphisms in anatomy are biased to form female phenotypes (Jost, 1970). This anatomical system was applied to behavioral sexual differentiation with masculinization resulting in display of male homotypical behavior (mounting-intromission, and ejaculation) and defeminization resulting in an inability to display female homotypical behavior (receptivity, lordosis in nonprimates). Studies demonstrated that masculinization and defeminization were separable processes that were independently expressed and could be affected by differences in the hormonal environment. The separability of these two processes made it possible to develop bisexuality. If a genetic male was masculinized by the hormonal environment, but not defeminized, he would exhibit bisexual behavior. Similarly, if a genetic female was exposed developmentally to a masculinizing hormonal environment, but one that did not defeminize her, then she would exhibit bisexuality. Some degree of bisexuality is common as described more fully below. Indeed, the only known species where males and females only express homotypical behavior (CIS males and CIS females, in current parlance) is the mythical Ramstergig (Beach, 1971). Bisexual behavior was sufficiently common that Young argued that female rodents were bisexual and males were not (Young, 1961b). Goy and Goldfoot (1975) found that there were species where females were not bisexual but the males of that species were. Furthermore, comparisons made across species revealed that one sex and only one sex exhibited bisexual behavior or bisexual potential (they just needed the appropriate sex-specific hormones administered; Goy & Goldfoot, 1975). In some species it was the female who showed bisexuality, while in others it was the male. No exception to this pattern of bisexuality has been reported in the 45 years since the publication of Goy & Goldfoot, 1975. Further evidence that this complementarity between bisexuality and CIS behavior comes from studies of inbred guinea pig strains 2 and 13. Gonadectomized and treated with the appropriate steroids, strain 2 males, but not females were bisexual, while in strain 13 females were bisexual, but males were not. Thus, the pattern of bisexuality appears to be heritable and to have a genetic basis possibly reflecting the expression of masculinization and defeminization. If masculinization is over expressed, it might not be detected in males as that is the way in which males are made, but it could result in bisexual females who were not defeminized, but masculinized. The opposite pattern would be seen if defeminization was over expressed. The notion that the two processes underlying behavioral sexual differentiation might account for bisexuality is one that Goy found of great interest, but which was not further explored.

Socialization and Sex Differences in Social Behavior

When Goy went to the Primate Lab in Madison, WI, to learn primate juvenile behavior, he studied juvenile monkeys who had been reared under a paradigm considered "normal" (Harlow, 1965). In this rearing system monkeys were housed with their mothers in single cages for the first 30–60 days of life and then housed singly for the rest of their childhood after removal from their mothers. During the first year of life, juvenile monkeys were put together in small groups where 5 days/week; they received 30 min/day of social interaction with 4–5 male and female peers (Harlow, 1965). Monkeys reared this way did not show the aberrant behavior displayed by monkeys reared in total social isolation, and thus this became the standard laboratory infant rearing condition (Wallen, 1996). What was not apparent until more monkeys were reared in this peer-access condition was that male monkeys were severely developmentally affected by the limited access to peers. Harlow's view of the adequacy of the Primate Lab rearing conditions reflected his view that juvenile play was a sign of adequate socialization and peer-reared monkeys showed high levels of play. What they didn't show was juvenile foot-clasp mounts where the juvenile males (and sometimes females) mimic the adult males' copulatory mount. Peer-reared males rarely if ever show this mount. The rarity of foot-clasp mounts among peer-reared males is probably not, as previously suggested (Harlow, 1965; Harlow & Lauersdorf, 1974), a normal developmental pattern but instead is characteristic of a socially deficient rearing environment. Goy encountered this negative attitude in the discussion of a paper he had given on animal models of human sexuality (Goy & Goldfoot, 1975). Robert Rose, who worked with monkeys in semi-natural social groups, stated: "… monkeys should be studied in a natural setting; caged monkeys are crazy." In typical fashion "… Goy replied that 'crazy' is perhaps an exaggeration, and the term 'legally insane' is more accurate" (Goy & Goldfoot, 1975). Rearing conditions had a profound effect on juvenile behavior. So, Goy stopped using the Primate Lab rearing method and reared all subsequent subjects in a mother-peer rearing condition in which 4–5 mother-infant pairs were continuously housed together during the infant's first year of life and then weaned at 1 year of age and singly housed, with daily 30 min interactions with the other group members. The change in the rearing had a profound effect on the juvenile monkey's behavior. Mother-peer reared males showed high levels of play as well as foot-clasp mounts. Interestingly, males and females did not differ in their threatening behavior, whereas when peer-reared there was a clear sex difference in threat, with males exhibiting four times the number of threats as did females. By contrast when the monkeys had continuous access to peers during their first year of life, males and females displayed an almost equal number of threats, with females displaying slightly more than did the males. Thus, rearing condition affected a variety of social behaviors and affected whether sex differences in social behavior were evident (Wallen, 1996; Wallen et al., 1981). What had started as a practical matter – why were our juvenile males not mounting – revealed the importance of behavior developing in a specific social context in order to see sex differences in juvenile behavior.

Timing of Androgen Exposure and Separation of Effects on Genitalia and Behavior

Probably one of Goy's most important papers is entitled *"Behavioral Masculinization Is Independent of Genital Masculinization in Prenatally Androgenized Female Rhesus Macaques."* When Goy published the lab's research showing that genetic female monkeys exposed fetally to androgens, either TP or DHTP, for long portions of pregnancy (50–75 days of gestation), their genitalia were masculinized as well as their juvenile behavior. Concerns were raised that because the treatment masculinized their genitals other monkeys might react to them as if they were males because they looked like males. While there is no reason to believe that monkeys have a notion of sex and that it is tied to how another monkey's genitals look, the possibility cannot be ruled out. The most direct way to address this issue is to find a prenatal treatment that modified the genitals, but does not modify behavior or modifies behavior, but does not modify the genitals. Goy et al. (1988) successfully found such treatments. They manipulated the timing and duration of treatment of pregnant females with TP. Treatments were either given early (first or second trimester) or late (third trimester). The duration of treatment was also varied so treatments were either 15 or 25 days. What did they find? Either 15- or 25-day early treatments masculinized genitals with female offspring of the longer treatment having more masculinized genitalia. Late treatments all occurred when genital differentiation was complete, so the had female-typical genitals and no masculinization for either short or long treatments. What about behavior? Two sexually dimorphic behaviors were collected, foot-clasp mounting (mounting) and rough and tumble play (play). Mounting was only increased by duration of treatment. 15-day TP treatments didn't increase mounting, whereas 25-day treatments either early or late did. Play was only increased by timing with early treatments not affecting mounting, but late treatment increasing play whether 15 or 25 days long. There was no consistent relationship between masculinization of genitals and masculinization of behavior. For mounting, two 25-day treatments increased mounting while one, 25-day Early treatment, masculinized both genitals and behavior, whereas the 25-day Late treatment masculinized behavior, but not genitalia. Thus, genital masculinization did not predict behavioral masculinization. A similar lack of relationship between anatomical and behavioral masculinization was found for play, where only Late treatments masculinized play with both 15- and 25-day treatments doing so. Both of these treatments resulted in androgen exposed females having female genitalia. The Early treatments all had masculinized genitalia, but not masculinized play. Thus, for play there was no treatment that masculinized both play and genitalia. All treatments that masculinized genitals did not masculinize play. For mounting, the 25 day early treatment masculinized both genitals and mounting, whereas the late treatment didn't masculinize genitals but did masculinize mounts.

These results do not support the socialization explanation for masculinizing behavior as there was no consistent relationship between genital and behavioral masculinization. These results also highlight why identifying organizational effects of hormones is so difficult. There appear to be different periods of sensitivity to the masculinizing effects of TP but that sensitivity varies with the behavior measured. Secondly, it is suggested that anatomical masculinization and behavioral masculinization occur at roughly different epochs of pregnancy with anatomical masculinization generally earlier than behavioral masculinization, but this is not a strict dichotomy. This could explain why girls with congenital adrenal hyperplasia have masculinized genitals, but little masculinized behavior and no apparent effect on gender identity (Meyer-Bahlburg et al., 2004).

Acknowledgments I thank my brother, Kurt Wallen, for the insightful comments on an earlier draft of this biography, which helped me greatly in developing the biography.

References

Allen, E. (1932). *Sex and internal secretions: A survey of recent research*. Williams & Wilkins Co.

Alsum, P., & Goy, R. W. (1974). Actions of esters of testosterone, dihydrotestosterone, or estradiol on sexual behavior in castrated male guinea pigs. *Hormones and Behavior, 5*(3), 207–217.

Baum, M. J. (1990). Frank Beach's research on the sexual differentiation of behavior and his struggle with the "organizational" hypothesis. *Neuroscience and Biobehavioral Reviews, 14*(2), 201–206. https://doi-org.proxy.library.emory.edu/10.1016/S0149-7634(05)80220-9

Baum, M. J., Valenstein, E. S., Phoenix, C., Gerall, A., Diamond, M., Feder, H. H. et al. (1999). In memoriam – A collection of remembrances of the life and career of Robert W. Goy (January 25, 1924–January 14, 1999). *36*(2), 187–211.

Beach, F. A. (1952). Sex and species differences in the behavioural effects of gonadal hormones. In *Ciba foundation symposium-hormones, psychology and behaviour (book I of colloquia on endocrinology)* (Vol. 3, pp. 3–17). Wiley.

Beach, F. A. (1971). Hormonal factors controlling the differentiation, development, and display of copulatory behavior in the ramstergig and related species. In *The biopsychology of development* (pp. 249–296).

Breland, K., & Breland, M. (1951). A field of applied animal psychology. *The American Psychologist, 6*(6), 202–204.

Breland, K., & Breland, M. (1961). The misbehavior of organisms. *American Psychologist, 16*(11), 681–6846.

Dantchakoff, V. (1938a). Role des hormones dans la manifestation des instincts sexuels. *Comptes Rendus Hebdomadaires des Séances de l'Académie des Sciences, 206*, 945–947.

Dantchakoff, V. (1938b). Sur les effets de l'hormone male dans un jeune cobaye femelle traite depuis un stade embryonnaire (inversions sexuelles). *Comptes Rendus des Seances de la Societe de Biologie et de Ses Filiales, 127*, 1255–1258.

Dempsey, E. W., Hertz, R., & Young, W. C. (1936). The experimental induction of oestrus (sexual receptivity) in the normal and ovariectomized guinea pig. *American Journal of Physiology-Legacy Content, 116*(1), 201–209.

Dewsbury, D. A. (2003). The conference on reproductive behavior: A history. *Hormones and Behavior, 43*(4), 465–473.

Goy, R. W. (1967). William Caldwell Young. *The Anatomical Record, 157*(1), 3–11.

Goy, R. W., & Goldfoot, D. A. (1975). Neuroendocrinology: Animal models and problems of human sexuality. *Archives of Sexual Behavior, 4*(4), 405–420.

Goy, R. W., & Jakway, J. S. (1959). The inheritance of patterns of sexual behaviour in female guinea pigs. *Animal Behaviour, 7*, 142–149.

Goy, R. W., & Wallen, K. (1979). Experiential variables influencing play, foot-clasp mounting and adult sexual competence in male rhesus monkeys. *Psychoneuroendocrinology, 4*(1), 1–12.

Goy, R. W., & Young, W. C. (1956). Strain differences in the behavioral responses of female guinea pigs to alpha-estradiol benzoate and progesterone. *Behaviour, 10*(3/4), 340–354.

Goy, R. W., & Young, W. C. (1957). Somatic basis of sexual behavior patterns in guinea pigs: Factors involved in the determination of the character of the soma in the female. *Psychosomatic Medicine, 19*(2), 144–151.

Goy, R. W., Hoar, R. M., & Young, W. C. (1957). Length of gestation in the guinea pig with data on the frequency and time of abortion and stillbirth. *The Anatomical Record, 128*(4), 747–757.

Goy, R. W., Bercovitch, F. B., & McBrair, M. C. (1988). Behavioral masculinization is independent of genital masculinization in prenatally androgenized female rhesus macaques. *Hormones and Behavior, 22*(4), 552–571.

Harlow, H. F. (1965). Sexual behavior in the rhesus monkey. In F. A. Beach (Ed.), *Sex and behavior* (pp. 234–265). Wiley.

Harlow, H. F., & Lauersdorf, H. E. (1974). Sex differences in passion and play. *Perspectives in Biology and Medicine, 17*(3), 348–360.

Hunt, H. F., Jernberg, P., & Lawlor, W. G. (1953). The effect of electroconvulsive shock on a conditioned emotional response; the effect of electroconvulsive shock under ether anesthesia. *Journal of Comparative and Physiological Psychology, 46*(1), 64–68.

Jost, A. (1953). Problems of fetal endocrinology: The gonadal and hypophyseal hormones. *Recent Progress in Hormone Research, 8*, 379–413.

Jost, A. (1970). Hormonal factors in the sex differentiation of the mammalian foetus. *Philosophical Transactions of the Royal Society of London. B, Biological Sciences, 259*(828), 119–131.

Martins, T., & Valle, J. R. (1948). Hormonal regulation of the micturition behavior of the dog. *Journal of Comparative and Physiological Psychology, 41*(5), 301–311.

McDonald, P., Beyer, C., Newton, F., Brien, B., Baker, R., Tan, H. S., et al. (1970). Failure of 5α-dihydrotestosterone to initiate sexual behaviour in the castrated male rat. *Nature, 227*(5261), 964–965.

Meyer-Bahlburg, H. F., Dolezal, C., Baker, S. W., Carlson, A. D., Obeid, J. S., & New, M. I. (2004). Prenatal androgenization affects gender-related behavior but not gender identity in 5–12-year-old girls with congenital adrenal hyperplasia. *Archives of Sexual Behavior, 33*(2), 97–104.

Phoenix, C. H. (1974). Effects of dihydrotestosterone on sexual behavior of castrated male rhesus monkeys. *Physiology & Behavior, 12*(6), 1045–1055.

Phoenix, C. H., Goy, R. W., Gerall, A. A., & Young, W. C. (1959). Organizing action of prenatally administered testosterone propionate on the tissues mediating mating behavior in the female guinea pig. *Endocrinology, 65*(3), 369–382.

Pomerantz, S. M., Roy, M. M., Thornton, J. E., & Goy, R. W. (1985). Expression of adult female patterns of sexual behavior by male, female, and pseudohermaphroditic female rhesus monkeys. *Biology of Reproduction, 33*(4), 878–889.

Price, D., Ortiz, E., & Zaaijer, J. J. (1967). Organ culture studies of hormone secretion in endocrine glands of fetal guinea pigs III. The relation of testicular hormone to sex differentiation of the reproductive ducts. *The Anatomical Record, 157*(1), 27–41.

Reddy, V. V., Naftolin, F., & Ryan, K. J. (1974). Conversion of androstenedione to estrone by neural tissues from fetal and neonatal rats. *Endocrinology, 94*(1), 117–121.

Sisk, C. L., & Zehr, J. L. (2005). Pubertal hormones organize the adolescent brain and behavior. *Frontiers in Neuroendocrinology, 26*(3–4), 163–174.

Thornton, J., & Goy, R. W. (1986). Female-typical sexual behavior of rhesus and defeminization by androgens given prenatally. *Hormones and Behavior, 20*(2), 129–147.

Thornton, J., Zehr, J. L., & Loose, M. D. (2009). Effects of prenatal androgens on rhesus monkeys: A model system to explore the organizational hypothesis in primates. *Hormones and Behavior, 55*(5), 633–644.

Valenstein, E. S., & Goy, R. W. (1957). Further studies of the organization and display of sexual behavior in male guinea pigs. *Journal of Comparative and Physiological Psychology, 50*(2), 115–119.

Valenstein, E. S., Riss, W., & Young, W. C. (1955). Experiential and genetic factors in the organization of sexual behavior in male guinea pigs. *Journal of Comparative and Physiological Psychology, 48*(5), 397–403.

Wallen, K. (1990). Desire and ability: Hormones and the regulation of female sexual behavio. *Neuroscience & Biobehavioral Reviews, 14*(2), 233–241.

Wallen, K. (1996). Nature needs nurture: the interaction of hormonal and social influences on the development of behavioral sex differences in rhesus monkeys. *Hormones and Behavior, 30*(4), 364–378.

Wallen, K. (2004, October 4). Gerall A A, interview.

Wallen, K. (2009). The organizational hypothesis: Reflections on the 50th anniversary of the publication of Phoenix, Goy, Gerall, and Young (1959). *Hormones and Behavior, 55*(5), 561–565.

Wallen, K. (2020). Hormones and behavior: An emergent journal. (editorial. *Hormones and Behavior, 121*, 104692.

Wallen, K., & Baum, M. J. (2002). Masculinization and defeminization in altricial and precocial mammals: Comparative aspects of steroid hormone action. In D. W. Pfaff (Ed.), *Hormones, brain, and behavior* (pp. 385–423). Academic Press.

Wallen, K., Goldfoot, D. A., & Goy, R. W. (1981). Peer and maternal influences on the expression of foot-clasp mounting by juvenile male rhesus monkeys. *Developmental Psychobiology, 14*(4), 299–309.

Whalen, R. E., & Luttge, W. G. (1971). Testosterone, androstenedione and dihydrotestosterone: Effects on mating behavior of male rats. *Hormones and Behavior, 2*(2), 117–112.

Wright, S. (1923). The effects of inbreeding on guinea-pigs. In *Scientific papers of the second international congress of eugenics held at American Museum of Natural History, New York, September 22–28, 1921*. Committee on Publication (Vol. 1, p. 266). Williams & Wilkins.

Young, W. C. (1961a). *Sex and internal secretions* (3rd ed.). Williams & Wilkins.

Young, W. C. (1961b). The hormones and mating behavior. In W. C. Young (Ed.), *Sex and internal secretions* (3rd ed., pp. 1173–1239). Williams and Wilkins.

Young, W. C., Goy, R. W., & Phoenix, C. H. (1964). Hormones and sexual behavior. *Science, 143*(3603), 212–218.

Charles H. Phoenix

11

Brishti A. White and Robin B. Oliverio

Abstract

Charles H. Phoenix was a scientist whose most notable work investigated the role of sex steroids on the organizational role of the central nervous system. Phoenix and his colleagues discovered the relationship between the organizational effect of androgens and the subsequent impact on downstream behaviors. In this work, he suggested behavior stems from neural changes caused by prenatal hormone administration. Although the mechanisms of this organization were not yet fully established, the work substantially provided a solid base for the field of behavioral neuroendocrinology.

Keywords

Sex differences · Androgens · Testosterone · Development · Organizational hypothesis · Activation · Sexual differentiation

Charles H. Phoenix was born in Webster, Massachusetts, but spent most of his early life in Connecticut ("Obituary", 2014). Following an honorable discharge from the US Army, Phoenix attended Boston University and received a Master of Arts degree before earning a psychology doctorate in 1954. Though Boston was the place Phoenix trained and met his wife, they soon moved to Lawrence, Kansas. Phoenix taught and conducted behavioral neuroendocrinology research at the University of

B. A. White · R. B. Oliverio (✉)
Department of Neuroscience, Rockefeller Neuroscience Institute, West Virginia University, Morgantown, WV, USA
e-mail: rboliverio@mix.wvu.edu

R. J. Nelson, Z. M. Weil (eds.), *Biographical History of Behavioral Neuroendocrinology*, https://doi.org/10.1007/978-3-031-12970-4_11

Kansas as a postdoctoral researcher. It was at this institution where Phoenix met his mentor, William C. Young, a professor of anatomy (Phoenix, 1999). Young was also the head of a research program investigating endocrinology and reproduction. Phoenix was so captivated by this research; he worked in Young's lab at night while continuing to teach during the day. Phoenix met future colleague, Robert (Bob) Goy, while engaging in this research. Following the birth of his children, their family packed up the car and headed west. Phoenix served as the assistant director of the Oregon Regional Primate Research Center from 1965 to 1982. Within this same period, Phoenix also taught in the Department of Medical Psychology at, what is now called, Oregon Health and Science University. Although a large portion of his career was spent as the director of the Oregon Regional Primate Research Center, his work with primates followed the guinea pig model used in his most notable contributions to behavioral neuroendocrinology.

Under the direction of Young, Phoenix and his colleagues began to investigate the effect of prenatal testosterone on mating behaviors. This series of studies was conducted in female guinea pigs and resulted in Phoenix's seminal publication, "Organizing action of prenatally administered testosterone propionate on the tissues mediating mating behaviors in the female guinea pig" in 1959. In 2009, reflecting on this influential publication, Phoenix stated that the novelty of these findings was not due to the male-typical mating behaviors the females displayed following testosterone administration (Phoenix, 2009). Indeed, this phenomenon had been reported as early as 1938, when Vera Dantchakoff discovered that prenatal testosterone exposure in female guinea pigs resulted in masculinized external genitalia as well as male-typical reproductive behaviors that were present in adulthood (Dantchakoff, 1938). However, the distinguishing feature of Phoenix's work lay in the attribution of these masculinized behaviors to a masculinized brain. Phoenix and his colleagues were the first to connect an alteration in sexual behaviors following prenatal androgen exposure to its influence on the central nervous system, a process they characterized as organizing the brain.

Phoenix and colleagues described four experiments from which the above conclusions were drawn (Phoenix et al., 1959). The first of these contained four different groups of guinea pigs. The first group was referred to as unmodified females and involved females which presented no notable abnormalities in the external genitalia after 1 mg of testosterone propionate had been administered to their mothers every 3–4 days between 10 and 27 days of the gestational period. The second group of nine females, referred to as hermaphrodites, was composed of females whose mothers had received 5 mg of testosterone propionate on day 10, 15, 18, or 24 after conception. These females presented with external genitalia which were similar to males. The third group was a control group consisting of females from mothers which had received no testosterone. Finally, the fourth group comprised males from mothers which had not been treated with testosterone. The hermaphrodites and most of the unmodified females were gonadectomized between postnatal days 80 and 150, except for a few unmodified females that were gonadectomized at postnatal day 45. Meanwhile, the males were gonadectomized early in life at postnatal day 21.

Following gonadectomy, three different concentrations of estradiol benzoate were administered subcutaneously, followed 36 h later by progesterone. The animals were then assessed for the presence of lordosis and mounting behavior. Phoenix and his colleagues found that lordosis was not related to the amount of estradiol administered. However, in about half of the animals which were exposed to the lower concentration of testosterone, and all of those with masculinized external genitalia, lordosis was reduced. Further, lordosis was reduced in all of the females which had been prenatally exposed to a larger concentration of testosterone and presented with masculinized external genitalia. Therefore, prenatal testosterone can alter adult mating behaviors without necessarily altering external genitalia. Furthermore, after estradiol and progesterone treatment, all animals displayed mounting behavior, while prior to adult hormone administration, only males and females with masculinized external genitalia mounted.

The second experiment was designed to test the persistence of the effects of prenatal testosterone exposure. This experiment included females with masculinized external genitalia some of which received daily testosterone treatments until adulthood. Again, females which had received a smaller concentration of testosterone and did not have masculinized genitalia were observed in this set of experiments, as well as females which had not been exposed to testosterone prenatally. Additionally, a new group was used in which females that had no prenatal exposure to testosterone received daily testosterone administration to assess the effect of this sex steroid later in life. Finally, mothers that had been injected with testosterone propionate and control females of the same age as the mothers were included to compare the impact of age at time of testosterone exposure. All animals were gonadectomized at similar ages as in the previous experiment, with the exclusion of the two final groups that were gonadectomized between 1.5 and 3 years of age.

As in the first experiment, mating behavior was observed following injections of estradiol benzoate and progesterone. Phoenix reported that the females which were treated daily with testosterone but had not been exposed to testosterone prenatally began to show lordosis following the cessation of testosterone treatment. In contrast, the animals which had received testosterone prenatally did not display lordosis. Similarly, mothers which had been injected with testosterone during the gestation period still displayed lordosis. Therefore, prenatal, but not postnatal testosterone exposure, permanently altered female mating behavior.

The third experiment investigated mating behavior in animals which were gonadectomized and received testosterone propionate in adulthood. The groups for this experiment were females that were exposed to testosterone prenatally and had masculinized external genitalia, control females which were gonadectomized in adulthood, and males which were gonadectomized at postnatal day 21. The gonadectomized females which had received testosterone prenatally displayed similar mating behavior to the gonadectomized males following an injection of testosterone propionate, and these behaviors differed significantly from the gonadectomized females which had not been exposed to testosterone prenatally. For instance, males and females from mothers that were treated with testosterone during gestation required less testosterone to begin mounting, and they exhibited more mounting compared to females with no prenatal exposure to testosterone. This suggested that prenatal testosterone was affecting tissue which mediated this masculine mating behavior and was thereby presenting an organizational function.

The final experiment sought to observe the mating behavior of males which were born to the mothers which received testosterone propionate injections during the gestation period. Three groups of males were used for this experiment: (1) males with mothers that had received testosterone treatment but received no subsequent exogenous hormone treatment, (2) males with mothers that had received testosterone during gestation and which were injected with testosterone after birth for 3 days then later in adulthood, and (3) a control group of males with mothers that were not treated with testosterone. Phoenix found that there was no significant difference in the mating behavior between groups. Thereby demonstrating that exogenous prenatal testosterone does not affect males in the same manner as females.

These experiments allowed Phoenix and his colleagues to assert that sex steroids organize the nervous system. He suggested that using behavior as a dependent variable and acknowledging that prenatal hormone administration affects this variable presents a case for the assumption that behavior arises from its neural correlates. In other words, from the experiments described above, Phoenix was laying the groundwork for understanding androgens as having an organizing effect on the central nervous system which could then be activated by hormones later in life. Although the specifics of this organization were not yet clear, the novelty in this very idea would allow the field of neuroendocrinology to advance greatly.

On the 50th anniversary of this paper, Phoenix stated that he found it unsurprising that this work still contained relevance in the field of endocrinology (Phoenix, 2009). However, he was amazed to find himself still around to see it, lamenting the passing that was all too soon of his younger colleague Bob Goy who could not witness this moment with him. Shortly before his own death, Phoenix had one request. He asked his friends and family not to perform any ceremonious mourning or even celebration of his life but to celebrate life itself.

References

Dantchakoff, V. (1938). Rôle des hormones dans la manifestation des instincts sexuels. *Comptes rendus de l'Académie des Sciences, 206*, 945–947.

"Obituary". 2014. Charles H. Phoenix Ph.D. Corvallis Gazette Times.

Phoenix, C. H. (1999). In memoriam: Robert W. Goy. *Archives of Sexual Behavior, 28*, 471–474.

Phoenix, C. H. (2009). Organizing action of prenatally administered testosterone propionate on the tissues mediating mating behavior in the female guinea pig. *Hormones and Behavior, 55*, 566.

Phoenix, C. H., Goy, R. W., Gerall, A. A., & Young, W. C. (1959). Organizing action of prenatally administered testosterone propionate on the tissues mediating mating behavior in the female guinea pig. *Endocrinology, 65*, 369–382.

Daniel Sanford Lehrman

12

Rae Silver

Abstract

Daniel Sanford Lehrman (1919–1972) was a well-rounded man. This biography highlights the legend of Lehrman that lives on in the minds of his students and their students, his scientific contributions, and his lasting and disappearing legacies. Lehrman's empirical and conceptual work covered the realms of development, neuroendocrinology of reproduction, learning, and motivation. His leap into the spotlight happened before he got his PhD in the postwar era of the early 1950s with a critique of the concepts of Konrad Lorenz. Lehrman's opus brought visibility in the English-speaking scientific communities to the research and ideas of Lorenz and other ethologists, including Tinbergen and eventually led to their Nobel prize. Lehrman worked on his critique for 3 years, and the pre-publication version was reviewed by world leaders of science. Unlike strategies that are broadly applied today, these reviewers urged Lehrman to remove the information showing that Lorenz findings on bird courtship were generalized to humans and were used to support the position of Nazis of that era. The application of politics to science merits continuing attention as the uses to which scientific discoveries can be put is worthy of deep scrutiny.

Keywords

Ethology · Neuroendocrinology · Reproduction · Development · Learning · Motivation · Avian · Courtship

R. Silver (✉)
Department of Psychology, Columbia University, New York, NY, USA

Department of Neuroscience, Barnard College, New York, NY, USA
e-mail: Rae.Silver@columbia.edu

R. J. Nelson, Z. M. Weil (eds.), *Biographical History of Behavioral Neuroendocrinology*, https://doi.org/10.1007/978-3-031-12970-4_12

Background of this Historical Biography

> My main orienting attitudes towards the study of animal behavior come from having been a bird watcher since I was 14 years old… Daniel S. Lehrman.

When I look at my office wall and see the beautiful face of Danny Lehrman in portraits that I have taken (Figs. 12.1 and 12.2), I experience the feelings he captured in explaining his love of animals "…curiosity and fascination, and of apprehension…" Lehrman wrote that "...Feelings of this kind have, for the observer, more in common with feelings involved in watching a sunset or reading a poem than they do with those involved in solving engineering problems, or in abstracting formalized general relationships from narrowly defined operations of experimenter and subject." For Lehrman, doing the research was about feelings. Lehrman once told me that there is something unique about people who turn to animals to explore their own feelings.

The Canadian author, Sheila Heti, offered a different angle. Heti wrote, in her novel *Pure Colour*, that humans are birds, fish, or bears. She suggests that people who are concerned with beauty and aesthetics are birds, those who focus on the common good are fish, and those who care most about their close relationships are bears. Daniel Sanford Lehrman was a giant of a man, physically and spiritually – he was a *bearbirdfish* or a *firdfarebish*. His enthusiasms knew no bounds and his passions escaped restraints. And he gave the same freedoms to his students and colleagues. I begin this historical biography with tales of Lehrman's astounding human qualities by considering some "Lehrman legends" gleaned from memories of his students. I then document his scientific work and lasting contributions. Thinking about Lehrman in preparation of this biography has been a pleasure. It has been

Fig. 12.1 Portrait of Daniel Lehrman (circa 1969). Photo: Portrait #1 attached

Fig. 12.2 Photo: Portrait #2 was taken on a bird-watching trip in Jamaica, Queens (circa 1970)

helped by many earlier discussions with Jay S. Rosenblatt, Lehrman's longtime colleague (Silver & Rosenblatt, 1987), by Rosenblatt's *Biographical Memoir written for the National Academy of Science (1995)),* and by the fond and funny recollections of friends and students including David Crews, Alison Fleming, Barry Komisaruk, and George Michel.

Lehrman Legends

Lehrman was an unusual man in his time, and that would be true even if he were here with us today. He was an extraordinary speaker. He knew what would interest each audience, and then he used that information to seduce his listeners, whatever the topic. Second, he had an independent streak with regard to academic rules and requirements, favoring his independent judgement of ability, motivation, and future capability. Finally, he was sensitive to the pressures faced by women in academia, perhaps because of his wife and daughters or perhaps because he was a firdfarebish. It was my impression that he struggled with how to implement fairness for the women who reported to him. Lehrman's qualities are reflected in the Lehrman legends we the students have in our heads. Here are a few of them.

Legend: Lehrman's Attitudes to Grades and Requirements

George Michel Danny did not do too well in City College (a "C" student) because he was spending all of his time at the Museum of Natural History. His mentor (Libbie Hyman) got him to work with Ernst Mayr who forced him to learn German by insisting that he read Lorenz in German. That subsequently led to the "Critique" paper in 1953. At that time, he was being mentored by T.C. Schneirla at New York University (NYU). They had accepted him in the graduate program only because Schneirla insisted.

David Crews When I applied to the Institute of Animal Behavior, we spent some time of my interview trading stories of "bad student done good." He showed me his NYU transcript where he was booted out for failing too many classes. (He was subsequently reinstated). I told him about my steady, steep rise in GPA in college and he seemed to find that a positive point.

David Crews Again Gladwyn Kingsley Noble was the head of the Department of Experimental Biology at the American Museum of Natural History in New York City and died in 1940. Danny had been a high school volunteer in Noble's lab. In fact, he published his first paper with Noble. After Noble's death, Frank Beach took over the position of department head. When I met Beach years later, he told me that Danny was a flake at the time and that he had fired him.

Rae Silver I met Danny when he did a site visit for NIH at H. Phillip Zeigler's lab. I had been working with Paul Witkovsky at the Columbia Medical school, trying to understand sensory processing in the pigeon brain, using electrophysiological techniques. Because I had been doing the work, Zeigler kindly asked me to present the results to the site visiting team. After the event, Danny took me aside and asked what I would be doing next, now that my PhD was done, as he was looking for someone who could work with Barry Komisaruk. I told him that I had just failed one of my MA comprehensive exams at City University of New York (CUNY) on the topic "Contribution of Modeling Work to Understanding Behavior." I had taken a relatively negative view of the status of that work at the time, much influenced by my husband Len Silver, who was working on his PhD in theoretical physics at Columbia University. Len had convinced me that there were too many uncontrolled parameters in psychology and biology, and the best strategy one could use in developing a model was to simplify the problem. (For example, to study neurons and circuits, first pretend a neuron is a perfect sphere, and that all spheres are equal). At the time, I did not see that progress was possible with any such strategies. I tried to make this argument in my MA comprehensive exam – and failed. When Danny heard this tale, he laughed and laughed and promptly invited me to be an "extra" graduate student in his Institute. Naturally, I expressed surprise, and so he told me that he did not seek folks who got straight A's, and that he had looked at the rest of my resume. From there, my academic career was launched and my personal and professional life changed course (Helmreich 2019).

Legend: Lehrman's Amazing Lecturing and Speaking Abilities

George Michel I enjoyed spending time with him at his home in the village and again at his home in Lodi, at his vacation place in Santa Fe, NM, and then hosting him at my home in Newton, MA, when he gave his talk at Harvard in 1971 or 1972. I remember when the Museum of Natural History held the Biopsychology of Development symposium, around 1969, and he had to speak last at the evening session for the general public. Margaret Mead spoke first (and she was prototypical Mead – powerful). Then Skinner spoke, and he was all about Walden Two and how we can create a better society. I thought that Danny was dead to be following two such crowd pleasers. He then gave his talk "Behavioral Science, Engineering and Poetry," and he captured the crowd. It was exhilarating to experience. As I noted – he was the most charismatic speaker I have ever known. His death really disrupted my life. I don't think that I ever got over it.

Barry Komisaruk One morning Danny walked in to where we graduate students were chatting and said "let's go to New York now. I have to give a talk." We said you never told us about it, and we were dressed in typical grad student garb...ragamuffins. He wasn't dressed too fine, either. (In all the years I knew and loved him, at fancy conferences, professional gatherings, etc., I only saw him wear a tie once... because the gambling casino at a "Foundations Fund Conference," a fancy Rockefeller Resort in Puerto Rico refused to let him in without one). But that morning Danny told us don't worry. Let's go. So we all got into his car and he drove to New York. Then we went into the Academy of Medicine building, up to a magnificent giant ballroom with brightly lit chandeliers and hundreds of people. Danny looked at it and said "Holy S...!...didn't expect THIS! I didn't prepare anything!" We sheepishly snuck in and sat down. Soon they called Danny. He told the audience, "I didn't bring any slides." Then proceeded to tell his "The Ring Dove Story." The audience was spellbound...it was one of the most brilliant, eloquent, talks I had ever heard Danny or anyone else give... up to then or ever since!

Lehrman's Mermaid and Minotaur

Rae Silver As a graduate student at CUNY, I had been advised by the graduate advisor to quit school as "you are a woman and women don't make it in Academia, and besides that you are married." As a student at the institute, I never ever felt demeaned by anyone. Danny became extraordinarily sensitive to the plight of women, possibly influenced by his wife, Dorothy Dinnerstein, author of the powerful vision of feminist theory, *The Mermaid and the Minotaur*. I did not know Dorothy very well and never really understood Danny's attitudes clearly. Danny had hired all male tenured or tenure-track faculty at the Institute (Colin Beer, Harvey Feder, Ernst Hansen, Barry Komisaruk, Jay Rosenblatt), and two women were hired in non-tenure-track jobs as research assistants (Mei Feng Cheng and Monika Impekoven). One day Danny admonished me to be more respectful and supportive of Mei Fang and Monika. I was pretty shocked and asked him how I could confer respect and wasn't that up to him? I asked what I could possibly do given that I was a mere graduate student, and he talked a bit about how respect is conferred by those around you. But looking back today, I feel him. We honor and elevate others in unspoken ways and perhaps Danny was struggling with his own positions. I think he was urging me to do my part. He may have spoken to everybody on this topic, but I never understood him well enough at the time to ask about his thinking on this question. Today, in these "#MeToo" days, it is clear that Danny was prescient.

Legendary Lehrman Meals

Barry Komisaruk Danny loved to describe memorable restaurant meals (e.g., a ridiculously expensive, one-time-in-his-lifetime, 5-star Michelin in France) he had in exquisite and prolonged (over an hour!) detail. One time as we were all (faculty, students, staff) sitting around the big seminar table in the Institute of Animal Behavior, Ralph Cooper, my graduate student, said, "Danny...tell us lunch."

Rae Silver I remember that Danny's descriptions of food were so vivid; I felt that I no longer had the need to eat after he told us about a meal. Once, Danny started a chocolate mousse contest with Steven Chernesky. In advance of the event, he bragged about his superior cooking skills and talked endlessly in anticipation of the contest. He lost the contest. Nevertheless, at the next opportunity, he described the flavors in such detail and with such love that I no longer wanted to eat chocolate mousse. The description was "filling" and could not be surpassed.

Legend: Lehrman Offered Freedom to Others

David Crews When I met with Danny and indicated that I was not interested in any of the animal model systems in the lab, he was not amused. However, I had a proposal for something different. I had read deeply into the beginnings of endocrinology (since I could read German, I was able to read Zondek's original papers). These early researchers were so creative. My original proposal was to study the green anole lizard. Danny then gave me, a first-year student, permission to work with the architects for the fourth floor, which still in planning stages at that time. I ordered two environmental chambers (which had to be hauled up by a crane on the side of the Institute building as the manufacturer had not given the correct widths of the units and they would not fit through the doorways!). To this day I cannot understand why Danny was so generous with his time (when he had it).

Lehrman's Legendary Loyalty

One of the questions Jay Rosenblatt addressed is why Lehrman established and kept the Institute of Animal Behavior at Rutgers University in Newark. Newark at that time was not a pleasant location and that branch of Rutgers was of no high status. As Rosenblatt tells it, Lehrman was repeatedly invited to move the Institute to the main Rutgers campus in New Brunswick. He also refused offers to move the Institute to other universities including Harvard University. Rosenblatt reported that he remained loyal to the place that supported him in the early years before he achieved fame and eminence.

I imagine that Lehrman must have had affection for public institutions of learning. He was educated in the Townsend Harris High School, a public high school in the NYC borough of Queens, and his undergraduate years were spent at City College of New York. At Rutgers Newark, he was awarded center support grants by the National Institute of Mental Health to support basic research and administration, training grants to support the training of students and postdoctoral fellows, and he himself was awarded a very prestigious lifetime Research Career Award from the National Institute of Mental Health, allowing him to pursue his research.

Lehrman's Scientific Contributions

Lehrman's empirical research encompassed four main themes: a developmental approach to behavioral analysis, evolutionary biology, neuroendocrinology, and the hormonal basis of various behaviors including reproductive behavior, learning, and motivation. He viewed each of these as a route to understanding many aspects of behavior change. For each, he was interested in what animals do when they live in their natural habitats rather than the arbitrary responses that investigators examine in artificial laboratory settings. He not only hired Colin Beer (a student of Tinbergen's) and facilitated setting up a field research site to study birds at Brigantine, but he also took us graduate students on bird-watching trips. One of his favorite local sites was the Jamaica Bay Wildlife Refuge in Queens in New York City. He could identify every bird there by its flight pattern and by its call. He patiently taught the naïve how to use binoculars, what clues to look for, what time of day to go birdwatching, how to set up a birding life list, etc. My second-most favorite picture of Lehrman (Fig. 12.2) is his response to a joke I made about a bird. The joke is lost, but "the look" remains.

There have been two prior tributes to Danny's contributions to science (Rosenblatt, 1995; Silver & Rosenblatt, 1987). These publications lay down the broad outlines of his conceptual, empirical, and administrative work. Instead of repeating that information, I provide a detailed overview of Lehrman's timeline of work through his publications, summarized in Table 12.1. The core of his scientific career was incredibly short, under 20 years from his PhD to his death. He worked on his critique of Konrad Lorenz and of ethology for three years and published it (Lehrman, 1953) prior to getting his PhD in 1954 (see Fig. 12.3 which shows a schematic of career timeline). His most visible work, as discussed below, was his pre-PhD critique of Konrad Lorenz. It was enabled by the fact that he had learned German to read the avian scientific literature and it was shaped by his experience as a cryptographer during World War II. In the postwar era, few English-speaking scientists read German and thus would not have known of Lorenz's research and theories. Lehrman was familiar with Lorenz's work on bird behavior and could read the original papers in German. Lehrman's critique played the important role of bringing ethological ideas to the attention of American psychologists and biologists. **Lorenz** and **Tinbergen** were awarded the Nobel Prize for Physiology and Medicine in 1973 for their wok in ethology. The problems tackled by ethologists, on the role of inherited aspects of behavior, remain a continuing realm of research. The molecular and genetic methods to explore these contributions have moved very far away from those of Lorenz and Tinbergen to the extent that researchers in this area of inquiry today may not necessarily see these Nobel prize winners as part of their intellectual history.

To understand Lehrman's contribution, it is helpful to know the context of his work, as it started in the postwar era. In the 1950s and the early 1960s, the study of behavioral organization was largely non-developmental and not brain based. Consider that it was only in 1963 that Tinbergen proposed that understanding behavioral development was important when analyzing the causes of behavior (Tinbergen, 1963). There was no tradition of trying to explore how adult behavior patterns might be understood through examination of their development. Very early on, Lehrman wrote about the difference between his approach to development and that of ethologists (Lehrman, 1953 pp 355–356). As we previously pointed out (Silver & Rosenblatt, 1987):

> Lehrman viewed the developmental approach to behavioral analysis as applicable in the usual sense to the ontogeny of behavior, but he also saw that it applied to adult patterns of behavior. It could provide a method for analyzing complex behavior patterns involving multiple stimuli and responses, interchanges between individuals, and sequential changes in behavior patterns over time, and could be applied to the analysis of the reproductive interaction between male and female ring dove as well as to the analysis of maternal and other behavior patterns…. [and]
>
> How does the reproductive cycle originate, and how is it regulated in pairs of doves, when it appears not to occur independently in either the male or the female? The experiments . . . point to the conclusion that changes in the activity of the endocrine system are induced or facilitated by stimuli coming from various aspects of the environment at different stages of the breeding cycle. Thus, participation in courtship appears to induce the secretion of hormones which facilitate the building of a nest; participation in nest building under these conditions contributes stimulation of the secretion of the hormone(s) which induce the birds to sit on eggs. Stimulation arising from the act of sitting on the eggs induces the further secretion of a hormone which (a) induces the birds to continue incubating, and (b) helps bring the birds into a condition of readiness to feed the young when they hatch. . . . the cycle appears to originate, and to be synchronized in male and female, because the

Table 12.1 Scientific publications and students

Publications of D.S. Lehrman		
1940	Egg Recognition by the Laughing Gull	G. K. Noble, D. S. Lehrman, The Auk, January; 57(1): 22–43. DOI: 10.2307/4078846
1953	A critique of Konrad Lorenz's theory of instinctive behavior	Lehrman DS. Q Rev Biol. Dec;28(4):337–63. DOI: 10.1086/399858
1955	The Physiological Basis of Parental Feeding Behavior in the Ring Dove (Streptopelia Risoria)	Lehrman DS. Behaviour. 7: 241-285. DOI: 10.1163/156853955X00094
1956	Comparative physiology (behavior)	Lehrman DS. Annual Review of Physiology.;18:527–42. DOI:10.1146/annurev.ph.18.030156.002523
1956	On the organization of maternal behavior and the problem of instinct	Lehrman, D. S. In Gresse, P. P. (Ed.), L'lnstinct duns le comportement des animaux et de l'homme. Paris: Masson, pp. 475–520. Cambridge: Cambridge University Press.
1957	Nurture, Nature, and Ethology.	Lehrman DS. Psyccritiques. 2. DOI: 10.1037/005505
1957	Oviduct response to estrogen and progesterone in the ring dove (Streptopelia risoria)	Lehrman DS, Brody P. Proc Soc Exp Biol Med. Jun;95(2):373–5. DOI: 10.3181/00379727-95-23226
1958	Induction of broodiness by participation in courtship and nestbuilding in the ring dove(Streptopella risoria)	Lehrman DS. J Comp Physiol Psychol. Feb;51(1):32–6. DOI: 10.1037/h0045891
1958	Effect of female sex hormones on incubation behavior in the ring dove (Streptopelia risoria)	Lehrman DS. J Comp Physiol Psychol. Apr;51(2):142–5. DOI: 10.1037/h0046502
1959	Hormonal responses to external stimuli in birds.	Lehrman, D. S. Ibis. 101: 478–496.
1959	On the origin of the reproductive behavior cycle in doves	Lehrman DS. Trans NY Acad Sci. Jun;21:682–8. DOI: 10.1111/j.2164-0947.1959.tb01714.x
1960	Previous breeding experience and hormone-induced incubation behavior in the ring dove	Lehrman DS, Wortis RP.Science. Dec 2;132(3440):1667–8. DOI: 10.1126/science.132.3440.1667
1961	Gonadotropin secretion in response to external stimuli of varying duration in the ring dove (Streptopelia risoria)	Lehrman DS, Wortis RP, Brody P. Proc Soc Exp Biol Med. Feb;106:298–300. DOI: 10.3181/00379727-106-26315
1961	Does prolactin induce incubation behavior in the ring dove?	Lehrman DS, Brody P.J Endocrinol. Jun;22:269–75. DOI: 10.1677/joe.0.0220269
1961	Ethology and psychology	Lehrman DS. Recent Adv Biol Psychiatry. 4:86–94. DOI: 10.1007/978-1-4684-8306-2_11
1961	Hormonal regulation of parental behavior in birds and infrahuman mammals	Lehrman DS. Sex and Internal Secretions, 3rd edition:1268–1382
1961	The presence of the mate and of nesting material as stimuli for the development of incubation behavior and for gonadotropin secretion in the ring dove (Streptopelia risoria)	Lehrman DS, Brody PN, Wortis RP.Endocrinology. Mar;68:507–16. DOI: 10.1210/endo-68-3-507

(continued)

Table 12.1 (continued)

1962	Varieties of learning and memory in animals	Lehrman, D. S. In Schmitt, F. 0. (Ed.), Macromolecular Specificity and Biological Memory. Cambridge: MIT Press, pp. 108–110.
1962	Ethology and psychology	Lehrman, D. S. Recent Adv. Biol. Psychiatry, 4: 86–94.
1963	On the initiation of incubation behaviour in doves	Lehrman DS. Animal Behaviour. 11: 433–438.
1963	Maternal behavior of the laboratory rat	Rosenblatt, J. S.. and Lehrman, D. S. In Rheingold, New York, Wiley, pp 8–57.
1964	The reproductive behavior of ring doves	Lehrman DS.Sci Am. Nov;211:48–54. DOI: 10.1038/scientificamerican1164-48
1964	Effect of prolactin on established incubation behavior in the ringdove	Lehrman DS, Brody PN.J Comp Physiol Psychol. Apr;57:161–5. DOI: 10.1037/h0046551
1964	Effect of castration of male ring doves upon ovarian activity of females	Erickson CJ, Lehrman DS.J Comp Physiol Psychol. Oct;58:164–6. DOI: 10.1037/h0038709
1964	Control of behavior cycles in reproduction	Lehrman, D. S. In (Ed.) W. Etkin. Social Behavior and Organization Among Vertebrates. The University of Chicago Press.
1965	Interaction between internal and external environments in the regulation of the reproductive cycle of the ring dove	Lehrman, D. S. In Beach, F. A. (Ed.), Sex and Behavior. New York: Wiley. pp. 335–380.
1966	Advances in the Study of Behavior	(Review by Scott JP of edited book) Lehrman DS, Hinde RA, Shaw E. American Midland Naturalist. 76: 253. DOI: 10.2307/2423255
1967	Selective inhibition by progesterone of androgen-induced behavior in male ring doves (Streptopelia risoria)	Erickson CJ, Bruder RH, Komisaruk BR, Lehrman DS. Endocrinology. Jul;81(1):39–44. DOI: 10.1210/endo-81-1-39
1967	Breeding experience and breeding efficiency in the ring dove	Lehrman DS, Wortis RP. Anim Behav. Apr-Jul;15(2):223–8. DOI: 10.1016/0003-3472(67)90003-6
1967	Role of the mate in the elicitation of hormone-induced incubation behavior in the ring dove	Bruder RH, Lehrman DS.J Comp Physiol Psychol. Jun;63(3):382–4. DOI: 10.1037/h0024634
1967	Exteroceptive stimulation of the reproductive system of the female ring dove (Streptopella risoria) by the mate and by the colony milieu	Lott D, Scholz SD, Lehrman DS. Anim Behav. Oct;15(4):433–7. DOI: 10.1016/0003-3472(67)90041-3
1968	Physiological conditions for the stimulation of prolactin secretion by external stimuli in the male ring dove	Friedman M, Lehrman DS. Anim Behav. Apr-Jul;16(2):233–7. DOI: 10.1016/0003-3472(68)90004-3
1969	Role of testosterone in progesterone-induced incubation behaviour in male ring doves (Streptopelia risoria)	Stern JM, Lehrman DS. J Endocrinol. May;44(1):13–22. DOI: 10.1677/joe.0.0440013
1969	Auditory stimulation of ovarian activity in the ring dove (Streptopelia risoria)	Lehrman DS, Friedman M. Anim Behav. Aug;17(3):494–7. DOI: 10.1016/0003-3472(69)90152-3

(continued)

Table 12.1 (continued)

1969	Advances in the Study of Behavior.	(Brief Preface to their edited book) Lehrman DS, Hinde RA, Shaw E. 2: vii. DOI: 10.1016/S0065-3454(08)60067-4
1970	Experiential background for the induction of reproductive behavior patterns by hormones	Lehrman, D. S. In Tobach, E., Aronson, L. R., and Shaw, E. (Eds.), Biopsychology of Development. New York: Academic, pp. 297–302.
1970	Semantic and conceptual issues in the nature-nurture problem	Lehrman, D. S. In: Development and Evolution of Behavior: Essays in Memory of T. C. Schnierla. L. A. Aronson, E. Tobach, D. S. Lehrman, and J. S. Rosenblatt (Eds.) San Francisco: W. H. Freeman.
1971	Behavioral science, engineering and poetry	Lehrman, D. S. In Tobach, E. Aronson, L. R. and Shaw, E. (Eds.), Biopsychology of Development. New York: Academic, pp. 459–471.
1971	The study of behavioral development	Lehrman, D. S., and Rosenblatt, J. S. In Moltz, H. E. (Ed.), Vertebrate Behavioral Development. New York: Academic, pp. 1–27.
1973	Situational and hormonal determinants of courtship, aggressive and incubation behavior in male ring doves (Streptopelia risoria)	Silver R, Feder HH, Lehrman DS. Hormones and Behavior. 4: 163–172. DOI: 10.1016/0018-506X(73)90026-3
1973	Relative effectiveness of diethylstilbestrol and estradiol benzoate in inducing female behavior patterns of ovariectomized ring doves (Streptopelia risoria)	Cheng MF, Lehrman DS. Hormones and Behavior. 4: 123–127. DOI: 10.1016/0018-506X(73)90022-6
1974	Effects of unseasonal environmental regime, group presence, group composition and males' physiological state on ovarian recrudescence in the lizard, Anolis carolinensis	Crews D, Rosenblatt JS, Lehrman DS. Endocrinology. Feb;94(2):541–7. DOI: 10.1210/endo-94-2-541
1974	Radioimmunoassay of plasma progesterone during the reproductive cycle of male and female ring doves (Streptopelia risoria)	Silver R, Reboulleau C, Lehrman DS, Feder HH. Endocrinology. Jun;94(6):1547–54. DOI: 10.1210/endo-94-6-1547.
1975	Gonadal hormone specificity in the sexual behavior of ring doves	Mei Fang Cheng, Lehrman D. Psychoneuroendocrinology. 1: 95–102. DOI: 10.1016/0306-4530(75)90026-8

behavior of the male and of the female, and the product of this behavior (nest, eggs, etc.), provide stimulation which influences the endocrine systems of the birds in ways which in turn contribute to the sequential changes in their behavior.

Neural and Neuroendocrine Basis of Reproductive Behavior

Imagine! A hot argument of the 1950s was whether the study of the brain could tell us anything about behavior. BF Skinner argued "no way." In 1955, Donald Hebb (a reviewer of Lehrman's critique of Lorenz) wrote an influential paper titled the

Fig. 12.3 Timeline of major events in Lehrman's life

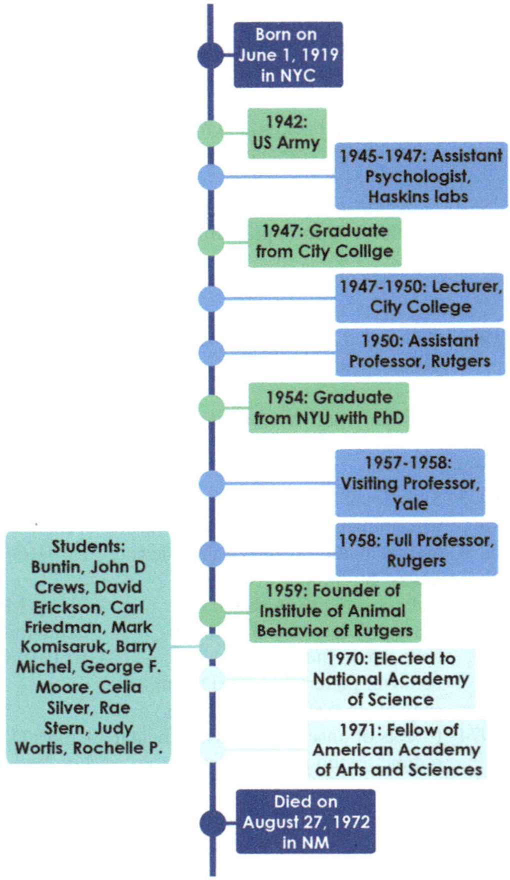

"Drives and the CNS" where CNS stood for the conceptual nervous system (Hebb, 1955). Here, Hebb made the case that studying the brain could in fact contribute to understanding behavior, countering arguments by BF Skinner that this could not be the case. Lehrman, from a different orientation than Hebb, shared the opinion that understanding the brain could help us to understand behavior. As a comparative psychologist Lehrman's interest centered on animal behavior. Lehrman was among the first to grasp the significance of the brain for the study of reproductive behavior (Lehrman, 1959, 1964b; Lehrman & Brody, 1964; Erickson & Lehrman, 1964; Hinde et al., 1965) (see Table 12.1). His starting point was Geoffrey Harris' discovery of the pituitary portal system demonstrating a vascular link between the hypothalamus and the anterior pituitary gland (Harris, 1955). This discovery enabled an understanding of how neurosecretory neurons might provide a link between the bird's response to a behavioral stimulus and hormone secretions. Vascular portal pathways provide a link between sensory and behavioral stimuli, endocrine secretions, and the physiological and behavioral consequences of hormonal secretions, a continuing realm of discovery (Yao et al., 2021). Today, this idea is so broadly accepted that it is considered self-evident – not a question. It is one of the building blocks of the field but no longer a contentious discussion point for researchers. The question of how behavior changes brain and hormones has not disappeared, but the level of analysis tackled by researchers today is much more reductionistic, aiming to map brain function neuron by neuron. Today's researchers often come from fields outside of behavioral research and for the most part, do not share a scientific history (e.g., Dulac et al., 2014; Moffitt et al., 2018).

Learning

The late 1940s and early 1950s was a period in which general theories of learning held sway over the field. The laboratory study of learning in this country was dominated by the single mechanism theories of Hull, Spence, and Tolman, the study of conditioning in the context of Pavlovian theory, and the newly formulated idea of operant conditioning proposed by Skinner. The goal of these students of behavior was a uniform theory of learning.

Lehrman's experience as a naturalist led him to see mysteries in the behavior of animals. He saw their behavior as more complex and opportunistic and motivated than the laboratory-based theorists of the day. An example often mentioned by Lehrman in discourse, writings, and teachings was the mystery of parental care of his ring doves as they successfully reared their young, even though they had no prior experience of squab (Lehrman, 1964a pp. 163–164.) The problem is still with us today, but the methods of analysis are new. As Lehrman wrote:

> . . . how does it happen that doves breeding for the first time manage to do these things appropriately? The answer lies in the fact that we must not think of "experience" as being limited to the effects of having performed the same behavior before. When a dove builds a nest and thus becomes attached to the nesting site so that she spends most of her time there, she is acquiring "experience" through which she becomes oriented to the nest in such a way that (a) the nest is the place where she is most likely to lay the egg, and (b) she is bound to come into contact with the egg after it appears even without having any intention to sit on it. Similarly, when doves sit on the eggs, they become attached to the nest-site even more, and they are there when the eggs hatch, so that they come into contact with the newly hatched squab. . . . the animal's experience during the early stages of the reproductive cycle may have an effect on its orientation to the world and upon its responses during later stages of the same cycle. This means that experience is playing a role in the succession from stage to stage even during the first breeding cycle.

Motivation

As is necessary in science, research focuses on accessible paradigms, and some phenomena are more accessible than others. Motivational theory of the 1950s was primarily laboratory based, mostly on rats, and was dominated by studies of hunger and thirst, as these were tractable. As a bird watcher and a naturalist, Lehrman had little interest in homeostatic systems. In his views, the behaviors necessary for successful reproduction, namely, courtship, nestbuilding, incubation, and feeding of the young were related to the secretion of gonadal and pituitary hormones. Lehrman focused on factors such as hormone-induced changes in peripheral tissue sensitivity and how that might affect behavior (Lehrman, 1964a). For example, he explored how hormone-mediated crop milk formation in male and female doves support the expression of regurgitation-feeding of squab. Such problems aligned with the notions of the day expressed in Frank Beach's famous comment on sex differences when he suggested that you can't be a carpenter without a hammer. Such problems, as related by Lehrman, took the behavior of feeding young to the higher-level question on central-peripheral-behavioral relationships that underlie all learning and motivation. Lehrman's brilliance lay in his ability to take specific instances of naturally occurring behavior and relate them to fundamental principles about complex behavior in his research and theorizing.

Current Status of Issues Addressed by Lehrman

Scientific consensus has been reached on a number of the issues addressed by Lehrman, and in many instances Lehrman's viewpoints have been assimilated into the general wisdom of the field. Behavioral endocrinology is seen as a specialty of comparative animal behavior. The general principles of behaviorally stimulated neuroendocrine secretion is a continuing area of analysis. Multi-site central and peripheral mechanisms of hormone action are no longer seen as alternative, but instead it is clear that hormones impact changes at multiple sites, at multiple levels, simultaneously in time.

The developmental approach to the analysis of adult behavior is accepted as a basic to behavioral analysis. Single process learning theories have virtually disappeared from the field. Hunger and thirst continue to be explored in sophisticated ways (Zimmerman, 2020) but are no longer considered a model of all motivational systems. The concepts of learned vs innate attributes of behavior no longer have conceptual power in theories about behavioral development, but the questions about how adult behavior emerges remain at the frontier. Today, Lehrman's strategy of observing naturally occurring behaviors is commonly used in many fields, including neuroscience as can be seen in studies of social behavior, curiosity, song learning and foraging, among many others.

Lasting Aspects of Lehrman's Work

Amazingly, he published his famous critique of Konrad Lorenz (Lehrman, 1953) before he got his PhD at NYU (1954). Prior to its publication, Lehrman had been working on this opus for about 3 years, and the paper was vetted by major thinkers of the era, including Karl Lashley, his student, Donald Hebb, and Hans Lucas Teuber. It was not just a scientific critique or a theory or a literature review; it was an emotionally laden document of the moment. To understand, consider the context in which he wrote. Lehrman had learned German when he worked at the American

Museum of Natural History, reading ornithology papers in German as a student of TC Schneirla. Lehrman served in the army (1942–1946) during World War II, working as a translator and cryptanalyst. Lehrman started to translate Lorenz's papers into English at around the end of the war, in 1946. In his original manuscript, Danny described the way in which Lorenz applied his concept of fixed action patterns and innate releasing mechanisms to human behavior. In the 1930s, based on his studies of interbreeding birds, Lorenz argued that it was necessary to "...avoid breeding between individuals and populations that differed significantly from the original population in order to prevent deterioration of genetically based behavior mechanisms" (Silver & Rosenblatt, 1987). This work had provided a biological basis for the concept of maintaining the genetic purity of an Aryan race and supported the ideology of the then ruling Nazi party. One can only imagine the emotional burden this carried for Lehrman. The reviewers objected to this discussion on the basis that it would detract from the science, and it was removed from the final paper. Whether or not this would be the decision made today, the outcome was that the debate between Lehrman and Lorenz on genetics and development continued on a scientific (apolitical) basis and was ever-increasingly fine-tuned until Lehrman's death (see Table 12.1 and (Lorenz, 1965, 2014)).

The issues raised in the Lehrman-Lorenz debates have not been resolved, have not faded away, and have not ramped down in emotional tone. What has changed is that it is no longer considered appropriate to hide them and to eliminate discourse. In fact, it is now considered appropriate to wax assertive and aggressive, as publicly as possible. As an example, one might look at the attacks, summarized in the New Yorker, in its unique journalistic style, on Kathryn Paigge Harden, a behavior geneticist (Lewis-Kraus, 2021). In that issue, the banner on the article by Gideon-Lewis Kraus reads "...on her left are those who assume that genes are irrelevant, on her right those who insist they are everything." The controversies and questions on the mechanisms underlying inheritance and development raised by Lehrman are with us today on both scientific and political levels.

Standing on the Shoulders of Giants

Scientists today build on the knowledge and discoveries made by others. In some cases, they continue and grow on the work of the scientists, like Lehrman, who have mentored and supervised them. In other cases, they build on prior discoveries with no knowledge of the shoulders on which they stand. In either case, many of the questions remain the same, but answers change over time, on both scientific and political aspects.

References

Dulac, C., O'connell, L. A., & Wu, Z. (2014). Neural control of maternal and paternal behaviors. *Science, 345*, 765–770.

Erickson, C. J., & Lehrman, D. S. (1964). Effect of castration of male ring doves upon ovarian activities of females. *Journal of Comparative and Physiological Psychology, 58*, 164.

Harris, G. W. (1955). *Neural control of the pituitary gland*. Edward Arnold.
Hebb, D. O. (1955). Drives and the C.N.S. (conceptual nervous system). *Psychological Review, 62*, 243–254.
Hinde, R. A., Lehrman, D. S., & Shaw, E. (1965). *Advances in the study of behavior*. Academic Press.
Lehrman, D. S. (1953). A critique of Konrad Lorenz's theory of instinctive behavior. *The Quarterly Review of Biology, 28*, 337–363.
Lehrman, D. S. (1959). On the origin of the reproductive behavior cycle in doves. *Transactions of the New York Academy of Sciences, 21*, 682–688.
Lehrman, D. S. (1964a). Control of behavior cycle in reproduction. In W. Etkin (Ed.), *Social behavior and organization among vertebrates*. University of Chicago Press.
Lehrman, D. S. (1964b). The reproductive behavior of ring doves. *Scientific American, 211*, 48–55.
Lehrman, D. S., & Brody, P. (1964). Effect of prolactin on established incubation behavior in the ring dove. *Journal of Comparative and Physiological Psychology, 57*, 161.
Lewis-Kraus, G. (2021). Can progressives be convinced that genetics matters? *The New Yorker*.
Lorenz, K. (1965). *Evolution and modification of behavior*. University of Chicago Press.
Lorenz, K. (2014). *Studies in animal and human behaviour*. Volume I (Translated by R. Martin). Harvard University Press.
Moffitt, J. R., Bambah-Mukku, D., Eichhorn, S. W., Vaughn, E., Shekhar, K., Perez, J. D., Rubinstein, N. D., Hao, J., Regev, A., Dulac, C., & Zhuang, X. (2018). Molecular, spatial, and functional single-cell profiling of the hypothalamic preoptic region. *Science, 362*.
Rosenblatt, J. S. (1995). Daniel Sanford Lehrman – June 1, 1919–August 27, 1972. *Biographical Memoirs. National Academy of Sciences (U.S.), 66*, 227–245.
Silver, R., & Rosenblatt, J. S. (1987). The development of a developmentalist: Daniel S. Lehrman. *Developmental Psychobiology, 20*, 563–570.
Tinbergen, N. (1963). On aims and methods of ethology. *Zeitschrift für Tierpsychologie, 20*, 410–433.
Yao, Y., Taub, A. B., Lesauter, J., & Silver, R. (2021). Identification of the suprachiasmatic nucleus venous portal system in the mammalian brain. *Nature Communications, 12*, 5643.
Zimmerman, C. A. (2020). The origins of thirst. *Science, 370*, 45–46.

Jay S. Rosenblatt

13

Alison S. Fleming and Harold I. Siegel

Abstract

Jay S. Rosenblatt was a developmental behavioral neuroendocrinologist who spent decades studying the neurobiological and endocrine bases for maternal behavior. Born in New York City, Rosenblatt was educated at New York University and the American Museum of Natural History before eventually joining the faculty at Rutgers University Institute of Animal Behavior. In addition to his work as a scientist, Rosenblatt was also a practicing psychoanalyst and an accomplished painter. His long scientific career was responsible for much of the current understanding of the regulation of maternal behavior, and his former trainees have continued those efforts.

Keywords

Maternal behavior · Parenting · Psychobiology · Psychoanalysis · Artist

Introduction

We are honored to have the opportunity to contribute to this volume and to provide our perspective on our mentor and friend, Jay S. Rosenblatt (Fig. 13.1), who has had such an impact on our own careers, who we have revered, and who we miss.

A. S. Fleming (✉)
Department of Psychology, University of Toronto at Mississauga, Mississauga, ON, Canada
e-mail: alison.fleming@utoronto.ca

H. I. Siegel
Department of Psychology, Rutgers University, Newark, NJ, USA
e-mail: hisiegel@psychology.rutgers.edu

R. J. Nelson, Z. M. Weil (eds.), *Biographical History of Behavioral Neuroendocrinology*, https://doi.org/10.1007/978-3-031-12970-4_13

Fig. 13.1 Celebrating his 89th birthday

Each of us has had our own unique and extensive opportunity to interact with and learn from Jay both as students and as scientific and professional colleagues. In preparing for an earlier version of this narrative (Fleming, 2007), "Three Faces of Jay S Rosenblatt," published in an issue of *Developmental Psychobiology* in Rosenblatt's honor, one of us (ASF) had the opportunity to talk with Rosenblatt over e-mail on numerous occasions where he permitted her to ask and he answered many questions, both of a personal and a professional nature. ASF also went to New York and interviewed Rosenblatt in Central Park, at his home in New Jersey, and over lunch on Amsterdam Avenue.

As a collaborator and on the same psychology faculty as Rosenblatt for decades after the university administration merged the Institute of Animal Behavior (IAB) and the Psychology Department, HIS was able to observe and participate as Rosenblatt continued to develop his views of his science, his world, and the academy. HIS also served as Chair of the Department for many years and he and Rosenblatt met with one another on a regular basis throughout that period. For us both, Rosenblatt also served as a confidante, colleague, and surrogate father. We both were fortunate to be able to visit with Rosenblatt during the weeks prior to his passing.

Rosenblatt provided us with a host of family photos and photos taken at various points in his life and career, with many taken by, and of, students and faculty at the IAB and the Psychology Department, Rutgers University-Newark Campus. In addition to our coveted time spent with Rosenblatt, we reread many of his articles and tried to understand the themes that had guided his thinking and his research and those of his students.

Before we start this biographical journey, why are we writing about the life and work of Jay S. Rosenblatt? Rosenblatt was a person who, in the modern period starting around 1960, virtually created a field of inquiry; the study of the psychobiology of mammalian maternal behavior. Rosenblatt provided the framework and the scaffolding for much of the work on maternal behavior that has taken place in the past six decades. He was among the first to study its humoral (Terkel & Rosenblatt, 1972, 1968) and hormonal (Bridges et al., 1977, 1978; Rosenblatt et al., 1998; Siegel & Rosenblatt, 1975a) bases; he explored the role of sensory factors, starting with somatosensory stimulation of mammary gland development during pregnancy derived through self-licking (Roth & Rosenblatt, 1966, 1967) and the role of chemosensory cues in its organization and regulation (Fleming & Rosenblatt, 1974a, b, c; Mayer & Rosenblatt, 1975, 1977, 1993). His work on experiential factors (which harks back to his studies on the role of experience in the mating behavior of male cats following castration) is legion and begins with his famous chapter with Daniel Lehrman (1963) and culminating in a progress report with Anne Mayer and Harold Siegel (Rosenblatt et al., 1979) describing the distinction between the onset of the behavior at parturition under hormonal control and its maintenance, through sensory factors and experience (Rosenblatt, 1970, 1975). Rosenblatt's students and postdocs also explored the role of neural and neurochemical factors. Michael Numan, Barry Komisaruk, and Rosenblatt were among the first to demonstrate the importance of the medial preoptic area (Numan et al., 1977). This work on neural control of maternal behavior, starting in 1975 (Fleming & Rosenblatt, 1974c, b), was very fruitfully followed up over the years by other students and postdocs, in collaboration with Joan Morrell and in the years that followed. The analysis of the neural and neurochemical mediation of maternal behavior produced as many as 25% of all research articles coming out of the IAB [to name a few (Felton et al., 1999, 1998, Giordano et al., 1989, 1991, Kalinichev et al., 2000, Komisaruk et al., 2000, Olazabal et al., 2004, 2002, Rosenblatt et al., 1996, Stern, 1990, Vernotica et al., 1999)].

Rosenblatt understood that with a social behavior like maternal behavior, there are two entities to deal with, the mother and the young, and the interaction and feedback between them, across development, which he labeled "behavioral synchrony." In part, this understanding derived from his studies of cats and their kittens, under Theodore C. Schneirla, analyzing the processes of their mutual regulation and modulation using brooder-reared kittens, isolated from the mother and littermates during various developmental periods, then returned to the mother. In kittens, he explored very early the olfactory, somatosensory, and thermal regulation of the offspring's response to its home environment and to its mother during nursing; and he discovered the establishment of individual nipple position preferences through learning, this at a time when many believed that learning in young mammals did not occur until well beyond the infantile period (Rosenblatt & Schneirla, 1962; Rosenblatt et al., 1961).

Finally, Rosenblatt did not restrict his work to rats but worked with multiple different species over his career, starting with cats (Freeman & Rosenblatt, 1978a, b; Rosenblatt & Schneirla, 1962; Rosenblatt et al., 1961), hamsters (Giordano et al., 1986; Siegel et al., 1979; Siegel & Rosenblatt, 1980), lizards (Crews et al., 1974), and, most productively, and in collaboration with Gaby Gonzalez-Mariscal and

Carlos Beyer, rabbits (Gonzalez-Mariscal et al., 2004, 1994, 2000; Gonzalez-Mariscal et al., 1998a, b). He even ventured into a few studies with humans (Rosenblatt, 1989b). His approach was to explore the similarities and differences in behavioral phenotypes across species; this comparative and evolutionary perspective, seen already in Rosenblatt (1989a), is best formulated in his article in the Scandinavian Journal of Psychology on the evolution of behavioral and nonbehavioral patterns of parental care in mammalian and avian species (Rosenblatt, 2003).

In short, Rosenblatt explored the psychobiology of maternal behavior and aspects of infant learning from many angles, at multiple analytic levels, using multiple technologies and techniques, as they became available and from a proximal, functional, developmental, comparative, and, more recently, evolutionary perspectives. He accomplished all this by working very hard, by giving his students guidance (but with a very long leash), by thinking deeply about issues, by being an excellent experimental scientist and methodologist, and by being inherently complex and seeing the world as complex. He was both willing to exploit unexpected, serendipitous findings and remain programmatic and somewhat dogged, in his pursuit of a problem. He was not a trend follower and rarely pursued the "sexy" ideas. Perhaps unbeknownst to him (or maybe he did know), he was a trend-setter himself; many careers have grown out of his conceptualizations and his work. Those of us who have had the opportunity to have worked with Rosenblatt and our students and theirs are very lucky indeed.

Three Faces

So, what about these "faces," and how have they contributed to the scientist and science? "Painting, researching, and psychoanalysis, they all expressed and satisfied different things in me," Rosenblatt acknowledged.

He was born in 1923 in the East Bronx of parents who were immigrants from Russia and Austria. His father was a furrier, a hard-working, quiet, but kind man. His mother was a "consummate housewife and mother," very warm and sociable, and a secure base. She came to the United States in the early twentieth century, to escape the Russian pogroms. His father arrived alone as a teenager, from Austria. Rosenblatt was the youngest of three children, with an older brother and sister. His childhood was a happy one, and his memory of his home life is very positive. He was admired and supported by his parents and his older sister, who was the intellectual of the family, not simply for being "cute," but for being very competent; he excelled in sports, was popular in school, had an active group of friends, did well academically, and showed real signs of talent in his painting. When I asked him what some positive memories from this early period were, in addition to his warm family life, he described listening to the stories of the Hebrews in biblical times (themes which come out in his paintings) told by young rabbis who also "took us on picnics when we played baseball." He also said, "when I was 16, I spent weekends and one-half day a week in his mentor's studio, painting and listening to classical music—a most enjoyable way of spending one's teens."

And negative memories? He had more trouble with this, but then said that times were very difficult during the depression and at that time his father became quite ill with blood poisoning and had frequent surgeries. "I continually worried about his life and the suffering he was enduring." His father was eventually cured, a miracle at that time. This brief personal biography does indeed explain the developmental antecedents of Rosenblatt, his work ethic, his self-confidence, his warmth, his interest in relationships, and the intellectual life that followed.

The Painter

It was in the context of his home that Rosenblatt began to paint. During his early teenage years, Rosenblatt painted on his own and did life drawing under the WPA program for artists and as apprentice in the studio of Ben Wilson, with the implicit and financial support of his father "who didn't really understand what (Jay) was doing," but supported him nonetheless. He was exceedingly grateful his entire life that his father was willing to pay for his painting lessons. However, it was his relationship with the painter, Ben Wilson, only 10 years his senior, that had the greatest impact on Rosenblatt. "I saw that he was thoughtful and deeply sincere in his work and that he would be supportive of my becoming a painter." Ben Wilson went on to become an established New York painter, influenced initially by the "great depicters of tragedy and morality—El Greco, Goya, and Daumier—and later by Kandinsky, Mondrian, and Max Weber." His work was primarily expressionistic and deeply moving.

Hence Rosenblatt's attachment to painting was also related to an attachment to Ben, with whom he kept in contact for 63 years until Ben died in 2001. Relationships and loyalty meant a lot to Rosenblatt. We are all pleased by that trait. In terms of the influences on Rosenblatt's own painting, Ben's style was one influence, but to that was added the influence of the French school, Braque, Cezanne, Picasso, perhaps more in terms of the form and esthetic and less in terms of the content. Rosenblatt's paintings show the strong influence of Judaism "because it provides a symbolism and setting within which to portray important contemporary feelings, stripped of their current images;" as well, he was influenced in his thinking by Marxism "—in its scientific, economic, philosophical, and political applications." His paintings are often political. They also show the horrors of war and of the holocaust; they show groupings of people who are survivors of social catastrophes such as the "Desaparecidos" in South America and of nuclear explosions. Some are representational; others are more expressionistic; others almost decorative and abstract. Most are somber; some are serene. Like the work of his mentor, Wilson, "he sought to infuse his work with a depth of emotion and to create a union of the cognitive and the expressive" (quote from a Wilson retrospective at the Blair Academy gallery www.blair.edu/Performing_Arts), a set of characteristics that also come out in Rosenblatt's approach to science.

Our separate visits to Rosenblatt's studio were revealing because unlike the prototype of the art studio, it was not a bright, open, environment. Instead, it was somewhat dark and close, down in the basement, occupied in the evenings. It is clear that he painted from his interior world and that he painted ideas and emotions. These are illuminated from within not without. Rosenblatt spoke about starting a new canvas on which he began the brushstrokes not knowing the path they would take; a process in contrast to carefully preparing the experimental steps required in a grant proposal. The paintings elicited a very mixed reaction of—dare to say it?—approach and withdrawal. On the one hand, they depict suffering or sorrow and the "apocalypse" and produce discomfort. On the other hand, most include relationships, contact, and interaction and nurturance and becoming comforting. Even those that describe the redemption ... but more especially, the many that show mothers and their children. The kids are not particularly cute or attractive to the outsider or viewer, but they clearly are to their mothers who are usually feeding and holding them. So here is one connection between Rosenblatt's science and his art, but his art traveled in other ways including but not limited to his feelings and experiences from the atrocities of World War II.

Rosenblatt did not become a professional painter, although Rutgers University mounted a one-man show of 56 of his paintings, drawings, and gouaches in the late 1990s, but painting was always important to him; he painted most of his life an activity which he shared during his later years with his second wife, Pat, who was a potter, sculpture, and painter in her own right (Fig. 13.2). Rosenblatt had quite the collection of drawings of people waiting for their flights at airports and train terminals. In later years, during psychology department meetings, Rosenblatt also drew faces, often of those around the table.

The Analyst

This is the most curious part of Rosenblatt and the least accessible. We recall as IAB students, we all became accustomed to Rosenblatt leaving work at 3 pm three days a week, and he would not deviate from this schedule—*even* if we needed him. It felt like he had another quite mysterious "other" life, a life that lasted 25 years. Why did Rosenblatt become an analyst? What did it do for him? Who influenced him? These questions were all put to Rosenblatt, and the answers were quite simple and not at all mysterious. Rosenblatt met Max Hertzman, a psychoanalyst and older colleague in the Psychology Department at CCNY where Rosenblatt taught as a TA and later Assistant Professor. Hertzman did a lot of pioneering work on Rorschach testing. "It was his thoughtfulness, originality, and honesty which attracted me to him. I felt close to him." Again, Rosenblatt became attached to a figure who served as a role model for things Rosenblatt himself wanted to do or be. Rosenblatt was attracted to psychoanalysis because he wanted to get down to the "individuality and ordinariness without (the) theory or generality" inherent in research and theoretical scientific work and to understand more about intrapsychic events both in his clients, some colleagues, and in himself. And the analytic approach was distinctly about development. In addition, Rosenblatt "wanted to see if his understanding of psychology related to the real world," which then led him to take a job at a Pediatric

Fig. 13.2 Paintings by Jay Rosenblatt with Mother and Child themes—themes he explores from both a subjective and a scientific perspective

Psychiatry Clinic at Brooklyn Jewish Hospital diagnosing emotionally disturbed, retarded, and cerebral palsied children. "I found that in fact my academic psychology background did help me to diagnose and understand these children." And then of course Rosenblatt's abiding interest in the child's mind also reflected his strong attachment to his wonderful children, Danny and Nina, his grandchildren, and his wife, Gilda, and who at the time also worked professionally with children.

Finally, Rosenblatt used his psychoanalytic training in his approach to his undergraduate clinical psychology course at Rutgers. Students prepared case studies of those in their own lives, and some of these cases became the focus of Rosenblatt's analysis. HS met several of these former students who described the course as unique, completely unforgettable, and at times surprising.

The Scientist

There were few indications from his earlier life that Rosenblatt would go into science. His favorite subjects in school were art and "the social sciences attracted (him) more than the sciences (which [he] feared a bit)." Between 1943 and 1945, Rosenblatt was in the army, stationed in various parts of Europe including several places in England, in Paris, and Liege, Belgium. One of his responsibilities was to design camouflage for military equipment. In 1946, he entered NYU to complete his training. The years between 1945 and 1953 were pivotal to Rosenblatt's scientific career. At NYU he met T.C. Schneirla whose work and theoretical position was clearly reflected in Rosenblatt's own subsequent work and thinking. Schneirla convinced Rosenblatt that having a "firm grounding in animal behavior (would provide) the broadest basis for understanding human behavior in addition to being of intrinsic interest." Rosenblatt describes a conversation between the two where Schneirla said, "well you are not going to get rich studying animal behavior"to which Rosenblatt responded "well, I was prepared not to get rich becoming a painter, so I might just as well not get rich studying animal behavior." And so, Rosenblatt started his PhD work nominally with Schneirla but also with Lester Aronson studying the role of hormones and experience in the organization of sexual behavior in male cats (Rosenblatt & Aronson, 1958b).

It was at the American Museum of Natural History (AMNH) on a grant with Schneirla that Rosenblatt began his first foray into the study of mother-young interactions in cats. During that time, he became very interested in early learning by the kittens and with Gerry Turkewitz who was a student, established that kittens develop home orientation by day 3–4; at the same time, he discovered that kittens develop nipple position preferences on day 1–2 of life. This work was important because it was counter to the belief based on more traditional conditioning paradigms that animals cannot learn at such young ages; it showed that by using "natural" species-characteristic situations, experience and learning can be shown to occur very early in life (Rosenblatt & Aronson, 1958a).

Earlier, as a teacher at NYU, later at City College of New York (CCNY) as a TA, and then as assistant professor at CCNY, Rosenblatt became close with Herbert Birch, a human developmental psychologist, who for a period constituted another important influence in Rosenblatt's intellectual development. Birch worked on development in newborn babies, providing Rosenblatt with an outlet for his interest in human development. It was during these formative years in New York City that Rosenblatt met Daniel S. Lehrman. Rosenblatt and Lehrman were graduate students together at NYU and also were colleagues at CCNY as assistant professors, as well as associates of the Department of Animal Behavior at AMNH where both of them did their doctoral research. Lehrman was the more advanced in animal behavior, as an expert ornithologist than Rosenblatt, had already published a research article at the age of 17, and was becoming world renowned because of his critique of Konrad Lorenz in 1953 (he received his PhD in 1954!). Says Rosenblatt of this relationship: "I learned a lot from Danny, but most of all we were warm friends who could talk to one another about the most important things in our lives." This affection is reflected in Rosenblatt's biography of Lehrman written for the National Academy of Sciences (Rosenblatt, 1995; Silver & Rosenblatt, 1987). Below we include a picture of Rosenblatt, Lehrman, and Robert Hinde (Fig. 13.3). Hinde became another important influence and close friend of Rosenblatt's, whose work on canaries provided Rosenblatt with yet another more "ethological" approach-from across the ocean

Fig. 13.3 Robert Hinde, Daniel S. Lehrman, and Jay S. Rosenblatt, Ethology Congress, 1955

(Oxford) in aid of understanding the endocrinology of another species-characteristic behavior.

The 1950s was a difficult period in US history, during the McCarthy era, and Rosenblatt and a number of his colleagues (Max Hertzman) were fingered as persons of interest by the House Un-American Activities Committee (HUAC), an "interest" that was conveyed to the CCNY administration which was followed by the nonrenewal of his contract. Rosenblatt soon thereafter moved to Rutgers University to join Lehrman in his newly created Institute of Animal Behavior (IAB) in Newark, New Jersey. Rosenblatt directed the IAB beginning at Danny's untimely death in 1972 and continued until 1989.

Rosenblatt began his work on rat maternal behavior after arriving at Rutgers. This work was what he spent most of his intellectual energy on for the next 45 years (Rosenblatt retired in 2005 although he continued to actively read and discuss the literature for another eight years). It is not possible to discuss each and every one of the 160 papers and chapters that Rosenblatt has written in his 60 years as a scientist. We will, however, describe some of the primary themes that have grown out of his work and the primary influences he has had on the field and on his students. As Rosenblatt himself pointed out, "most theorists are known for one or two ideas that formed the framework of their thinking." In Rosenblatt's case there are two and, maybe, three, principal ideas that have come to be identified with his perspective and have had an impact on the field in general and on the direction of own work, in particular (for overview, see (Corter & Fleming, 2002, Fleming & Li, 2002, Moore, 1995, Numan et al., 2006, Rosenblatt, 1995, Silver & Rosenblatt, 1987, Stern, 1990)]. We feel forever indebted to Rosenblatt for these ideas.

The first major contribution to the field was his recognition that given the right ecologic and naturalistic context, one can demonstrate learning by the neonate at ages that are considerably younger than believed possible at the time this work was first done in 1960s (Rosenblatt et al., 1969). The sensory modalities recruited for this early learning were initially single modalities, thermal, and tactile but, through associative processes, came to depend heavily on the olfactory sense and eventually became truly multimodal. Rosenblatt's early kitten work, illustrating these processes and described above, provided the groundwork for an entire field of study that focuses on learning within naturalistic contexts and by neonatal animals (Rosenblatt et al., 1969; Rosenblatt, 1971). This idea, tested well before its time, is reflected in more recent work called "the constraints on learning." Although Rosenblatt is most famous for his work on rat maternal behavior, which he started when he joined Lehrman's new Institute of Animal Behavior in 1958, his first work on kittens (started in 1954 at the AMNH) really epitomizes his approach to the organization of behavior and the influence of Schneirla's Approach/Withdrawal theory (Schneirla & Rosenblatt, 1961; Schneirla, 1952), namely, that to understand development one must understand the transition between dependence on basic sensory-motor reflexes which occurs first, to the development of affectively based perceptual-motivational relations. Hence, through learning, simple responses to primary stimuli (thermal and tactile) that vary along the intensity dimension become approach or active withdrawal from affectively laden multimodal stimuli (often with the addition of olfactory information) (Rosenblatt, 1971).

When ASF asked Rosenblatt what was the most exciting personal moment in his career (a rather unfair question, realized after asking), he described it was these kitten discoveries, which were in fact, quite serendipitous. While weighing kittens on a daily basis, he had noted and later set Gerry Turkewitz to study, that whenever kittens were replaced back into their home environments (after removal of the mother), the kittens would return to the home corner and do so more rapidly each day. Since the kittens seemed to know where to go, despite having no vision, this could only be based on early learning of the olfactory characteristics of the area where mother normally nurses her young (Rosenblatt, 1983; Rosenblatt et al., 1969).

He said of this work and the kitten work that followed, "It seemed to me all the problems of early development could be studied in the development of home orientation of kittens-development of sensory capacities and developmental transition in the use of sensory systems, development of motor capabilities, and the effect on sensory system use (when kittens rise off the floor, crawling to walking) they cannot use olfactory stimuli as well or as continuously and need to shift to vision with olfactory support, etc.), transition from use of socially conditioned stimuli (nest odors) to social stimuli (mother), development of learning and cognitive structures (internalization of path taking) and of course, their emotional development indicated by their distressed vocalization and its termination" (Rosenblatt, 1983, Rosenblatt et al., 1969).

The second set of ideas or themes that have come to be identified with the Rosenblatt framework are reflected in the concepts of behavioral transitions in the maternal behavior cycle and behavioral synchrony between mother and young. Since these concepts have had such an impact on our own work and are by now part of the vocabulary of the study of maternal behavior ("onset vs. maintenance"), they will now be discussed in some detail (Rosenblatt, 1970).

Although the study of maternal behavior is not primarily about early development, Rosenblatt treated the phases in the maternity cycle much as he would any developmental problem, as a series of developmental transitions. He became very interested in the phenomenology and then the mechanisms that mediated the development of maternal responsiveness from mating through pregnancy, to pregnancy termination, through the postpartum period, and into and through weaning. Each of these phases was characterized, for each he established the role of sensory factors, the associated physiological changes, the feedback effects of behavior, the role of endocrine factors, and then the role of shifts in hedonic and affective mechanisms. For each phase the mother undergoes, there occurs a synchronized set of behavioral and physiological changes in the offspring. While in his rodent work, the emphasis was on the mother, the developmental status, and needs of the offspring changed accordingly, so that mother and offspring were mutually adapted in their behavior to one another.

According to this view maternal care begins before the young appear, with conception. Once mating and pregnancy has taken place, the mother-to-be experiences endocrine changes that alter her behavior and psychology. She shows enhanced nesting and self-grooming behaviors, she comes to restrict her movements and reduce activity level, focusing instead on a particular nest site; she changes her eating preferences and behavior, with increases in preference for needed nutrients [63]; and she shows variations in emotional behavior, changes in perception, and comes to attend to some cues over others (Lott & Rosenblatt, 1969; Rosenblatt, 1980).

The next transition in the maternity cycle that Rosenblatt and his students have studied most intensively occurs at the end of pregnancy and was first described in a chapter by Rosenblatt and Lehrman (1963). "The transition between the onset of maternal behavior and its maintenance is a powerful one; with onset being mediated by the hormonal changes occurring towards the end of pregnancy and at parturition." While Rosenblatt did not know which hormones "turned on" maternal behavior initially, his 1972 studies with Joseph Terkel on the role of blood borne factors in the functional parabiotic manipulation showed that the relevant factors were present during the last 48 h of gestation (Terkel & Rosenblatt, 1968, 1972). These studies with Terkel were followed up by studies by other students and postdocs working at the IAB. These described the pregnancy effect in which mothers undergo elevations in maternal responsiveness across pregnancy, peaking close to parturition.

Pregnancy termination, removal of the uterus with or without the ovaries, became an important experimental procedure associated with the Rosenblatt lab. Hysterectomy during latter stages of pregnancy resulted in short latency maternal behavior shown to young foster pups while the combination of hysterectomy and ovariectomy did not. What were the ovaries doing to stimulate maternal care? Soon after displaying maternal care, the hysterectomized animals also showed estrous behavior; under conditions of normal termination of pregnancy by parturition, females become receptive as a function of postpartum estrous permitting them to become pregnant again after a short delay (Siegel & Rosenblatt, 1975b; Bridges,

1975). These studies led to a recognition that estrogen priming is also essential to the later discovered activational effects of oxytocin and prolactin in the initiation of maternal responsiveness.

The third phase in the maternity cycle, "the maintenance phase," has, until recently, received very little attention. However, again Rosenblatt did some landmark experiments on the role of experiences acquired during the postpartum on the subsequent expression of the behavior at a time when the parturitional hormones were no longer playing a role (Lott & Rosenblatt, 1969). These experiments suggested the existence of a sensitive period for the long-term effectiveness of a postpartum experience with young; it also suggested a role for different sensory modalities in the experience.

Work initially by Bridges (Bridges, 1975; Bridges et al., 1977; Cohen & Bridges, 1981; Lott & Rosenblatt, 1969) and then by our laboratory (summarized in Fleming and Li (2002)) and Numan et al. (2006) followed up on these early studies (see also (Stern, 1989, 1990, Stern & Keer, 1999, Stern et al., 2002), and we began to explore the role of the expression of the behavior per se and of somatosensory versus olfactory experiences with the pups in the maintenance of the motivation to mother. We illustrated the importance of the timing and duration of the maternal experience, and, as with other forms of learning, the importance of the interval during which pups are not present on responsiveness at test (the retention interval). These studies started by Rosenblatt show that reproductive behaviors like maternal behavior may be species-typical and relatively stereotyped in form, but they are nevertheless subject to the influences of experience and learning and exhibit considerable flexibility in when they are expressed and with what intensity. These behavioral flexibilities are mimicked by flexibility in brain function and structure.

The end of the maternity cycle, that of weaning, was also of considerable importance to Rosenblatt but it remains to this day the least studied phase (Stern & Keer, 1999). Rosenblatt's early work on weaning by mother cats of her kittens is a true classic and describes behavioral synchronies between mother and kittens that are all too easy to translate into a human experience. The imagery of the analogy is palpable. The "approach and withdrawal" tendencies that both we, as parents, and our teenage children experience in relation to one another is strangely analogous to the behavior shown by the mother cat who actively withdraws from her growing and proactive litter prior to weaning (Reisbick et al., 1975).

Finally, in collaboration with his students and postdocs, Rosenblatt also explored in depth the role of neural and neurochemical factors in the maternal behavior cycle. Led by Michael Numan and Barry Komisaruk, Rosenblatt's lab was among the first to demonstrate the importance of the medial preoptic area in the regulation of maternal behavior (Numan et al., 1977); see (Numan & Insel, 2003). Work by Joan Morrell and others showed the importance of pathways between the MPOA and other limbic sites and their neurochemical connections to the expression of maternal behavior (Giordano et al., 1989; Lonstein & Stern, 1997; Mayer & Rosenblatt, 1993; Numan & Insel, 2003; Olazabal et al., 2004).

Finally, what consumed Rosenblatt in his later years was less experimental in nature and more theoretical—that is to understand the evolution of parental behavior (PB). This interest only became consuming once Rosenblatt had essentially left the lab and was based primarily on an in-depth analysis of the extant literatures. Of this undertaking he wrote:

I had an insight about how to organize the material of my review and my thinking about the evolution. Instead of directly characterizing PB as evolutionary sequences, I now believe that the backbone of the evolution consists of the evolution of the reproductive physiology (sexual and parental) in the various vertebrate classes of which there are basically three patterns found in the fishes-amphibia, reptiles-birds, and monotremes-marsupials-eutherian mammals. The physiology provided the platform on the basis of which PB initially arose in the bony fishes as an aspect of mating behavior (egg-scatterers and related pattern). Further evolution of PB required elaboration of adaptations to specific features of the breeding habitat in each of the clades of fishes. The evolutionary sequences can be traced in detail in many fish species, and this is the only true direct analysis of the evolution of PB. All other "evolutionary analyses' are derived but no less valid if recognized as such." Although he was not able to finish this analysis, he invited anyone who was interested, to take on the task—a tall order!

As important as the science itself, how Rosenblatt approached his science was fun to watch and to emulate. His basic artistic nature contributed to his creative methodological designs which he worked through with his students. To study kitten approach to mothers, he created a thermal gradient in the cage; to study nipple position preference, he applied different colors to the nipples which could identify where the kitten had suckled; to study the role of olfaction in nursing, he exploited techniques to produce temporary anosmia; to study the role of maternal self-stimulation and licking, he used a rat collar to prevent females from performing the behaviors. This creative flair in design and methodology was designed to test hypotheses that were equally as intriguing or subtle. One study that he suggested was to assess brain estrogen and oxytocin receptors in virgins at different points in the sensitization process, to test our assumption (based squarely on Rosenblatt's earlier assertion) that virgin maternal behavior is entirely nonhormonal in origin. He was interested in testing this idea and going further to assess at what point in the sensitization process the relevant substrate to the hormone's changes. In fact, he was collaborating on this idea when he became ill. This is the kind of question Rosenblatt liked to ask, and it is the kind of question that often turned into the PhD theses of his many, many students and colleagues.

Professional and Personal Recognition and Appreciation

Rosenblatt received significant professional recognition for his thinking and work on rat maternal behavior. His research was federally funded for over 40 years. Simply put, he provided the model for all work in this area in all species that have been studied. The following are representative of his well-deserved awards and accomplishments: Senior Investigator Award from the International Society for Developmental Psychobiology, Daniel S. Lehrman Lifetime Achievement Award from the Society of Behavioral Neurobiology, Fellow of the American Association for the Advancement of Science, and Honorary Degrees from Goteborg University (Sweden) and National University of Education at a Distance (Spain). Rosenblatt's home Department of Psychology at Rutgers University—Newark established an annual Jay S. Rosenblatt Scholarly Prize for the best publication by a graduate student.

He is also the "father" or "uncle" to many of his students, his postdocs, and his younger colleagues. In the mid 1980s, the alumni and then-current members of the IAB organized a conference to celebrate Rosenblatt's 60th birthday and to honor the 25th anniversary of the founding of the IAB; the scientific contributions resulted in a unique volume published by the New York Academy of Sciences (Komisaruk et al., 1986). Over the last 45 years, Rosenblatt has trained at least as many students and postdocs and influenced countless others. We have benefited enormously from his example in his science and his creative mind, as have our students. Many of us would simply not be doing what we do today had we not met Rosenblatt, read his work, or studied under his supervision (Fig. 13.4). He easily and frequently expressed positive thoughts and encouragement to his many students, and there are clear cases of Rosenblatt serving as a model for other mentors and for serving as a mentor for other advisors' students. During our combined almost century of knowing and loving Rosenblatt, he never disappointed and he often surprised.

Fig. 13.4 Jay among his many students and colleagues, 1986

References

Bridges, R. S. (1975). Long-term effects of pregnancy and parturition upon maternal responsiveness in rat. *Physiology & Behavior, 14*, 245–249.

Bridges, R. S., Feder, H. H., & Rosenblatt, J. S. (1977). Induction of maternal behaviors in primigravid rats by ovariectomy, hysterectomy, or ovariectomy plus hysterectomy – effect of length of gestation. *Hormones and Behavior, 9*, 156–169.

Bridges, R. S., Rosenblatt, J. S., & Feder, H. H. (1978). Stimulation of maternal responsiveness after pregnancy termination in rats – effect of time of onset of behavioral testing. *Hormones and Behavior, 10*, 235–245.

Cohen, J., & Bridges, R. S. (1981). Retention of maternal-behavior in nulliparous and primiparous rats – effects of duration of previous maternal experience. *Journal of Comparative and Physiological Psychology, 95*, 450–459.

Corter, C. M., & Fleming, A. S. (2002). Psychobiology of maternal behavior in human beings. In M. Bornstein (Ed.), *Handbook of parenting* (Second ed.). L. Erlbaum Associates.

Crews, D., Rosenblatt, J. S., & Lehrman, D. S. (1974). Effects of unseasonal environmental regime, group presence, group composition and males' physiological state on ovarian recrudescence in the lizard, Anolis carolinensis. *Endocrinology, 94*, 541–547.

Felton, T. M., Linton, L., Rosenblatt, J. S., & Morrell, J. I. (1998). Intact neurons of the lateral habenular nucleus are necessary for the nonhormonal, pup-mediated display of maternal behavior in sensitized virgin female rats. *Behavioral Neuroscience, 112*, 1458–1465.

Felton, T. M., Linton, L., Rosenblatt, J. S., & Morell, J. I. (1999). First and second order maternal behavior related afferents of the lateral habenula. *Neuroreport, 10*, 883–887.

Fleming, A. S. (2007). The three faces of Jay S. Rosenblatt. *Developmental Psychobiology, 49*, 2–11.

Fleming, A. S., & Li, M. (2002). Psychobiology of maternal behavior and its early determinants in nonhuman mammals.

Fleming, A. S., & Rosenblatt, J. S. (1974a). Maternal behavior in the virgin and lactating rat. *Journal of Comparative and Physiological Psychology, 86*, 957–972.

Fleming, A. S., & Rosenblatt, J. S. (1974b). Olfactory regulation of maternal behavior in rats. I. Effects of olfactory bulb removal in experienced and inexperienced lactating and cycling females. *Journal of Comparative and Physiological, 86*, 221–232.

Fleming, A. S., & Rosenblatt, J. S. (1974c). Olfactory regulation of maternal behavior in rats. II. Effects of peripherally induced anosmia and lesions of the lateral olfactory tract in pup-induced virgins. *Journal of Comparative and Physiological, 86*, 233–246.

Freeman, N. C. G., & Rosenblatt, J. S. (1978a). Interrelationship between thermal and olfactory stimulation in the development of home orientation in newborn kittens. *Developmental Psychobiology, 11*, 437–457.

Freeman, N. C. G., & Rosenblatt, J. S. (1978b). Specificity of litter odors in the control of home orientation among kittens. *Developmental Psychobiology, 11*, 459–468.

Giordano, A. L., Siegel, H. I., & Rosenblatt, J. S. (1986). Intrasexual aggression during pregnancy and the estrous-cycle in golden-hamsters (Mesocricetus-Auratus). *Aggressive Behavior, 12*, 213–222.

Giordano, A. L., Siegel, H. I., & Rosenblatt, J. S. (1989). Nuclear estrogen receptor binding in the preoptic area and hypothalamus of pregnancy-terminated rats: Correlation with the onset of maternal behavior. *Neuroendocrinology, 50*, 248–258.

Giordano, A. L., Siegel, H. I., & Rosenblatt, J. S. (1991). Nuclear estrogen receptor binding in microdissected brain regions of female rats during pregnancy: Implications for maternal and sexual behavior. *Physiology & Behavior, 50*, 1263–1267.

Gonzalez-Mariscal, G., Diaz-Sanchez, V., Melo, A. I., Beyer, C., & Rosenblatt, J. S. (1994). Maternal-behavior in New-Zealand white-rabbits – quantification of somatic events, motor patterns, and steroid plasma-levels. *Physiology & Behavior, 55*, 1081–1089.

Gonzalez-Mariscal, G., Cuamatzi, E., & Rosenblatt, J. S. (1998a). Hormones and external factors: Are they "on/off" signals for maternal nest-building in rabbits? *Hormones and Behavior, 33*, 1–8.

Gonzalez-Mariscal, G., Melo, A. I., Chirino, R., Jimenez, P., Beyer, C., & Rosenblatt, J. S. (1998b). Importance of mother/young contact at parturition and across lactation for the expression of maternal behavior in rabbits. *Developmental Psychobiology, 32*, 101–111.

Gonzalez-Mariscal, G., Melo, A. I., Parlow, A. F., Beyer, C., & Rosenblatt, J. S. (2000). Pharmacological evidence that prolactin acts from late gestation to promote maternal behaviour in rabbits. *Journal of Neuroendocrinology, 12*, 983–992.

Gonzalez-Mariscal, G., Chirino, R., Beyer, C., & Rosenblatt, J. S. (2004). Removal of the accessory olfactory bulbs promotes maternal behavior in virgin rabbits. *Behavioural Brain Research, 152*, 89–95.

Kalinichev, M., Rosenblatt, J. S., & Morrell, J. I. (2000). The medial preoptic area, necessary for adult maternal behavior in rats, is only partially established as a component of the neural circuit that supports maternal behavior in juvenile rats. *Behavioral Neuroscience, 114*, 196–210.

Komisaruk, B. R., Siegel, H. I., Cheng, M.-F., & Feder, H. H. (1986). *Reproduction: A behavioral and neuroendocrine perspective*. New York Academy of Sciences.

Komisaruk, B. R., Rosenblatt, J. S., Barona, M. L., Chinapen, S., Nissanov, J., O'bannon, R. T., Johnson, B. M., & Del Cerro, M. C. (2000). Combined c-fos and 14C-2-deoxyglucose method to differentiate site-specific excitation from disinhibition: Analysis of maternal behavior in the rat. *Brain Research, 859*, 262–272.

Lonstein, J. S., & Stern, J. M. (1997). Somatosensory contributions to c-fos activation within the caudal periaqueductal gray of lactating rats: Effects of perioral, rooting, and suckling stimuli from pups. *Hormones and Behavior, 32*, 155–166.

Lott, D. F., & Rosenblatt, J. S. (1969). Development of maternal responsiveness during pregnancy in the rat. In B. Foss (Ed.), *Determinants of infant behavior*. Methuen.

Mayer, A. D., & Rosenblatt, J. S. (1975). Olfactory basis for delayed onset of maternal-behavior in virgin female rats – experiential effects. *Journal of Comparative and Physiological Psychology, 89*, 701–710.

Mayer, A. D., & Rosenblatt, J. S. (1977). Effects of intranasal zinc sulfate on open field and maternal behavior in female rats. *Physiology & Behavior, 18*, 101–109.

Mayer, A. D., & Rosenblatt, J. S. (1993). Persistent effects on maternal aggression of pregnancy but not of estrogen/progesterone treatment of nonpregnant ovariectomized rats revealed when initiation of maternal behavior is delayed. *Hormones and Behavior, 27*, 132–155.

Moore, C. L. (1995). Maternal contributions to mammalian reproductive development and the divergence of males and females. In P. J. B. Slater, J. S. Rosenblatt, C. T. Snowdon, & M. Milinski (Eds.), *Advances in the study of behavior*. Academic Press.

Numan, M., & Insel, T. R. (2003). *Hormonal and nonhormonal basis of maternal behavior*. Springer-Verlag.

Numan, M., Rosenblatt, J. S., & Komisaruk, B. R. (1977). Medial preoptic area and onset of maternal behavior in the rat. *Journal of Comparative and Physiological Psychology, 91*, 146–164.

Numan, M., Fleming, A. S., & Levy, F. (2006). Maternal behavior. In J. D. Neill (Ed.), *Knobil and Neill's physiology of reproduction*. Elsevier.

Olazabal, D. E., Kalinichev, M., Morrell, J. I., & Rosenblatt, J. S. (2002). MPOA cytotoxic lesions and maternal behavior in the rat: Effects of midpubertal lesions on maternal behavior and the role of ovarian hormones in maturation of MPOA control of maternal behavior. *Hormones and Behavior, 41*, 126–138.

Olazabal, D. E., Abercrombie, E., Rosenblatt, J. S., & Morrell, J. I. (2004). The content of dopamine, serotonin, and their metabolites in the neural circuit that mediates maternal behavior in juvenile and adult rats. *Brain Research Bulletin, 63*, 259–268.

Reisbick, S., Rosenblatt, J. S., & Mayer, A. D. (1975). Decline of maternal behavior in the virgin and lactating rat. *Journal of Comparative and Physiological Psychology, 89*, 722–732.

Rosenblatt, J. S. (1970). Views on the onset and maintenance of maternal behavior in the rat. In L. E. T. Aronson, J. S. Rosenblatt, & D. S. Lehrman (Eds.), *Development and evolution of behavior: Essays in memory of T.C. Schneirla*. Freeman.

Rosenblatt, J. S. (1971). Suckling and home orientation in the kitten: A comparative developmental study. In E. Tobach, L. Aronson, & E. Shaw (Eds.), *The biopsychology of development*. Academic Press.

Rosenblatt, J. S. (1975). Selective retrieving by maternal and nonmaternal female rats. *Journal of Comparative and Physiological Psychology, 88*, 678–686.

Rosenblatt, J. S. (1980). Hormonal and non-hormonal regulation of maternal-behavior - a theoretical survey. *Reproduction Nutrition Development, 20*, 791–800.

Rosenblatt, J. S. (1983). Olfaction mediates developmental transition in the altricial newborn of selected species of mammals. *Developmental Psychobiology: The Journal of the International Society for Developmental Psychobiology, 16*, 347–375.

Rosenblatt, J. S. (1989a). The physiological and evolutionary background of maternal responsiveness. *New Directions for Child Development*, 15–30.

Rosenblatt, J. S. (1989b). The physiological and evolutionary background of maternal responsiveness among humans. In M. Bornstein (Ed.), *Mammalian responsiveness: Characteristics and consequences*. Jossey Bass.

Rosenblatt, J. S. (1995). Daniel Sanford Lehrman – June 1, 1919–August 27, 1972. *Biographical memoirs. National Academy of Sciences (U.S.), 66*, 227–245.

Rosenblatt, J. S. (2003). Outline of the evolution of behavioral and nonbehavioral patterns of parental care among the vertebrates: Critical characteristics of mammalian and avian parental behavior. *Scandinavian Journal of Psychology, 44*, 265–271.

Rosenblatt, J., & Aronson, L. (1958a). The influence of experience on behavioural effects of androgen in prepuberally castrated male cats. *Animal Behaviour, 6*, 171–182.

Rosenblatt, J. S., & Aronson, L. R. (1958b). The influence of experience on the behavioural effect of androgen in prepuberally castrated male cats. *Animal Behaviour, 6*, 171–182.

Rosenblatt, J. S., & Lehrman, D. S. (1963). Maternal behavior in the laboratory rat. In H. L. Rheingold (Ed.), *Maternal behavior in mammals*. Wiley.

Rosenblatt, J. S., & Schneirla, T. C. (1962). The behavior of cats. In E. Hafez (Ed.), *The behavior of domestic animals*. Balliere, Tendall and Cox.

Rosenblatt, J. S., Turkewitz, G., & Schneirla, T. (1961). Early socialization in the domestic cat as based on feeding and other relationships between female and young. In B. Foss (Ed.), *Determinants of infant behavior*. Methuen.

Rosenblatt, J. S., Turkewitz, G., & Schneirla, T. C. (1969). Development of home orientation in newly born kittens. *Transactions of the New York Academy of Sciences, 31*, 231–250.

Rosenblatt, J. S., Siegel, H. I., & Mayer, A. D. (1979). Progress in the study of maternal behavior in the rat: Hormonal, nonhormonal, sensory, and developmental aspects. In J. S. Rosenblatt, R. A. Hinde, C. Beer, & M. C. Busnel (Eds.), *Advances in the study of behavior*. Elsevier.

Rosenblatt, J. S., Hazelwood, S., & Poole, J. (1996). Maternal behavior in male rats: Effects of medial preoptic area lesions and presence of maternal aggression. *Hormones and Behavior, 30*, 201–215.

Rosenblatt, J. S., Olufowobi, A., & Siegel, H. I. (1998). Effects of pregnancy hormones on maternal responsiveness, responsiveness to estrogen stimulation of maternal behavior, and the lordosis response to estrogen stimulation. *Hormones and Behavior, 33*, 104–114.

Roth, L. L., & Rosenblatt, J. S. (1966). Mammary glands of pregnant rats: Development stimulated by licking. *Science, 151*, 1403–1404.

Roth, L. L., & Rosenblatt, J. S. (1967). Changes in self-licking during pregnancy in rat. *Journal of Comparative and Physiological Psychology, 63*, 397–400.

Schneirla, T. C. (1952). A consideration of some conceptual trends in comparative psychology. *Psychological Bulletin, 49*, 559–597.

Schneirla, T. C., & Rosenblatt, J. S. (1961). Animal research – panel, 1960. 1. Behavioral organization and genesis of the social bond in insects and mammals. *American Journal of Orthopsychiatry, 31*, 223–253.

Siegel, H. I., & Rosenblatt, J. S. (1975a). Estrogen-induced maternal behavior in hysterectomized-overiectomized virgin rats. *Physiology & Behavior, 14*, 465–471.
Siegel, H. I., & Rosenblatt, J. S. (1975b). Hormonal basis of hysterectomy-induced maternal behavior during pregnancy in the rat. *Hormones and Behavior, 6*, 211–222.
Siegel, H. I., & Rosenblatt, J. S. (1980). Hormonal and behavioral aspects of maternal care in the hamster: A review. *Neuroscience and Biobehavioral Reviews, 4*, 17–26.
Siegel, H. I., Cohen, P., & Rosenblatt, J. S. (1979). The effect of hysterectomy on hormone-induced lordosis behavior in hamsters. *Physiology & Behavior, 23*, 851–853.
Silver, R., & Rosenblatt, J. S. (1987). The development of a developmentalist: Daniel S. Lehrman. *Developmental Psychobiology, 20*, 563–570.
Stern, J. M. (1989). Maternal behavior: Sensory, hormonal, and neural determinants. In F. R. Brush & S. Levine (Eds.), *Psychoendocrinology*. Elsevier.
Stern, J. M. (1990). Multisensory regulation of maternal behavior and masculine sexual behavior: A revised view. *Neuroscience and Biobehavioral Reviews, 14*, 183–200.
Stern, J. M., & Keer, S. E. (1999). Maternal motivation of lactating rats is disrupted by low dosages of haloperidol. *Behavioural Brain Research, 99*, 231–239.
Stern, J. M., Yu, Y. L., & Crockett, D. P. (2002). Dorsolateral columns of the spinal cord are necessary for both suckling-induced neuroendocrine reflexes and the kyphotic nursing posture in lactating rats. *Brain Research, 947*, 110–121.
Terkel, J., & Rosenblatt, J. S. (1968). Maternal behavior induced by maternal blood plasma injected into virgin rats. *Journal of Comparative and Physiological Psychology, 65*, 479–482.
Terkel, J., & Rosenblatt, J. S. (1972). Humoral-factors underlying maternal behavior at parturition – cross transfusion between freely moving rats. *Journal of Comparative and Physiological Psychology, 80*, 365–371.
Vernotica, E. M., Rosenblatt, J. S., & Morrell, J. I. (1999). Microinfusion of cocaine into the medial preoptic area or nucleus accumbens transiently impairs maternal behavior in the rat. *Behavioral Neuroscience, 113*, 377–390.

Peter Robert Marler

14

John C. Wingfield and Marilyn Ramenofsky

Abstract

Peter R. Marler (1928–2014) began his scientific career at the University College of London and at Cambridge University steeped in natural history and with a deep interest in animals and plants in general, but particularly birds. At Cambridge he gained a broad knowledge of the fast-growing field of ethology. This focus plus his own observations of local dialects in the songs and calls of a local songbird precipitated a life long and distinguished career on song learning and communication. His research gained international attention not only because of the parallels with human speech (also learned) but also the intersections of fields from natural history and ethology to neurobiology and neuroendocrine aspects of behavior. Marler's scientific acumen and insights percolated many biological disciplines including behavioral neuroendocrinology. Many of his former associates have gone on to distinguished academic careers leaving an astonishing legacy for future generations.

Keywords

Song · Learning · Semantic communication · Birds · Primates · Endocrinology

J. C. Wingfield (✉) · M. Ramenofsky
Department of Neurobiology, Physiology and Behavior, University of California, Davis, CA, USA
e-mail: jcwingfield@ucdavis.edu; mramenofs@ucdavis.edu

R. J. Nelson, Z. M. Weil (eds.), *Biographical History of Behavioral Neuroendocrinology*, https://doi.org/10.1007/978-3-031-12970-4_14

Peter and Judith Marler. (With permission)

After receiving a B.Sc. degree from the University College, University of London in 1948, a young graduate student became involved with surveys of vegetation and potential nature reserves. The surveys took him to the field in the UK, France, and the Azores, where he became fascinated by geographic variation in songs of a widespread European songbird, the chaffinch, *Fringilla coelebs*. Although conducted in his spare time, his studies suggested that chaffinches had diverse dialects of songs and calls spanning geographic scales rather like dialects in human speech. This student from the Department of Botany, University College, University of London, then began a Ph.D. project on plant ecology in bogs. His doctoral field work across the UK provided more observations that the calls and songs of chaffinches showed regional differences.

That student was Peter R. Marler who went on to build on his observations in natural history to show that dialects of birds were learned, as is human speech, while becoming one of the ethologists of his time who defined the modern era of animal behavior studies at the intersections with ecology, evolution, and neurobiology, including endocrinology and neuroendocrinology (Ball & Dooling, 2017). He nurtured these interfaces and commented on other fundamental biological processes integrating behavior with physiological systems. For example, the role of biological clocks in animal behavior as emphasized by a meeting in 1960 that brought together researchers in the field. Marler understood how biological clocks play a fundamental role in the expression of behavioral traits at appropriate times. Even then, his insight was far reaching and integrative and his historical perspective inspirational (Ball & Dooling, 2017; Marler, 2005).

Marler was born in the southern town of Slough, just west of London, on 24 February 1928. He married Judith Gallen, also from Slough, in Edinburgh on 1 September 1954. Marler was already involved in his second graduate degree, a D.Phil., at Cambridge when he met Judith. They visited Kew Gardens, but never entered as Marler was distracted by pigeon behavior which he explained to Judith with much enthusiasm. He was engrossed with birds from as early as 8 years old. At that time, he raised a pet rook (*Corvus frugilegus*) that developed a habit of entering neighbors houses through open windows and bringing home anything bright and shiny. Marler dutifully returned jewelry and other items the rook had pilfered. His fascination with birds remained unabated throughout his life.

The Marlers had three children, Christopher, Catherine, and Marianne. One of them, Catherine A. Marler, is a professor of psychology and an accomplished behavioral neuroendocrinologist at the University of Wisconsin. She has contributed widely to the field. Her research generally is in the biology of brain and behavior in deer mice, *Peromyscus californicus*. She has explored the hormonal mechanisms underlying agonistic and reproductive behavior in both the field and laboratory. Christopher directs the multimedia national school in Kigali, Rwanda, and serves as a consultant to the government. Marianne lives in Davis, California.

After graduating with a Ph.D. in botany in 1952, Marler received a fellowship from the Nature Conservancy to pursue a D.Phil. this time in Zoology, at the University of Cambridge graduating in 1954. He joined the Madingley Ornithological Station at Cambridge and under the guidance of Professor William H. Thorpe and Robert Hinde – who had recently completed his D.Phil. Now starting out in a new discipline, Marler had much to learn about the growing field of ethology. His first research projects included studies of aggression and agonistic behavior in captive birds, but he never forgot his strong interest in animal communication and vocal behavior, particularly of songbirds.

Because Thorpe focused on song, Marler turned more to the extensive vocal repertoire of the chaffinch. With a fellowship from Jesus College at Cambridge, Marler took advantage of access to the recently developed sound spectrograph producing a visual display of the frequency modulation of a vocalization that allowed him to produce a functional analysis of the entire repertoire of a focal animal. This was perhaps the first time such a complete analysis had been accomplished.

Robert Hinde's work on courtship behavior of cardueline finches presented an opportunity for Marler to compare vocal repertoires among species documenting evidence for evolution of shared vocal signals across species. This caused much debate as to whether such shared signals existed across species and were not just arbitrary sounds, which was the firmly held belief at the time. Marler's "clashes with authority" as he called them, "which he attributed in retrospect more to youthful exuberance than intellectual acumen" (quoted from P.R. Marler autobiography, Department of Neurobiology, Physiology and Behavior, College of Biological Sciences, University of California, Davis), set the stage for many studies of vocal repertoires across vertebrates and eventually for the development of semantic communication, such as predator alerts, within and among species. Additionally, during his fellowship at Cambridge Marler was introduced to academic meetings including international scientists and thus was fully engaged in discourse with other ethologists. His presentations indicated the breadth of his thinking as he emphasized the roles of the multiple peripheral and central levels of sensory processing at which specificity of responsiveness can be imposed. Marler never lost sight of the mechanistic underpinnings of ethology and its importance to understanding biological processes in breadth.

Throughout his career Marler retained interest in vocal communication, mostly of birds, but also in other animals that would become research foci later. His studies of free-living, local bird species near Cambridge showed that louder vocal signals communicated across distances appeared more species specific, especially vocalizations related to reproduction. In contrast, quieter signals communicated over much smaller distances and often related to alarm calls in social species, appeared to be much more similar across species (Marler, 1957). This paper became a citation classic in *Current Contents*, 1985.

As his fellowship at Cambridge was ending, he accepted a faculty position at the University of California, Berkeley. He moved there in 1957 with Judith. The position at Berkeley was advertised as one would expect, but Marler was actually second in line for the job. The department's first choice turned the job down, and Marler was then offered the position and accepted it. Dick Aiken was chair of the department at Berkeley and wanted to strengthen behavioral biology in their faculty. They chose the right person! The journey from the UK to California was quite eventful and while transiting the Panama Canal, Marler would rush from one side of the boat to the other trying to identify the tropical birds in the forest that surrounded them.

After arrival in Berkeley, Marler began exploring the California countryside searching for a model species on which to continue and expand his research ideas on vocalizations and their development. Several species already had preliminary data on vocalizations and the one he chose was the coastal, white-crowned sparrow, *Zonotrichia leucophrys nuttalli*, which is abundant, adapts well to captivity, is non-migratory and importantly, has well defined song dialects. Marler was able to show that song dialects of white-crowned sparrow were learned. Although young birds had the capacity to learn many different sounds in their environment, they strongly preferred songs of their own species. The concept of "instinct to learn" began to develop with this interplay of innate preferences and learned songs. Further research showed that song learning was restricted to a short but relatively flexible sensitive period early in life. The parallels of song learning and human speech were very compelling and Marler's work gained national and international attention. Marler had outstanding colleagues at Berkeley including Frank Beach (hormones and behavior) and Howard Bern (comparative endocrinology). Students at Berkeley soon focused on Marler's research programs and early graduate students included John Eisenberg, Mark Konishi, and Fernando Nottebohm. Others followed. These colleagues presented Marler with a very broad intellectual scope to pursue his own research and also gain background for future investigations on a much wider front (Ball & Dooling, 2017).

During his early years at Berkeley, Marler realized that a textbook on animal behavior was badly needed for an increasing population of undergraduate and graduate students who were becoming very interested in the field. He published a much needed, immensely popular, and detailed textbook of animal behavior together with his close friend and colleague, William Hamilton (Marler & Hamilton, 1966). It turned out to be a 5 year project, but true to form, it focused on mechanisms and included work on invertebrates as well as vertebrates. Hamilton graduated from Cornell University in mammalogy before conducting graduate work under Starker Leopold at Berkeley in the Museum of Natural History. He later joined the faculty at the University of California, Davis in 1963. Hamilton's wide-ranging research on birds complemented Marler's expertise. Their combined and extensive experience with ethology and ethologists at the interfaces with so many other disciplines gave them a unique opportunity to bring together a very intensive but often scattered literature into a single volume. It was published after Marler had moved to Rockefeller University although work on it began while in Berkeley. The textbook became a reference for all areas of animal behavior that inspired many students and professionals alike to pursue careers in this field. Many of these later pioneered investigations at the interfaces of animal behavior with endocrinology and neuroendocrinology (Ball & Dooling, 2017).

After 9 years at Berkeley, Marler began to be restless for new opportunities, and in 1966 he was persuaded to move across the continent to the Rockefeller University in New York City. The then President, Detlev Bronk, was expanding the university's expertise to include animal behavior. Together with Don Griffin, Marler built a thriving Rockefeller University Field Research Center (RUFRC) near the village of Millbrook in the Hudson Valley. The center included 1000 hectares of eastern forest, streams, and juniper prairies about 130 km north of the "Rock" main campus which was on the Upper East Side of Manhattan. Marler continued his seminal studies on song learning but initially had to find, again, a new model species. At first, he imported canaries from Belgium. These had been selected for song and his group was able to show that song was learned by young canaries while "white noise" interfered with learning adult songs, and experimental deafening resulted in primitive songs produced by young birds as they matured. He also chose a local wild songbird, the song sparrow, *Melospiza melodia*, that has a complex song and the congeneric swamp sparrow, *M. georgiana*, that had a simpler song. Experiments by outstanding students, postdoctoral fellows and assistant professors recruited by Marler to explore the mechanisms of song learning in these species established a very productive and inspirational atmosphere with facilities to pursue these investigations. The bucolic surroundings at the RUFRC nurtured many discussions and weekly seminars attracted leaders of various fields of ethology and neurobiology resulting in very animated discussions. The intellectual climate was outstanding and inspiring (Ball & Dooling, 2017). Additionally, there were socials held at RUFRC folks' homes nearby the Field Research Station where lively discussions continued.

Song learning experiments involved finding nests of the focal species in the field and bringing nestlings into captivity where they were hand-raised under various conditions to tease apart the mechanisms and characteristics of the learning process. The bulk of this work fell to Judith who along with Marler devised an exacting and exhaustive scheme for rearing hatchlings to full sized fledglings which took at least 20 days. Actually, hand rearing expertise began while the Marlers were at Madingley where they raised jackdaws, *Corvus monedula*. When the Marlers moved to Berkeley by boat (freighter out of Glasgow and through the Panama Canal to San Francisco), they brought the jackdaws along (12 housed in a huge cage)! Judith Marler became a legendary expert, indeed the authority, on hand-raising songbird nestlings that allowed critical experimental investigations of the processes of vocal learning. Such studies were clearly not possible in other organisms such as humans. Hand raising also required a 100%-time commitment to collecting nestlings and feeding them at regular intervals from early morning to late night. This exhausting commitment was mercifully restricted to spring and early summer months, but without Judith's care and knowledge, many critical experiments could not have been performed with such effectiveness.

Investigations of song learning and production were conducted in Millbrook assisted by an outstanding Research Scientist, Susan Peters, who worked in the field and the laboratory with Marler from 1973 to 1989. Susan proved to be exceptionally qualified for this work and together both Marler and Susan unraveled many intricacies of song development.

The work on canaries attracted Marler's former graduate student, Fernando Nottebohm, from Berkeley to RUFRC where he continued his landmark research on song control systems in songbirds and the neurobiology of song. Nottebohm discovered that neurogenesis occurred in adult songbirds each year as they learned new songs and crystallized the final song. This discovery was at first considered a "heresy" to infer that neurogenesis could occur in adult brains. Since then, neurogenesis in adult vertebrate brains has been widely demonstrated in many contexts. Nottebohm's mechanistic approaches and those of Mark Konishi (deceased 2020, formally of Caltech), who worked with owls and songbirds stimulated many other investigations across the globe including exploring the role of endocrinology and neuroendocrinology. These results were presented at ever growing numbers of symposia and large poster sessions at meetings such as the Society for Neuroscience held each autumn.

While in Berkeley, Marler became increasingly interested in research on primates, particularly in Africa. Essentially this began when he was invited to review social communication at a meeting of primatologists at Stanford University in 1964. The meeting piqued Marler's interest in vocal repertoires of primates, and he took a sabbatical year with his family at Makerere University in Uganda, Africa, from 1965 to 1966. This heralded the start of his primate work where he defined in detail the vocal repertoires of several forest living primates under natural conditions. Later, he worked with Jane Goodall at Gombe National Park in Tanzania and documented the full repertoire of the chimpanzee, *Pan troglodytes* (see also Seyfarth & Cheney, 2015). Soon thereafter Marler conducted the first "play back" experiments on Vervet monkeys, *Chlorocebus pygerythrus*, opening the door to so many experimental studies on vocalizations and social interactions in other vertebrate species. These studies also laid the foundations for further seminal investigations on primates by former graduate students and postdoctoral fellows such as John Oates,

Tom Struhsaker, John Mitani, Dorothy Cheney, and Robert Seyfarth as well as many others (see Seyfarth & Cheney, 2015).

Throughout his career, Marler espoused the advantages that vocalizations models presented for underlying neurobiological mechanisms, including neuroendocrinology. In 1988, Marler was invited to write a review chapter (Wingfield & Marler, 1988) on the endocrine basis of communication for Knobil and Neill's monumental treatise on the physiology of reproduction. An even more extensive updated version followed (Wingfield et al., 1994). At this point it was abundantly clear the impact Perter Marler had had on ethology in general, but also his insight into underlying neurobiology and neuroendocrinology. It is clear how influential Marler has been in the integration of these fields including field and laboratory work. Although he would never have considered himself a "card-carrying" behavioral neuroendocrinologist, he did have a very firm grasp of important questions and penetrating insight relating to mechanisms, central and regulatory such as hormonal control (see also Marler, 2005 for a historical perspective). Marler's group also experimentally demonstrated a role for sex steroids in song learning (Marler et al., 1988). His influence, then, has been critical in developing the field of behavioral neuroendocrinology from the perspective of linking the biology of animals in their natural world with mechanisms at the reductionist levels. In 2004 he gave a plenary lecture at the meeting of the Society for Behavioral Neuroendocrinology in Lisbon, Portugal, resulting in a paper on the topic of ethology and the origins of behavioral endocrinology (Marler, 2005). Other examples of his influence are apparent in the seminal books on hormones and social behavior (Adkins-Regan, 2005) and behavioral endocrinology in general (Nelson & Kriegsfeld, 2018; Pfaff & Joëls, 2017) and others.

After many highly productive years at the RUFRC, Marler made it known that he was available for a possible move and he was attracted, over several other offers, to move back across country to the University of California Davis. Both Marler and Judith considered California as home and were delighted to be back, especially in the golden hills and savannah of central California. Marler had a hand in developing his own program at UC Davis as well as forming the Department of Neurobiology, Physiology and Behavior. A very logical and creative name again involving mechanisms of behavior at so many levels.

By the time he retired Marler had advised 17 graduate students and 28 postdoctoral fellows. After retirement he edited a book (Marler & Slabbekoorn, 2004) on nature's music and then launched another big editorial project on the neurobiology of bird song from ethology to neurobiology and neuroendocrinology (Zeigler & Marler, 2008). Even after retirement Marler was still addressing integrative aspects of behavior incorporating anatomy, developmental, learning, evolution, physiology, genetics, and cell and molecular processes. The 2008 book also included chapters on "a personal note" about William Thorpe (by R. A. Hinde), a personal journey with birdsong by Mark Konishi, and the remarkable discovery of replaceable neurons by Fernando Nottebohm. Both of the latter two authors were former students of Marler's, and were instrumental in working out the songbird neural circuits that regulated song production, perception, and development. Such fundamental discoveries drove a tsunami of research to the present day.

Marler received many accolades and honors (Ball & Dooling, 2017). Foremost was his election as a member of the USA National Academy of Sciences in 1971; member of the American Academy of Arts and Sciences in 1970 and Honorary Foreign Fellow of the Royal Society, UK, in 2008. He was an elected fellow and president of many behavior-related societies, recipient of numerous awards, and member of numerous boards and so on.

Peter R. Marler died on 5 July 2014 of pneumonia soon after he and his family were evacuated to the town of Winters from the path of one of California's notorious wildfires – the Monticello Fire. His life and research accomplishments still guide us as we now strive to understand changing behavioral processes and their control, and evolution, in a world ravaged by climate change. Pollution and loss of habitat all have their own unique ways of affecting individuals, populations, and ecological communities from expression of behavioral traits to their neurobiological and neuroendocrine controls. We miss Peter Marler's insight, scientific acumen bolstered by his sheer joy observing animals in their natural world as we try to make sense of what is ahead.

Acknowledgments We are grateful for detailed discussions with Judith Marler about the life and times of Peter and their journey through life from England to North America and the wilds of Africa. We also thank Catherine A. Marler for her comments.

References

Adkins-Regan, E. (2005). *Hormones and social behavior*. Princeton University Press. https://doi.org/10.1515/9781400850778

Ball, G. F., & Dooling, R. J. (2017). Peter R Marler, 1928–2014. *The Auk, 134*, 932–934.

Marler, P. (1957). Specific distinctiveness in the communication signals of birds. *Behaviour, 11*, 13–39.

Marler, P. R. (2005). Ethology and the origins of behavioral endocrinology. *Hormones and Behavior, 47*, 493–502.

Marler, P. R., & Hamilton, W. J. (1966). *Mechanisms of animal behavior*. Wiley.

Marler, P. R., & Slabbekoorn, H. (Eds.). (2004). *Nature's music. The science of bird song* (p. 513). Elsevier Academic Press.

Marler, P. R., Peters, S., Ball, G. F., Dufty, A. M., Jr., & Wingfield, J. C. (1988). The role of sex steroids in the acquisition and production of bird song. *Nature, 336*, 770–772.

Nelson, R. F., & Kriegsfeld, L. (2018). *Introduction to behavioral endocrinology*. Sinauer Press.

Pfaff, D. W., & Joëls, M. (Eds.). (2017). *Hormones, brain and behavior* (3rd ed.). Academic Press. Editors-in-chief

Seyfarth, R. M., & Cheney, D. L. (2015). Obituary: Peter Marler (February 24, 1928–July 5, 2014). *International Journal of Primatology, 36*, 14–17.

Wingfield, J. C., & Marler, P. R. (1988). Endocrine basis of communication: Reproduction and aggression. In E. Knobil & J. D. Neill (Eds.), *The physiology of reproduction* (pp. 1647–1677). Raven Press.

Wingfield, J. C., Whaling, C. S., & Marler, P. R. (1994). Communication in vertebrate aggression and reproduction: The role of hormones. In E. Knobil & J. D. Neill (Eds.), *Physiology of reproduction* (2nd ed., pp. 303–342). Raven Press.

Zeigler, H. P., & Marler, P. R. (Eds.). (2008). *Neuroscience of birdsong* (p. 550). Cambridge University Press.

Julian M. Davidson

15

Randy J. Nelson, Benjamin D. Sachs, and Irving Zucker

Abstract

Julian M. Davidson (15 April 1931–31 December 2001) grew up in Scotland, attended university in Israel, and completed graduate and postdoctoral studies in the United States. He spent his scientific career in the United States. He trained in neuroendocrinology with PhD mentor, William Ganong, and postdoctoral mentor, Charles Sawyer, and became interested in behavior during his postdoctoral studies with Frank Beach. Davidson's work focused on male sexual behavior; his contributions were both broad and deep. For example, Davidson studied the behavioral neuroendocrinology underlying sexual behavior in male rodents, nonhuman primates, and humans. His studies of humans included the role of hormones in sexual arousal and behavior in pre- and postmenopausal women, the effects of aging and diabetes in sexual functioning in men, the influence of pharmacologic agents on sexual functioning, and the relationship of oxytocin to orgasm (Morrissette & Myers, 2002).

Keywords

Sex behavior · Male sexual behavior · Testosterone · Penile reflexes · Neuroendocrinology · Hyenas · Human sexual function · Sexology

R. J. Nelson (✉)
Department of Neuroscience, Rockefeller Neuroscience Institute, West Virginia University, Morgantown, WV, USA
e-mail: Randy.Nelson@hsc.wvu.edu

B. D. Sachs
Department of Psychological Sciences, University of Connecticut, Storrs, CT, USA

I. Zucker
Department of Psychology and Integrative Biology, University of California, Berkeley, CA, USA

R. J. Nelson, Z. M. Weil (eds.), *Biographical History of Behavioral Neuroendocrinology*, https://doi.org/10.1007/978-3-031-12970-4_15

Early Years

Julian Mordecai Davidson, the third of four sons, was born in 1931 in Dublin, Ireland, to an Orthodox Jewish family. At the age of four, his family moved to Glasgow, Scotland, where his father established a knitwear manufacturing company. He passed the Scottish Higher Examinations at 16. He yearned to help settle Israel by moving there to be a farmer. However, the kibbutzim preferred only experienced farmers, so he worked on a model farm in southern England that was managed by the Zionist Youth Movement, Bnei Akiva (Bloch & Davidson, 1967). According to all reports, Davidson was an enthusiastic, but terrible, farmer. Nonetheless, he achieved his dream and moved to a kibbutz in Israel when he was 19 but left to pursue his education a year later. Davidson graduated from Hebrew University in Jerusalem in 1955 with BS and MS degrees in agriculture. He moved to the United States and studied with preeminent neuroendocrinologists.

Academics

Davidson first enrolled in the Physiology PhD program at UC San Francisco. He earned his PhD in 1959 with William Ganong, then completed postdoctoral studies, first with Charles Sawyer at UCLA (1959–1960), and then with Frank Beach (1962–1963) at Berkeley. He married his wife of 41 years, Ann, in 1959, and they had three accomplished children and several grandchildren. In 1963, Davidson joined the Department of Physiology (since renamed the Department of Cellular and Molecular Physiology) at Stanford University, where he spent his entire career with a few sojourns. For example, in 1970–1971 he was a Guggenheim Fellow at Oxford; he was a visiting scholar at the Battelle Seattle Research Center (1974–1975), a visiting professor at the University of Athens Medical School (1978–1979), and a guest researcher at the US NIH (1985–1986). His first papers were published in 1960, and over his 37 year publishing career, more than 170 papers were authored or co-authored by Davidson (Morrissette & Myers, 2002).

Behavioral neuroendocrinology comprises many examples of behavioral biologists learning neuroendocrinology and incorporating these techniques into their behavioral studies. But there are few examples of individuals trained by preeminent neuroendocrinologists pivoting to study behavior. But that was the path of Davidson. During his postdoc in Beach's lab, he befriended and collaborated with Gordon Bermant and Stephen Glickman examining the role of the limbic system on male rat mating behavior (Bermant et al., 1968; Sachs, 2003). Later during his career, Davidson collaborated with Glickman on the hyena projects (e.g., (Glickman et al., 1987b, Smith et al., 1987a, b). He also collaborated with Bermant later in his career to produce a book (Bermant & Davidson, 1974).

Behavioral Neuroendocrinology: Rodents

Most of Davidson's early papers focused on endocrine feedback mechanisms in the hypothalamic-pituitary-adrenal (HPA) and hypothalamic-pituitary-gonadal (HPG) axes. In the early sixties, he published a paper with Ganong on the effects of LSD on adrenal function in dogs (Ganong et al., 1961). In the mid-1960s, Davidson published several papers detailing the progressive waning of sexual behavior in rats after castration (e.g., (Davidson, 1966b)). Shortly after Heimer and Larsson (1964) reported that an intact MPOA was necessary for male copulatory behavior in rats, Davidson published a paper describing a study wherein implants of testosterone into the MPOA restored copulation in castrated rats (Davidson, 1966a). His team extended this research a decade later (Cheung & Davidson, 1977; Damassa et al., 1977). Davidson also demonstrated that implants of cortisone into the hypothalamus inhibited ACTH secretion (Davidson et al., 1963) and that intra-hypothalamic implants of cyproterone (an antiandrogen) stimulated male reproductive function; he hypothesized that decreased testosterone stimulation in the median eminence activated the HPG axis (Bloch & Davidson, 1967). This work was followed up with his discovery that estradiol maintained male rat mating behaviors. Throughout the 1970s, Davidson also published papers detailing regulation of female rat mating behavior and regulation of estrous cycles (e.g., (Davidson & Smith, 1974, Gray et al., 1978, Smith et al., 1973)). One important study (Damassa et al., 1977) established that differences in circulating testosterone (T) are not significant factors in determining variance in sexual performance among individual male rats. The results of this study indicate that, among typical adult male rats, neither quantitative nor qualitative aspects of sexual behavior are determined by circulating testosterone concentrations. By treating castrated rats with testosterone implants, they observed that blood testosterone concentrations adequate to restore full copulatory behavior were less than 10% of concentrations present in intact males, suggesting that all males, including those that fail to copulate, secrete more than adequate amounts of androgens to sustain vigorous male sex behavior; failure to do so resides in properties of relevant tissue substrates responsive to androgens, presumably in the CNS. Their group also reported that sexual motivation was increased in male rats by treatment of yohimbine, an α2-adrenoceptor antagonist (Clark et al., 1984). His lab was joined in the early 1970s by postdoctoral scholar, Erla Smith. She remained in Davidson's lab throughout his career and served as the "rock of the lab," chief-of-staff, and manager and organized counterpoint to him (Erla Smith quoted in Sachs, 2003). She was also coauthor to more than one-third of his papers.

The 1970s also provoked Davidson's interests in altered states of consciousness. He studied the physiological responses associated with meditation and the so-called mystical states (e.g., (Davidson, 1976, 1980). This work led to his exploration of human orgasm as the most common and accessible (and legal) altered state of consciousness, and he published several papers on this topic (e.g., (Davidson, 1980; Davidson & Davidson, 1980)).

Behavioral Neuroendocrinology: Humans and Hyenas

In addition to studies of the altered states of human orgasms, Davidson became interested in human sex behaviors and other responses. He traveled to Edinburgh to work with John Bancroft to learn human sex-research techniques (Morrissette & Myers, 2002). Davidson became such a sexology expert that many traveled to his lab to learn these techniques. Throughout the 1980s, Davidson published increasingly more on human and nonhuman primate sexual responses (e.g., (Chen et al., 1981, Coe et al., 1981, Davidson & Myers, 1988, Davidson et al., 1983, Herdt & Davidson, 1988, Davidson, 1984)). In one paper he detailed oxytocin release in response to male orgasms (Carmichael et al., 1987). His work with squirrel monkeys was an extended collaboration with his Stanford collaborator, Seymour Levine. Davidson continued to make major contributions to understanding the mechanisms underlying sexual behavior using nonhuman animal models to the end of his career.

Davidson loved the Bay Area and he loved to travel. He traveled to New Guinea to study a population of androgen-insensitive men who were sexed as females early in life but transformed into male phenotypes after puberty (Herdt & Davidson, 1988). In Africa, he joined with his former postdoctoral colleague, Stephen Glickman, and Glickman's research associate, Laurence Frank, to study the endocrine mechanisms underlying the atypical anatomy and social organization of spotted hyenas, and he continued this collaboration with the captive population of hyenas housed in the Berkeley hills (Frank et al., 1985a, b; Glickman et al., 1987a).

He trained a number of outstanding PhD students including George Bloch, David Damassa, Marcia Stefanick, and John Clark, postdoctoral fellows (e.g., Lin Myers, Diane Morrissette, and Marie Carmichael), and visiting scholars (e.g., Ray Rosen, Manuel Mas, C. Sue Carter, and Benjamin Sachs) (Sachs, 2003).

Davidson was a dedicated supporter of social justice. This may have reflected in part the loss of his father's family during the Holocaust. His wife Ann and he were involved in many political marches and other activities associated with human rights and social justice. He was a kind and generous man with a fine sense of humor. During the early 1990s, at the age of 59, he was diagnosed with early-onset Alzheimer's Disease (AD). This diagnosis prematurely ended his academic career. It was painful to observe the gradual decline that robbed him of his acute intelligence and personhood. He was honored for his significant and substantial contributions to the field of sexology during the 1993 meeting of the International Academy of Sex Research (Morrissette & Myers, 2002). In her heartwarming book, *Alzheimer's, A Love Story: One Year in My Husband's Journey*, Ann Davidson chronicles a year in their lives living with AD (Davidson, 1997). Ann was a remarkable loving caretaker. Davidson passed away at the age of 70 from complications of AD leaving behind a legacy of excellent science, many friends, former trainees, and colleagues who helped bring rigorous neuroendocrinology to the field of behavioral neuroendocrinology. He is missed to this day by the many whose paths he crossed (Fig. 15.1).

Fig. 15.1 Julian Davidson

JULIAN M. DAVIDSON
(1931–2001)

References

Bermant, G., & Davidson, J. (1974). *Biological bases of sexual behavior*. Harper and Row.

Bermant, G., Glickman, S. E., & Davidson, J. M. (1968). Effects of limbic lesions on copulatory behavior of male rats. *Journal of Comparative and Physiological Psychology, 65*, 118–125.

Bloch, G. J., & Davidson, J. M. (1967). Antiandrogen implanted in brain stimulates male reproductive system. *Science, 155*, 593–595.

Carmichael, M. S., Humbert, R., Dixen, J., Palmisano, G., Greenleaf, W., & Davidson, J. M. (1987). Plasma oxytocin increases in the human sexual response. *The Journal of Clinical Endocrinology & Metabolism, 64*, 27–31.

Chen, J. J., Smith, E. R., Gray, G. D., & Davidson, J. M. (1981). Seasonal changes in plasma testosterone and ejaculatory capacity in squirrel monkeys (Saimiri sciureus). *Primates, 22*, 253–260.

Cheung, C. Y., & Davidson, J. M. (1977). Effects of testosterone implants and hypothalamic lesions on luteinizing hormone regulation in the castrated male rat. *Endocrinology, 100*, 292–302.

Clark, J. T., Smith, E. R., & Davidson, J. M. (1984). Enhancement of sexual motivation in male rats by yohimbine. *Science, 225*, 847–849.

Coe, C. L., Chen, J., Lowe, E. L., Davidson, J. M., & Levine, S. (1981). Hormonal and behavioral changes at puberty in the squirrel monkey. *Hormones and Behavior, 15*, 36–53.

Damassa, D. A., Smith, E. R., Tennent, B., & Davidson, J. M. (1977). The relationship between circulating testosterone levels and male sexual behavior in rats. *Hormones and Behavior, 8*, 275–286.

Davidson, J. M. (1966a). Activation of the male rat's sexual behavior by intracerebral implantation of androgen. *Endocrinology, 79*, 783–794.

Davidson, J. M. (1966b). Characteristics of sex behaviour in male rats following castration. *Animal Behaviour, 14*, 266–272.
Davidson, J. M. (1976). The physiology of meditation and mystical states of consciousness. *Perspectives in Biology and Medicine, 19*, 345–379.
Davidson, J. M. (1980). The psychobiology of sexual experience. In *The psychobiology of consciousness*. Springer.
Davidson, J. M. (1984). Response to "hormones and human sexual behavior" by John Bancroft, MD. *Journal of Sex & Marital Therapy, 10*, 23–27.
Davidson, A. (1997). *Alzheimer's disease: A love story*. Birch Lane Press.
Davidson, R. J., & Davidson, J. M. (1980). *The psychobiology of consciousness*. Plenum.
Davidson, J., & Myers, L. (1988). Endocrine factors in sexual psychophysiology. In R. C. Rosen & J. G. Beck (Eds.), *Patterns of sexual arousal*. Guilford.
Davidson, J. M., & Smith, E. R. (1974). Gonadotropin release as a function of mating and steroid feedback in the female rat. *Hormones and Behavior, 5*, 163–174.
Davidson, J. M., Feldman, S., & Conforti, N. (1963). Cerebral involvement in the inhibition of ACTH secretion by hydrocortisone. *Endocrinology, 72*, 936–946.
Davidson, J. M., Chen, J. J., Crapo, L., Gray, G. D., Greenleaf, W. J., & Catania, J. A. (1983). Hormonal changes and sexual function in aging men. *The Journal of Clinical Endocrinology & Metabolism, 57*, 71–77.
Frank, L., Davidson, J., & Smith, E. R. (1985a). Androgen levels in the spotted hyaena Crocuta crocuta: The influence of social factors. *Journal of Zoology, 206*, 525–531.
Frank, L., Smith, E., & Davidson, J. (1985b). Testicular origin of circulating androgen in the spotted hyaena Crocuta crocuta. *Journal of Zoology, 207*, 613–615.
Ganong, W. F., Goldfien, A., Halevy, A., Davidson, J. M., & Boryczka, A. (1961). Effect of lysergic acid diethylamide on adrenocortical and adrenal medullary function in the dog. *Acta Endocrinologica, 37*, 583–588.
Glickman, S. E., Frank, L. G., Davidson, J. M., Smith, E. R., & Siiteri, P. (1987a). Androstenedione may organize or activate sex-reversed traits in female spotted hyenas. *Proceedings of the National Academy of Sciences, 84*, 3444–3447.
Glickman, S. E., Frank, L. G., Davidson, J. M., Smith, E. R., & Siiteri, P. K. (1987b). Androstenedione may organize or activate sex-reversed traits in female spotted hyenas. *Proceedings of the National Academy of Sciences of the United States of America, 84*, 3444–3447.
Gray, G., Södersten, P., Tallentire, D., & Davidson, J. (1978). Effects of lesions in various structures of the suprachiasmatic-preoptic region on LH regulation and sexual behavior in female rats. *Neuroendocrinology, 25*, 174–191.
Heimer, L., & Larsson, K. (1964). Drastic changes in the mating behaviour of male rats following lesions in the junction of diencephalon and mesencephalon. *Experientia, 20*, 460–461.
Herdt, G. H., & Davidson, J. (1988). The Sambia "turnim-man": Sociocultural and clinical aspects of gender formation in male pseudohermaphrodites with 5-alpha-reductase deficiency in Papua New Guinea. *Archives of Sexual Behavior, 17*, 33–56.
Morrissette, D., & Myers, L. S. (2002). In memoriam: Julian M. Davidson. *Archives of Sexual Behavior, 31*, 311–318.
Sachs, B. D. (2003). Julian M. Davidson (April 15, 1931–December 31, 2001). *Hormones and Behavior, 43*, 265–280.
Smith, E. R., Bowers, C. Y., & Davidson, J. M. (1973). Circulating levels of plasma gonadotropins in 4- and 5-day cycling rats. *Endocrinology, 93*, 756–758.
Smith, E. R., Lee, R. L., Schnur, S. L., & Davidson, J. M. (1987a). Alpha2-adrenoceptor antagonists and male sexual behavior: I. Mating behavior. *Physiology & Behavior, 41*, 7–14.
Smith, E. R., Lee, R. L., Schnur, S. L., & Davidson, J. M. (1987b). Alpha2-adrenoceptor antagonists and male sexual behavior: II. Erectile and ejaculatory reflexes. *Physiology & Behavior, 41*, 15–19.

Stephen E. Glickman

16

Christine M. Drea and Zuleyma Tang-Martínez

Abstract

Stephen E. Glickman (1933–2020) was an American comparative psychologist and scholar of the history of psychology, who contributed over 100 publications relevant to the study of animal behavior, cognitive and behavioral neuroscience, reproductive neuroendocrinology and anatomy, and integrative and evolutionary biology. His early research career, spanning roughly 26 years, was dominated by investigation of the neurological substrates of learning and arousal, and by the comparative study of curiosity. In his later research career, spanning roughly 36 years, Glickman was best known for his illuminating work on the sexual differentiation and development of the spotted hyena (*Crocuta crocuta*) – a species that came to be renowned for the female's highly unusual suite of 'masculinized' traits. Glickman can be credited with unraveling many of this species' mysteries, including by establishing, at the University of California, Berkeley, the only captive hyena colony worldwide and assembling a team of highly specialized collaborators who provided unparalleled research synergy. In honor of his scientific contributions and the creation of this unique intellectual environment, the field station of UC Berkeley was renamed, in 2020, the "Stephen Glickman Field Station for the Study of Behavior, Ecology and Reproduction."

Keywords

Comparative psychology · Curiosity · Female masculinization · Reinforcement · Sexual differentiation · Spotted hyena

C. M. Drea (✉)
Department of Evolutionary Anthropology, Duke University, Durham, NC, USA
e-mail: cdrea@duke.edu

Z. Tang-Martínez
Department of Biology, University of Missouri – St. Louis, St. Louis, MO, USA
e-mail: zuleyma@umsl.edu

R. J. Nelson, Z. M. Weil (eds.), *Biographical History of Behavioral Neuroendocrinology*, https://doi.org/10.1007/978-3-031-12970-4_16

Stephen Glickman holding a young spotted hyena cub at the U.C. Berkeley hyena colony. Photo: C.M. Drea

Early Influences and Education

Stephen E. Glickman was born on 17 January 1933, in the Bronx, New York. His parents had wanted his middle name simply to be the letter "L," but at his birth, the attendant, who was of Latina heritage, misunderstood and recorded the middle name as "El," which translates from Spanish as "Stephen The Glickman." The name stuck and this vignette perhaps engendered the whimsical sense of humor for which Glickman was later known. Along with a sense of humor, Glickman had a life-long fascination for animal behavior and attributed the genesis of his career trajectory to two familial influences (The UC Berkeley Emeriti Association's Legacy Project, 2019). Firstly, his attraction to animals began in childhood, spurred by frequent visits to the Bronx Zoo, where his nanny's brother was keeper at the rhinoceros house. Hours spent staring at a rhinoceros convinced young Glickman that the rhinoceros was communicating with him, igniting in him a passion for discovery that would fuel his later studies and research. Secondly, both of his parents were teachers – his mother taught piano, in affiliation with The Julliard School, and his father taught junior high school mathematics. Together, they instilled in him a deep and lasting sense of the importance of teaching and learning, both of which he viewed as gifts. Thus began a life of connecting to animals, and of educating himself and others about them.

Glickman earned a B.S. in Psychology at Brooklyn College, in New York, in 1954, and a Ph.D. in Psychology at McGill University, in Montreal, in 1959. Supervised at McGill by Peter M. Milner, a pioneer in neuroscience, and Donald O. Hebb, the "father of neuropsychology," Glickman began his neuroscience career standing on the shoulders of giants. He greatly admired his graduate mentors, fondly crediting Milner for tutoring him in Hebbian theory (Glickman, 2012). In his dissertation research on the "Reinforcing properties of arousal," Glickman addressed novelty-seeking behavior or curiosity and showed that the reticular activating system mediates sensory reward (Glickman, 1958). Two years later, he published his first landmark paper, on the neural processes involved in consolidating memory (Glickman, 1961), followed by a coedited volume on cognitive processes and the brain (Milner & Glickman, 1965). This important, early work presaged the explosive growth in cognitive neuroscience.

Junior Professorships

Glickman accepted an Assistant Professorship in Psychology at Northwestern University, in Evanston, Illinois, and remained in this position from 1958 to 1965. In 1959, he also served as a temporary instructor at the University of New Mexico, in Albuquerque, and during the summers of 1959–1960, received postdoctoral training in Physiology at the University of Washington, in Seattle. During this period, Glickman continued the line of work initiated during his graduate studies, on neural processes and arousal or exploration, still primarily in laboratory rodents [e.g., (Glickman & Feldman, 1961)]; however, he increasingly adopted a more broadly comparative perspective in regard to his study species (Glickman & Hartz, 1964; Routtenberg & Glickman, 1964). With this burgeoning interest in comparative psychology also came an interest in the behavior of animals living under more natural conditions (i.e., outside of a laboratory setting). Adopting a greater breadth of approaches to studying the behavior of diverse species, across environments, would become a recurring theme during Glickman's career.

Most notably, starting in 1959, Glickman, along with his graduate student, Richard W. Sroges, ran a comparative study of object exploration in which they compared the responses of over 200 captive zoo animals, representing roughly 100 species, to the presentation of a series of novel objects scaled to the "handling" capacity of the animals: Deceptively simple in its design, this large-scale undertaking would become another benchmark study (Glickman & Sroges, 1966). The subjects included primates from Glickman's beloved Bronx Zoo, but the study was primarily conducted at the Lincoln Park Zoo, in Chicago, Illinois, where it was facilitated by the zoo's then Director, R. Marlin Perkins. Best known for his role as host of the television program Mutual of Omaha's Wild Kingdom, Perkins scrutinized Glickman during the first 6 months of his after-hours data collection and ultimately entrusted Glickman with keys to the mammal and reptile houses, essentially providing free rein (The UC Berkeley Emeriti Association's Legacy Project, 2019). Glickman recalled this period with great fondness and was grateful to Perkins for providing such an unparalleled opportunity, but had not truly anticipated that this work would become a major part of his legacy. The main finding, well ahead of its time within comparative psychology, was the discovery of a relationship between an animal's curiosity, as revealed by its manipulative object exploration, its natural feeding ecology, and its relative brain size (Glickman & Sroges, 1966): Animals that relied on narrow diets readily available for consumption would lack curiosity, whereas those (particularly primates) that had varied diets requiring manipulatory behavior would be intensely curious and would also show greater encephalization. Reframed years later within the context of the evolution of intelligence, such relationships would be further recognized and more formally examined [e.g., (Clutton-Brock & Harvey, 1980)], and remain particularly germane and current (DeCasien et al., 2017).

The original findings on curiosity would be written up during a period, from 1962 to 1964, while Glickman was a Fellow of the Miller Institute for Basic Research in Science, at the University of California in Berkeley. UC Berkeley Professor Frank A. Beach sponsored this fellowship, having recognized in Glickman a kindred spirit. In a classic paper, entitled "The snark was a boojum," Beach (1950) had criticized the field of comparative psychology, both for its overreliance on the rat as a study species (accounting for 70% of all subjects) and on conditioned learning as the experimental paradigm of choice. Beach had argued, compellingly, for the reinstatement of the comparative, evolutionary method in American psychology and, as Glickman noted, had "remained an outspoken champion of careful behavioral description as a prerequisite for physiological analysis" [(Glickman, 1994) p. 162]. These tenets just as readily could have described Glickman's own research. A keen observational eye, coupled with a broad-based scientific interest, and an ability to capture, synthesize, and distill the relevant details into a cogent question or explanation, were all hallmark qualities in Glickman's intellectual arsenal.

Glickman accepted a position as Associate Professor of Psychology at the University of Michigan, in Ann Arbor, from 1965 to 1968. Here, he and his graduate student, Bernard B. Schiff, wrote their classic and widely cited review on the biological theory of reinforcement, in which they theorized about neural mechanisms of learning and suggested that reinforcement evolved as a positive feedback mechanism to ensure species-typical approach or avoidance responses to appropriate stimuli (Glickman & Schiff, 1967). This work related to growing appreciation at the time of a species "biological preparedness" (Garcia & Koelling, 1966), establishing that there were, in fact, constraints on what animals could learn, based on what responses could be reasonably reinforced [e.g., (Glickman, 1973)]. These ideas were revolutionary at the time because they challenged the preeminence of strict behaviorism, which had emphasized "General Laws of Learning" and the "Principles of Equipotentiality" (Thorndike, 1911). According to established doctrine, the learned behavioral responses of animals should not vary with the use of different stimuli or reinforcers – dogma that had justified focusing on the laboratory rat as a general model for understanding learning in all species. Glickman and Schiff's (1967) review would eventually be recognized as foundational in advancing the then nascent field of animal cognition; the concepts of species-specific responses and biological preparedness are now accepted tenets in virtually all modern research on animal cognition.

Life and Professorship in Berkeley

Glickman became a Fellow of the Center for Advanced Study in the Behavioral Sciences, in Palo Alto, California, from 1968 to 1969, while transitioning to his position as Professor of Psychology at UC Berkeley – a position that he held from 1968 to 2008. The latter was a return home, of sorts, as it was here, during his earlier

Miller Fellowship, that Glickman had met Krista Zimmermann, a Berkeley native, who became his wife (in 1964) and life-long partner, and with whom he would raise a family and come to share a passion for "all things hyena." Outside of academia, Glickman developed another passion – for stock car racing – which he indulged at the Sear's Point Raceway in Sonoma, California. Thankfully, he approached his hobby with the same methodical care and level-headedness that characterized his science. After racing, it was drinks at Sam's Anchor Café in Tiburon, overlooking the bay and Angel Island, as Glickman was a man of tradition and gradually amassed a collection of local haunts and habits. Many informal, but often highly insightful and productive, scientific discussions were held at Berkeley restaurants, where menus were optional, the staff bringing Glickman his favorite meals without ever a distracting decision having to be made.

From 1972 to 1977, Glickman succeeded Beach in serving as Director of UC Berkeley's Field Station for Behavioral Research (FSBR), located in the Berkeley hills, adjacent to Tilden Park. Originally home to beagles – the subjects of Beach's seminal behavioral endocrine work – it would later become home to the world's only captive spotted hyena colony (see below). Here, Glickman put into practice his and Beach's shared appreciation for a truly comparative approach to psychological research. Recognizing the need to study animals in more salient environments (Glickman & Caldwell, 1994), Glickman encouraged moving research animals from their cages in Psychology's Tolman Hall to more naturalistic enclosures at the FSBR, where the animals' lives would be enriched. Along with collaborators, postdoctoral associates, and graduate and undergraduate students, Glickman conducted behavioral, neural, anatomical, and hormonal research on a wide array of species (including skunks, kangaroo rats, woodrats, gerbils, grasshopper mice, ground squirrels, moles, and lizards) using an integrative mix of approaches applied in laboratory, seminatural, and field settings [e.g., (Baran & Glickman, 1970; Caldwell et al., 1984; Glickman et al., 1970)]. Particularly admiring the skunks' tenacity, upon completion of his behavioral observations, he arranged for their release back to Tilden Park (The UC Berkeley Emeriti Association's Legacy Project, 2019). Glickman's fascination with animals was such that he even housed a rescued sloth, named Fast Eddie, in one of his home closets.

In 1996, Glickman additionally became Professor of Integrative Biology (also at UC Berkeley) and resumed his directorship position at the FSBR. He became Professor Emeritus in 2009 and remained active in research until his death. Steve Glickman passed on 22 May 2020, at his home in Berkeley; Krista Glickman died the following year. Their passing left a monumental void to family, friends, former students, colleagues, and the Berkeley community.

Establishing the Berkeley Hyena Project: A Collaborative Network for an "Experiment of Nature"

Glickman is perhaps best recognized for his role in establishing a hyena colony at UC Berkeley and directing a long-term program of research on this fascinating species. In the early 1980s, Laurence G. Frank, a postdoctoral fellow at UC Berkeley, and Julian M. Davidson, a reproductive endocrinologist at Stanford University, approached Glickman with a proposition to collaborate on an integrated study of the spotted hyena (*Crocuta crocuta*). Frank had been studying a wild population – the Talek clan – in the Masai Mara of Kenya and had enlisted Davidson to help him better understand the species' reproductive physiology. The unique reproductive anatomy of the female had perplexed evolutionary biologists since the time of Aristotle! The proposal hinged on bringing spotted hyenas into captivity, as Frank and Davidson needed more controlled access to animals than was possible in nature. To explore this possibility, they invited Glickman to visit the animals and field site in Africa. Although Glickman was initially noncommittal about embarking on such an undertaking, his wife, an inveterate nature lover, insisted, saying, "Let me understand this; Laurence Frank has invited us to Africa, to live in a tent on the banks of the Talek River, surrounded by elephants, and zebras, and giraffes, and hyenas, and you don't know whether we should go? Why are we married?" (The UC Berkeley Emeriti Association's Legacy Project, 2019). So persuaded, the Glickmans set out to the Masai Mara, where the Kenyan Wildlife Service and local Masai tribesmen were conducting a hyena cull. The arrangement was to spare cubs, raise them by hand, and allow for their transport to the USA in two cohorts of ten, one in 1984, the other the following year. Funded initially and for the next 22 years by the National Institute of Mental Health (and, ultimately, by the National Science Foundation and UC Berkeley), Glickman redesigned Beach's dog enclosures to make them suitable and hyena proof – for animals that can chew through metal, this was no small task! Thus began Glickman's most prized scientific adventure.

The spotted hyena emerged as Glickman's "experiment of nature" for pursuing questions about the role of naturally circulating androgens on sex differences in behavior and morphology: "We hoped that through the use of an unusual female mammal we would challenge the adequacy of current understanding of the process of sexual differentiation" [(Glickman et al., 1992a) p. 138]. Indeed, spotted hyenas challenge most norms. The female of the species is the most highly "masculinized" of extant female mammals. To the untrained eye, the external genitalia of both sexes seem indistinguishable, with females converging on the male form; nevertheless, morphological sex differences become apparent upon careful scrutiny, particularly when animals bear erections (Frank et al., 1990). The female lacks a vagina and is instead endowed with a peniform clitoris through which she urinates, copulates, and gives birth. Additionally, within natural social groups or clans, adult females, which are larger and heavier than adult males (Swanson et al., 2013), can be exceptionally aggressive, socially dominating adult males in most behavioral interactions (Glickman et al., 1993). How did such an unusual species come to be?

During its over 30-year tenure, the Berkeley Hyena Project contributed countless new insights into this enigmatic species. Glickman's vision for and execution of a successful project entailed the recruitment, over many years, of a diverse team of uniquely qualified collaborators, representing biochemists, obstetricians and gynecologists, reproductive and behavioral endocrinologists, behavioral neuroscientists, urogenital anatomists and pediatric urologists, sensory specialists, and molecular biologists, among others. Throughout, under careful project management and colony stewardship by Mary L. Weldele, Glickman maintained close coordination with field ecologists, initially with Frank (Frank et al., 1989), to whom he remained forever indebted, and ultimately, with his former graduate student, Kay Holekamp (Michigan State University), who took over the long-term field study of the Talek clan (Dloniak et al., 2004; Wahaj et al., 2007). So compelling was Glickman's multifaceted network that the National Science Foundation funded a sociologist of science, Elihu M. Gerson, to study the mechanisms of collaboration that had come to characterize the Berkeley Hyena Project. Gerson had noted that the uniqueness was reflected not only in the number of specialties represented, but in the "different specialties making technical connections that almost never get made."

The Ontogeny of Sex Differences in Spotted Hyenas

Because contemporary field researchers of spotted hyenas could not reliably distinguish the sexes, the full suite of unusual female traits had yet to be characterized. The foundational studies at the colony were thus aimed at describing hyena ontogeny in relation to the emergence of sex differences, and at identifying the hormonal correlates of social behavior and reproductive anatomy. The unifying hypothesis was that costs of morphological masculinization would be offset by benefits of female social dominance.

Some of the earliest discoveries were documented in a series of remarkable video clips that Glickman relished showing and narrating. Accordingly, viewers could witness, somewhat aghast, a primiparous female giving birth through a peniform clitoris. The female's urogenital opening, despite increased elasticity relative to the male's (Steinetz et al., 1997), is still too small to allow passage of a relatively large, precocial cub. Her first born thus inevitably becomes lodged in the clitoris, until the structure eventually tears open (Frank & Glickman, 1994; Frank et al., 1995). Delayed parturition, coupled with an umbilical cord that is too short for the journey, results in placental detachment, fetal hypoxia or anoxia, and hence a preponderance of stillbirths in first-time mothers, unambiguously establishing a reproductive cost to females.

The next clip would show live births in a multiparous female, wherein the first cub delivered (far more expeditiously) gained a home-court advantage over later-arriving siblings and fought with litter mates upon their emergence (Frank et al., 1991). Born precocial, with eyes open and teeth erupted, the cub's stereotypical bite-shakes could be seen puncturing and removing the placenta that still covered newborns; singletons could be seen showing the same aggression toward inanimate objects. Such early aggression establishes a life-long, rank relation between twins and may contribute to reduction of larger litters (Frank et al., 1991; Wahaj et al., 2007). Thereafter, neonatal aggression is gradually substituted by prosociality and play, which are likely necessary for the cubs' social integration at the communal den (Drea et al., 1996).

Viewers would then be regaled by this lighter side of hyenas: Whereas immature males could be tentatively incited to play, female peers would play with reckless abandon (Pedersen et al., 1990). The last clips would further illustrate the reversal of behavioral sex differences, as well as the prominent role of social facilitation in hyena life. Notably, the primate-like (albeit uncontestedly female-dominant), hierarchical social structure of spotted hyenas, which emerged de novo in captivity (Jenks et al., 1995), is based on redirected aggression, recruitment, and coalition formation (Zabel et al., 1992). The pressure to "do what others do" is such a powerful force in hyena life that it also entices repeat turn-taking in scent marking (or "pasting") (Woodmansee et al., 1991), communal drinking, urination, defecation, and "head-to-tail" greeting ceremonies (following even the most minor of social separations), and even overcomes learned food aversions (Glickman et al., 1997). The hyenas' cohesive social structure is further evidenced by their complex, individualized olfactory and vocal communicatory systems (Drea et al., 2002b; Mathevon et al., 2010) and their impressive cooperative skills and social intelligence (Drea & Carter, 2009; Drea & Frank, 2013).

Identifying the Mechanisms of Sex Differences in Unusual Species

The only theory available at the time relevant to explaining the unusual features of female spotted hyenas was Alfred Jost's theory of mammalian sexual differentiation, involving androgen exposure during critical periods of development (Jost, 1953). Glickman's complementary endocrine studies of spotted hyena anatomy and behavior were thus aimed, initially, at identifying a source of androgens in females, the timing of fetal exposure to these androgens, and the extent to which different female traits were masculinized. Subsequently, they were aimed at testing the putative androgenic mechanism through a combination of approaches (e.g., gonadectomies, hormone treatments, histological and molecular studies).

With early collaborators, including Pentti K. Siiteri (UC San Francisco), Gerry R. Cunha (UCSF), and Paul Licht (UC Berkeley), Glickman identified the maternal ovary (versus adrenal) as a source of androgens and the role of placental enzymes in converting ovarian androstenedione to testosterone (Glickman et al., 1987; Glickman et al., 1992b; Licht et al., 1992, 1998; Yalcinkaya et al., 1993). High concentrations of maternal testosterone would thereby be available to developing fetuses of both sexes throughout gestation, with presumably masculinizing effects on daughters. Indeed, antiandrogens administered to pregnant females throughout gestation partially confirmed this route for masculinizing aspects of hyena genitalia: Maternal treatment feminized external features of the "phallus" in the offspring of both sexes (i.e., making it shorter, thicker, and more rounded in contour) and also enlarged the urogenital meatus (Drea et al., 1998) such that, ultimately, the reproductive cost of clitoral delivery in primiparous daughters was eliminated (Drea et al., 2002a). Most notably, however, retention of the basic phallic structure in all fetuses and surviving offspring of treated dams also suggested that gross genital development was independent of maternal androgens (Drea et al., 1998). A role for androgen-independent postnatal genital development was likewise supported by data from gonadectomized animals (Glickman et al., 1998). These findings suggested multiple routes of sexual differentiation, leaving open the possibility of a role for early estrogens (Place & Glickman, 2004) and/or hormone-independent genetic programming.

Later collaborations, including with Alan J. Conley (UC Davis) and Larry Baskin (UCSF), further confirmed androgen-independence of gross genital development, refining the developmental timing in steroidogenesis and illuminating the intricacies of urogenital formation (Browne et al., 2006; Cunha et al., 2003, 2005), while comparative molecular investigations across hyena species, in collaboration with Geoffrey L. Hammond (University of British Columbia) and Michael J. McPhaul (University of Texas Southwestern), clarified the specific roles of sex hormone binding globulin and the androgen receptor (Cunha et al., 2014; Hammond et al., 2012).

Additional anatomical studies were facilitated by (1) access to brain and other central nervous system (CNS) tissues obtained primarily from hyenas culled in Northern Kenya (requiring an expedition on which Glickman had to be accompanied by armed guards during a politically unstable period) and (2) auditory studies of control, antiandrogen-treated, and gonadectomized animals living at the FSBR. Collaboration with Nancy G. Forger and Geert J. De Vries (then at the University of Massachusetts, Amherst) characterized traditional sex differences in perineal muscles and motoneurons (Forger et al., 1996), a diminished sex difference in the hypothalamus (Fenstermaker et al., 1999), and absent or possibly reversed sex differences in arginine vasopressin innervation of the forebrain (Rosen et al., 2006); collaboration with Dennis McFadden (University of Texas, Austin) confirmed female masculinization of otoacoustic emissions (McFadden et al., 2006). Female spotted hyenas thus present a mosaic in which some neuroanatomical traits are also remarkably masculinized, whereas others are not.

This same mosaic had always been evident in hyena behavior, in which female aggression and rough play appear masculinized, but sexual and maternal behavior do not. A remaining puzzle presented by female behavior is that aggression and social dominance over males exist despite traditional, adult sex differences in testosterone concentrations (at least outside of pregnancy), suggesting possibly more potent organizational than activational androgenic effects. Nonetheless, both activational and organizational effects on female aggression are implicated by reduction, respectively, of adult female aggression following ovariectomy and of infant female aggression following prenatal antiandrogen treatment [reviewed in (Conley et al., 2020)]. Whereas the study of adult aggression identified an ovarian source of masculinizing hormones, the study of infant aggression more specifically identified a role for androgens. Likewise, in nature, offspring aggression correlates with the concentration of maternal fecal androgen metabolites during late gestation (Dloniak et al., 2006). Together, these findings suggest that late gestational androgens must masculinize certain neural structures underlying female behavior.

Lastly, true to his comparative and collaborative tradition, Glickman also investigated questions of development in other unusual species, including moles (Rubenstein et al., 2003; Whitworth et al., 1999) elephants, and wallabies (Glickman et al., 2005), to test the generalizability of androgen-dependent and/or androgen-independent mechanisms, and encouraged others to follow suite (e.g., in lemurs (Drea & Weil, 2008) and mongooses (Drea et al., 2021). Such comparative studies further confirm female development as an active process, identify a broader role for androgens in females than previously recognized, and illustrate the multiple mechanisms (and timelines) at play in producing species-typical patterns of anatomical and/or behavioral sex differences.

Glickman's breadth and depth of research coverage on the topic of female masculinization defy brevity. In condensing these and other findings, he succinctly and humbly noted that a review paper in *Trends in Endocrinology and Metabolism* (Glickman et al., 2006) "conveys the revisions in our understanding of sexual differentiation, in all mammalian species (including humans), that are necessitated by our studies of the spotted hyena." Throughout, Glickman graciously credited the numerous collaborators, whose different areas of expertise, including on human reproductive development and fertility (Baskin et al., 2006; Cunha et al., 2020), helped "translate basic science into useful medical discoveries" (The UC Berkeley Emeriti Association's Legacy Project, 2019).

Colony Closure

Unfortunately, what was clearly an advantage with regard to the advancement of knowledge increasingly became a disadvantage with regard to reductionist approaches required by funding institutions. Support for an expensive, integrative project, with a colony large enough to meet scientific rigor, could not be met, logistically, by acquiring individual support for its component research questions. Glickman's final effort to secure NIH funds to advance hyena research, during a recession, met with limited enthusiasm from a program officer who anticipated that such a proposal "would not be quantitatively competitive with proposals involving mice and zebra fish." Thus, in an ironic twist of fate, lack of appreciation for the comparative perspective ultimately led to termination of this remarkable project and closure of the colony. Gradually, through Weldele's Herculean efforts, the remaining hyenas would join the others previously placed in zoos or sanctuaries around the globe; the last hyenas left the Berkeley Hills in 2014, silencing the evocative long-distance "whoops" that could be heard and enjoyed all the way in neighboring Walnut Creek. It is unlikely that such a project will ever be renewed.

Service and Teaching

A strong supporter of equity and inclusion in science, Glickman applied the same philosophy evident in his research to his teaching and service. At UC Berkeley, he served on several Academic Senate committees, including the Berkeley Division Committee on Budget and Interdepartmental Relations, Committee on Teaching, and the Animal Care and Use Committee, but he was particularly proud to have served on the inaugural Title IX Committee, and made significant inroads to promote diversity during his Chairmanship of the Department of Psychology (1977–1982). One of his greatest challenges in this position was to unify a large faculty around a common theme (The UC Berkeley Emeriti Association's Legacy Project, 2019). Toward that end, he invited influential speakers, including Robert V. Guthrie, an African-American scholar who authored the book, *Even the Rat Was White: A Historical View of Psychology* (Guthrie, 1976). Such events helped profile the significant, yet often overlooked, contributions of minority psychologists. During his chairmanship, the department also tenured three women, two on the same day (The UC Berkeley Emeriti Association's Legacy Project, 2019). Although Glickman appropriately credited the women faculty for their accomplishments, for him this was a particularly satisfying achievement.

As a new faculty member at UC Berkeley, Glickman initially developed and taught courses in Physiological Psychology and Animal Behavior. A notable characteristic, which carried over from his own early work on learning and reinforcement, is that Glickman encouraged his students to think outside the box and question existing paradigms. When, in the early 1980s, a cohort of African American students requested that he offer a seminar on sickle cell anemia – a topic clearly outside his wheelhouse – Glickman accepted, contingent on broadening the topic and engaging the students in independent research (The UC Berkeley Emeriti Association's Legacy Project, 2019). Together, they tackled various aspects of institutional racism, including the related nature/nurture debate missing from Arthur R. Jensen's study on heredity and IQ. The students conducted hypothesis-driven research on attitudinal impacts of different housing conditions, thereby illustrating the profound role of environmental factors; the seminar was fondly recalled by Glickman as one of his most memorable.

Glickman also had a profound interest in the origins of ideas and how they developed over time. He thus developed another course on the History of Psychology – a topic on which he also published (Glickman, 1985) – where he explored the backgrounds of prominent scholars (such as Luigi Galvani, Thaddeus L. Bolton, Sigmund Freud, and William James) to relate their lives to their intellectual products (The UC Berkeley Emeriti Association's Legacy Project, 2019). He brought such topics to life through his gift for story-telling. Much beloved by his students, Glickman was the recipient, in 1975, of UC Berkeley's Distinguished Teaching Award. Likewise, among his graduate students, postdocs, and colleagues, he was considered a superlative role model, on both academic and personal fronts. He earned the Berkeley Citation from the chancellor for "exceeding the standards of excellence" in a field.

Other Contributions

Increasingly captivated by historical antecedents, Glickman was also drawn to discover the origins of the spotted hyena's less-than-stellar reputation, as these animals continued to be unjustly maligned for their purported hermaphroditism and scavenging habits. He researched attitudes towards spotted hyenas, tracing their evolution from Aristotle, in 384 BC, through twelfth-century bestiaries, to contemporary authors and artists. In a delightfully titled paper, "The spotted hyena from Aristotle to the Lion King: Reputation is everything," Glickman debunked many a fallacy about spotted hyenas (Glickman, 1995). Relatedly, hyenas at the FSBR had served as models for Disney's 1994 film *The Lion King*, a point that Glickman would later describe with amused pride. He credited the animators' attention to detail for inadvertently depicting these bone-crushing carnivores with teeth that were all of equal size – an artifact of the canines of founding FSBR animals having been blunted to reduce the potential of severe injuries arising from intraspecific aggression.

Glickman's historical recreations extended to sociocultural perspectives. He was thus also conversant with feminist approaches to animal behavior, with women who made transformative contributions to the field, and with how views about sex and sexuality have changed since the early origins of the discipline. His take on these subjects is expounded upon in a paper titled: "Culture, Disciplinary Tradition, and the Study of Behavior: Sex, Rats, and Spotted Hyenas" (Glickman, 2000), published as part of a Wenner-Gren conference volume on the theme of science, gender, and society. Glickman's breadth of interests and interdisciplinary knowledge played an important role during the conference, as a bridge between those participants who were practicing scientists (e.g., primatologists, biologists, comparative psychologists) and those who were philosophers, historians, and sociologists of science.

Lastly, a classic debate in the origins of ideas that most fascinated Glickman was the one between Charles Robert Darwin and Alfred Russel Wallace over the theory of evolution through natural selection. During a visit to the UK, Glickman poured over the original correspondences between these two men (while a library attendant in white gloves turned the pages for him); he particularly relished the authors' handwritten notes in the margins of shared drafts. Glickman admired how Wallace, a poor, self-educated man, made himself into a serious biologist – one who represented such a threat to Darwin's primacy. Of particular interest was the contrast between Darwin and Wallace in their views on the evolution of the human brain (Glickman, 2009). So deep was Glickman's interest in Wallace, that in 1997, he spent a sabbatical traveling with his wife to Indonesia to retrace Wallace's journey through the Malay Archipelago. Glickman simply wanted to "see what could still be seen" in remote locales (The UC Berkeley Emeriti Association's Legacy Project, 2019). Ever inquisitive, his own curiosity about the world around him, about key thinkers and the evolution of their trade, no doubt fueled his passion for understanding the inner workings of other animals, from his early friend the rhinoceros to the spotted hyenas to which he dedicated his career.

References

Baran, D., & Glickman, S. E. (1970). "Territorial marking" in the Mongolian gerbil: A study of sensory control and function. *Journal of Comparative and Physiological Psychology, 71*, 237–245.

Baskin, L. S., Yucel, S., Cunha, G. R., Glickman, S. E., & Place, N. J. (2006). A neuroanatomical comparison of humans and spotted hyena, a natural animal model for common urogenital sinus: Clinical reflections on feminizing genitoplasty. *The Journal of Urology, 175*, 276–283.

Beach, F. A. (1950). The snark was a boojum. *American Psychologist, 5*, 115–124.

Browne, P., Place, N. J., Vidal, J. D., Moore, I. T., Cunha, G. R., Glickman, S. E., & Conley, A. J. (2006). Endocrine differentiation of fetal ovaries and testes of the spotted hyena (*Crocuta crocuta*): Timing of androgen-independent versus androgen-driven genital development. *Reproduction, 132*, 649–659.

Caldwell, G. S., Glickman, S. E., & Smith, E. R. (1984). Seasonal aggression independent of seasonal testosterone in wood rats. *Proceedings of the National Academy of Sciences of the United States of America, 81*, 5255–5257.

Clutton-Brock, T. H., & Harvey, P. H. (1980). Primates, brains and ecology. *Journal of Zoology, 190*, 309–323.

Conley, A., Place, N. J., Legacki, E. L., Hammond, G. L., Cunha, G. R., Drea, C. M., Weldele, M. L., & Glickman, S. E. (2020). Spotted hyaenas and the sexual spectrum: Reproductive endocrinology and development. *The Journal of Endocrinology, 247*, R27–R44.

Cunha, G. R., Wang, Y., Place, N. J., Liu, W., Baskin, L., & Glickman, S. E. (2003). Urogenital system of the spotted hyena (*Crocuta crocuta* Erxleben): A functional histological study. *Journal of Morphology, 256*, 205–218.

Cunha, G. R., Place, N. J., Baskin, L., Conley, A., Weldele, M., Cunha, T. J., Wang, Y. Z., Cao, M., & Glickman, S. E. (2005). The ontogeny of the urogenital system of the spotted hyena (*Crocuta crocuta* Erxleben). *Biology of Reproduction, 73*, 554–564.

Cunha, G. R., Risbridger, G., Wang, H., Place, N. J., Grumbach, M., Cunha, T. J., Weldele, M., Conley, A. J., Barcellos, D., Agarwal, S., Bhargava, A., Drea, C., Hammond, G. L., Siiteri, P., Coscia, E. M., McPhaul, M. J., Baskin, L. S., & Glickman, S. E. (2014). Development of the external genitalia: Perspectives from the spotted hyena (*Crocuta crocuta*). *Differentiation, 87*, 4–22.

Cunha, G. R., Liu, G., Sinclair, A., Cao, M., Glickman, S., Cooke, P. S., & Baskin, L. (2020). Androgen-independent events in penile development in humans and animals. *Differentiation, 111*, 98–114.

Decasien, A. R., Williams, S. A., & Higham, J. P. (2017). Primate brain size is predicted by diet but not sociality. *Nature ecology & evolution, 1*, 1–7.

Dloniak, S. M., French, J. A., Place, N. J., Weldele, M. L., Glickman, S. E., & Holekamp, K. E. (2004). Non-invasive monitoring of fecal androgens in spotted hyenas (*Crocuta crocuta*). *General and Comparative Endocrinology, 135*, 51–61.

Dloniak, S. M., French, J. A., & Holekamp, K. E. (2006). Rank-related maternal effects of androgens on behaviour in wild spotted hyaenas. *Nature, 440*, 1190–1193.

Drea, C. M., & Carter, A. N. (2009). Cooperative problem solving in a social carnivore. *Animal Behaviour, 78*, 967–977.

Drea, C. M., & Frank L. G. (2013). The social complexity of spotted hyenas. In F. B. M. de Waal & P. L. Tyack (Eds.), *Animal social complexity* (pp. 121–148). Cambridge (MA): Harvard University Press.

Drea, C. M., & Weil, A. (2008). External genital morphology of the ring-tailed lemur (*Lemur catta*): Females are naturally "masculinized". *Journal of Morphology, 269*, 451–463.

Drea, C. M., Hawk, J. E., & Glickman, S. E. (1996). Aggression decreases as play emerges in infant spotted hyaenas: Preparation for joining the clan. *Animal Behaviour, 51*, 1323–1336.

Drea, C. M., Weldele, M. L., Forger, N. G., Coscia, E. M., Frank, L. G., Licht, P., & Glickman, S. E. (1998). Androgens and masculinization of genitalia in the spotted hyaena (*Crocuta crocuta*). 2. Effects of prenatal anti-androgens. *Journal of Reproduction and Fertility, 113*, 117–127.

Drea, C. M., Place, N. J., Weldele, M. L., Coscia, E. M., Licht, P., & Glickman, S. E. (2002a). Exposure to naturally circulating androgens during foetal life incurs direct reproductive costs in female spotted hyenas, but is prerequisite for male mating. *Proceedings of the Royal Society B-Biological Sciences, 269*, 1981–1987.

Drea, C. M., Vignieri, S. N., Kim, H. S., Weldele, M. L., & Glickman, S. E. (2002b). Responses to olfactory stimuli in spotted hyenas (*Crocuta crocuta*): II. Discrimination of conspecific scent. *Journal of Comparative Psychology, 116*, 342–349.

Drea, C. M., Davies, C., Greene, L. K., Mitchell, J., Blondel, D. V., Shearer, C. L., Feldblum, J. T., Dimac-Stohl, K. A., Smyth-Kabay, K. N., & Clutton-Brock, T. (2021). An intergenerational androgenic mechanism of female intrasexual competition in the cooperatively breeding meerkat. *Nature Communications, 12*, 7332.

Fenstermaker, S. B., Zup, S. L., Frank, L. G., Glickman, S. E., & Forger, N. G. (1999). A sex difference in the hypothalamus of the spotted hyena. *Nature Neuroscience, 2*, 943–945.

Forger, N. G., Frank, L. G., Breedlove, S. M., & Glickman, S. E. (1996). Sexual dimorphism of perineal muscles and motoneurons in spotted hyenas. *The Journal of Comparative Neurology, 375*(2):333–343.

Frank, L., & Glickman, S. (1994). Giving birth through a penile clitoris: Parturition and dystocia in the spotted hyaena (*Crocuta crocuta*). *Journal of Zoology, 234*, 659–665.

Frank, L. G., Glickman, S. E., & Zabel, C. J. (1989). Ontogeny of female dominance in the spotted hyaena: Perspectives from nature and captivity. *Symposia of the Zoological Society of London, 61*, 127–146.

Frank, L. G., Glickman, S. E., & Powch, I. (1990). Sexual dimorphism in the spotted hyaena (*Crocuta crocuta*). *Journal of Zoology, 221*, 308–313.

Frank, L. G., Glickman, S. E., & Licht, P. (1991). Fatal sibling aggression, precocial development, and androgens in neonatal spotted hyenas. *Science, 252*, 702–704.

Frank, L. G., Weldele, M. L., & Glickman, S. E. (1995). Masculinization costs in hyaenas. *Nature, 377*, 584–585.

Garcia, J., & Koelling, R. A. (1966). Relation of cue to consequence in avoidance learning. *Psychonomic Science, 4*, 123–124.

Glickman, S. E. (1958). *Reinforcing properties of arousal*. McGill University.

Glickman, S. E. (1961). Perseverative neural processes and consolidation of the memory trace. *Psychological Bulletin, 58*, 218.

Glickman, S. E. (1973). Responses and reinforcement. In R. A. Hinde & J. Stevenson-Hinde (Eds.), *Constraints on learning: Limitations and predispositions* (pp. 207–241). Academic Press.

Glickman, S. E. (1985). Some thoughts on the evolution of comparative psychology. In S. Koch & D. Leary (Eds.), *A century of psychology as science* (pp. 738–782). McGraw Hill.

Glickman, S. E. (1994). Frank A. Beach (13 April 1911-15 June 1988). *Proceedings of the American Philosophical Society, 138*, 160–164.

Glickman, S. E. (1995). The spotted hyena from aristotle to the lion king: Reputation is everything. *Social Research*, 501–537.

Glickman, S. E. (2000). Culture, disciplinary tradition, and the study of behavior: Sex, rats, and spotted hyenas. In S. Strum & L. Fedigan (Eds.), *Primate encounters: Models of science, gender and society* (pp. 275–295). University of Chicago Press.

Glickman, S. E. (2009). Charles Darwin, Alfred Russel Wallace and the creation/evolution of the human brain and mind. *Gayana (Concepción), 73*, 32–41.

Glickman, S. E. (2012). Donald Olding Hebb: Returning the nervous system to psychology. *Portraits of Pioneers in Psychology: Volume II, 2*, 227–244.

Glickman, S. & Caldwell, G. (1994). Studying natural behavior in artificial environments: The problem of "salient elements". In *Naturalistic environments in captivity for animal behavior research* (pp. 197–216). State University of New York Press.

Glickman, S. E., & Feldman, S. M. (1961). Habituation of the arousal response to direct stimulation of the brainstem. *Electroencephalography and Clinical Neurophysiology, 13*, 703–709.

Glickman, S. E., & Hartz, K. E. (1964). Exploratory behavior in several species of rodents. *Journal of Comparative and Physiological Psychology, 58*, 101–104.

Glickman, S. E., & Schiff, B. B. (1967). A biological theory of reinforcement. *Psychological Review, 74*, 81–109.

Glickman, S. E., & Sroges, R. W. (1966). Curiosity in zoo animals. *Behaviour, 26*, 151–188.

Glickman, S. E., Higgins, T. J., & Isaacson, R. L. (1970). Some effects of hippocampal lesions on the behavior of Mongolian gerbils. *Physiology & Behavior, 5*, 931–938.
Glickman, S. E., Frank, L. G., Davidson, J. M., Smith, E. R., & Siiteri, P. K. (1987). Androstenedione may organize or activate sex-reversed traits in female spotted hyenas. *Proceedings of the National Academy of Sciences of the United States of America, 84*, 3444–3447.
Glickman, S. E., Frank, L. G., Licht, P., Yalcinkaya, T., Siiteri, P. K., & Davidson, J. (1992a). Sexual differentiation of the female spotted hyena. One of nature's experiments. *Annals of the New York Academy of Sciences, 662*, 135–159.
Glickman, S. E., Frank, L. G., Pavgi, S., & Licht, P. (1992b). Hormonal correlates of 'masculinization' in female spotted hyaenas (*Crocuta crocuta*). 1. Infancy to sexual maturity. *Journal of Reproduction and Fertility, 95*, 451–462.
Glickman, S. E., Frank, L. G., Holekamp, K. E., Smale, L., & Licht, P. (1993). Costs and benefits of 'androgenization' in the female spotted hyena: The natural selection of physiological mechanisms. *Perspectives in Ethology, 10*, 87–117.
Glickman, S. E., Zabel, C. J., Yoerg, S. I., Weldele, M. L., Drea, C. M., & Frank, L. G. (1997). Social facilitation, affiliation, and dominance in the social life of spotted hyenas. In C. S. Carter, I. I. Lederhendler, & B. Kirkpatrick (Eds.), *The integrative neurobiology of affiliation*. The MIT Press.
Glickman, S. E., Coscia, E. M., Frank, L. G., Licht, P., Weldele, M. L., & Drea, C. M. (1998). Androgens and masculinization of genitalia in the spotted hyaena (*Crocuta crocuta*). 3. Effects of juvenile gonadectomy. *Journal of Reproduction and Fertility, 113*, 129–135.
Glickman, S. E., Short, R. V., & Renfree, M. B. (2005). Sexual differentiation in three unconventional mammals: Spotted hyenas, elephants and tammar wallabies. *Hormones and Behavior, 48*, 403–417.
Glickman, S. E., Cunha, G. R., Drea, C. M., Conley, A. J., & Place, N. J. (2006). Mammalian sexual differentiation: Lessons from the spotted hyena. *Trends in Endocrinology and Metabolism, 17*, 349–356.
Guthrie, R. V. (1976). *Even the rat was white: A historical view of psychology*. Harper & Row.
Hammond, G. L., Miguel-Queralt, S., Yalcinkaya, T. M., Underhill, C., Place, N. J., Glickman, S. E., Drea, C. M., Wagner, A. P., & Siiteri, P. K. (2012). Phylogenetic comparisons implicate sex hormone-binding globulin in "masculinization" of the female spotted hyena (*Crocuta crocuta*). *Endocrinology, 153*, 1435–1443.
Jenks, S. M., Weldele, M. L., Frank, L. G., & Glickman, S. E. (1995). Acquisition of matrilineal rank in captive spotted hyaenas – emergence of a natural social system in peer-reared animals and their offspring. *Animal Behaviour, 50*, 893–904.
Jost, A. (1953). Problems of fetal endocrinology – the gonadal and hypophyseal hormones. *Recent Progress in Hormone Research, 8*, 379–418.
Licht, P., Frank, L. G., Pavgi, S., Yalcinkaya, T. M., Siiteri, P. K., & Glickman, S. E. (1992). Hormonal correlates of 'masculinization' in female spotted hyaenas (*Crocuta crocuta*). 2. Maternal and fetal steroids. *Journal of Reproduction and Fertility, 95*, 463–474.
Licht, P., Hayes, T., Tsai, P., Cunha, G., Kim, H., Golbus, M., Hayward, S., Martin, M. C., Jaffe, R. B., & Glickman, S. E. (1998). Androgens and masculinization of genitalia in the spotted hyaena (*Crocuta crocuta*). 1. Urogenital morphology and placental androgen production during fetal life. *Journal of Reproduction and Fertility, 113*, 105–116.
Mathevon, N., Koralek, A., Weldele, M., Glickman, S. E., & Theunissen, F. E. (2010). What the hyena's laugh tells: Sex, age, dominance and individual signature in the giggling call of *Crocuta crocuta*. *BMC Ecology, 10*, 1–16.
Mcfadden, D., Pasanen, E. G., Weldele, M. L., Glickman, S. E., & Place, N. J. (2006). Masculinized otoacoustic emissions in female spotted hyenas (*Crocuta crocuta*). *Hormones and Behavior, 50*, 285–292.
Milner, P. M., & Glickman, S. E. (1965). *Cognitive processes and the brain: An enduring problem in psychology: Selected readings*. D. Van Nostrand.
Pedersen, J. M., Glickman, S. E., Frank, L. G., & Beach, F. A. (1990). Sex-differences in the play-behavior of immature spotted hyenas, *Crocuta crocuta*. *Hormones and Behavior, 24*, 403–420.

Place, N. J., & Glickman, S. E. (2004). Masculinization of female mammals: Lessons from nature. *Hypospadias and Genital Development, 545*, 243–253.
Rosen, G. J., De Vries, G. J., Villalba, C., Weldele, M. L., Place, N. J., Coscia, E. M., Glickman, S. E., & Forger, N. G. (2006). Distribution of vasopressin in the forebrain of spotted hyenas. *The Journal of Comparative Neurology, 498*, 80–92.
Routtenberg, A., & Glickman, S. E. (1964). Visual cliff behavior in undomesticated rodents, land and aquatic turtles, and cats (Panthera). *Journal of Comparative and Physiological Psychology, 58*, 143–146.
Rubenstein, N. M., Cunha, G. R., Wang, Y. Z., Campbell, K. L., Conley, A. J., Catania, K. C., Glickman, S. E., & Place, N. J. (2003). Variation in ovarian morphology in four species of New World moles with a peniform clitoris. *Reproduction, 126*, 713–719.
Steinetz, B. G., Randolph, C., Weldele, M., Frank, L. G., Licht, P., & Glickman, S. E. (1997). Pattern and source of secretion of relaxin in the reproductive cycle of the spotted hyena (*Crocuta crocuta*). *Biology of Reproduction, 56*, 1301–1306.
Swanson, E. M., McElhinny, T. L., Dworkin, I., Weldele, M. L., Glickman, S. E., & Holekamp, K. E. (2013). Ontogeny of sexual size dimorphism in the spotted hyena (*Crocuta crocuta*). *Journal of Mammalogy, 94*, 1298–1310.
The UC Berkeley Emeriti Association's Legacy Project. (2019). *Irving Zucker interview of Stephen Glickman*. The University of California.
Thorndike, E. L. (1911). *Animal intelligence*. Macmillan.
Wahaj, S. A., Place, N. J., Weldele, M. L., Glickman, S. E., & Holekamp, K. E. (2007). Siblicide in the spotted hyena: Analysis with ultrasonic examination of wild and captive individuals. *Behavioral Ecology, 18*, 974–984.
Whitworth, D. J., Licht, P., Racey, P. A., & Glickman, S. E. (1999). Testis-like steroidogenesis in the ovotestis of the European mole, *Talpa europaea*. *Biology of Reproduction, 60*, 413–418.
Woodmansee, K. B., Zabel, C. J., Glickman, S. E., Frank, L. G., & Keppel, G. (1991). Scent marking (pasting) in a colony of immature spotted hyenas (*Crocuta crocuta*): A developmental study. *Journal of Comparative Psychology, 105*, 10–14.
Yalcinkaya, T. M., Siiteri, P. K., Vigne, J. L., Licht, P., Pavgi, S., Frank, L. G., & Glickman, S. E. (1993). A mechanism for virilization of female spotted hyenas in utero. *Science, 260*, 1929–1931.
Zabel, C. J., Glickman, S. E., & Frank, L. G. (1992). Coalition formation in a colony of prepubertal spotted hyenas (*Crocuta crocuta*). In A. H. Harcourt & F. B. M. De Waal (Eds.), *Coalitions and alliances in humans and other animals* (pp. 113–134). Oxford University Press.

Mary F. Dallman

17

Seema Bhatnagar and Claire-Dominique Walker

Abstract

Mary Fenner Dallman was a pioneer in neuroendocrinology and neuroscience making seminal discoveries in negative feedback control of the hypothalamic-pituitary adrenal axis and in regulation of adrenal function and in identifying the relationship between metabolic functions and the stress response. She received her BS in Chemistry from Smith College, her PhD in Physiology from Stanford University, and her postdoctoral fellowship from the University of California at San Francisco (UCSF). These accomplishments occurred in an era when few women were expected or encouraged to become faculty members and independent scientists. She became the first female faculty member in the basic sciences at UCSF and stayed there until her retirement in 2007. She trained generations of scientists active today and influenced countless others. She appreciated and respected the past, but her excitement for the questions that remained to be asked and answered was contagious. She would have been excited to see what lies ahead.

Keywords

Stress · Glucocorticoids · Hypothalamic-pituitary adrenal · Negative feedback · Facilitation · Comfort food

---

S. Bhatnagar (✉)
Department of Anesthesiology and Critical Care, Children's Hospital of Philadelphia, The Perelman School of Medicine at the University of Pennsylvania, Philadelphia, PA, USA
e-mail: bhatnagars@chop.edu

C.-D. Walker
Department of Psychiatry, McGill University, Douglas Institute, Montreal, QC, Canada

R. J. Nelson, Z. M. Weil (eds.), *Biographical History of Behavioral Neuroendocrinology*, https://doi.org/10.1007/978-3-031-12970-4_17

Introduction

Mary Dallman was a pioneer, a maverick, and a genuinely unique thinker (Fig. 17.1). It is fair to say that her research identified and established many features of hypothalamic-pituitary adrenal (HPA) axis function that we know today. Her research spanned multiple generations of trainees, spanned waves of technical innovations, and advanced our understanding of the HPA axis in fundamental ways. Throughout her career, she stayed true to herself – only pursuing those questions that truly interested her. Below, we have tried to provide an overview of her research career by focusing on what we believe to be three major contributions of her research, as we see it and as we experienced it.

Mary Fenner Dallman was born in 1935 in New York City. She did part of her schooling in New Jersey (Ridgewood High School) although she thought of herself as a New Yorker in origin. She attended Smith College starting in 1952 and graduated with a degree in Chemistry in 1956. Dallman received her PhD from Stanford University (with Dr. Gene Yates) and completed a post-doctoral fellowship year in Stockholm with Dr. Bengt Andersson and then a longer fellowship, at UCSF with Dr. Fran Ganong in the Department of Physiology. Dallman joined the UCSF faculty in 1970 becoming the first female faculty member in basic science at UCSF. Indeed, she was among the first female scientists working in neuroscience in academia. She rose to professor and vice chair of the Department of Physiology in the 1980s. Dallman would stay at UCSF for the remainder of her career. She retired in 2007 and then became Professor Emerita until her death, which occurred in December 2021 during the time of this writing.

It is hard to fully convey Dallman's spirit, "joie de vivre," and sharp wit. She was extremely generous in sharing her thoughts and ideas with others. She never worried about what others were working on; she only wanted to advance the science whether that was through her work or the work of others. She took much joy and pleasure from her work, whether because of a successful experiment, a small set of

Fig. 17.1 Mary Dallman

data, the work of her trainees, or from larger accomplishments. She was a generous lifelong mentor to all who were fortunate to have joined her lab and her mentorship extended well beyond the lab. She was also an adamant supporter of women in science, endocrinology, and neuroendocrinology. Indeed, Dallman was the president of "Women in Endocrinology," served on the Endocrine Society Council, and was an associate editor of "Endocrinology."

Her research career, encompassing more than 240 publications, can be roughly divided into three phases. In her early work, she established unique time domains through which glucocorticoids exerted their negative feedback effects. A second phase of her career examined the neural substrates of facilitation of stress responses that can be seen in repeated and chronic stress conditions and a third phase focused on the interplay of glucocorticoids and metabolism. More specifically, she discovered the central role of glucocorticoids to mediate the action of "comfort food" on the stress response. The second and third phases overlapped in time. Below we present brief overviews of the main contributions that Dallman's research made in these phases.

Glucocorticoid Negative Feedback

Dallman's interest in negative feedback control within the HPA axis dates primarily to her dissertation work with Gene Yates in the 1960s at Stanford University. In those early studies, she focused on mathematical modeling of feedback, identifying fundamental properties of negative feedback function of the HPA axis including showing that there were multiple corticosterone-sensitive inputs to CRF (corticotropin-releasing factor)-secreting neurons (Yates et al., 1961; Dallman & Yates, 1967, 1969). Her work showed that there was regulation of adrenocortical function by "its own recent past history" (Dallman & Yates, 1967) and early on established that there is very rapid feedback inhibition of ACTH by corticosterone that occurs in the order of minutes (Dallman et al., 1972). This body of work on negative feedback functions within the HPA axis culminated in a series of elegant reviews in which she and colleagues laid out the evidence for multiple time courses for corticosteroid inhibition of ACTH (adrenocorticotropic hormone) secretion, identifying fast, delayed, and slow feedback domains (Keller-Wood & Dallman, 1984; Dallman et al., 1987a, b). Earlier work on the specific role of glucocorticoids in responses of the HPA axis to hypoglycemia (Keller-Wood et al., 1983b) and hypoxia (Raff et al., 1983) in dogs or hypovolemia in rats (Darlington et al., 1989) and dogs (Wood et al., 1982) has been important to refine the dose and time domains of glucocorticoid requirements under various physiologically stressful conditions. Fast feedback inhibition occurs within seconds/minutes of rises in corticosterone and is a function of the rate of rise of corticosterone concentrations. Delayed feedback occurs within 2 h or so of elevations in corticosterone, the time frame of inhibition that would occur in response to stress. In contrast, slow feedback occurs within many hours or days and likely occurs during pathological conditions. These time domains, in our view, remain an underappreciated aspect of negative glucocorticoid feedback because they suggest a high degree of complexity in the control of the HPA response to stress that is still not fully understood: that the HPA response to stress is controlled both by *ongoing* inhibitory and stimulatory signals as well as by *past* inhibitory and stimulatory signals.

Facilitation of Stress Responses

The prescient question raised by Dallman and colleagues (Dallman et al., 1972) was why, with the clear inhibition of ACTH by corticosteroids, repeated stress (using a homotypic stressor) does not always evoke progressively smaller responses. This question paved the way toward a better understanding of how prior stress, either acute or chronic, influences reactivity of the HPA axis to subsequent novel/heterotypic stressors. This led to the concept of facilitation first described in publications with Mortyn Jones starting in 1973 (Dallman & Jones, 1973) and in her work in rats (Akana et al., 1996; Akana et al., 1992) and in dogs (Wood et al., 1981; Raff et al., 1983). They defined facilitation as the response to a novel challenge after a prior acute or chronic stress. Despite the negative feedback produced by glucocorticoids released by the prior stress, the response to a novel stress was maintained at a normal (acute stress-matched) level or was even higher than the response to the acute novel stress alone (Dallman & Jones, 1973). This ability to supersede glucocorticoid feedback allows the individual to remain responsive to novel challenges despite the negative feedback from previous stressors. There were also studies showing that chronically stressed rats exhibited reduced sensitivity to fast negative feedback (Young et al., 1990). A series of studies were undertaken to better understand facilitation and its neural substrates. These led to the discovery that the paraventricular thalamic nucleus, particularly its posterior division, was a key regulator of the facilitated HPA response in chronically stressed animals (Bhatnagar & Dallman, 1998, 2000b) with significant effects of corticosterone in the prefrontal cortex, amygdala, and hippocampus (Akana et al., 2001; Bradbury et al., 1993) that differ from those classically observed at the level of the hypothalamic paraventricular nucleus (PVN). These studies provided the fundamental mechanistic bases of ongoing research investigating the multiple consequences of chronic stress exposure during several life periods.

Metabolic Interactions with the Stress Axis

Dallman's interest in examining the interactions between metabolic and stress regulation has been quite present throughout her work, although it is only in the second half of her career that she fully articulated the novel concept that glucocorticoids could potentiate the effects of comfort food to reduce central hypothalamic CRH (corticotropin-releasing hormone) and adrenocortical responses to stress. This important paradigm shift in the conception of stress regulation not only placed metabolic regulation at the center of homeostatic stress responses in preclinical models, but this concept also helped design new interventions in clinical population highly vulnerable to chronic stress. Recent clinical studies in collaboration with the Center for Health and Community at UCSF demonstrated, for instance, that long-term adaptations to chronic stress when caloric intake and visceral fat are high lead to *reduced* basal and stress-induced cortisol secretion (Tomiyama et al., 2011), similar to the situation described in animal models.

How did the concept develop along Dallman's scientific contributions? A first influential review written by Dallman in the mid-1980s explored the possibility that one of the important roles of the ventromedial hypothalamus (VMH) was to inhibit adrenocortical activity, insulin secretion, and food intake during the inactive phase of the diurnal cycle and that both signals from the periphery (insulin, fat stores, glucocorticoids) and circadian inputs amplified by the VMH (Choi et al., 1998) ensured adequate feeding in the active phase of the cycle, overriding VMH inhibitory inputs (Dallman, 1984). When interactions between these peripheral signals are disrupted in the case of diabetic rats (Scribner et al., 1991), PVN-lesioned rats (King et al., 1992), or genetically obese Zucker rats with hyperinsulinemia (Walker et al., 1992), specific regulatory components of the HPA axis are altered, in particular rhythmic adrenocortical activity also determined by caloric intake (Dallman et al., 1993; Hanson et al., 1994; Akana et al., 1994). Although the link among caloric intake, energy signals, and the HPA axis was known for some time, a first report in 1997 suggested that availability of excess calories in the form of sucrose (30% sucrose for 10 days) was able to limit acute stress responses in normal rats (Strack et al., 1997). This was followed by a number of studies in adrenalectomized (ADX) rats that demonstrated the profound effect of sucrose ingestion to restore normal CRH expression in the hypothalamus (Laugero et al., 2001) as well as metabolic deficits (Bell et al., 2000) caused by the lack of circulating corticosterone. Interestingly, central infusion of corticosterone in ADX rats drinking sucrose blocked the restorative effects of sucrose on basal metabolism and HPA activity (Laugero et al., 2002), suggesting a distinct effect of peripheral vs. central corticosterone to mediate sucrose effects. In intact rats, corticosterone can increase the willingness to consume sweet solutions (Bhatnagar et al. 2000a) as well as lard (la Fleur et al., 2004; Pecoraro et al., 2004), both nutrients that result in diverse metabolic signals reducing acute HPA activity.

Although this line of work was developed mostly using effects on acute stress responses, a parallel interest in Dallman's lab was to examine neural pathways leading to the altered homeostatic and behavioral phenotype under chronic stress conditions (see above sections). Given the strong evidence for metabolic signals to regulate acute HPA activity, it was natural to examine whether these interactions were changed under chronic stress conditions. For instance, studies using cold stress conditions (4 days) have shown that the increased caloric intake of sucrose and chow combined with clamped stress levels of corticosterone was able to maintain normal activity in the HPA axis even under chronic stress conditions (Bell et al., 2002), demonstrating the potent effect of nutrient-induced metabolic signals to dampen stress activation under both acute and chronic stress conditions (Dallman et al., 2003a). A landmark paper published in 2003 (Dallman et al., 2003b) established the model by which "comfort food" and the increased intake of palatable nutrients such as glucose and lard provide metabolic feedback signals to reduce HPA activity under chronic stress conditions. In this model (which requires at least 24 h increases in glucocorticoids), the action of glucocorticoids on the CNS is mainly stimulatory, in particular on the CRH neurons of the central amygdala that are activated under chronic stress conditions and provide increased drive to the hypothalamic PVN neurons. The increased glucocorticoid concentrations augment the salience of palatable food and intake of "comfort food" (Pecoraro et al., 2004; la Fleur et al., 2005) as well as enhance fat deposition peripherally. As a result, the increased metabolic signals from enlarged energy stores now represent the major players to dampen HPA activity because sensitivity of the hypothalamic PVN (Fig. 17.2) and other nuclei to glucocorticoid-mediated *negative* feedback is considerably reduced under chronic stress conditions.

Highlighting the physiological importance of metabolic signals in conditions of chronic stress has led to several elegant studies to clarify the hierarchical nature of these signals and their interactions with glucocorticoids. Among the multitude of potential signals (e.g., leptin, ghrelin, opioids, and insulin), most studies have

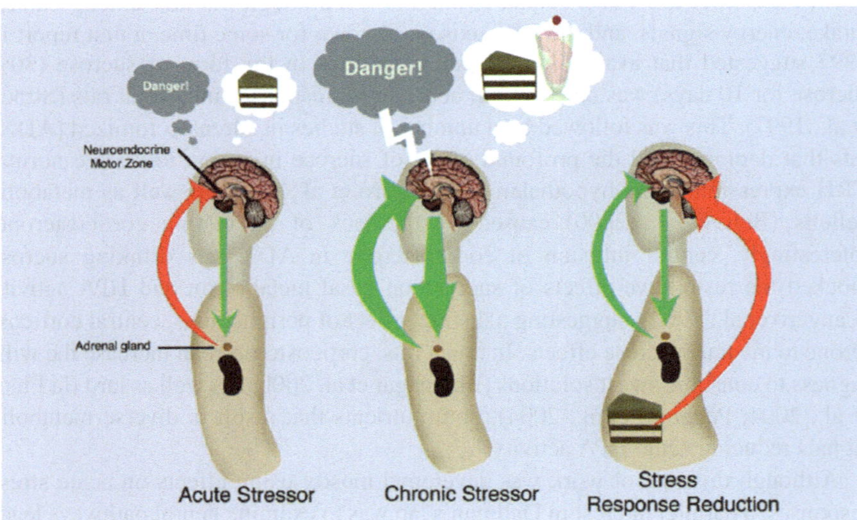

Fig. 17.2 Schematic of regulation of feeding and the HPA axis with acute (*left*) and chronic (*middle*) stressors and the increased signal for stress reduction in the brain that comes from increased intra-abdominal fat stores (*right*). Acute stressors provoke transient increases in HPA activity that are self-limited because of rapid feedback effects of glucocorticoids on the motor output of the axis. Additionally, acute stressors alter behavior and may, through elevated glucocorticoids, enhance the motivation to eat high-sweet, fatty foods (*left*). With a chronic stressor, the elevated glucocorticoid signal acts positively on the brain to promote further activation of the chronic stress response system. Glucocorticoids and insulin further augment the drive for and hedonic response to high-sweet and fatty foods (*middle*). When a combination of elevated glucocorticoids and insulin has acted to increase intra-abdominal caloric storage, an unidentified signal from these stores acts on the brain to reduce the overall level of activity of the chronic stress response network (*right*). (Reprinted from Dallman et al. (2004))

focused on insulin because this hormone increases both the choice of lard (la Fleur et al., 2004) or sugar consumption (incentive salience) and abdominal fat depots in combination with increasing concentrations of glucocorticoids observed in chronic stress and obesity conditions. In the absence of insulin, increased glucocorticoids do not drive high palatable food intake (Dallman et al., 2004). Insulin-mediated actions are observed throughout the brain and affect several areas of the limbic system in addition to vagal afferent nerves to regulate food intake and motivation to eat. The integration of the metabolic feedback loop into the larger context of the "emotional brain" might serve the purpose of relieving stress-induced discomfort. However, when this strategy is used repeatedly instead of a conscious thought about ways of coping with a stressor (using prefrontal executive functions), intake of comfort food can become habitual and lead to obesity (Dallman, 2010). The elegant and intelligent research work of Dallman and her trainees has provided a unique and sophisticated window to explain these mechanisms.

Other Major Contributions

During the course of studying negative feedback, her lab established protocols for glucocorticoid replacement in adrenalectomized rats. This work showed that adrenalectomy followed by replacement of corticosterone that achieved 5–7 μg/dl was adequate to normalize ACTH and thymus gland weight (Akana et al., 1985). Her work also revealed characteristics of adrenal function, including regulation of adrenal hypertrophy and compensatory growth (Engeland & Dallman, 1976). Holzwarth and Dallman (1979) showed that adrenal sensitivity to ACTH varies diurnally (Kaneko et al., 1981; Akana et al., 1986) and that circadian variation was critical for regulation of the response to stress (Jacobson et al., 1988). An additional important set of findings showed that the adrenal gland not only proportionally responded to ACTH but could also integrate the ACTH signal over time, producing a glucocorticoid signal that substantially outlasted the ACTH signal (Keller-Wood et al., 1983a), providing another layer of complexity in the impact of stress on HPA activity.

Conclusions

Dallman's elegant research contributed in many fundamental ways to our sophisticated knowledge of the regulation of the stress axis and glucocorticoid feedback. Her contributions were well ahead of their time, and some of her findings only gained widespread acceptance years later. While it is impossible to summarize the many contributions Dallman made to endocrinology, neuroendocrinology, and neuroscience, we have tried to provide a picture of Dallman's prolific work and the indelible mark that it has left on current stress neurobiology research. Two of Dallman's qualities were her enthusiasm and determination that percolated throughout all her life and scientific activities; she knew the past and was resolutely embracing the future. She would have been excited to see what lies ahead.

Acknowledgments We thank Dr. Hershel Raff for his valuable comments on this chapter.

References

Akana, S. F., Cascio, C. S., Shinsako, J., & Dallman, M. F. (1985). Corticosterone: Narrow range required for normal body and thymus weight and ACTH. *The American Journal of Physiology, 249*(5 Pt 2), R527–R532. https://doi.org/10.1152/ajpregu.1985.249.5.R527. PMID: 2998210.

Akana, S. F., Cascio, C. S., Du, J. Z., Levin, N., & Dallman, M. F. (1986). Reset of feedback in the adrenocortical system: An apparent shift in sensitivity of adrenocorticotropin to inhibition by corticosterone between morning and evening. *Endocrinology, 119*(5), 2325–2332. https://doi.org/10.1210/endo-119-5-2325. PMID: 3021439.

Akana, S. F., Dallman, M. F., Bradbury, M. J., Scribner, K. A., Strack, A. M., & Walker, C. D. (1992). Feedback and facilitation in the adrenocortical system: Unmasking facilitation by partial inhibition of the glucocorticoid response to prior stress. *Endocrinology, 131*(1), 57–68. https://doi.org/10.1210/endo.131.1.1319329. PMID: 1319329.

Akana, S. F., Strack, A. M., Hanson, E. S., & Dallman, M. F. (1994). Regulation of activity in the hypothalamo-pituitary-adrenal axis is integral to a larger hypothalamic system that determines caloric flow. *Endocrinology, 135*(3), 1125–1134. https://doi.org/10.1210/endo.135.3.8070356. PMID: 8070356.

Akana, S. F., Hanson, E. S., Horsley, C. J., Strack, A. M., Bhatnagar, S., Bradbury, M. J., Milligan, E. D., & Dallman, M. F. (1996). Clamped corticosterone (B) reveals the effect of endogenous B on both facilitated responsivity to acute restraint and metabolic responses to chronic stress. *Stress, 1*(1), 33–49. https://doi.org/10.3109/10253899609001094. PMID: 9807060.

Akana, S. F., Chu, A., Soriano, L., & Dallman, M. F. (2001). Corticosterone exerts site-specific and state-dependent effects in prefrontal cortex and amygdala on regulation of adrenocorticotropic hormone, insulin and fat depots. *Journal of Neuroendocrinology, 13*(7), 625–637. https://doi.org/10.1046/j.1365-2826.2001.00676.x. PMID: 11442777.

Bell, M. E., Bhatnagar, S., Liang, J., Soriano, L., Nagy, T. R., & Dallman, M. F. (2000). Voluntary sucrose ingestion, like corticosterone replacement, prevents the metabolic deficits of adrenalectomy. *Journal of Neuroendocrinology, 12*(5), 461–470. https://doi.org/10.1046/j.1365-2826.2000.00488.x. PMID: 10792586.

Bell, M. E., Bhargava, A., Soriano, L., Laugero, K., Akana, S. F., & Dallman, M. F. (2002). Sucrose intake and corticosterone interact with cold to modulate ingestive behaviour, energy balance, autonomic outflow and neuroendocrine responses during chronic stress. *Journal of Neuroendocrinology, 14*(4), 330–342. https://doi.org/10.1046/j.1365-2826.2002.00784.x. PMID: 11963830.

Bhatnagar, S., & Dallman, M. (1998). Neuroanatomical basis for facilitation of hypothalamic-pituitary-adrenal responses to a novel stressor after chronic stress. *Neuroscience, 84*(4), 1025–1039. https://doi.org/10.1016/s0306-4522(97)00577-0. PMID: 9578393.

Bhatnagar, S., Bell, M. E., Liang, J., Soriano, L., Nagy, T. R., & Dallman, M. F. (2000a). Corticosterone facilitates saccharin intake in adrenalectomized rats: Does corticosterone increase stimulus salience? *Journal of Neuroendocrinology, 12*(5), 453–460. https://doi.org/10.1046/j.1365-2826.2000.00487.x. PMID: 10792585.

Bhatnagar, S., Viau, V., Chu, A., Soriano, L., Meijer, O. C., & Dallman, M. F. (2000b). A cholecystokinin-mediated pathway to the paraventricular thalamus is recruited in chronically stressed rats and regulates hypothalamic-pituitary-adrenal function. *Journal of Neuroscience, 20*(14), 5564–5573. https://doi.org/10.1523/JNEUROSCI.20-14-05564.2000. PMID: 10884340; PMCID: PMC6772345.

Bradbury, M. J., Strack, A. M., & Dallman, M. F. (1993). Lesions of the hippocampal efferent pathway (fimbria-fornix) do not alter sensitivity of adrenocorticotropin to feedback inhibition by corticosterone in rats. *Neuroendocrinology, 58*(4), 396–407. https://doi.org/10.1159/000126569. PMID: 8284025.

Choi, S., Wong, L. S., Yamat, C., & Dallman, M. F. (1998). Hypothalamic ventromedial nuclei amplify circadian rhythms: Do they contain a food-entrained endogenous oscillator? *The Journal of Neuroscience, 18*(10), 3843–3852. https://doi.org/10.1523/JNEUROSCI.18-10-03843.1998. PMID: 9570813.

Dallman, M. F. (1984). Viewing the ventromedial hypothalamus from the adrenal gland. *American Journal of Physiology-Regulatory, Integrative and Comparative Physiology, 246*(1 Pt 2), R1–R12. https://doi.org/10.1152/ajpregu.1984.246.1.R1. PMID: 6320667.

Dallman, M. F. (2010). Stress-induced obesity and the emotional nervous system. *Trends in Endocrinology and Metabolism, 21*(3), 159–165. https://doi.org/10.1016/j.tem.2009.10.004. Epub 2009 Nov 18. PMID: 19926299.

Dallman, M. F., & Jones, M. T. (1973). Corticosteroid feedback control of ACTH secretion: Effect of stress-induced corticosterone secretion on subsequent stress responses in the rat. *Endocrinology, 92*(5), 1367–1375. https://doi.org/10.1210/endo-92-5-1367. PMID: 4348647.

Dallman, M. F., & Yates, F. E. (1967). Anatomical and functional mapping of central neural input and feedback pathways of the adrenocortical system. *Proceedings of the Royal Society of Medicine, 60*(9), 904–905. PMID: 4293423; PMCID: PMC1902040.

Dallman, M. F., & Yates, F. E. (1969). Dynamic asymmetries in the corticosteroid feedback path and distribution-metabolism-binding elements of the adrenocortical system. *Annals of the New York Academy of Sciences, 156*(2), 696–721. https://doi.org/10.1111/j.1749-6632.1969.tb14008.x. PMID: 4309115.

Dallman, M. F., Jones, M. T., Vernikos-Danellis, J., & Ganong, W. F. (1972). Corticosteroid feedback control of ACTH secretion: Rapid effects of bilateral adrenalectomy on plasma ACTH in the rat. *Endocrinology, 91*(4), 961–968. https://doi.org/10.1210/endo-91-4-961. PMID: 4340612.

Dallman, M. F., Akana, S. F., Cascio, C. S., Darlington, D. N., Jacobson, L., & Levin, N. (1987a). Regulation of ACTH secretion: Variations on a theme of B. *Recent Progress in Hormone Research, 43*, 113–173. https://doi.org/10.1016/b978-0-12-571143-2.50010-1. PMID: 2819993.

Dallman, M. F., Akana, S. F., Jacobson, L., Levin, N., Cascio, C. S., & Shinsako, J. (1987b). Characterization of corticosterone feedback regulation of ACTH secretion. *Annals of the New York Academy of Sciences, 512*, 402–414. https://doi.org/10.1111/j.1749-6632.1987.tb24976.x. PMID: 2831781.

Dallman, M. F., Strack, A. M., Akana, S. F., Bradbury, M. J., Hanson, E. S., Scribner, K. A., & Smith, M. (1993). Feast and famine: Critical role of glucocorticoids with insulin in daily energy flow. *Frontiers in Neuroendocrinology, 14*(4), 303–347. https://doi.org/10.1006/frne.1993.1010. PMID: 8258378.

Dallman, M. F., Akana, S. F., Laugero, K. D., Gomez, F., Manalo, S., Bell, M. E., & Bhatnagar, S. (2003a). A spoonful of sugar: Feedback signals of energy stores and corticosterone regulate responses to chronic stress. *Physiology & Behavior, 79*(1), 3–12. https://doi.org/10.1016/s0031-9384(03)00100-8. PMID: 12818705.

Dallman, M. F., Pecoraro, N., Akana, S. F., La Fleur, S. E., Gomez, F., Houshyar, H., Bell, M. E., Bhatnagar, S., Laugero, K. D., & Manalo, S. (2003b). Chronic stress and obesity: A new view of "comfort food". *Proceedings of the National Academy of Sciences of the United States of America, 100*(20), 11696–11701. https://doi.org/10.1073/pnas.1934666100. Epub 2003 Sep 15. PMID: 12975524.

Dallman, M. F., la Fleur, S. E., Pecoraro, N. C., Gomez, F., Houshyar, H., & Akana, S. F. (2004). Minireview: Glucocorticoids–food intake, abdominal obesity, and wealthy nations in 2004. *Endocrinology, 145*(6), 2633–2638. https://doi.org/10.1210/en.2004-0037. Epub 2004 Mar 24. PMID: 15044359.

Darlington, D. N., Keil, L. C., & Dallman, M. F. (1989). Potentiation of hormonal responses to hemorrhage and fasting, but not hypoglycemia in conscious adrenalectomized rats. *Endocrinology, 125*(3), 1398–1406. https://doi.org/10.1210/endo-125-3-1398. PMID: 2667956.

Engeland, W. C., & Dallman, M. F. (1976). Neural mediation of compensatory adrenal growth. *Endocrinology, 99*(6), 1659–1662. https://doi.org/10.1210/endo-99-6-1659. PMID: 1001260.

Hanson, E. S., Bradbury, M. J., Akana, S. F., Scribner, K. S., Strack, A. M., & Dallman, M. F. (1994). The diurnal rhythm in adrenocorticotropin responses to restraint in adrenalectomized rats is determined by caloric intake. *Endocrinology, 134*(5), 2214–2220. https://doi.org/10.1210/endo.134.5.8156924. PMID: 8156924.

Holzwarth, M. A., & Dallman, M. F. (1979). The effect of hypothalamic hemi-islands on compensatory adrenal growth. *Brain Research, 162*(1), 33–43. https://doi.org/10.1016/0006-8993(79)90753-4. PMID: 216464.

Jacobson, L., Akana, S. F., Cascio, C. S., Shinsako, J., & Dallman, M. F. (1988). Circadian variations in plasma corticosterone permit normal termination of adrenocorticotropin responses to stress. *Endocrinology, 122*(4), 1343–1348. https://doi.org/10.1210/endo-122-4-1343. PMID: 2831028.

Kaneko, M., Kaneko, K., Shinsako, J., & Dallman, M. F. (1981). Adrenal sensitivity to adrenocorticotropin varies diurnally. *Endocrinology, 109*(1), 70–75. https://doi.org/10.1210/endo-109-1-70. PMID: 6263595.

Keller-Wood, M. E., & Dallman, M. F. (1984). Corticosteroid inhibition of ACTH secretion. *Endocrine Reviews, 5*(1), 1–24. https://doi.org/10.1210/edrv-5-1-1. PMID: 6323158.

Keller-Wood, M. E., Shinsako, J., & Dallman, M. F. (1983a). Integral as well as proportional adrenal responses to ACTH. *The American Journal of Physiology, 245*(1), R53–R59. https://doi.org/10.1152/ajpregu.1983.245.1.R53. PMID: 6307068.

Keller-Wood, M. E., Shinsako, J., & Dallman, M. F. (1983b). Feedback inhibition of adrenocorticotropic hormone by physiological increases in plasma corticosteroids in conscious dogs. *The Journal of Clinical Investigation, 71*(4), 859–866. https://doi.org/10.1172/jci110840. PMID: 6300189.

King, B. M., Zansler, C. A., Richard, S. M., Gutierrez, C., & Dallman, M. F. (1992). Paraventricular hypothalamic obesity in rats: Role of corticosterone. *Physiology & Behavior, 51*(6), 1207–1212. https://doi.org/10.1016/0031-9384(92)90310-x. PMID: 1641422.

la Fleur, S. E., Akana, S. F., Manalo, S. L., & Dallman, M. F. (2004). Interaction between corticosterone and insulin in obesity: Regulation of lard intake and fat stores. *Endocrinology, 145*(5), 2174–2185. https://doi.org/10.1210/en.2003-1359. Epub 2004 Feb 12. PMID: 14962993.

la Fleur, S. E., Houshyar, H., Roy, M., & Dallman, M. F. (2005). Choice of lard, but not total lard calories, damps adrenocorticotropin responses to restraint. *Endocrinology, 146*(5), 2193–2199. https://doi.org/10.1210/en.2004-1603. Epub 2005 Feb 10. PMID: 15705773.

Laugero, K. D., Bell, M. E., Bhatnagar, S., Soriano, L., & Dallman, M. F. (2001). Sucrose ingestion normalizes central expression of corticotropin-releasing-factor messenger ribonucleic acid and energy balance in adrenalectomized rats: A glucocorticoid-metabolic-brain axis? *Endocrinology, 142*(7), 2796–2804. https://doi.org/10.1210/endo.142.7.8250. PMID: 11415998.

Laugero, K. D., Gomez, F., Manalo, S., & Dallman, M. F. (2002). Corticosterone infused intracerebroventricularly inhibits energy storage and stimulates the hypothalamo-pituitary axis in adrenalectomized rats drinking sucrose. *Endocrinology, 143*(12), 4552–4562. https://doi.org/10.1210/en.2002-220613. PMID: 12446582.

Pecoraro, N., Reyes, F., Gomez, F., Bhargava, A., & Dallman, M. F. (2004). Chronic stress promotes palatable feeding, which reduces signs of stress: Feedforward and feedback effects of chronic stress. *Endocrinology, 145*(8), 3754–3762. https://doi.org/10.1210/en.2004-0305. Epub 2004 May 13. PMID: 15142987.

Raff, H., Shinsako, J., Keil, L. C., & Dallman, M. F. (1983). Vasopressin, ACTH, and corticosteroids during hypercapnia and graded hypoxia in dogs. *The American Journal of Physiology, 244*(5), E453–E458. https://doi.org/10.1152/ajpendo.1983.244.5.E453. PMID: 6303130.

Scribner, K. A., Walker, C. D., Cascio, C. S., & Dallman, M. F. (1991). Chronic streptozotocin diabetes in rats facilitates the acute stress response without altering pituitary or adrenal responsiveness to secretagogues. *Endocrinology, 129*(1), 99–108. https://doi.org/10.1210/endo-129-1-99. PMID: 1647314.

Strack, A. M., Akana, S. F., Horsley, C. J., & Dallman, M. F. (1997). A hypercaloric load induces thermogenesis but inhibits stress responses in the SNS and HPA system. *The American Journal of Physiology, 272*(3 Pt 2), R840–R848. https://doi.org/10.1152/ajpregu.1997.272.3.R840. PMID: 9087645.

Tomiyama, A. J., Dallman, M. F., & Epel, E. S. (2011). Comfort food is comforting to those most stressed: Evidence of the chronic stress response network in high stress women. *Psychoneuroendocrinology, 36*(10), 1513–1519. https://doi.org/10.1016/j.psyneuen.2011.04.005. Epub 2011 Sep 8. PMID: 21906885.

Walker, C. D., Scribner, K. A., Stern, J. S., & Dallman, M. F. (1992). Obese Zucker (fa/fa) rats exhibit normal target sensitivity to corticosterone and increased drive to adrenocorticotropin during the diurnal trough. *Endocrinology, 131*(6), 2629–2637. https://doi.org/10.1210/endo.131.6.1332842. PMID: 1332842.

Wood, C. E., Shinsako, J., Keil, L. C., Ramsay, D. J., & Dallman, M. F. (1981). Hormonal and hemodynamic responses to 15 ml/kg hemorrhage in conscious dogs: Responses correlate to body temperature. *Proceedings of the Society for Experimental Biology and Medicine, 167*(1), 15–19. https://doi.org/10.3181/00379727-167-41117. PMID: 6262832.

Wood, C. E., Shinsako, J., Keil, L. C., Ramsay, D. J., & Dallman, M. F. (1982). Apparent dissociation of adrenocorticotropin and corticosteroid responses to 15 ml/kg hemorrhage in conscious dogs. *Endocrinology, 110*(4), 1416–1421. https://doi.org/10.1210/endo-110-4-1416. PMID: 6277605.

Yates, F. E., Leeman, S. E., Glenister, D. W., & Dallman, M. F. (1961). Interaction between plasma corticosterone concentration and adrenocorticotropin releasing stimuli n the rat: Evidence for the reset of an endocrine feedback control. *Endocrinology, 69*(1), 67–80. https://doi.org/10.1210/endo-69-1-67

Young, E. A., Akana, S., & Dallman, M. F. (1990). Decreased sensitivity to glucocorticoid fast feedback in chronically stressed rats. *Neuroendocrinology, 51*(5), 536–542. https://doi.org/10.1159/000125388. PMID: 2162011.

Benjamin D. Sachs

18

Robert L. Meisel and Michael J. Baum

Abstract

Ben Sachs is one of the world's experts on the neuroendocrine mechanisms that control sexual arousal, penile erectile function, and the expression of mating behavior in male rodents. His research laboratory produced experimental work on the respective roles of testosterone and/or its neural metabolite estradiol, acting perinatally to organize circuits that control male reproductive behaviors in adult male rats and acting postpubertally to activate these same circuits and the resultant expression of reproductive behaviors. Sachs was a master at integrating experimental animal and human clinical research findings in numerous review papers published over the course of his career that advanced our understanding of how hormones act in the brain to control male reproductive behaviors. He was also a generous mentor and friend to numerous students, postdoctoral fellows, and academic colleagues from around the world (including the two authors of this biographical sketch).

Keyword

Neuroendocrinology · Mating behavior · Sex steroids · Copulation

R. L. Meisel (✉)
Department of Neuroscience, University of Minnesota, Minneapolis, MN, USA
e-mail: meisel@umn.edu

M. J. Baum
Department of Biology, Boston University, Boston, MA, USA
e-mail: baum@bu.edu

R. J. Nelson, Z. M. Weil (eds.), *Biographical History of Behavioral Neuroendocrinology*, https://doi.org/10.1007/978-3-031-12970-4_18

Sachs' Early Years

Benjamin Sachs (Fig. 18.1) was born in 1936 in Madrid, Spain. His parents met at the University of Berlin where they were both students earning their PhDs in Philology. Given the political situation in Germany, Sachs' parents moved to Spain where his father was offered a position at the Complutense University of Madrid. The threat of civil war in Spain led the family to emigrate to the United States in 1937, settling in New York City. Sachs' father died in 1939 following a burst appendix. Sachs' mother worked for the International Auxiliary Language Association after which she taught at Hunter College (Spanish) and at the Walden School (German), where Sachs attended high school. She resigned her position at the Walden School after accepting a position teaching at the Sarah Lawrence College (German).

Preparation for Graduate School?

After high school, Sachs enrolled at the City College of New York in 1953. He then started graduate school at Duke University before dropping out and joining the Army. In the Army, Sachs was stationed in Newark, NJ, where he administered the Army's entrance examinations. Upon his discharge from the Army, Sachs returned to City College for a master's degree in School Psychology. As part of his undergraduate education he took classes taught by Daniel Lehrman and Jay Rosenblatt, who seeded his interest in comparative animal behavior.

Fig. 18.1 Benjamin Sachs circa 1980s

Graduate and Postgraduate Years

After earning his MS in Education from City College, Sachs went to the University of California Berkeley to study comparative animal behavior with Frank Beach. At that time, Beach had an interest in quail photoperiod and social behaviors, with the quail housed in the Berkeley Field Station at Strawberry Canyon. A storage shed was cleaned out to house the quail, and Sachs began his studies at about the same time that Beach lost interest in the project. But Sachs persevered, and that research was the subject of his dissertation (PhD awarded in 1966) and the basis of papers in *Science* and in the first issue of a new journal, *Hormones and Behavior.*

Not Sold in Storrs

After two postdoctoral years at Lehrman's Institute of Animal Behavior at Rutgers in Newark, Sachs and his wife Jacqueline accepted faculty positions at the University of Connecticut in 1968. Jacqueline had a notable research career studying language development. Depending on one's point of view, Storrs is the best of worlds or the worst of worlds. Storrs is in a typically idyllic location in the rolling, wooded hills of eastern Connecticut. People in this part of Connecticut are committed to maintaining this iconic beauty at the expense of life's daily conveniences (e.g., supermarkets). "Not sold in Storrs," a joke to some and the mantra of others, was a tribute to television ads proclaiming products that were only available through the TV offering and "not sold in stores."

The University of Connecticut was a wonderful backdrop for Sachs' research interests and personal life. He and Jacqueline found a house named Tranquility with a large barn that served as a two-car garage and, over the years, as a home for bats, mice, barn swallows, house sparrows, and several cats. For a short time, the barn also housed a pair of horses which Sachs and his daughter Naomi (now a faculty member at the University of Maryland) would ride on the nearby Nipmuck Trail. The rural serenity provided quiet two-lane roads along which Sachs would jog through much of the year. The Agricultural College of the University of Connecticut campus was an outlet for Sachs' interests in animal behavior. The open barn policy was a wonderful resource that led to a descriptive study of lamb play behavior with Sachs' graduate student, Valerie Harris (Sachs & Harris, 1978). Paradoxically, Sachs' primary research focus on rodent sexual behavior confined him to the windowless basement of the Psychology Building, with cinderblock and cement walls providing the only scenery.

The Making of a Fondateur (or Slaying the Dragon)

French language WikipédiA has a comprehensive and extremely well-produced overview of the field of sexology (Fig. 18.2). Scrolling down to the bottom of the page (https://fr.wikipedia.org/wiki/Portail:Sexualit%C3%A9_et_sexologie), there

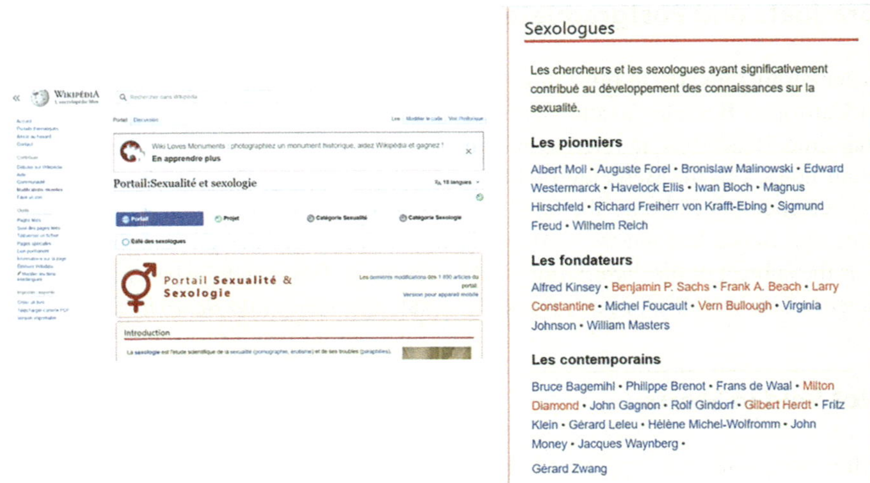

Fig. 18.2 A French language WikipédiA site listing Ben Sachs as a founder of the field of sexuality (https://fr.wikipedia.org/wiki/Portail:Sexualit%C3%A9_et_sexologie)

is a listing of the pioneers, founders, and contemporaries among sexology scholars. Listed among these individuals are Freud, Kinsey, Masters, Johnson, and Beach, familiar and notable names. Included among the founders (along with Frank Beach) is Ben Sachs.

How did Sachs' name emerge among these other notables? We do not know of course, but our speculation is that it follows from an influential chapter Sachs wrote entitled "The Physiology of Male Sexual Behavior." In 1988, Ernst Knobil and Jimmy Neill edited "The Physiology of Reproduction," a two-volume collection of chapters that provided comprehensive coverage of (as the title suggests) all things related to reproduction. Donald Pfaff was a section editor and asked Sachs to write the chapter on male sexual behavior (Sachs & Meisel, 1988). Sachs agreed and in turn graciously asked one of the authors of this biography (RLM) to be a coauthor. Sachs was an extremely efficient and effective writer, and with his knowledge of the field needed no help with the chapter. The decision to include a coauthor reflected Sachs' kindness and support of the career of an assistant professor.

There were many fits and starts and with the deadline for submission looming Sachs lamented that the writing of the chapter was dragging on. Finally, he proclaimed that we were going to slay this "drag-on," and a frenzied effort was put forth to produce an inclusive overview of the research on male sexual behavior that could be a valuable resource for both the experts and individuals new to the field. That goal was overwhelmingly met with the two editions of the chapter referenced over 1000 times! The first and second (published in 1994) editions of this chapter were each more than 300 manuscript pages, with over 1000 references and roughly 20 figures/tables. Perhaps this effort does not make Sachs one of the founders of the field of sexology; nonetheless his impact through this chapter alone made him a monumental contributor to the field.

You Can't Be a Carpenter Without a Hammer

The saying "You can't be a carpenter without a hammer" as it relates to sexual behavior is often attributed to Frank Beach (Thornton et al., 2009). In the context of sexual behavior, it refers to hypotheses regarding the development of different sexual behavior patterns between males and females. Why male rats mount females with pelvic thrusting is attributed here to differences in physical morphology. That is, males have a penis and females do not. Beach's view ran counter to hormone-centric hypotheses which asserted (and still do) that the brain is masculinized during a sensitive period in prenatal/early postnatal development and that neural pathways underlying mounting and thrusting by males result from this early hormonal exposure.

Along with his graduate student Donna Emery, Sachs conceived of a study in which typical adult female rats were given extended exposure to estradiol through a subcutaneous pellet (Emery & Sachs, 1975). After several months of exposure, these estradiol-treated females were paired with other female rats that were in heat (sexually responsive). A portion of the long-term estradiol-treated females mounted the female rats in heat. Now, male rats have distinct patterns of mounting that correspond to mounts with penile insertion (intromission) or mounts with insertion that include seminal emission (ejaculation). A dramatic finding from the Emery and Sachs study was that some of the females also showed the mounting pattern associated with ejaculation in male rats. Sachs had previously published observations that painful stimulation sexually aroused non-copulating male rats, inducing them to mount females (Barfield & Sachs, 1968). Pinching the tails of the estradiol-treated females induced mounting, including the ejaculatory pattern, in an additional cohort of females tested. Based on these results, Sachs called for a refinement of the prevailing theory of hormonally based sexual differentiation of copulation. Indeed, it was later found that fetal female rats are exposed perinatally to surprisingly high (almost male-like) concentrations of testosterone which partially masculinize neural circuits to allow the expression of male-typical patterns of mating later in life (Baum, 2009).

Penile Responses and Copulation

It is interesting that much of Sachs' career following the Emery and Sachs' paper was devoted to studying the neural regulation of penile responses in male rats, along with their role as a behavioral component of copulation. This programmatic approach began with behavioral description that led to a conceptual view of the functional roles the behavioral components held during copulation (Sachs, 1978).

We already noted the primary copulatory motor movements of male rats during copulation (mounts, mounts with intromission, and mounts with intromission and ejaculation). A different testing paradigm was developed to identify the penile movements associated with male copulation. This approach involved eliciting penile responses outside of the context of copulation. Here, male rats were put on their backs with their head and upper torso partially restrained in a plastic cup. The experimenter held the lower limbs to prevent them from moving and then used a thin dowel to retract the penile sheath. After a few minutes, the rat would exhibit clusters of responses that included three types of movements of the penis: erection (penile tumescence), flips (dorsiflexions of the penis), and cups (a flaring of the tip of the penis).

During copulation, these movements of the penis are obscured. Sachs developed a clever way to observe the penile responses during copulation by testing the animals in a clear bottom glass aquarium that was suspended over a mirror tilted at a 45-degree angle. This arrangement made it possible to obtain a ventral view of the male and female during copulation and correlate penile responses with the male's motor copulatory patterns. What further aided this analysis was the surprising occurrence of copulatory movements that were accompanied by extravaginal penile responses. With this approach, it appeared that erection primarily accompanied mounts without insertion, mounts with intromission were linked to the flips, and cups were involved with seminal emission during ejaculation.

How the specific penile responses are integrated with the rat's mounting patterns is still not understood. However, it appears that the coordination of the different movements occurs at different levels of the central nervous system. With the tests of penile responses independent of copulation, it takes several minutes for penile responses to be initiated. This latency seems to be controlled through inhibition from higher levels of the nervous system as rats with lower thoracic spinal transection initiate these penile responses in a few seconds with similar clusters of responses.

Sachs was interested in behavioral and neural mechanisms of male copulatory behavior from a comparative perspective, though most of the work in his laboratory focused on rats. Male rats initially sniff and investigate females for a period of time after which the males initiate bouts of overt copulatory activity. Males will mount females and initiate a few seconds of thrusting that may or may not terminate in intromission. There are several series of these mounts/intromissions that lead up to an intromission that includes ejaculation. After the male ejaculates, it ignores the female for 5–10 min before initiating another copulatory bout. This quiescent period (termed the postejaculatory interval) initially includes the male vocalizing at a frequency of 22 kHz (the absolute refractory period), with the vocalizations ending for the latter part of the postejaculatory interval (relative refractory period). Sachs noted the array of measures that different laboratories recorded during tests of male copulatory behavior, including latencies to initiate events (e.g., mount and intromission latencies), intervals between events (e.g., interintromission intervals), and numbers of events. The expectation was that these measures mapped onto conceptualizations of behavioral mechanisms underlying male copulatory behavior (e.g., arousal and performance measures). Sachs used normative data sets to calculate the intercorrelations among individual components of male copulatory behavior through factor analyses to identify the minimal number of mechanisms (factors) that accounted for male copulatory behavior. Based on these analyses, Sachs identified four factors that encompassed the variance among the measures of copulation. The four factors were an *copulatory rate factor* (comprising the interintromission interval, latency from the first intromission to the first ejaculation, duration of the postejaculatory vocalizations, and latency from ejaculation to the first intromission in the next copulatory bout), an *initiation factor* (latency to the first mount, latency to the first intromission, and relative refractory period), a *hit rate factor* (the proportion of mounts that included intromission), and an *intromission count factor* (number of intromissions as well as the postejaculatory interval). Sachs noted both that "Much of nature is notable for its lack of parsimony, and it can be deceptive to make things easy by leaving out the hard parts" (Sachs, 1978, pp. 275–276).

The other element of identifying behavioral mechanisms was the idea that they would be valuable parts of developing a neurobiological understanding of the control of male copulatory behavior. With his students Valerie Harris and Donna Emery (Harris & Sachs, 1975; Emery & Sachs, 1976), lesions were made in the corticomedial amygdala or bed nucleus of the stria terminalis (a region receiving afferents from the corticomedial amygdala) in male rats. Common feature of each of these lesions was a dramatic increase in the numbers of intromissions preceding ejaculation compared with control males (intromission count factor) and an increase in the interintromission intervals (copulatory rate factor). More variable effects were seen in measures linked to the initiation factor. Sachs proposed both that it might be possible to map the behavioral mechanisms underlying copulation on to specific neural circuits and that the behaviors might also emerge from feedback loops among these circuits.

Organic Versus Psychogenic Erections

Sachs and colleagues noted that the incidence of erection (seen via a ventral view of free moving male rats) was greatly increased when males were downwind (vs. upwind) of an estrous female rat. Sachs initially labeled these as "psychogenic" erections that did not follow from any direct physical manipulation of the penis (see Sachs (2003) for his later views on psychogenic erection). He suggested that the neural circuitry controlling such psychogenic erections in rats may resemble the occurrence of erections that occur in men while they fantasize about a sexual partner and/or while they dream. Sachs and colleagues proceeded to study the role of the olfactory-pheromone processing circuitry (e.g., olfactory bulb, medial amygdala, bed nucleus of the stria terminalis, and the nucleus paragigantocellularis), as well as the steroidal regulation of noncontact erection in male rats. In addition to publishing numerous experimental articles on the neuroendocrine regulation of noncontact (psychogenic) erectile function, Sachs published several reviews that integrated experimental animal and clinical work on the control of penile erection (e.g., Sachs (2003)). Sachs and his trainees regularly presented their research findings on the neuroendocrine regulation of penile erection in rats and mice at annual meetings of the Society for Behavioral Neuroendocrinology as well as the International Academy of Sex Research—the premier association for reporting research findings related to all aspects of human sexuality.

Conclusions: A Scholar's Life

Sachs' start in his career was as a student of animal behavior. This broad interest extended into his knowledge of sexual behavior that included insect mating behavior, including the adaptive significance of the vast array of penile morphologies. This scholarship made Sachs a valuable resource in the field. His approach to understanding male sexual behavior in rodents from the conceptual level, to the hormonal and neural control, and to physiological mechanisms underlying the control of penile responses including the integration of all of these elements still makes him unique in the field. Sachs was intellectually rigorous (including his use of language) and was never one to leave out the "hard parts" in developing and testing hypotheses. With this rigor of thought (that he expected in others as well), his scholarship meant that he enjoyed the process of science and was always interested in the research of others in the field. Sachs' scientific rigor and interest in the ideas of others were the bases for his wonderful mentorship of students and postdocs, who he saw as colleagues in his research endeavor. We do not know why WikipédiA labeled Ben Sachs as a "fondateur" among researchers of sexology, though his contributions certainly make him deserving of this distinction.

References

Barfield, R. J., & Sachs, B. D. (1968). Sexual behavior: Stimulation by painful electrical shock to skin in male rats. *Science, 161*, 392–395.

Baum, M. J. (2009). Sexual differentiation of pheromone processing: Links to male-typical mating behavior and partner preference. *Hormones and Behavior, 55*, 579–588.

Emery, D. E., & Sachs, B. D. (1975). Ejaculatory pattern in female rats without androgen treatment. *Science, 190*, 484–486.

Emery, D. E., & Sachs, B. D. (1976). Copulatory behavior in male rats with lesions in the bed nucleus of the stria terminalis. *Physiology & Behavior, 17*, 803–806.

Harris, V. S., & Sachs, B. D. (1975). Copulatory behavior in male rats following amygdaloid lesions. *Brain Research, 86*, 514–518.

Sachs, B. D. (1978). Conceptual and neural mechanisms of masculine copulatory behavior. In T. E. McGill, D. A. Dewsbury, & B. D. Sachs (Eds.), *Sex and behavior: Status and prospectus* (pp. 267–295). Plenum Press.

Sachs, B. D. (2003). The false organic-psychogenic distinction and related problems in the classification of erectile dysfunction. *International Journal of Impotence Research, 15*, 72–78.

Sachs, B. D., & Harris, V. S. (1978). Sex differences and developmental changes in selected juvenile activities (play) of domestic lambs. *Animal Behaviour, 26*, 678–684.

Sachs, B. D., & Meisel, R. L. (1988). The physiology of male sexual behavior. In E. Knobil & J. D. Neill (Eds.), *The physiology of reproduction* (Vol. 2, pp. 1393–1485). Raven Press.

Thornton, J., Zehr, J. L., & Loose, M. D. (2009). Effects of prenatal androgens on rhesus monkeys: A model system to explore the organizational hypothesis in primates. *Hormones and Behavior, 55*, 633–645.

Ronald J. Barfield

19

Gregory F. Ball and Jacques Balthazart

Abstract

Ronald (Ron) Barfield was born in Detroit Michigan on 25 July 1936 and devoted his entire career to the study of the endocrine control of behavior. His research was heavily influenced by the ethological tradition that he was exposed to through his PhD advisor Nicholas Collias. After postdoctoral studies at the Institute of Animal Behavior at Rutgers-Newark, his entire career took place at the New Brunswick, NJ, campus of Rutgers University where he arrived as assistant professor in 1967 and was promoted full professor in 1974. Barfield's major research impact relates to two interrelated lines of research. His careful stereotaxic implantations of sex steroids in the brain provided foundational information on the central sites of action of steroid hormones in relation to the activation of male and female sexual behavior and aggression. Barfield also discovered in 1972 that male rats produce ultrasonic vocalizations during the post-ejaculatory refractory period. A series of studies investigating these calls in several different contexts related to sexual, social, and parental behavior resulted in many publications spanning more than two decades.

Keywords

Sex steroid hormones · Brain implants · Ultrasonic vocalization · Ejaculation · Refractory period

G. F. Ball (✉)
Department of Psychology, University of Maryland, College Park, MD, USA
e-mail: gball@umd.edu

J. Balthazart
University of Liège, GIGA Neurosciences, Research Group in Behavioral Neuroendocrinology, Liège, Belgium

R. J. Nelson, Z. M. Weil (eds.), *Biographical History of Behavioral Neuroendocrinology*, https://doi.org/10.1007/978-3-031-12970-4_19

Ronald "Ron" Barfield was a pioneer in the development of the modern field of behavioral neuroendocrinology. He played a crucial role in the identification of brain sites where sex steroids act to activate sexual behaviors in both birds and rodents. He also was at the origin of the discovery of the ultrasonic vocalizations produced by male rats during the post-ejaculatory refractory period, and this initiated another fruitful line of research. In addition to his significant scientific legacy, Barfield was an active leader in the field through his participation in many national and international conferences and societies that sponsored research in behavioral neuroendocrinology and his service on editorial boards for significant journals such as *Behavioral Neuroscience* and *Hormones and Behavior*. He served as a program officer at the National Science Foundation both in the Animal Behavior program and in the Neuroendocrinology program from 1994 to 1997, where he was able to support research important to the field.

Barfield was born in Detroit Michigan on 25 July 1936, and he died on 29 September 2015 in Palm Beach Gardens, Florida. He received all of his university degrees at the University of California, Los Angeles (UCLA). He graduated with an AB in Zoology in 1959 and then completed a master's degree in 1962 and PhD in 1965. His doctoral advisor was Nicholas Collias, a prominent ornithologist/ethologist at UCLA, who is perhaps best known for his studies of nest building in weaver birds (*Textor cucullatus*). Collias, as many mechanistically oriented ethologists, had conducted studies on the hormonal control of nest building and other behaviors related to reproduction in his primary study species, the weaver bird, and Barfield collaborated with him in the early 1960s on this project (reviewed in Collias and Collias (1984)). This work planted the seed for Barfield to pursue studies of hormones and behavior in birds. Barfield's time with Collias also inculcated in him an ethological perspective to the study of animal behavior that remained with Ron for his entire career.

An emerging concept in the 1960s was that steroids act directly in the brain to facilitate the activation of adult behavior. It was reasoned that in gonadectomized animals, small implants of crystalline steroid placed directly into select brain areas should activate behavior and would help build an understanding as to how hormones modulate neural circuits to activate behavior. Robert Lisk at Princeton pioneered the stereotaxic technique allowing to administer centrally steroids such as estradiol and to investigate neuroendocrine and behavioral effects (Lisk, 1960). This approach was in the zeitgeist of the mid-1960s, Barry Komisaruk, for example, working with Daniel Lehrman at Rutgers simultaneously initiated research projects in birds examining the central effects of steroids on behavior. Komisaruk investigated progesterone in relation to the onset of parental behavior in ring doves (Komisaruk, 1967).

For his thesis, Barfield studied the effects of intracranial implants of androgen on sexual and aggressive behavior in capons (castrated roosters (Barfield, 1965a, b)). He found that copulatory behavior was activated by implants of testosterone in the preoptic area (POA) but not in other brain areas. Interestingly, these implanted birds exhibited no aggressive or courtship behavior suggesting that these behaviors are activated by androgens acting at other brain sites. Other experiments described in his PhD thesis (Barfield, 1965b) and published in abstract form only (Barfield, 1965a) suggested that some aggressive behavior was activated in capons that had received an implant of testosterone propionate in the lateral forebrain, more specifically the paleostriatum (now renamed basal ganglia (Reiner et al., 2004)), a brain region that was implicated by earlier studies in the control of aggression (Phillips, 1964; Putkonen, 1967). By combining androgen implants in the preoptic area and in the lateral forebrain, Barfield was then able to test in a direct manner the older ethological theory postulating that courtship involves a conflict between sexual and aggressive motivations. Waltzing, a courtship behavior directed at females, was indeed observed in a few of the dual implanted capons providing some support for this hypothesis, but there were not enough subjects in this experiment that also lacked proper controls. This ethological concept is nowadays largely abandoned, but this work clearly shows the influence that ethological thinking had on Barfield's research.

Barfield had two famous neuroendocrinologists on his committee – Richard Whalen and Charles Sawyer. While approving Barfield's proposed topic during his thesis proposal meeting, Sawyer raised a number of practical issues with which Barfield had not considered. One was how he was going to stereotaxically implant androgens in the chickens. In his proposal to investigate central effects of androgen on aggression and copulation in capons, Barfield had not realized that stereotaxic devices were not readily available for studies in birds. At Sawyer's recommendation, Barfield went to see David Kopf who had recently started his Los Angeles-based Kopf instrument company and who was among the first to produce high-quality stereotaxic devices. Kopf asked for a few rooster heads and personally modified a stereotaxic device to be compatible with work in chickens so that Barfield could conduct his dissertation research.

After UCLA, Barfield moved to the Institute of Animal Behavior at Rutgers, Newark, to pursue postdoctoral studies with Daniel Lehrman. He conducted work on ring doves with Lehrman performing brain implant studies in ring doves and studying the effects of female stimuli on gonadotrophic activity in males (Barfield, 1967, 1971a, b). He also initiated seminal studies on rats with Benjamin (Ben) Sachs, and rats turned out to be his primary study species for the rest of his career. After his 2 years of postdoctoral research at Rutgers, Newark, in 1967, he joined the Zoology Department as an assistant professor at Douglas College on the New Brunswick campus of Rutgers. This position initiated a series of appointments that he held at Rutgers until his retirement in 1999. In 1970, he joined the Biology Department of Livingston College as an associate professor, and he was promoted to Full Professor in 1974 and to Professor II in 1981. The biological sciences at New Brunswick were amalgamated into a single department in 1974, and Barfield's appointment over the remainder of his career was in that single Department of Biological Sciences.

Barfield's research impact can be appreciated based on an examination of the interrelated lines of research that he pursued nearly throughout his career. The hallmark of this work was his ability to perform careful stereotaxic implantations that provided foundational information on the sites of action of steroid hormones in relation to the activation of sex and aggression. Work from Barfield's lab played a key role in the development of a consensus concerning where steroid hormones act to regulate male-typical and female-typical sexual behaviors (Davis & Barfield, 1979a, b; Etgen & Barfield, 1986; Glaser et al., 1985, 1987; Rubin & Barfield, 1980, 1983). He also studied the issue of steroid specificity identifying the importance of estrogenic metabolites of testosterone acting in the POA in the regulation of many male-typical behaviors (Christie & Barfield, 1979; Davis & Barfield, 1979b). In 1972, Barfield discovered that male rats produce ultrasonic vocalizations during the post-ejaculatory refractory period (Barfield & Geyer, 1972, 1975). He then conducted a series of studies investigating these ultrasonic calls in several different contexts related to sexual, social, and parental behavior, which resulted in a large number of publications spanning more than two decades. This work was summarized in a general review in 1986 (Barfield & Thomas, 1986). He also investigated reproductive behavior in relation to ultrasonic vocalization in male rats and combined studies of central and systemic administration of sex steroids. Barfield eventually ventured into studying mice, because beginning in the 1990s, they were considered as a potentially better model for behavioral neuroendocrinologists than rats because of the ability to conduct in various mouse strain studies of transgenic mice (Matochik et al., 1994; Nyby et al., 1992; White et al., 1998).

Over the course of a 30-year publishing career (from 1968 to 1998), Barfield published about 100 referred journal articles in addition to an influential set of book chapters and other reviews. His impact is apparent by measures of his citations such as his h index of 44 based on 5278 total citations (web of knowledge January 2022). These numbers will of course continue to grow as Barfield's work continues to be cited.

Barfield always considered himself as an ethologist studying the impact of hormones on spontaneous motivated behaviors (one would have said instinctive behavior in these early days). When one of us (JB) was the president of the Society for Behavioral Neuroendocrinology and thus had the privilege to organize the presidential symposium during the annual meeting in 2004 in Lisbon, Portugal, we (JB and GFB) both decided that it would be a good idea to devote this symposium to the "Ethological Roots of Behavioral Neuroendocrinology." The idea was to highlight the role played by European ethologists in the development of behavioral neuroendocrinology and then illustrate the more recent developments as well as conceptual continuity of the field based largely on studies in birds that were traditionally used extensively as subjects in ethology. We had the good fortune to be able on this occasion to gather for the last time in a symposium session Peter Marler and Robert Hinde who described with their first-hand knowledge the interactions that took place during the 1950s and 1960s between European ethologists and the founders of behavioral neuroendocrinology in the United States. The third talk was then appropriately given by Ron Barfield who described the convergence of interests between ethologists, behavioral endocrinologists, and neuroendocrinologists and promoted the development of studies on the identification of the sites of steroid action in the brain. The final talk in this historical part of the symposium was then presented by Greg Ball who explained how ideas originally developed by Daniel Lehrman in his work on ring doves still continue to influence current research on the interplay between hormones, the brain, and behavior. A picture of the first three speakers taken on this occasion is shown in Fig. 19.1.

Barfield mentored a series of students while at Rutgers who went on to make contributions to the field of behavioral neuroendocrinology and other fields of science such as Tracy McIntosh, Mary Erskine, Beverly Rubin, John Matochik, Eric Pleim, Lynette Geyer, and Nicholas White. Moreover, he was supportive of many young scientists in our field that he encountered in a variety of places. Barfield also was a spokesperson for the field often facilitating conversations at conferences by asking questions or making comments related to hormones and behavior in the broader contexts of ethology and neuroscience. He also was, at an informal level, contributing to establish the "memory of the field" by taking countless pictures of participants at all meetings he was attending, in particular the Conference of Reproductive Behavior that is at the origin of the current annual meeting of the Society for Behavioral Neuroendocrinology. One of us (JB) organized in 2014 a meeting on the occasion of his official retirement, and Barfield who attended had prepared an extensive PowerPoint file containing more than 100 photos (only a small part of his collection) summarizing the last 50 years of the behavioral neuroendocrinology community. We were able to project these photos on a wall during several coffee breaks and other social events to the great satisfaction of all participants.

On a more personal note, Barfield was intellectually generous to both of us during our entire careers. He enjoyed talking with us in part because our work is in birds, and we built on his thesis work in chickens. He was always a great sounding board for new ideas and he would take the time to discuss them. One of us (JB) personally encountered Barfield's notable reactions at the end of talks during

Fig. 19.1 The speakers and the organizer of the symposium on "Ethological Roots of Behavioral Neuroendocrinology" in Lisbon during the meeting of the Society for Behavioral Neurobiology (2014). From left to right: Peter Marler, Jacques Balthazart (organizer), Robert Hinde, and Ron Barfield. (Photo by GFB)

conferences at his first oral presentation. It was in 1975 during the 14th International Ethology Conference in Parma, Italy. At the time, JB was working on the endocrine control of reproductive behavior in male ducks and had the privilege of presenting his results in front of a large audience including very prominent members of the field such as Konrad Lorenz, Uli Weidman, Frank McKinney and Irenaus Eibl-Eibesfedlt. As a young PhD student who was a native Francophone, at that time, he had a less than optimal command of English. Needless to say, there was much stress when the talk was completed based on the fear that questions would come afterward in English that would not be understandable. However, at the end of the presentation, Barfield, who JB did not know at the time, stood up and started commenting about the presentation. The comments were mostly positive even if they were hard to follow for a native French speaker. Barfield spoke about as long as the length of the original talk so that at the end of his comments there was no time for other questions, much to JB's relief!

After retirement, Barfield enjoyed an active life in Florida involving sailing and sports such as softball with his wife, Paula Davis. Ron is survived by Paula and his two children, David and Rachel. Although he has passed, his scientific legacy lives on.

References

Barfield, R. J. (1965a). Induction of aggressive and courtship behavior by intracerebral implants of androgen in capons. *American Zoologist, 5*, 203.

Barfield, R. J. (1965b). *Induction of sexual and aggressive behavior by intracerebral implants of androgen in the capon* (p. 118). University of California.

Barfield, R. J. (1967). Activation of sexual and aggressive behavior by androgen implants in the brain of the male Ring Dove. *American Zoologist, 7*, 800.

Barfield, R. J. (1971a). Activation of sexual and aggressive behavior by androgen implanted into the male ring dove brain. *Endocrinology, 89*, 1470–1476.

Barfield, R. J. (1971b). Gonadotrophic hormone secretion in the female ring dove in response to visual and auditory stimulation by the male. *The Journal of Endocrinology, 49*, 305–310.

Barfield, R. J., & Geyer, L. A. (1972). Sexual behavior: Ultrasonic postejaculatory song of the male rat. *Science, 176*, 1349–1350.

Barfield, R. J., & Geyer, L. A. (1975). The ultrasonic postejaculatory vocalization and the postejaculatory refractory period of the male rat. *Journal of Comparative and Physiological Psychology, 88*, 723–734.

Barfield, R. J., & Thomas, D. A. (1986). The role of ultrasonic vocalizations in the regulation of reproduction in rats. *Annals of the New York Academy of Sciences, 474*, 33–43.

Christie, M. H., & Barfield, R. J. (1979). Effects of aromatizable androgens on aggressive behaviour among rats (Rattus norvegicus). *The Journal of Endocrinology, 83*, 17–26.

Collias, N. E., & Collias, E. C. (1984). *Nest building and bird behavior*. Princeton University Press.

Davis, P. G., & Barfield, R. J. (1979a). Activation of feminine sexual behavior in castrated male rats by intrahypothalamic implants of estradiol benzoate. *Neuroendocrinology, 28*, 228–233.

Davis, P. G., & Barfield, R. J. (1979b). Activation of masculine sexual behavior by intracranial estradiol benzoate implants in male rats. *Neuroendocrinology, 28*, 217–227.

Etgen, A. M., & Barfield, R. J. (1986). Antagonism of female sexual behavior with intracerebral implants of antiprogestin RU 38486: Correlation with binding to neural progestin receptors. *Endocrinology, 119*, 1610–1617.

Glaser, J. H., Etgen, A. M., & Barfield, R. J. (1985). Intrahypothalamic effects of progestin agonists on estrous behavior and progestin receptor binding. *Physiology & Behavior, 34*, 871–877.

Glaser, J. H., Etgen, A. M., & Barfield, R. J. (1987). Temporal aspects of ventromedial hypothalamic progesterone action in the facilitation of estrous behavior in the female rat. *Behavioral Neuroscience, 101*, 534–545.

Komisaruk, B. R. (1967). Effects of local brain implants of progesterone on reproductive behavior in ring doves. *Journal of Comparative and Physiological Psychology, 64*, 219–224.

Lisk, R. D. (1960). Estrogen-sensitive centers in the hypothalamus of the rat. *The Journal of Experimental Zoology, 145*, 197–207.

Matochik, J. A., Sipos, M. L., Nyby, J. G., & Barfield, R. J. (1994). Intracranial androgenic activation of male-typical behaviors in house mice: Motivation versus performance. *Behavioural Brain Research, 60*, 141–149.

Nyby, J., Matochik, J. A., & Barfield, R. J. (1992). Intracranial androgenic and estrogenic stimulation of male-typical behaviors in house mice (Mus domesticus). *Hormones and Behavior, 26*, 24–45.

Phillips, R. E. (1964). Wildness in the mallard duck. Effects of brain lesions and stimulation on escape behavior and reproduction. *Journal of Comparative Neurology, 122*, 139–155.

Putkonen, P. T. S. (1967). Electrical stimulation of the avian brain. Behavioral and autonomic reactions from the archistriatum, ventromedial forebrain and the diencephalon in the chicken. In *Annales Academiae Scientiarum Fennicae. Ser. A. 5, Medica* (Vol. 130, pp. 9–95).

Reiner, A. D., Perkel, J., Bruce, L., Butler, A., Csillag, A., Kuenzel, W., Medina, L., Paxinos, G., Shimizu, T., Striedter, G., Wild, M., Ball, G. F., Durand, S., Güntürkün, O., Lee, D. W., Mello, C. V., White, S. A., Hough, G., Kubikova, L., Smulders, T. V., Wada, K., Dugas-Ford,

J., Husband, S., Yamamoto, K., & Yu, J. (2004). Revised nomenclature for avian telencephalon and some related brainstem nuclei. *The Journal of Comparative Neurology, 473*, 377–414.

Rubin, B. S., & Barfield, R. J. (1980). Priming of estrous responsiveness by implants of 17 beta-estradiol in the ventromedial hypothalamic nucleus of female rats. *Endocrinology, 106*, 504–509.

Rubin, B. S., & Barfield, R. J. (1983). Induction of estrous behavior in ovariectomized rats by sequential replacement of estrogen and progesterone to the ventromedial hypothalamus. *Neuroendocrinology, 37*, 218–224.

White, N. R., Prasad, M., Barfield, R. J., & Nyby, J. G. (1998). 40- and 70-kHz vocalizations of mice (Mus musculus) during copulation. *Physiology & Behavior, 63*, 467–473.

Lynwood George Clemens

20

Casey L. Henley and Jennifer A. Cummings

Abstract

Lynwood G. Clemens was a pioneer in the field of behavioral neuroendocrinology. Trained by Frank Beach, Robert Goy, and Roger Gorski, a strong foundation was set for Clemens to establish himself as an impactful researcher. Combined with his innate curiosity and imagination, Clemens made many significant discoveries about the roles of organizational and activational gonadal hormone exposure on adult sexual behavior in rodents, using novel techniques and innovative experimental designs. Clemens' enduring impact on the field does not stop with his academic contributions; he has inspired generations of behavioral neuroendocrinologists either directly through his extraordinary mentoring or indirectly through his founding of the *Conference on Reproductive Biology*, which would eventually transform into the annual meeting for the *Society for Behavioral Neuroendocrinology*.

Keywords

Activational effects · Aromatase · Behavioral neuroendocrinology · Gonadal hormones · Mentoring · Neurotransmitters · Organizational effects · Rodent behavior · Sexual behavior

Followed dreams, not rules. No regrets. –Lynwood G. Clemens

C. L. Henley
Neuroscience Program, Physiology Department, Michigan State University, East Lansing, MI, USA
e-mail: mcgove14@msu.edu

J. A. Cummings (✉)
Department of Psychology, University of Michigan, Ann Arbor, MI, USA
e-mail: jacummin@umich.edu

R. J. Nelson, Z. M. Weil (eds.), *Biographical History of Behavioral Neuroendocrinology*, https://doi.org/10.1007/978-3-031-12970-4_20

In the Beginning

Lynwood (Lyn) Clemens enrolled at Pennsylvania State University with the plan of becoming a journalist. Fortunately for the field of neuroendocrinology, science called him, and he earned his BS in Psychology in 1960. A faculty mentor at Penn State, who recognized Clemens' inquisitive mind and propensity for comparative analysis, suggested that Clemens apply to graduate school at Berkeley to work with a professor studying comparative psychology. Clemens moved across the country and earned his PhD training with a pioneer in the emerging field of behavioral endocrinology, Frank A. Beach. After completing his postdoctoral training with Robert Goy at the Oregon Primate Research Center and Roger Gorski at UCLA (University of California, Los Angeles), he joined the faculty in the Zoology Department at Michigan State University in 1968 where he stayed for the remainder of his career. Clemens' 48 years as a Spartan were highlighted by a history of influential contributions to the field of hormones and behavior, a record of continuous federal funding, the minting of two dozen PhDs, and a reputation as an inspirational teacher and a mentor of his students.

His resourceful nature was demonstrated early in graduate school. Clemens thought he was moving to Berkeley to study comparative psychology but soon discovered that his laboratory was "just working with rats." Undeterred and unwilling to accept this fate and abandon a comparative approach, he pushed ahead and introduced a new animal model to the Beach's laboratory: deer mice. Clemens proceeded to characterize the sexual behavior of *Peromyscus maniculatus* males and wrote a review comparing the ejaculatory response characteristics of multiple mammals with the hope that interspecies comparisons would lead to a better understanding of human sexual functions (Beach et al., 1966). As usual, Clemens' instincts were sound: his first publication from graduate school in 1966 is still being cited, highlighting the importance and continuing impact of his early work.

After completing his PhD in 1966, Clemens first moved north to conduct his postdoctoral training with Robert Goy at the Oregon National Primate Research Center, followed by training with Roger Gorski at UCLA. His postdoctoral work with Goy and Gorski broadened Clemens' research perspectives, bridging his work examining hormones and behavior with integrative neuroscience. During this time, Clemens transitioned from using ablation-and-replacement techniques to more elegant experiments delivering minute quantities of steroid hormones directly into the brain via intracranial implants. Bringing his surgical skills to the Gorski laboratory, Clemens followed up on a previous study by Beach (1944) that examined the inhibitory role of the cerebral cortex in the sexual behavior of female rats. To address this hypothesis, Clemens and his colleagues used a novel method of delivering potassium chloride to the brain to depress cortical activity in a complete, but reversible, manner (Clemens et al., 1967). This study – published in *Science* and the first behavioral work to come out of the Gorski laboratory – demonstrated that transiently inhibiting the cerebral cortex would facilitate female sexual behavior in a manner similar to hormone priming. This publication was one of the first reports of tonic inhibition of female sexual behavior by a neural mechanism.

Clemens' research as a postdoctoral fellow set him on a future path that integrated his passions for behavior and endocrinology with his emerging interest in neuroscience, a field just beginning to take shape but commonly denoted at the time as physiological psychology. The *Society for Neuroscience* did not yet exist, but Clemens and others were forging the way, applying novel techniques and combining fields of study to test a seemingly endless array of new hypotheses. With a strong background in behavioral analysis and newly learned skills in neuroscience techniques, Clemens transitioned to his faculty position at Michigan State University.

Organizational Effects of Gonadal Hormones

Upon his arrival in East Lansing, Clemens embarked on the study of the organizational effects of gonadal hormones on developing animals, complementing his previous interests in the activational effects of these compounds in adults. Specifically, his early work at Michigan State focused on exploring the organizational effects of testosterone on male and female sexual behavior in hamsters (Carter et al., 1972; Coniglio & Clemens, 1976; Doty et al., 1971) and determining the roles of the testosterone metabolites, estradiol, and non-aromatizable androgens, in the development of later sexual behaviors (Coniglio et al., 1973a, b; Paup et al., 1972, 1974). These were among the earliest reports to implicate the conversion of testosterone to estradiol during early development as a key factor in masculinization and defeminization of adult sexual behavior in rodents.

Although the importance of exogenous perinatal hormones on adult sexual behavior in rodents had been established by his laboratory and other investigators, Clemens continued to wonder if naturally occurring variations in hormone levels during prenatal life also could affect adult behavior. He had previously determined that females undergo some defeminization in utero because the female offspring of pregnant rats treated gestationally with either anti-androgens or aromatase inhibitors showed increased female sexual behavior in adulthood (Clemens & Gladue, 1978; Gladue & Clemens, 1978). Applying an inventive approach to identify litter positions by Caesarean delivery, Clemens and his students discovered that female rats located next to at least one male in the uterus during gestation were more likely to show male-typical sexual behaviors as adults when tested with sexually active stimulus females. Additionally, these changes in behavior could be blocked with prenatal treatment of the mother with an anti-androgen (Clemens et al., 1978). In a separate study, ovariectomized pregnant females were treated with oil vehicle, estradiol, or an aromatase inhibitor. The female offspring that were not exposed to estradiol during gestation (ovariectomized plus oil or aromatase inhibitor) showed increased female sexual behavior compared to the groups exposed to estradiol (Witcher & Clemens, 1987). Thus, this body of work confirmed that not only can the administration of exogenous hormones during development alter sexual behavior as an adult but demonstrated that naturally occurring variations in hormone exposure in utero have an important role in programming adult sexual behavior.

Clemens continued to examine the effects of perinatal gonadal hormones throughout his career, expanding on his early work to include end points beyond sexual behavior. Focusing on effects on anatomy, Clemens' students examined the development of the sexually dimorphic spinal nucleus of the bulbocavernosus (SNB), a cluster of cell bodies in the lumbar spinal cord linked to sexual reflexes in male rodents (Wagner & Clemens, 1989a; Wee & Clemens, 1987). The development of the SNB in mice was found to be dependent on exposure to gonadal hormones during gestation (Wagner & Clemens, 1989b). Additionally, the neural circuit from the brain to the spinal cord also showed sensitivity to steroid hormones, with estrogen-concentrating neurons in the periventricular nucleus of the hypothalamus projecting to the androgen-sensitive SNB (Wagner et al., 1993), indicating that neural circuits have location-specific sensitivities to hormones, reflecting the intricate relationships between hormones and behavior.

Clemens and his students also studied the effects of perinatal exposure to the pervasive environmental contaminants, polychlorinated biphenyls (PCBs), on sexual behaviors in rats. Perinatal exposure to PCBs, which can exert both estrogenic and antiestrogenic effects, was found to alter female sexual behavior (Chung et al., 2001; Wang et al., 2002), female partner preference (Cummings et al., 2008), and maternal behavior (Cummings et al., 2005; Simmons et al., 2005). This line of inquiry provided further evidence of the potential of compounds in the environment to interact with reproductive processes in mammalian species. Interestingly, a cross-fostering research design and a focus on maternal behavior indicated that some perturbations in the adult behavior of offspring were due in part to the altered maternal care provided by the exposed mother, rather than direct exposure of the offspring to the contaminant.

In later years, Clemens returned to investigating the effects of perinatal exposure to exogenous gonadal hormones on adult behavior, examining partner preference behavior in male and female rats. Similar to sexual behavior, exposure to estradiol during early postnatal development masculinized partner preference in female rats (Henley et al., 2009). Additionally, postnatal exposure to exogenous testosterone in intact males increased the amount of time males spent with stimulus males while decreasing sexual behavior with stimulus females (Henley et al., 2010). These results confirmed the roles of gonadal hormones on the development of partner preference, a critical component of mating behavior with implications for sexual orientation.

Activational Effects of Gonadal Hormones on Sexual Behavior

Whereas many laboratories limited the scope of their research to examining the role for hormones on behavioral output at a particular period in an animal's lifetime, Clemens was never one to stick to traditions. Along with pursuing studies on hormone actions during development, his relentless curiosity inspired a return to his earlier interests during his graduate and postdoctoral years on hormones acting during adulthood. To this end, Clemens' laboratory began to examine the role of testosterone and estradiol in the activation of male sexual behavior in adulthood in hamsters (Christensen et al., 1973; De Bold & Clemens, 1978), rats (Christensen & Clemens, 1974, 1975), and several strains of mice (Clemens et al., 1988; Wee et al., 1988; Wee & Clemens, 1989). Like the organizational effects of estradiol, the Clemens' laboratory reported that the aromatization of testosterone into estradiol was necessary to activate male-typical behaviors in adulthood, particularly in rats. These studies added continued support for the aromatization hypothesis. They also represented some of the earliest studies to examine the direct effects of hormones on the brain since most hormonal studies up to this point used systemic injections to test the effects of steroids on behavior. Evidence suggested that the preoptic area (POA) of the hypothalamus was a critical neural site for the regulation of male sexual behavior in rats. Christensen and Clemens (1974, 1975), therefore, decided to use an innovative technique and deliver steroids directly into the POA via indwelling cannulae. They found that estradiol alone in the POA was able to induce mounts, intromissions, and ejaculations in castrated animals and that the effects of testosterone could be blocked by coadministration of an aromatase enzyme inhibitor. These studies indicated that aromatization of testosterone was taking place directly within the POA and that estradiol was a metabolite necessary for the activation of male sexual behavior in rats.

Clemens and his students had demonstrated that steroid hormones administered directly into the brain could activate sexual behavior, but brain circuits rely on the actions of neurotransmitters, not just hormones, for communication. Therefore, Clemens expanded his work to analyze how neurotransmitter systems mediate the effects of gonadal hormones on sexual behaviors. Through a series of pharmacological experiments, acetylcholine, acting via muscarinic receptors, was found to play an important role in facilitating both female and male sexual behaviors, whereas norepinephrine, possibly mediated by $\alpha 2$ receptor action, was found to have an inhibitory effect on lordosis. In females, intracerebral administration of muscarinic agonists increased the display of lordosis, whereas muscarinic antagonists reduced display of the behavior (Clemens et al., 1980, 1981, 1989; Clemens & Dohanich, 1980; Dohanich et al., 1984; Richmond & Clemens, 1986a, b). Additionally, administration of estradiol increased muscarinic receptor binding in the POA and medial basal hypothalamus (Dohanich et al., 1982). Infusions of norepinephrine directly to the POA of hormone-primed females, however, reduced lordosis levels (Caldwell & Clemens, 1986). Similar cholinergic influences on male sexual behavior were reported in which blockade of muscarinic receptors in the POA decreased the number of males that displayed sexual behavior (Hull et al., 1988).

Having carefully explored the roles of gonadal hormones and the importance of neurotransmitter systems on sexual behavior, Clemens again extended his reach to investigate other implications of hormone exposure during adult life. In one line of research, his laboratory reported that ~40% of male mice of the B6D2F1 genotype continued to show ejaculatory reflexes as long as 25 weeks after castration (Clemens et al., 1988; Wee et al., 1988; Wee & Clemens, 1989). Follow-up studies indicated that these so-called "continuers" did not show any differences in serum hormone concentrations, aromatase activity, or estrogen receptor levels compared to "noncontinuers" or those males that stopped showing copulatory behavior after castration (Sinchak et al., 1996). The continuer males may have had differential exposure to steroid hormones during perinatal development, leading to adult variation in their responsiveness to circulating nongonadal hormones or other signaling molecules.

Finally, Clemens' laboratory also explored factors that account for individual differences in female sexual behavior. Although activational hormone exposure did alter female behavior (Yang & Clemens, 1997), other components such as the behavior of stimulus males could also affect the expression of female behavior (Yang & Clemens, 1996, 1997, 1998). Additionally, characterization of female behavior led to the hypothesis that female-female mounting should not simply be viewed as a male trait seen in masculinized females but might serve to establish and maintain a social hierarchy among females living in groups (Fang & Clemens, 1999).

Clemens' many and varied research contributions played an important role in setting the foundation for the emerging field of behavioral neuroendocrinology and created new paths for other scientists to follow and new questions for them to explore. Much of what was learned about the organizational and activational roles of gonadal hormones on mammalian behavior was inspired by Clemens' curiosity and ingenuity.

Advancing the Field Through Networking and Mentoring

Although Clemens' research contributions to the field of behavioral neuroendocrinology have been significant, he also had a remarkable influence on the social culture of the field. Clemens' desire to provide opportunities for networking and collaboration, as well as his strong emphasis on the importance of mentoring, shaped the lives of generations of trainees, both directly and indirectly.

Perhaps, his most impressive impact on the field was the founding of the *Eastern Regional Conference on Reproductive Behavior* in 1969, initially a small annual meeting designed to bring together scientists who were studying the physiological bases of reproductive behavior. The meeting expanded in popularity and membership over the years, gathering like-minded scientists from trainees to senior investigators, allowing for discourse ranging from the basic science of reproductive biology to application of knowledge to clinical populations. As the conference grew, the name changed, first becoming the *Conference on Reproductive Behavior* (CRB) and later the *Society for Behavioral Neuroendocrinology* (SBN). The opportunities for informal discussions brought about by the nature of the conference in its various reincarnations paved the way for behavioral endocrinologists to share their nascent ideas, advancing the study of reproductive psychobiology immeasurably.

Importantly, many of the collaboration opportunities at the conference were driven by Clemens' commitment to create a more inclusive social structure and his passion for mentorship as he added new activities to the conference program, such as "trainee" and "meet-the-professor" lunches. A tradition developed in which many graduate students would give their first presentation on their dissertation work at the meeting, providing an introduction to their future colleagues (Dewsbury, 2003).

Further evidence of Clemens' impact on the social culture of the field can be seen in his reorganization of the CRB's conference program upon its 25th anniversary meeting in 1993. Clemens and Tony Nunez – that year's meeting cohost and Clemens' long-time friend and research collaborator – completely revamped the organization and structure of the meeting by removing individual poster presentations in favor of 2- or 3-h symposia organized by a selection of the field's leading younger and more diverse scientists. Though some participants questioned the appropriateness of this new format, its success was undeniable. Indeed, the symposia structure has been adopted by a variety of organizations for their annual meetings, highlighting its effectiveness in engaging attendees.

To provide an additional avenue for increased interaction of behavioral neuroendocrinologists at a larger conference, Clemens organized social hours during the annual *Society for Neuroscience* conference beginning in 1982. These informal social gatherings were incredibly successful, allowing for – among other things – greater networking between the field's younger scientists with more established researchers. These socials were hosted by Clemens for many years, driven by his enjoyment for the informal discourse that would take place at the events, as well as his passion for increasing opportunities for the field's next generation of scientists. Now, organized socials of this nature are an integral piece of the program at the annual *Society for Neuroscience* meeting, spanning a wide range of area-specific neuroscience topics.

Finally, a discussion about Clemens' indelible impact on the field would not be complete without mention of his exceptional mentoring of students and postdoctoral fellows. Anyone sufficiently fortunate to have worked with Clemens knows that he not only set high standards for his students but also provided them with the room they require to make mistakes, learn, and grow. With over five decades of research success, Clemens contributed to the scientific careers of dozens of trainees, from undergraduate students to technicians, graduate students, postdocs, and collaborators. His impact continues to affect the field of behavioral endocrinology as those individuals who passed through his laboratory continue to shape the field, undoubtedly using lessons learned from Clemens in their own work in research, academia, government, and industry.

Conclusion

In acknowledgment of his profound and lasting influence on the field of hormones and behavior, Clemens was awarded the *Daniel S. Lehrman Lifetime Achievement Award* in Behavioral Endocrinology at the *Society for Behavioral Neuroendocrinology* annual meeting in 2009. After Clemens' passing in 2016, the *Society for Behavioral Neuroendocrinology* also established the *Lyn Clemens Award*, which recognizes a postdoctoral scholar who shows

Fig. 20.1 Lynwood G. Clemens

talent and promise for contributions to the field of behavioral endocrinology. An award supporting an up-and-coming investigator by providing the opportunity to deliver an invited address at the annual meeting is an ideal way to honor Clemens' place in the field.

Clemens' long and productive career is marked by milestones of advancements in knowledge regarding the neural and hormonal control of reproductive behavior. His contributions to the field of behavioral neuroendocrinology are not limited to published findings, however. The importance Clemens placed on the value of mentorship was evident in his actions both within and beyond his laboratory. The profound impact Clemens has had on the field of hormones and behavior has been felt for decades and will continue to live on in his work and those he influenced (Fig. 20.1).

Acknowledgments We would like to offer our special thanks to Dr. Gary Dohanich for his valuable contributions and constructive suggestions. We also wish to thank Dr. Tony Nunez and Dr. Linda Coniglio for sharing their stories and providing historical context for much of the work. We also must express our great appreciation to Dr. Lynwood Clemens for being an outstanding mentor and role model.

References

Beach, F. A. (1944). Effects of injury to the cerebral cortex upon sexually-receptive behavior in the female rat. *Psychosomatic Medicine, 6*, 40–55.

Beach, F. A., Westbrook, W. H., & Clemens, L. G. (1966). Comparisons of the ejaculatory response in men and animals. *Psychosomatic Medicine, 28*(5), 749–763. https://doi.org/10.1097/00006842-196609000-00006

Caldwell, J. D., & Clemens, L. G. (1986). Norepinephrine infusions into the medial preoptic area inhibit lordosis behavior. *Pharmacology Biochemistry & Behavior, 24*, 1015–1023.

Carter, C. S., Clemens, L. G., & Hoekema, D. J. (1972). Neonatal androgen and adult sexual behavior in the golden hamster. *Physiology & Behavior, 9*(7), 89–95.

Christensen, L. W., & Clemens, L. G. (1974). Intrahypothalamic implants of testosterone or estradiol and resumption of masculine sexual behavior in long-term castrated male rats. *Endocrinology, 95*(4), 984–990. https://doi.org/10.1210/endo-95-4-984

Christensen, L. W., & Clemens, L. G. (1975). Blockade of testosterone-induced mounting behavior in the male rat with intracranial application of the aromatization inhibitor, androst-l,4,6-triene-3,17-dione. *Endocrinology, 97*(6), 1545–1551.

Christensen, L. W., Coniglio, L. P., Paup, D. C., & Clemens, L. G. (1973). Sexual behavior of male golden hamsters receiving diverse androgen treatments. *Hormones and Behavior, 229*, 223–229.

Chung, Y.-W., Nunez, A. A., & Clemens, L. G. (2001). Effects of neonatal polychlorinated biphenyl exposure on female sexual behavior. *Physiology & Behavior, 74*, 363–370.

Clemens, L. G., & Dohanich, G. P. (1980). Inhibition of lordotic behavior in female rats following intracerebral infusion of anticholinergic agents. *Pharmacology Biochemistry & Behavior, 13*, 89–95.

Clemens, L. G., & Gladue, B. A. (1978). Feminine sexual behavior in rats enhanced by prenatal inhibition of androgen aromatization. *Hormones and Behavior, 11*, 190–201.

Clemens, L. G., Wallen, K., & Gorski, R. A. (1967). Mating behavior: Facilitation in the female rat after cortical application of potassium chloride. *Science, 157*(3793), 1208–1209. https://doi.org/10.1126/science.157.3793.1208

Clemens, L. G., Gladue, B. A., & Coniglio, L. P. (1978). Prenatal endogenous androgenic influences on masculine sexual behavior and genital morphology in male and female rats. *Hormones and Behavior, 10*, 40–53.

Clemens, L. G., Humphrys, R. R., & Dohanich, G. P. (1980). Cholinergic brain mechanisms and the hormonal regulation of female sexual behavior in the rat. *Pharmacology Biochemistry & Behavior, 13*, 81–88.

Clemens, L. G., Dohanich, G. P., & Witcher, J. A. (1981). Cholinergic influences on estrogen-dependent sexual behavior in female rats. *Journal of Comparative and Physiological Psychology, 95*, 763–770.

Clemens, L. G., Wee, B. E. F., Weaver, D. R., Roy, E. J., Goldman, B. D., & Rakerd, B. (1988). Retention of masculine sexual behavior following castration in male B6D2F1 mice. *Physiology & Behavior, 42*, 69–76.

Clemens, L. G., Barr, P. J., & Dohanich, G. P. (1989). Cholinergic regulation of female sexual behavior in rats demonstrated by manipulation of endogenous acetylcholine. *Physiology & Behavior, 45*, 437–442.

Coniglio, L. P., & Clemens, L. G. (1976). Period of maximal susceptibility to behavioral modification by testosterone in the golden hamster. *Hormones and Behavior, 7*, 267–282.

Coniglio, L. P., Paup, D. C., & Clemens, L. G. (1973a). Hormonal factors controlling the development of sexual behavior in the male golden hamster. *Physiology & Behavior, 10*(6), 1087–1094. https://doi.org/10.1016/0031-9384(73)90192-3

Coniglio, L. P., Paup, D. C., & Clemens, L. G. (1973b). Hormonal specificity in the suppression of sexual receptivity of the female golden hamster. *Journal of Endocrinology, 57*(1), 55–61. https://doi.org/10.1677/JOE.0.0570055

Cummings, J. A., Nunez, A. A., & Clemens, L. G. (2005). A cross-fostering analysis of the effects of PCB 77 on the maternal behavior of rats. *Physiology & Behavior, 85*, 83–91. https://doi.org/10.1016/j.physbeh.2005.04.001

Cummings, J. A., Clemens, L. G., & Nunez, A. A. (2008). Exposure to PCB 77 affects partner preference but not sexual behavior in the female rat. *Physiology & Behavior, 95*, 471–475. https://doi.org/10.1016/j.physbeh.2008.07.016

de Bold, J. F., & Clemens, L. G. (1978). Aromatization and the induction of male sexual behavior in male, female, and androgenized female hamsters. *Hormones and Behavior, 11*(3), 401–413. https://doi.org/10.1016/0018-506X(78)90040-5

Dewsbury, D. A. (2003). The conference on reproductive behavior: A history. In *Hormones and behavior* (Vol. 43, Issue 4, pp. 465–473). Academic Press. https://doi.org/10.1016/S0018-506X(03)00026-6

Dohanich, G. P., Witcher, J. A., Weaver, D. R., & Clemens, L. G. (1982). Alteration of muscarinic binding in specific brain areas following estrogen treatment. *Brain Research, 241*(2), 347–350.

Dohanich, G. P., Barr, P. J., Witcher, J. A., & Clemens, L. G. (1984). Pharmacological and anatomical aspects of cholinergic activation of female sexual behavior. *Physiology & Behavior, 32*, 1021–1026.

Doty, R. L., Carter, C. S., & Clemens, L. G. (1971). Olfactory control of sexual behavior in the male and early-androgenized female hamster. *Hormones and Behavior, 2*, 325–335.

Fang, J., & Clemens, L. G. (1999). Contextual determinants of female–female mounting in laboratory rats. *Animal Behaviour, 57*, 545–555. http://www.idealibrary.comon

Gladue, B. A., & Clemens, L. G. (1978). Androgenic influences on feminine sexual behavior in male and female rats: Defeminization blocked by prenatal antiandrogen treatment. *Endocrinology, 103*(5), 1702–1709.

Henley, C. L., Nunez, A. A., & Clemens, L. G. (2009). Estrogen treatment during development alters adult partner preference and reproductive behavior in female laboratory rats. *Hormones and Behavior, 55*(1), 68–75.

Henley, C. L., Nunez, A. A., & Clemens, L. G. (2010). Exogenous androgen during development alters adult partner preference and mating behavior in gonadally intact male rats. *Hormones and Behavior, 57*(4–5), 488–495. https://doi.org/10.1016/j.yhbeh.2010.02.007

Hull, E. M., Pehek, E. A., Bitran, D., Holmes, G. M., Warner, R. K., Band, L. C., Bazzett, T., & Clemens, L. G. (1988). Brain localization of cholinergic influence on male sex behavior in rats: Antagonists. *Pharmacology Biochemistry & Behavior, 31*, 175–178.

Paup, D. C., Coniglio, L. P., & Clemens, L. G. (1972). Masculinization of the female Golden hamster by neonatal treatment with androgen or estrogen. *Hormones and Behavior, 3*, 123–131. https://doi.org/10.1016/0018-506X(72)90014-1

Paup, D. C., Coniglio, L. P., & Clemens, L. G. (1974). Hormonal determinants in the development of masculine and feminine behavior in the female hamster. *Behavioral Biology, 10*(3), 353–363. https://doi.org/10.1016/S0091-6773(74)91944-0

Richmond, G., & Clemens, L. G. (1986a). Cholinergic mediation of feminine sexual receptivity: Demonstration of progesterone Independence using a progestin receptor antagonist. *Brain Research, 373*, 159–163.

Richmond, G., & Clemens, L. G. (1986b). Evidence for involvement of midbrain central gray in cholinergic mediation of female sexual receptivity in rats. *Behavioral Neuroscience, 100*(3), 376–380.

Simmons, S. L., Cummings, J. A., Clemens, L. G., & Nunez, A. A. (2005). Exposure to PCB 77 affects the maternal behavior of rats. *Physiology & Behavior, 84*, 81–86. https://doi.org/10.1016/j.physbeh.2004.10.022

Sinchak, K., Roselli, C. E., & Clemens, L. G. (1996). Levels of serum steroids, aromatase activity, and estrogen receptors in preoptic area, hypothalamus, and amygdala of B6D2F1 male house mice that differ in the display of copulatory behavior after castration. *Behavioral Neuroscience, 110*(3), 593–602. https://doi.org/10.1037/0735-7044.110.3.593

Wagner, C. K., & Clemens, L. G. (1989a). Anatomical organization of the sexually dimorphic perineal neuromuscular system in the house mouse. *Brain Research, 499*, 93–100.

Wagner, C. K., & Clemens, L. G. (1989b). Perinatal modification of a sexually dimorphic motor nucleus in the spinal cord of the B6D2F1 house mouse. *Physiology & Behavior, 45*, 831–835.
Wagner, C. K., Sisk, C. L., & Clemens, L. G. (1993). Neurons in the paraventricular nucleus of the hypothalamus that project to the sexually dimorphic lower lumbar spinal cord concentrate 3H-estradiol in the male rat. *Journal of Neuroendocrinology, 5*(5), 545–551. https://doi.org/10.1111/J.1365-2826.1993.TB00520.X
Wang, X. Q., Fang, J., Nunez, A. A., & Clemens, L. G. (2002). Developmental exposure to polychlorinated biphenyls affects sexual behavior of rats. *Physiology & Behavior, 75*, 689–696.
Wee, B. E. F., & Clemens, L. G. (1987). Characteristics of the spinal nucleus of the bulbocavernosus are influenced by genotype in the house mouse. *Brain Research, 424*, 305–310.
Wee, B. E. F., & Clemens, L. G. (1989). Environmental influences on masculine sexual behavior in mice. *Physiology & Behavior, 46*, 867–872.
Wee, B. E. F., Weaver, D. R., & Clemens, L. G. (1988). Hormonal restoration of masculine sexual behavior in long-term castrated B6D2F1 mice. *Physiology & Behavior, 42*, 77–82.
Witcher, J. A., & Clemens, L. G. (1987). A prenatal source for defeminization of female rats is the maternal ovary. *Hormones and Behavior, 21*, 36–43.
Yang, L.-Y., & Clemens, L. G. (1996). Relation of intromissions to the female's postejaculatory refractory period in rats. *Physiology & Behavior, 60*(6), 1505–1511. https://doi.org/10.1016/S0031-9384(96)00314-9
Yang, L., & Clemens, L. G. (1997). Function of intromissions on intromission-return latency of female rats during paced sexual behavior. *Physiology & Behavior, 61*(6), 889–894.
Yang, L.-Y., & Clemens, L. G. (1998). Influence of male-related stimuli on female postejaculatory refractory period in rats. *Physiology & Behavior, 63*(4), 675–682.

Bruce S. McEwen

21

Ilia N. Karatsoreos

Abstract

From his early work in deciphering how adrenal hormones can enter the brain and impact cellular function to groundbreaking findings uncovering the mechanisms by which steroid hormones affected neuronal structure, Bruce S. McEwen (1938–2020) made an indelible mark on the field of behavioral neuroendocrinology. His work helped to clarify how "good stress" and "bad stress" could affect both the brain and body, particularly the factors that led to allostatic load and overload which can have significant health consequences. In later years, McEwen made important contributions at the national and international levels, as a champion of appreciating the impact of stressful early childhood experiences on adult mental and physical health and understanding how socioeconomic status and related stressors could affect the brain and general health. Both a deep and broad thinker, applying tools from the level of molecules to the level of sociology, McEwen helped to shape modern behavioral neuroendocrinology.

Keywords

Steroid hormones · Stress · Sex differences · Brain · Plasticity

Early Life and Scientific Foundations

Bruce Sherman McEwen (1938–2020) was a true scientific "renaissance man," who made significant and lasting contributions in many areas of science, from basic cell biology, to neuroendocrinology, to neuroscience, and even to the social sciences. It

I. N. Karatsoreos (✉)
Neuroscience and Behavior Program, Center for Neuroendocrinology, Department of Psychological and Brain Sciences, University of Massachusetts, Amherst, MA, USA
e-mail: ikaratsoreos@umass.edu

R. J. Nelson, Z. M. Weil (eds.), *Biographical History of Behavioral Neuroendocrinology*, https://doi.org/10.1007/978-3-031-12970-4_21

is safe to say that his findings have laid the foundation for many aspects of modern neuroscience and endocrinology, with significant implications for mental and physical health, as well as policy. This chapter mostly highlights his work in the realm of behavioral neuroendocrinology, as a full accounting of his accomplishments in all areas of neuroscience would likely fill a volume of their own.

Born in Fort Collins, Colorado, in the winter of 1938, McEwen would become a giant in several different fields of scientific inquiry. After graduating from University High School in Ann Arbor, Michigan, he obtained his undergraduate degree in chemistry at Oberlin College, in Oberlin, Ohio. From there, McEwen pursued his graduate work in cell biology in the lab at Alfred Mirsky at the Rockefeller University (then Institute) in New York City. As described by McEwen, while working with Mirsky and Vincent Allfrey, he was exposed to their interests in gene expression. McEwen particularly found Mirsky's insistence that gene X environment interactions were critical in understanding biology formative in his own thinking as he developed as a biologist, though he did not support Mirsky's past attempts to discredit Oswald Avery's work on DNA and heredity (personal communication). It is fitting that in 1999, he was named the Alfred E. Mirsky Professor at the Rockefeller, likely in part due to the decades of work aimed at understanding the long-term impact of life experiences on the structure and function of the brain. This intellectual exposure and the mastering of fundamental biochemical and cell biology methods were both central in the work to follow. Indeed, it is sometimes surprising to those familiar with the work of McEwen in adrenal and gonadal hormones to learn that his training was that of a biochemist and that his early work was focused on understanding the intricate biochemical machinery by which cells made and transported proteins.

After brief forays in Goteborg Sweden as a USPHS NIH postdoctoral fellow and then as an assistant professor of Zoology at the University of Minnesota, McEwen returned to Rockefeller but this time as an assistant professor working alongside Dr. Neal Miller. It is during this period in the late 1960s that McEwen made some of his most substantial findings; findings that would determine the trajectory of not only his research group, but of entire fields.

The work of Berthold launched the field of behavioral neuroendocrinology, demonstrating that "gonadal secretions" were responsible for a great many of the physical, physiological, and behavioral traits observed in a variety of species. Eventually, the steroid hormones responsible were identified and isolated, including the isolation of the adrenal steroidal compounds A, B, C, and D (B being corticosterone) by Kendall in the 1930s (Simoni et al., 2002). However, it is also safe to say that until the work of McEwen and his coworkers, the mechanism by which these steroid hormones could act to change the structure and function of the brain, or indeed that actually even enter the brain from the periphery (a result taken as fact in the present day), was a complete mystery.

Understanding How Adrenal and Gonadal Hormones Can Impact the Brain

Eisenfeld and Axelrod provided compelling evidence that estradiol from the periphery could bind to several brain sites (Eisenfeld, 1969). Using similar approaches in 1968, McEwen and coworkers Jay Weiss and Leslie Schwartz were able to document that peripherally delivered radiolabeled corticosterone could be detected in several brain regions including the hypothalamus, cortex, amygdala, and hippocampus, with particularly high affinity (McEwen et al., 1968). They went further to document the comparative timecourse of this radioactivity in the blood and the different brain regions, which gradually reduced over the course of 6 h. This gave insight to the potential dynamics of stress-induced release of adrenal steroids, and the potential temporal windows for their acute effects on brain function. Though limited to a characterization of the location and timecourse of corticosterone in the brain, this work set the stage for understanding the potential mechanisms by which it could act. Indeed, the discussion of their landmark 1968 *Nature* paper presciently noted, "This control might operate at the genetic level, for there is evidence for the accumulation and action of steroid hormones in cell nuclei of target tissues" (McEwen et al., 1968). In 1969 and 1970, bringing to bear the biochemical techniques he and his lab had been honing since the early 1960s, the McEwen, Schwartz, and Weiss trio published two key additional findings. The first explored differences between the hippocampus and the septum and how hydrocortisone and dexamethasone competed (McEwen et al., 1969). Extending this work, by using a method to highly purify cell nuclei from tissues, they were able to demonstrate that in the hippocampus, over 36% of peripherally injected radiolabeled corticosterone was concentrated in cell nuclei (McEwen et al., 1970). This finally set the stage for work that McEwen did with Linda Plapinger, which identified the interactions of corticosterone with macromolecules extracted from nuclei of brain tissues (Mcewen & Plapinger, 1970). Combined with work of other groups around the same time, McEwen and Plapinger made the observation that "From these results and those from other laboratories on nuclear hormone binding factors, it seems highly probable that the binding substance is a protein of a type which is believed to have a regulatory function in the cell nucleus over RNA and subsequent protein synthesis" (Mcewen & Plapinger, 1970). This protein type would eventually be revealed as the glucocorticoid receptor and lead to the discovery of the mechanisms of genomic actions of steroid hormones.

Of course, although part of a wider set of findings by researchers working on similar problems, these findings heralded the rapid development of the field interested in how adrenal "stress" hormones could affect cellular function in the brain. McEwen and his team went on to use autoradiographic techniques to demonstrate binding of corticosterone to specific cell populations, as well as particular cellular compartments (Gerlach & McEwen, 1972). His groundbreaking work on corticosterone in the limbic system not only provided the foundation for understanding how stress can affect brain and behavior (for good or for ill) but helped to launch our understanding of the fundamental mechanisms by which steroid hormones could have impacts on gene transcription and eventually to the genomic vs. non-genomic dichotomy of steroid hormone effects more generally.

Though groundbreaking, the work of McEwen and his team on adrenal hormones and their effects in the brain were not the only major contributions he made to the fields of endocrinology, behavioral neuroendocrinology, and neuroscience. McEwen had a long-standing scientific partnership with Donald Pfaff, a fellow Rockefeller investigator, that started in the early 1970s and lasted for nearly 50 years. Both in his own lab and in collaboration with Pfaff, McEwen and his team made foundational discoveries about the effects of estrogens on the brain. Using similar approaches as had been employed with corticosterone, McEwen and his collaborators made significant findings on the concentration of estradiol in the brain (McEwen & Pfaff, 1970; McEwen et al., 1970a, b; Zigmond & McEwen, 1970), its subcellular localization (Plapinger & McEwen, 1973), and its role in sexual behavior and sexual differentiation of the nervous system (McEwen et al., 1975), to name but a few. Given the renewed interest in the importance of understanding sex as a biological variable in biomedical sciences, revisiting and understanding these discoveries is no less important now as it was when they were originally documented.

The Impact of the Environment on the Brain

A theme that winds its way through McEwen's work is an interest in understanding the bidirectional interaction of the brain and body. He was particularly interested in the complexity of these interactions beyond the classic endocrine feedback models. Indeed, when McEwen started his work in this area, the role of the steroid hormones on the brain was mostly considered as the mechanisms by which they could regulate their own secretion. However, subsequent decades of work, largely spearheaded by McEwen and his ever-growing family tree of trainees, would provide a detailed picture of how the steroid hormones could fundamentally alter brain circuits, and the molecular underpinnings of complex behaviors, beyond the hypothalamus.

The major area of work that McEwen is perhaps best known for is understanding how stress, as first articulated by Hans Selye (Selye, 1936), can have both good (i.e., eustress) and bad (i.e., distress) effects on brain, behavior, and general physiology. McEwen's lab helped to pioneer the understanding of how chronic stress could lead to atrophy of the cells of the hippocampus (Sapolsky et al., 1985a; Sapolsky, 1992) with significant impacts on behavior. Work by graduate students and fellows in the McEwen lab helped to characterize the role of glucocorticoids in the brain and their development over time (Meaney et al., 1985a, b; Sapolsky et al., 1985b). In sum, McEwen and his lab helped to establish that adrenal hormones could contribute to the negative outcomes observed in aging, ischemia, and stress (Sapolsky & Pulsinelli, 1985; Sapolsky et al., 1986; Sapolsky, 1999) but that glucocorticoids could affect neural plasticity in the hippocampus and also affect the process of hippocampal neurogenesis (which was a novel finding in its own right) (McEwen et al., 1991; Gould et al., 1991, 1997; Watanabe et al., 1992; Cameron et al., 1998). Work from his group led to the notion that chronic stress could leave a significant fingerprint on neural structure, even following recovery from stress, and that early life experiences could significantly impact hippocampal plasticity and aging; effects also observed in humans (Lupien et al., 1998, 1999, 2000, 2009).

The brain-body interactions in which McEwen was always so interested are captured in his work on "allostatic load," a term that he developed with Elliot Stellar (McEwen & Stellar, 1993) and based on the ideas of Sterling and Eyer with respect to blood pressure (Sterling & Eyer, 1988). This concept was built upon work from his own lab, and many others, that repeated activation of the physiologic systems that respond to stressors in the environment, their inappropriate termination, or their failure to engage properly could lead to "wear and tear" on the brain and body (McEwen, 1998; McEwen & Wingfield, 2003; Karatsoreos & McEwen, 2011). This could eventually lead to allostatic overload, a situation in which the body's capacity to respond to additional stressors is not only exhausted but actually starts to cause significant negative effects. This includes increased inflammation and altered immune responses, as well as negative cardiometabolic effects, and neurobehavioral function. For a far more thorough accounting of these contributions, readers are pointed to reviews published by McEwen and his laboratory on the neurobiology of stress (McEwen, 2005; McEwen et al., 2015) as well as a popular science book (McEwen & Lasley, 2002). That chronic stress could leave an indelible mark on future behaviors and capacity to respond to stressors in the environment remains a cornerstone of our understanding of the links between neuropsychiatric disease and chronic stress. It also set the stage for his contributions as an advocate and scientific champion on understanding how early life stress and socioeconomic status can affect the developing child.

Long-Term Consequences of Early Childhood Experiences

In more recent years, as part of a growing professional partnership with his brother, sociologist Craig McEwen, McEwen became very involved in national and international efforts that bridged his fundamental findings in neuroendocrinology and the consequences of stress and development on a societal level (McEwen & McEwen, 2017). His work with the MacArthur Foundation Research Network on Socioeconomic Status and Health and the National Council on the Developing Child helped to further the ideas of "biological embedding" that adverse child experiences could lead leave a molecular and physiological fingerprint, significantly affecting these individuals in adulthood, as well as the cumulative costs of stress (Lupien et al., 2000; Seeman et al., 2001; Shonkoff et al., 2009; McEwen & Gianaros, 2010; McEwen, 2012; McEwen & McEwen, 2017). He was a major supporter for continued work to understand the biological consequences of early life experiences and how this could affect the trajectory of development. It is fitting that McEwen's scientific contributions, particularly in adrenal and gonadal hormone biology, can be imagined as their own "feedback loop," where environmental challenges lead to changes in the brain and behavior, which can then go on to further impact individuals later in life and how they cope with subsequent psychological and physical stressors.

Conclusions

McEwen was that rare combination of both a broad and a deep thinker, whose creativity and persistence were rivaled only by his generosity. The work of McEwen and his collaborators helped to establish the "brain as an endocrine organ" (McEwen, 1974), and his work articulating the central role of the brain as both a driver and target of stress responses (McEwen, 2006, 2007) is a cornerstone of modern stress research. He trained countless scientists who have gone on to become major thought leaders in their own right – a fact documented by the massive "scientific family tree" that was a gift from the lab to McEwen on the occasion of his 70th birthday, a multi-square meter record that quite literally filled an entire wall in the laboratory. Importantly, McEwen was a firm believer in both top-down and bottom-up approaches in science, and he was particularly wary of becoming wedded to hypotheses or possessing illusory knowledge. The door to his office bore a quote popularized by the historian Daniel J. Boorstin, "The greatest obstacle to discovery is not ignorance—it is the illusion of knowledge." McEwen clearly embraced this philosophy in his work and instilled it in the extensive network of trainees who walked by the quote every time they visited to discuss data, even when they returned decades later as established investigators. The work of McEwen's laboratory was not only critical in the development of the field of behavioral neuroendocrinology but also had significant consequences for our understanding of the two-way street that is brain-body interactions.

Acknowledgments The author would like to thank Bruce for being a one-of-a-kind scientist and mentor to dozens of graduate students and postdoctoral fellows. It was impossible to highlight the contributions of all of his mentees throughout the years given the constrained nature of this chapter.

References

Cameron, H. A., Tanapat, P., & Gould, E. (1998). Adrenal steroids and N-methyl-D-aspartate receptor activation regulate neurogenesis in the dentate gyrus of adult rats through a common pathway. *Neuroscience, 82*, 349–354. S0306452297003035 [pii].

Eisenfeld, A. J. (1969). Hypothalamic oestradiol-binding macromolecules. *Nature, 224*, 1202–1203. https://doi.org/10.1038/2241202a0

Gerlach, J. L., & McEwen, B. S. (1972). Rat brain binds adrenal steroid hormone: Radioautography of hippocampus with corticosterone. *Science, 175*, 1133–1136. https://doi.org/10.1126/science.175.4026.1133

Gould, E., Woolley, C. S., & McEwen, B. S. (1991). Adrenal steroids regulate postnatal development of the rat dentate gyrus: I. Effects of glucocorticoids on cell death. *The Journal of Comparative Neurology, 313*, 479–485. https://doi.org/10.1002/cne.903130308

Gould, E., McEwen, B. S., Tanapat, P., et al. (1997). Neurogenesis in the dentate gyrus of the adult tree shrew is regulated by psychosocial stress and NMDA receptor activation. *The Journal of Neuroscience, 17*, 2492–2498.

Karatsoreos, I. N., & McEwen, B. S. (2011). Psychobiological allostasis: Resistance, resilience and vulnerability. *Trends in Cognitive Sciences, 15*, 576–584. https://doi.org/10.1016/j.tics.2011.10.005

Lupien, S. J., de Leon, M., de Santi, S., et al. (1998). Cortisol levels during human aging predict hippocampal atrophy and memory deficits. *Nature Neuroscience, 1*, 69–73. https://doi.org/10.1038/271

Lupien, S. J., Nair, N. P., Brière, S., et al. (1999). Increased cortisol levels and impaired cognition in human aging: Implication for depression and dementia in later life. *Reviews in the Neurosciences, 10*, 117–139. https://doi.org/10.1515/revneuro.1999.10.2.117

Lupien, S. J., King, S., Meaney, M. J., & McEwen, B. S. (2000). Child's stress hormone levels correlate with mother's socioeconomic status and depressive state. *Biological Psychiatry, 48*, 976–980. https://doi.org/10.1016/s0006-3223(00)00965-3

Lupien, S. J., McEwen, B. S., Gunnar, M. R., & Heim, C. (2009). Effects of stress throughout the lifespan on the brain, behaviour and cognition. *Nature Reviews. Neuroscience, 10*, 434–445. https://doi.org/10.1038/nrn2639

McEwen, B. S. (1974). Why the brain is truly an endocrine organ. In L. Mastroianni (Ed.), *Clinician* (pp. 13–21). Springer.

McEwen, B. S. (1998). Stress, adaptation, and disease. Allostasis and allostatic load. *Annals of the New York Academy of Sciences, 840*, 33–44.

McEwen, B. S. (2005). Glucocorticoids, depression, and mood disorders: Structural remodeling in the brain. *Metabolism, 54*, 20–23.

McEwen, B. S. (2006). Protective and damaging effects of stress mediators: Central role of the brain. *Dialogues in Clinical Neuroscience, 8*, 367–381.

McEwen, B. S. (2007). Physiology and neurobiology of stress and adaptation: Central role of the brain. *Physiological Reviews, 87*, 873–904. 87/3/873 [pii] 10.1152/physrev.00041.2006.

McEwen, B. S. (2012). Brain on stress: How the social environment gets under the skin. *Proceedings of the National Academy of Sciences of the United States of America, 109*(Suppl 2), 17180–17185. https://doi.org/10.1073/pnas.1121254109

McEwen, B. S., & Gianaros, P. J. (2010). Central role of the brain in stress and adaptation: Links to socioeconomic status, health, and disease. *Annals of the New York Academy of Sciences, 1186*, 190–222. NYAS5331 [pii] 10.1111/j.1749-6632.2009.05331.x.

McEwen, B. S., & Lasley, E. N. (2002). *The end of stress as we know it*. National Academies Press.

McEwen, C., & McEwen, B. S. (2017). *Social structure, adversity, toxic stress, and intergenerational poverty: An early childhood model*. Social Science Research Network.

McEwen, B. S., & Pfaff, D. W. (1970). Factors influencing sex hormone uptake by rat brain regions. I. Effects of neonatal treatment, hypophysectomy, and competing steroid on estradiol uptake. *Brain Research, 21*, 1–16. https://doi.org/10.1016/0006-8993(70)90016-8

Mcewen, B. S., & Plapinger, L. (1970). Association of 3H corticosterone-1,2 with macromolecules extracted from brain cell nuclei. *Nature, 226*, 263–265. https://doi.org/10.1038/226263a0

McEwen, B. S., & Stellar, E. (1993). Stress and the individual: Mechanisms leading to disease. *Archives of Internal Medicine, 153*, 2093–2101.

McEwen, B. S., & Wingfield, J. C. (2003). The concept of allostasis in biology and biomedicine. *Hormones and Behavior, 43*, 2–15. S0018506X02000247 [pii].

McEwen, B. S., Weiss, J. M., & Schwartz, L. S. (1968). Selective retention of corticosterone by limbic structures in rat brain. *Nature, 220*, 911–912.

McEwen, B. S., Weiss, J. M., & Schwartz, L. S. (1969). Uptake of corticosterone by rat brain and its concentration by certain limbic structures. *Brain Research, 16*, 227–241. https://doi.org/10.1016/0006-8993(69)90096-1

McEwen, B. S., Weiss, J. M., & Schwartz, L. S. (1970). Retention of corticosterone by cell nuclei from brain regions of adrenalectomized rats. *Brain Research, 17*, 471–482. https://doi.org/10.1016/0006-8993(70)90254-4

McEwen, B. S., Pfaff, D. W., & Zigmond, R. E. (1970a). Factors influencing sex hormone uptake by rat brain regions. II. Effects of neonatal treatment and hypophysectomy on testosterone uptake. *Brain Research, 21*, 17–28. https://doi.org/10.1016/0006-8993(70)90017-x

McEwen, B. S., Pfaff, D. W., & Zigmond, R. E. (1970b). Factors influencing sex hormone uptake by rat brain regions. 3. Effects of competing steroids on testosterone uptake. *Brain Research, 21*, 29–38. https://doi.org/10.1016/0006-8993(70)90018-1

McEwen, B. S., Pfaff, D. W., Chaptal, C., & Luine, V. N. (1975). Brain cell nuclear retention of [3H]estradiol doses able to promote lordosis: Temporal and regional aspects. *Brain Research, 86*, 155–161. https://doi.org/10.1016/0006-8993(75)90649-6

McEwen, B. S., Coirini, H., Westlind-Danielsson, A., et al. (1991). Steroid hormones as mediators of neural plasticity. *The Journal of Steroid Biochemistry and Molecular Biology, 39*, 223–232.

McEwen, B. S., Bowles, N. P., Gray, J. D., et al. (2015). Mechanisms of stress in the brain. *Nature Neuroscience, 18*, 1353–1363. https://doi.org/10.1038/nn.4086

Meaney, M. J., Sapolsky, R. M., & McEwen, B. S. (1985a). The development of the glucocorticoid receptor system in the rat limbic brain. I. Ontogeny and autoregulation. *Developmental Brain Research, 18*, 159–164.

Meaney, M. J., Sapolsky, R. M., & McEwen, B. S. (1985b). The development of the glucocorticoid receptor system in the rat limbic brain. II. An autoradiographic study. *Brain Research, 350*, 165–168. https://doi.org/10.1016/0165-3806(85)90260-3

Plapinger, L., & McEwen, B. S. (1973). Ontogeny of estradiol-binding sites in rat brain. I. Appearance of presumptive adult receptors in cytosol and nuclei. *Endocrinology, 93*, 1119–1128. https://doi.org/10.1210/endo-93-5-1119

Sapolsky, R. M. (1992). *Stress, the aging brain and the mechanisms of neuron death*. MIT Press.

Sapolsky, R. M. (1999). Glucocorticoids, stress, and their adverse neurological effects: Relevance to aging. *Experimental Gerontology, 34*, 721–732.

Sapolsky, R. M., & Pulsinelli, W. A. (1985). Glucocorticoids potentiate ischemic injury to neurons: Therapeutic implications. *Science, 229*, 1397–1400.

Sapolsky, R. M., Krey, L. C., & McEwen, B. S. (1985a). Prolonged glucocorticoid exposure reduces hippocampal neuron number: Implications for aging. *The Journal of Neuroscience, 5*, 1222–1227.

Sapolsky, R. M., Meaney, M. J., & McEwen, B. S. (1985b). The development of the glucocorticoid receptor system in the rat limbic brain. III. Negative-feedback regulation. *Brain Research, 350*, 169–173. https://doi.org/10.1016/0165-3806(85)90261-5

Sapolsky, R. M., Krey, L. C., & McEwen, B. S. (1986). The neuroendocrinology of stress and aging: The glucocorticoid cascade hypothesis. *Endocrine Reviews, 7*, 284–301. https://doi.org/10.1210/edrv-7-3-284

Seeman, T. E., McEwen, B. S., Rowe, J. W., & Singer, B. H. (2001). Allostatic load as a marker of cumulative biological risk: MacArthur studies of successful aging. *Proceedings of the National Academy of Sciences of the United States of America, 98*, 4770–4775. https://doi.org/10.1073/pnas.081072698

Selye, H. (1936). A syndrome produced by diverse nocuous agents. *Nature, 138*, 32–32. https://doi.org/10.1038/138032a0

Shonkoff, J. P., Boyce, W. T., & McEwen, B. S. (2009). Neuroscience, molecular biology, and the childhood roots of health disparities: Building a new framework for health promotion and disease prevention. *JAMA, 301*, 2252–2259. https://doi.org/10.1001/jama.2009.754

Simoni, R. D., Hill, R. L., & Vaughan, M. (2002). The isolation of thyroxine and cortisone: The work of Edward C. Kendall. *Journal of Biological Chemistry, 277*, 21–22. https://doi.org/10.1016/S0021-9258(20)85219-3

Sterling, P., & Eyer, J. (1988). Allostasis: A new paradigm to explain arousal pathology. In S. Fisher & J. Reason (Eds.), *Handbook of life stress, cognition and health* (pp. 629–649). Wiley.

Watanabe, Y., Gould, E., & McEwen, B. S. (1992). Stress induces atrophy of apical dendrites of hippocampal CA3 pyramidal neurons. *Brain Research, 588*, 341–345. 0006-8993(92)91597-8 [pii].

Zigmond, R. E., & McEwen, B. S. (1970). Selective retention of oestradiol by cell nuclei in specific brain regions of the ovariectomized rat. *Journal of Neurochemistry, 17*, 889–899. https://doi.org/10.1111/j.1471-4159.1970.tb02242.x

Donald W. Pfaff

22

Zachary M. Weil

Abstract

Donald Wells Pfaff is a pioneering neuroscientist who has spent his career elucidating the neural and neuroendocrine mechanisms that regulate behavior. He was trained at Harvard and MIT (Massachusetts Institute of Technology) before becoming a faculty member at Rockefeller University in New York where he spent his entire career. Beginning as a graduate student, Pfaff used steroid receptor autoradiography to uncover the binding sites of estrogens in the brain. From those early data, he extensively mapped the neural circuitry responsible for female mating posture, lordosis, elucidating the sensory, neuroanatomical, neuroendocrine, and molecular events that were required for the expression of this critical behavioral outcome. He also made seminal discoveries on gonadotropin-releasing hormone ontogeny and function. Pfaff's productivity and scientific and popular writing have had a significant impact on the field.

Keywords

Neuroendocrinology · Lordosis · Behavior · Circuitry

What do you call a woman who uses the rhythm method for birth control???? Mom! –Don Pfaff c. 2009.

The memorable aspect of the telling of this old joke was not so much that a scientist who had done more than anyone to uncover the neural and endocrine mechanisms of sexual behavior was telling it, but rather that he couldn't get through it without repeatedly bursting into uncontrolled (and apparently uncontrollable) laughter.

Z. M. Weil (✉)
Department of Neuroscience, Rockefeller Neuroscience Institute, West Virginia University, Morgantown, WV, USA
e-mail: Zachary.weil@hsc.wvu.edu

R. J. Nelson, Z. M. Weil (eds.), *Biographical History of Behavioral Neuroendocrinology*, https://doi.org/10.1007/978-3-031-12970-4_22

Donald Wells Pfaff was born on 9 December 1939 in Rochester, New York, and was a standout student, serving among other things as an eighth grade class president, sports editor of his high school newspaper, and a National Honor Society senior standard bearer. Although always interested in the biological sciences, Pfaff studied math and physics at Harvard and graduated *magna cum laude* in 1961. He had narrowly decided against enrolling in medical school and becoming a surgeon like his father and decided to pursue graduate training. One factor that influenced that decision was his experience volunteering in the Massachusetts state mental hospital and his feeling that psychiatric medicine in the early 1960s badly trailed other medical disciplines. Pfaff then entered the Brain and Cognitive Sciences program at MIT after being assured that he could still enroll in classes at the Harvard Medical School.

At MIT, Pfaff joined Joseph Altman's lab. Altman was a relatively junior investigator having joined the faculty at MIT only a few years earlier but was in the midst of publishing a tremendous series of papers describing adult neurogenesis (Altman, 1962a, b; Altman & Das, 1965, 1967). Altman had been using the then relatively new tool, tritiated thymidine autoradiography, to show strong evidence that new cells were being born in the olfactory bulb, dentate gyrus, and cerebral cortex of the adult rat and cat. Although these papers were published in high-profile journals, it took decades to overcome the dogma that new neurons are not born in adulthood (Gross, 2000). In fact, it would take the work of Pfaff's future Rockefeller colleague Fernando Nottebohm to overcome this barrier decades later (Goldman & Nottebohm, 1983).

Pfaff, never one to simply go with the flow, mentioned to Altman that he would be interested in pursuing a project with more direct links to behavior. In a response that Pfaff would recall verbatim (it was only seven words long!) to generations of his trainees and would launch a research program that made fundamental discoveries about the role of the neuroendocrine system in the control of the brain and behavior, Altman replied "I'd like to do something with hormones."

At the time, the dependence of female mating behavior, and in particular the control of lordosis posture, had been shown to be estrogen-dependent in ovariectomized rodents (Boling & Blandau, 1939; Beach, 1948). Moreover, intracranial implants of estrogens had suggested that the basal hypothalamus was a site necessary and sufficient for the activation of lordosis behavior (Dörner et al., 1968; Lisk, 1962). However, there were many fundamental questions regarding the regulation of this behavior that remained unspecified. Although behavioral circuitry (sensory inputs, central processing, and motor outflows) was being mapped, for instance, by fellow New Yorker Eric Kandel (e.g., Kandel and Tauc (1964)), this was being done in a simple model system, the sea slug *Aplysia californicus*. Pfaff saw the opportunity to take the inverse approach, that is, to thoroughly investigate a simple behavioral process in a relatively complex organism, laboratory rats.

Pfaff spent much of the next 4 years optimizing the conditions for steroid autoradiography. The technique was challenging because the specific activity of the tritiated steroids was low, and it was difficult to keep the bound steroids in place for the extended time necessary to image them. This work, it is worth noting, required a

9-month exposure of radioligand-treated brains to the emulsion! Despite these challenges, Pfaff was able to map out the localization of estrogen binding sites and noted that there were particularly strong sites in the preoptic area, hypothalamus, and limbic structures and in the midbrain central gray (Pfaff, 1968a, b; Pfaff & Keiner, 1973). Notable from this early work was that the ventromedial hypothalamus (VMH) exhibited a particularly strong accumulation of the radiolabeled estrogen while the cells which would later be identified as gonadotropin-releasing hormone neurons did not. On the strength of these and other early findings, Pfaff earned his PhD in 1965 and after attending the Woods Hole course in electrophysiology arrived at Rockefeller University where he would remain for the rest of his long and fruitful career.

Rockefeller in the mid-1960s was in the process of transitioning from a private research institution into a degree-granting university. To that end, the institute's president, Detlev Bronk, had recruited Carl Pfaffmann from Brown as a faculty member and vice president and charged him with developing the biobehavioral group. Pfaffmann was successful in recruiting accomplished scientists from high-profile institutions including Neal Miller, William Estes, George Miller, Peter Marler, and Donald Griffin (Riggs, 1997). Moreover, this group was also attracting younger talent, including Pfaff and his future collaborator Bruce McEwen.

Pfaff would join Pfaffmann's laboratory as an NSF-funded postdoc but quickly earned a faculty appointment and by 1978 was a full professor and the head of the Laboratory of Behavioral Neuroscience. In those early years, Pfaff, often in collaboration with the nearby McEwen lab, worked to fully define the circuitry underlying female mating behavior as well as more generally to understand the role of steroid hormone regulation of the brain and behavior (McEwen et al., 1979). Pfaff's work demonstrated that estrogen treatment of the ventromedial hypothalamus was both necessary and sufficient for the induction of lordosis behavior (Davis et al., 1979; Pfaff, 1979; Meisel & Pfaff, 1984). He took the position that to consider the circuitry "mapped" that the entire system would have to be interrogated, from the sensory inputs that initiate the response through the central processing and coordination with internal endocrine states and the motor outputs (Pfaff, 1979).

It was critical that if more general rules of neuroendocrine coordination were to be drawn from this work, Pfaff utilize a comparative approach and as he put it examine the estrogen accumulating cells from "fish to philosopher" (Pfaff, 1979; Morrell et al., 1975a). To that end, much work in the 1970s involved extending his initial rat findings into species across taxa including songbirds, fish, nonhuman primates, and reptiles (Morrell et al., 1975b; Davis et al., 1977; Morrell et al., 1979).

Over the ensuing years, the Pfaff's lab methodically used his original data on estrogen accumulation as well as subsequent biochemical and tract tracing techniques to delineate the anatomical system that controlled lordosis both up and down the neuraxis. Over time, he would identify a series of neuronal "modules," in the spinal cord, lower brain stem, midbrain, hypothalamus, and forebrain. These structures were synaptically interconnected, their activity was enhanced by estrogen (and progesterone) stimulation, and together they facilitated the induction of lordosis in response to appropriate sensory stimulation (Pfaff, 1979, 1999).

The Pfaff's lab, spearheaded by his longtime collaborator Lee-Ming Kow (who joined the lab in the 1970s and was still there four decades later!), also utilized electrophysiological techniques to detect the effects of estrogens and various neurotransmitter systems on the activity of cells in the lordosis circuitry. Particularly notable among these findings was the discovery that estrogens expanded the receptive fields of cutaneous receptors that innervated the flanks of female rodents (Kow & Pfaff, 1973). Thus, the pressure induced by a male mounting and placing his paws on her back was significantly more likely to induce lordosis when estrogen concentrations were high (Kow et al., 1976).

There was also mounting evidence that estrogenic induction of lordosis behavior required new gene expression. In the late 1970s and early 1980s, the laboratory reported that estrogens could induce new proteins in the VMH and that blockade of protein synthesis in that hypothalamic structure could abrogate lordosis (Mobbs et al., 1988; Parsons et al., 1982). In the ensuing years, Pfaff hunted down numerous endocrine and neurotransmitter systems that were both upregulated by estrogen and facilitated lordosis behavior, including progesterone receptors, adrenergic and muscarinic receptors, oxytocin and its cognate receptor, thyroid hormones, enkephalins, and many more (summarized in Kow et al., 2016, Pfaff, 2017a). As the field evolved, Pfaff was able to keep step with the emerging technology and continued to use his combination of anatomical, biochemical, electrophysiological, and now molecular biological techniques to drill down into the interlocking systems underlying female reproductive behavior.

Pfaff was quite proud of his finding in the early 1970s that gonadotropin-releasing hormone (GnRH, then called luteinizing hormone factor [LRF]) could facilitate the induction of estrogen-primed lordosis behavior even in the absence of pituitary hormones (Pfaff, 1973). His pride, in retrospect, emerged not only from the novel finding or that it was published in *Science* but that he was able to conceive the experimental approach one evening, order the animals the next morning, and conduct the experiment the next week (Pfaff, 2017b). That story was always available whenever regulatory processes delayed experiments. Pfaff's lab later demonstrated that LRF/GnRH infused into the midbrain central gray could facilitate lordosis while neutralizing antibodies against LRF could abolish it (Sakuma & Pfaff, 1980; Sakuma & Pfaff, 1983). Thus, this provided an additional way in which gonadal status could be reflected in behavioral outputs.

Pfaff's experience with Altman and neuronal migration in graduate school helped him make a fundamental discovery regarding GnRH neuron biology. Marlene Schwanzel-Fukuda, a postdoc in the Pfaff's lab, was studying GnRH neurons that were found in the nervus terminalis, part of the accessory olfactory system. During development, GnRH immunoreactivity is detectable outside of the CNS (central nervous system) before it is apparent in the brain. Schwanzel-Fukuda and Pfaff therefore were examining GnRH immunoreactivity in fetal mice across development when Pfaff realized that these neurons appeared to be migrating (Schwanzel-Fukuda & Pfaff, 1989). Indeed, they would later show that the GnRH neurons are born in the olfactory placode before migrating into the brain during development.

Failure of this process results in hypogonadotropic hypogonadism (Kallmann's syndrome). Similar ontogenesis of GnRH neurons occurs across taxa.

Throughout his long career, Pfaff has accomplished feats that are unique and uniquely challenging for academic scientists. He has remained focused on the reproductive behavior and circuitry that has dominated his laboratories' activities for more than half a century. Yet as science has evolved over that time, Pfaff has always been at the forefront, eager to use the new tools to address the old problems. He was quick to adopt viral vectors, transgenic mice, stem cell technology, single cell genomics, and much more (Kaplitt et al., 1994; Martin et al., 2011; Magariños et al., 2018; Geary et al., 2001). Additionally, this focus has not deterred him from making discoveries beyond the lordosis circuitry as evidenced by his remarkable discovery of the ontogeny of the GnRH neurons and later forays into brain arousal, among others. Finally, Pfaff was singular in that even as a world-class scientist with a remarkable number of issues competing for his time, he still made it a priority (and a time when we as members of his laboratory were told to try hard not to bother him!) to block out one day a week where he could run his own electrophysiological experiments.

The scope of accomplishments achieved by Pfaff is truly remarkable. As of this writing (in 2021), Pfaff has published over 960 scientific papers and more than 35 books! His restless intellect and tremendous discipline have allowed him to run a superbly productive research laboratory during the day but also extend his ideas far beyond traditional neurobiological questions. In recent years, he has written books on human socialization (Pfaff, 2020), altruism (Pfaff & Sherman, 2015), fairness (Pfaff, 2007), and my personal favorite, robots (Pfaff, 2015)! One particularly worthy accomplishment in recent years is the publication of an extensive neuroscience textbook (edited along with Nora Volkow) that has been made available free to students in developing countries (Pfaff & Volkow, 2016). Pfaff was continuously NIH funded for more than 40 years and even managed to hold an active R01 from 1969 to 2013. He was elected to the American Academy of Arts and Sciences in 1992 and 2 years later to the National Academy of Sciences. In 2010, Bruce McEwen and Pfaff shared the Ipsen Neuronal Plasticity Award with Tom Insel, then the director of the National Institutes of Mental Health. He was also awarded the 2011 Daniel Lehrman award for lifetime achievement in the field of behavioral neuroendocrinology (Choleris et al., 2012).

Remarkably an offhand comment from his PhD advisor set Pfaff on the path for a half century of work. Don Pfaff has been central to the development of the field of behavioral neuroendocrinology from its earliest days. In that time, he has methodically pursued his research aims, trained generations of younger scientists, and done as much to reveal the neurobiology of behavior as any modern scientist.

Acknowledgments My sincere appreciation to Richard Zigmond for an entertaining conversation about Pfaff's early years at Rockefeller.

References

Altman, J. (1962a). Are new neurons formed in the brains of adult mammals? *Science, 135*, 1127–1128.

Altman, J. (1962b). Autoradiographic study of degenerative and regenerative proliferation of neuroglia cells with tritiated thymidine. *Experimental Neurology, 5*, 302–318.

Altman, J., & Das, G. D. (1965). Autoradiographic and histological evidence of postnatal hippocampal neurogenesis in rats. *The Journal of Comparative Neurology, 124*, 319–335.

Altman, J., & Das, G. D. (1967). Postnatal neurogenesis in the guinea-pig. *Nature, 214*, 1098–1101.

Beach, F. A. 1948. Hormones and behavior, a survey of interrelationships between endocrine secretions and patterns of overt response., P.B. Hoeber.

Boling, J. L., & Blandau, R. J. (1939). The estrogen-progesterone induction of mating responses in the spayed female rat. *Endocrinology, 25*, 359–364.

Choleris, E., Mccarthy, M. M., & Kavaliers, M. (2012). Daniel S. Lehrman Lifetime Achievement Award. *Hormones and Behavior, 61*, 1–3.

Davis, R. E., Morrell, J. I., & Pfaff, D. W. (1977). Autoradiographic localization of sex steroid-concentrating cells in the brain of the teleost Macropodus opercularis (Osteichthyes: Belontiidae). *General and Comparative Endocrinology, 33*, 496–505.

Davis, P. G., Mcewen, B. S., & Pfaff, D. W. (1979). Localized behavioral effects of tritiated estradiol implants in the ventromedial hypothalamus of female rats. *Endocrinology, 104*, 898–903.

Dörner, G., Döcke, F., & Moustafa, S. (1968). Differential localization of a male and a female hypothalamic mating centre. *Reproduction, 17*, 583–586.

Geary, N., Asarian, L., Korach, K. S., Pfaff, D. W., & Ogawa, S. (2001). Deficits in E2-dependent control of feeding, weight gain, and cholecystokinin satiation in ER-alpha null mice. *Endocrinology, 142*, 4751–4757.

Goldman, S. A., & Nottebohm, F. (1983). Neuronal production, migration, and differentiation in a vocal control nucleus of the adult female canary brain. *Proceedings of the National Academy of Sciences of the United States of America, 80*, 2390–2394.

Gross, C. G. (2000). Neurogenesis in the adult brain: Death of a dogma. *Nature Reviews. Neuroscience, 1*, 67–73.

Kandel, E. R., & Tauc, L. (1964). Mechanism of prolonged heterosynaptic facilitation. *Nature, 202*, 145–147.

Kaplitt, M. G., Kwong, A. D., Kleopoulos, S. P., Mobbs, C. V., Rabkin, S. D., & Pfaff, D. (1994). Preproenkephalin promoter yields region-specific and long-term expression in adult brain after direct in vivo gene transfer via a defective herpes simplex viral vector. *Proceedings of the National Academy of Sciences, 91*, 8979–8983.

Kow, L. M., & Pfaff, D. W. (1973). Effects of estrogen treatment on the size of receptive field and response threshold of pudendal nerve in the female rat. *Neuroendocrinology, 13*, 299–313.

Kow, L. M., Malsbury, C. W., & Pfaff, D. W. (1976). Lordosis in the male golden hamster elicited by manual stimulation: Characteristics and hormonal sensitivity. *Journal of Comparative and Physiological Psychology, 90*, 26–40.

Kow, L. M., Lee, A., Klinge, C., Warner, M., Gustafsson, J.-A., & Pfaff, D. (2016). *Genetic and epigenetic mechanisms in neural and hormonal controls over female reproductive behaviors*. Elsevier.

Lisk, R. D. (1962). Diencephalic placement of estradiol and sexual receptivity in the female rat. *American Journal of Physiology-Legacy Content, 203*, 493–496.

Magariños, A. M., Pedron, S., Creixell, M., Kilinc, M., Tabansky, I., Pfaff, D. W., & Harley, B. A. (2018). The feasibility of encapsulated embryonic medullary reticular cells to grow and differentiate into neurons in functionalized gelatin-based hydrogels. *Frontiers in Materials, 5*, 40.

Martin, E. M., Devidze, N., Shelley, D. N., Westberg, L., Fontaine, C., & Pfaff, D. W. (2011). Molecular and neuroanatomical characterization of single neurons in the mouse medullary gigantocellular reticular nucleus. *The Journal of Comparative Neurology, 519*, 2574–2593.

Mcewen, B. S., Davis, P. G., Parsons, B., & Pfaff, D. W. (1979). The brain as a target for steroid hormone action. *Annual Review of Neuroscience, 2*, 65–112.

Meisel, R. L., & Pfaff, D. W. (1984). RNA and protein synthesis inhibitors: Effects on sexual behavior in female rats. *Brain Research Bulletin, 12*, 187–193.

Mobbs, C. V., Harlan, R. E., Burrous, M. R., & Pfaff, D. W. (1988). An estradiol-induced protein synthesized in the ventral medial hypothalamus and transported to the midbrain central gray. *The Journal of Neuroscience, 8*, 113–118.

Morrell, J., Kelley, D., & Pfaff, D. (1975a). *Sex steroid binding in the brains of vertebrates. Brain–endocrine interactions II* (pp. 230–256). Karger.

Morrell, J. I., Kelley, D. B., & Pfaff, D. W. (1975b). Autoradiographic localization of hormone-concentrating cells in the brain of an amphibian, Xenopus laevis. II. Estradiol. *The Journal of Comparative Neurology, 164*, 63–77.

Morrell, J. I., Crews, D., Ballin, A., Morgentaler, A., & Pfaff, D. W. (1979). 3H-estradiol, 3H-testosterone and 3H-dihydrotestosterone localization in the brain of the lizard Anolis carolinensis: An autoradiographic study. *The Journal of Comparative Neurology, 188*, 201–223.

Parsons, B., Rainbow, T. C., Pfaff, D. W., & Mcewen, B. S. (1982). Hypothalamic protein synthesis essential for the activation of the lordosis reflex in the female rat. *Endocrinology, 110*, 620–624.

Pfaff, D. W. (1968a). Autoradiographic localization of radioactivity in rat brain after injection of tritiated sex hormones. *Science, 161*, 1355–1356.

Pfaff, D. W. (1968b). Uptake of 3H-estradiol by the female rat brain. An autoradiographic study. *Endocrinology, 82*, 1149–1155.

Pfaff, D. W. (1973). Luteinizing hormone-releasing factor potentiates lordosis behavior in hypophysectomized ovariectomized female rats. *Science, 182*, 1148–1149.

Pfaff, D. W. (1979). *Estrogens and brain function*. Springer.

Pfaff, D. W. (1999). *Drive: Neurobiological and molecular mechanisms of sexual motivation*. MIT press.

Pfaff, D. W. (2007). *The neuroscience of fair play*. Dana Press.

Pfaff, D. W. (2015). *A neuroscientist looks at robots*. World Scientific.

Pfaff, D. (2017a). *How the vertebrate brain regulates behavior*. Harvard University Press.

Pfaff, D. (2017b). *How the vertebrate brain regulates behavior: Direct from the lab*. Harvard University Press.

Pfaff, D. W. (2020). *Origins of human socialization*. Academic Press.

Pfaff, D., & Keiner, M. (1973). Atlas of estradiol-concentrating cells in the central nervous system of the female rat. *The Journal of Comparative Neurology, 151*, 121–158.

Pfaff, D. W., & Sherman, S. (2015). *The altruistic brain: How we are naturally good*. Oxford University Press.

Pfaff, D. W., & Volkow, N. D. (2016). *Neuroscience in the 21st century: From basic to clinical*. Springer.

Riggs, L. A. (1997). Carl Pfaffman. *Biographical Memoirs, 71*, 71.

Sakuma, Y., & Pfaff, D. W. (1980). LH-RH in the mesencephalic central grey can potentiate lordosis reflex of female rats. *Nature, 283*, 566–567.

Sakuma, Y., & Pfaff, D. W. (1983). Modulation of the lordosis reflex of female rats by LHRH, its antiserum and analogs in the mesencephalic central gray. *Neuroendocrinology, 36*, 218–224.

Schwanzel-Fukuda, M., & Pfaff, D. W. (1989). Origin of luteinizing hormone-releasing hormone neurons. *Nature, 338*, 161–164.

Eberhard Gwinner

23

Barbara Helm, Michaela Hau, and Wolfgang Goymann

Abstract

Eberhard (Ebo) Gwinner was a German ornithologist and chronobiologist. Following his doctorate in classical ethology, further formative experiences included postdoctoral training in biological rhythms and behavioral endocrinology. Gwinner combined these backgrounds to coin his trademark, integrative research on biological timekeeping under both natural and experimental conditions. He is most recognized for his seminal work and persistently authoritative monography on circannual rhythms. Gwinner also elaborated contributions of reproductive and adrenocortical hormones to avian annual cycle behavior, and his studies of multiple pacemaker interactions and the role of melatonin contributed majorly to understanding circadian systems. Gwinner pioneered many fields that unfolded beyond his lifetime, for example, research on light pollution and urbanization in wild organisms. Across subjects, much of his research revolved around bird migration, a topic that fascinated him throughout his lifetime and that he studied in daring and persistent experiments across continents.

Keywords

Circannual · Circadian · Multi-pacemaker · Internal resonance · Melatonin · Migration · Behavior · Bird

B. Helm (✉)
Department for Bird Migration, Swiss Ornithological Institute, Sempach, Switzerland
e-mail: barbara.helm@vogelwarte.ch

M. Hau
Max Planck Institute for Ornithology, Eberhard-Gwinner-Straße, Seewiesen, Germany

University of Konstanz, Department of Biology, Konstanz, Germany

W. Goymann
Max Planck Institute for Ornithology, Eberhard-Gwinner-Straße, Seewiesen, Germany

R. J. Nelson, Z. M. Weil (eds.), *Biographical History of Behavioral Neuroendocrinology*, https://doi.org/10.1007/978-3-031-12970-4_23

Eberhard Gwinner was born in 1938 in Stuttgart, Germany. Early on, he felt a strong pull to the outdoors and was keen to watch birds near the family home and on the grandparental farm. Encouraged by supportive teachers and amateur ornithologists, his studies soon became systematic. Through a series of natural history publications in ornithological journals, started as a 17-year-old, Gwinner quickly demonstrated his observational and analytical skills. He then enrolled to study zoology, botany, and physiology at the Universities of Freiburg and Tübingen. Soon afterward, he agreed to work toward a doctorate with the navigation researcher and designated Max Planck director Gustav Kramer (Schwartz & Daan, 2017). However, right at the start of Gwinner's thesis project in 1959, Kramer tragically died during fieldwork. Gwinner then found a new mentor in the Nobel laureate Konrad Lorenz, then director at the Max Planck Institute for Behavioural Physiology. Through Lorenz, Gwinner was trained in classical ethology, although he carried out his studies in far-reaching independence, studying ravens (*Corvus corax*) at a remote field station. His thesis focused on "expressive and social behavior of the raven" (Gwinner, 1964) and was published in *Zeitschrift für Tierpsychologie* (now *Ethology*), but he covered many additional aspects of raven behavior.

Upon completion of his thesis in 1964, Gwinner continued at the Max Planck Institute for Behavioural Physiology as a postdoc. He joined the recently established department of Jürgen Aschoff, who pioneered the nascent field of chronobiology (Schwartz & Daan, 2017), in the Bavarian village of Erling-Andechs near Munich. Little did Gwinner know at this time that he would continue to work there and eventually lead the Department for Biological Rhythms and Behaviour for most of his life. Although Aschoff's own work focused on circadian rhythms (Aschoff, 1967), he was intrigued by the possibility of long-term, annual rhythms (Aschoff, 1955). Like Rowan (1926), Aschoff suspected that these rhythms were particularly important for migratory birds, whose movements imposed broad changes in photoperiod, but experimental support for these ideas was lacking. This was the main call for the young Gwinner (Fig. 23.1), next to early circadian studies (in particular (Gwinner, 1966a, b)).

Gwinner set out to explore, describe, test, and summarize evidence for circannual rhythms. He carried out remarkable circannual studies on migratory songbirds, including initial demonstration of circannual rhythms in the African winter quarters and under constant photoperiods in the laboratory (Gwinner, 1968a, b, 1967). This phase marked a breakthrough in Gwinner's scientific career, making him a figurehead of the budding circannual research field which also included Pengelley's findings on hibernating mammals (Pengelley & Kelly, 1966) and evidence from carpet beetles (Blake, 1958). It also marked a milestone in his personal life when he met and eventually married fellow biologist Helga. The two of them shared a wide range of interests centered on the natural world. Together, they raised three children.

Two study visits to the United States widened the scope of Gwinner's research, one with the avian endocrinologist Donald Farner at the University of Washington (1969–1970) and the other with the chronobiologist Colin Pittendrigh at Stanford University (1970–1971). Gwinner picked up further skills, particularly in physiology, and extended his pioneering research to include neuroendocrinology. Gwinner's

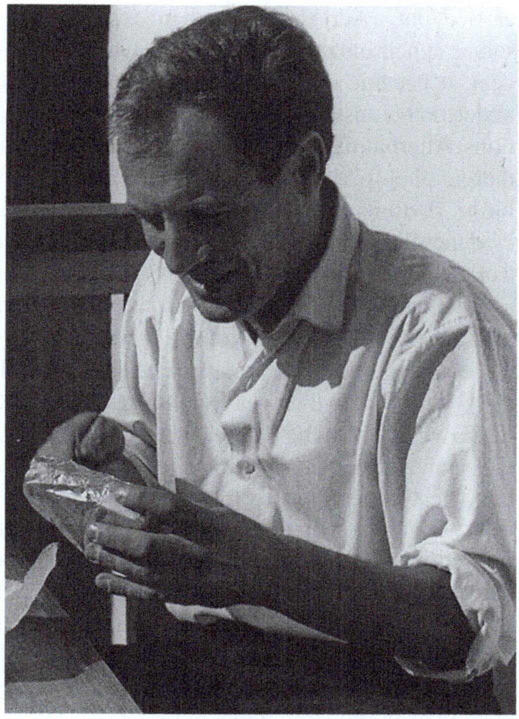

Fig. 23.1 Eberhard Gwinner as a young postdoc, ca. 1967. (Courtesy Helga Gwinner)

integrative views also took shape, whereby he increasingly synthesized circannual and circadian insights from mechanistic, evolutionary, and ecological perspectives. Untiringly, Gwinner kept adding novel aspects, mainly through strategic engagement of young researchers with state-of-the-art skills, thereby safeguarding that his group stayed on the frontline of research (e.g., Brandstätter et al. (2000)). Throughout his career, Gwinner was equally respected in the communities of behavioral physiology, ornithology, and chronobiology, as evidenced by the multiple roles he held and honors that were awarded to him.

Below, we summarize selected contributions to behavioral neuroendocrinology and point readers to original sources that may not be evident to international researchers. Early works in particular were mostly published in German.

Circannual Research

As a postdoctoral researcher, Gwinner initiated his first long-term experiment to examine rhythmicity of seasonal physiology (molt, body mass cycles, nocturnal migratory restlessness or *Zugunruhe*) in migratory birds. It required substantial ornithological skills to collect and hand-raise nestlings of the chosen species, the tiny willow warbler (*Phylloscopus trochilus* (Gwinner, 1967, 1968a, b)) which weighs only 10 g. Gwinner succeeded, and while waiting for the seasons to pass, he carried out a series of circadian experiments, including important evidence for social synchronization via song (Gwinner, 1966a). Meanwhile, within roughly a year, it became clear that the seasonal physiology of the warblers was rhythmic under both constant 12 h (Light:Dark 12:12 h) days and natural day-length conditions. Alternating phases of nocturnal migratory restlessness and molt demonstrated that no photoperiodic change was needed for rhythmicity. The observation that the birds' rhythms had period lengths that differed from 12 months consolidated the evidence for an endogenous circannual clock, rather than entrainment to uncontrolled annual synchronizing cues (i.e., *Zeitgeber* (Gwinner, 1967)).

Gwinner reasoned that the circannual rhythms he had discovered might be an artefact since the birds were kept from reaching their equatorial winter quarters. To test this hypothesis, he translocated small warblers to the Democratic Republic of the Congo, where their seasonal processes were studied in outdoor and indoor aviaries (Gwinner, 1968b). The translocated birds behaved similarly to those recorded in Germany. In particular, the birds continued to show autumnal migratory restlessness in Africa, indicating that the migratory drive was not triggered by northerly locations. Gwinner speculated that the birds' circannual rhythms encoded an inherited migration program that enabled them to reach their wintering and breeding grounds by traveling for a fixed amount of time in a fixed direction (Gwinner, 1996a).

During the following decades, Gwinner elaborated and tested this hypothesis. The main expansions of his circannual research were as follows. Firstly, the hypothesis of an inherited migration program would require that the birds also show seasonally changing directional preferences of their movements, even if sheltered from environmental information. Gwinner proceeded to prove this spatiotemporal aspect for some species in collaboration with orientation researcher Wolfgang Wiltschko (e.g. Wiltschko (1974), Gwinner and Wiltschko (1980)). Secondly, the hypothesis would require large differences in the program between species and local populations to match their geographically distinct migration behaviors. Gwinner also found evidence for inferred local adaptation, for example, in studies of warblers, flycatchers, and stonechats (*Saxicola torquatus*; e.g., Gwinner (1968a), Berthold et al. (1971), Gwinner and Schwabl-Benzinger (1982), Helm et al., 2009)). These comparative studies identified differences in the timing and length of different phases of seasonal processes. Moreover, they also indicated that the robustness of the circannual rhythm and the permissive conditions under which it is expressed differed by population or species (e.g., Gwinner (1989b)). For example, two flycatcher species required different constant day lengths for rhythmicity to persist, depending on wintering latitude, and stonechats that continuously lived under constant equatorial day lengths showed more robust rhythmicity than higher-latitude congeners (Gwinner & Dittami, 1990). These observations kindled Gwinner's keen interest in the interaction between circannual rhythms and photoperiodism (i.e., the control of annual events in response to day length, see below), as well as with other potential *Zeitgeber* (e.g., Gwinner and Scheuerlein (1998), Scheuerlein and Gwinner (2002), Gwinner (1975)).

A third expansion of Gwinner's circannual studies involved greater attention to the reproductive system, thus completing his annual cycle approach. This involved repeatedly measuring the reproductive organs (testes in males, follicles in females) through surgical opening and subsequent sealing of the abdominal cavity (laparotomy (Berthold et al. (1972)). Gwinner and colleagues also measured reproductive hormones such as LH (luteinizing hormone), GnRH (gonadotropin-releasing hormone), and testosterone. In vivo and in vitro findings revealed that even under constant photoperiods, these hormones rise from low winter levels to elevated summer levels (Bluhm et al., 1991) and that LH and testosterone show circannual rhythms (e.g., Dittami and Gwinner (1987), Gwinner et al. (1995)). Remarkably, circannual rhythms in LH persisted even in castrated males (Dittami & Gwinner, 1987).

A final, major expansion of Gwinner's circannual research was a broad comparative approach. He keenly collected any information on circannual rhythms in living organisms, from algae to mammals, and engaged with conceptualizations and potential mechanisms of these enigmatic clocks. In 1986, he published a monograph which still stands as the authoritative text on the subject (Gwinner, 1986).

Photoperiodic and Circadian Contributions to Annual Timing

Gwinner's research consistently addressed photoperiodism. The novel findings on the endogenous nature of circannual rhythms faced challenges similar to those to circadian rhythms some decades earlier (Schwartz & Daan, 2017). Like for circadian rhythms, various researchers questioned whether an endogenous driver existed (reviewed in Wikelski et al. (2008)) in addition to a well-established photic mechanism (e.g., Rowan (1926)). Throughout his career, Gwinner responded to such criticism by thoughtful experiments (Gwinner & Wozniak, 1982). Being well aware of the importance of photic conditions, he viewed circannual rhythms as connected with photoperiodism and circadian rhythms (Gwinner, 1973). In analogy to circadian rhythms, Gwinner expected circannual rhythms to have evolved in association with *Zeitgeber* to which they would entrain and which could modulate them (Gwinner, 1989b). Furthermore, in the wake of findings by Hamner, Follett, and Farner, he was interested in the role of circadian rhythms for photoperiodism as well as for circannual rhythms.

Gwinner's earliest circannual experiments on warblers evaluated developmental programming effects, for example, via day length experienced at hatching. Soon after, he began to experimentally study responses of birds to various photoperiods as a tool to clarify interactions between circannual rhythms, circadian rhythms, and photoperiodism. For these studies, he adopted another model species, the highly photoperiodic European starling (*Sturnus vulgaris*). His work incorporated novel neuroendocrine insights on avian deep brain photoreception and theoretical circadian work on phase response curves. He found that the circadian system was affected by the annual cycle (Gwinner & Turek, 1971), for example, by demonstrating that the observed lengthening and splitting of activity were due to seasonal increase of testosterone levels (Gwinner, 1974). Conversely, his experiments also showed that photoperiodic responses involved circadian time measurement (Gwinner & Eriksson, 1977). It seemed plausible that like for mammals, the duration of the nocturnal rise in melatonin provided the crucial link between photoperiodic time measurement and annual timing. Therefore, Gwinner and colleagues conducted a series of experiments in which they manipulated melatonin secretion, mainly through lesioning of its primary production site, the pineal gland. These studies showed clearly that pinealectomy compromised the circadian system of birds (see below), but that annual cycles, circannual rhythms, and photoperiodic entrainment were hardly affected (Gwinner & Dittami, 1982).

Nonetheless, in subsequent, elegant studies on house sparrows (*Passer domesticus*), Gwinner and colleagues showed that the pineal gland retained photoperiodic information (Brandstätter et al., 2000). The avian pineal gland is itself a circadian oscillator, which in vitro continues to rhythmically secrete melatonin. Gwinner's team first compared the behavior and melatonin profiles between sparrows exposed to short days versus long days. They then released the birds into constant darkness, where their behavior and melatonin profiles continued to reflect the previous photoperiod. The researchers hypothesized that the birds' photoperiodic memory was located in the pineal gland. When they removed and cultured pineal glands to measure melatonin release under constant conditions, they found that features of the previous photoperiodic exposure were indeed retained on a petri dish. While Gwinner thus confirmed circadian time measurement and photoperiodic modulation of the circadian system, direct circadian effects on annual timing were not supported (such as counting of days, see Wikelski et al. (2008)).

Gwinner also used starlings to explore interactions between circannual rhythms and photoperiodism (Gwinner, 1973). He showed that the birds' annual cycle readily entrained to day length, even when being exposed to five simulated photoperiodic cycles within 1 year (Gwinner, 1977). However, the phase of entrainment (i.e., the timing of annual cycle events relative to photoperiod) changed systematically, indicating contributions from an underlying pacemaker. Series of experiments further demonstrated that starlings only express circannual rhythms under a narrow set of permissive photoperiods, which was in stark contrast to the wide range of entrainments in some migratory songbirds (Gwinner, 1989b; Gwinner & Wozniak, 1982). Starlings exposed to constant photoperiods slightly shorter than 12 h in autumn grow and then permanently retain fully developed testes. Conversely, starlings exposed to photoperiods slightly longer than 12 h go through one gonadal cycle and thereafter remain in regressed state. The gonadal arrest under long days, also referred to as photorefractoriness, can be broken by interim exposure to short days (Gwinner & Wozniak, 1982). Gwinner further showed that the regressed state is based on an arrest of the underlying circannual system. This was demonstrated by transferring starlings from 13 h days to 12 h days after 10, 14, and 20 months (Gwinner et al., 1989). The dynamics of the subsequent testicular and molt cycles were similar between groups, indicating that the circannual oscillator had not advanced. Gwinner concluded from the starling experiments that the circannual oscillator in this species requires phase-specific photoperiodic conditions. Under constant 12 h days, the endogenous changes in the photoperiodic response system are sufficient to permit expression of the underlying free-running rhythm (Gwinner & Wozniak, 1982).

Comparative Studies of Photoperiodism and Annual Cycles

Unlike starlings, Gwinner's initial circannual models, long-distance migrants, generally showed robust circannual rhythms that were expressed under a wide range of constant photoperiods (e.g., from LD 10:14 to 16:8 h, see Berthold et al. (1972)). He thus resumed work with long-distance migrants to further examine photoperiodic responses. Garden warblers (*Sylvia borin*) were particularly promising because their wide wintering range extends across the equator (Gwinner, 1987). Thereby, garden warblers in early winter would naturally either experience shortening photoperiod or constant 12 h days at the equator or increasing photoperiod on the southern hemisphere. Nonetheless, all would need to be back for breeding in the following spring. Gwinner demonstrated that appropriate timing was facilitated by strong dependence of the photoperiodic response on the phase of the annual cycle (summarized in Gwinner (1996b)). In early winter, responses to natural austral photoperiod or to constant 15 h days were minute. While being potentially slightly head-started, the birds were prevented from obtaining breeding condition in the winter quarters. Later in winter, birds responded to the same photoperiods with rapid, full gonadal growth and marked increase in LH (Bluhm et al., 1991; Gwinner et al., 1988). Gwinner concluded that the timing and requirements of reproductive arrest (i.e., photorefractoriness) were finely adjusted to the photoperiods a bird would naturally experience. Long-distance migrants, who experience a broad range of winter photoperiods, generally showed delayed reproductive responsiveness, but even closely related species differed in details of the timing program, apparently linked to wintering latitude (Gwinner & Schwabl-Benzinger, 1982; Gwinner, 1989b).

Gwinner's comparative interest in photoperiodism and circannual rhythms also focused on birds that lived permanently under the unchanging day lengths of equatorial regions, as their timing programs would be expected to differ substantially from starlings. The main taxon was the stonechat, which at the time was considered the songbird with the largest north-south breeding range (but is now considered a species complex). Gwinner and colleagues monitored the annual cycle of stonechats in different locations, including in equatorial East Africa (Dittami & Gwinner, 1985; Helm & Gwinner, 1999). These birds might have been predicted to show free-running rhythms or breed continuously or opportunistically depending on weather. Instead, pairs defended territories year-round but displayed strictly annual cycles of gonadal recrudescence and regression, followed by molt. In the following decades, Gwinner bred and studied African stonechats in southern Germany alongside congeners from higher latitudes.

Like Aschoff (1955) before, Gwinner surmised that cycles under constant day length were based on particularly robust circannual rhythms. He subsequently confirmed this hypothesis by demonstrating free-running rhythms over 10 years, including in birds that hatched under constant 12 h days (Gwinner & Dittami, 1990; Gwinner, 1996b). These free-running rhythms however contrasted with the birds' strictly annual breeding. Thus, Gwinner and colleagues tested multiple potential *Zeitgeber*, for example, non-photoperiodic photic cues and food availability (Gwinner & Scheuerlein, 1998; Scheuerlein & Gwinner, 2002). When Gwinner's team examined photoperiodism, they surprisingly discovered some responses that were similar to those of northern congeners (Helm & Gwinner, 1999, 2006; Gwinner & Scheuerlein, 1999). Comparative studies among stonechats became a major research line in the Gwinner group, with parallel work in the field and captivity, incorporating annual cycles, endocrinology, and behavior (Helm et al., 2009; Goymann et al., 2006; Raess & Gwinner, 2005)). Gwinner was also interested in rhythmicity in other equatorial species (Dittami & Gwinner, 1990). Late in life, he studied annual cycles of Darwin's finches on Galapagos (Hau et al., 2004). There, he discovered an equatorial lifestyle that was apparently non-annual and primarily opportunistic.

Extending the Scope of Annual Physiology: Hormones, Metabolism, and Cognition

Gwinner's search for mechanisms that underlie circannual rhythms, paired with his keen interest in understanding birds under natural conditions, continuously broadened the scope of his research. After a link between annual cycles and melatonin was not supported in birds, Gwinner also examined contributions of other hormones. His initial work focused on testosterone, whose interactions with the circadian system were already confirmed (see above and Turek and Gwinner (1982)). He found direct, delaying effects of implanted testosterone on molt in starlings (Schleussner et al., 1985). He also discovered that during molt, levels of testosterone were reduced, whereas thyroxine was elevated. Gwinner tested the hypothesis that in contrast to their delaying effects on the end of reproduction, reproductive hormones early in the season could advance, or at least fine-tune, reproductive timing. He tested ideas mainly indirectly, via behavioral mechanisms that affected hormones. For example, he investigated whether circannual cycles of breeding and LH were affected by pair bond (e.g., Gwinner et al. (1995)), and whether availability of mates or nesting opportunities advanced reproductive hormone levels and gonadal recrudescence (Dittami et al., 1987). Overall, these studies identified small, augmenting effects of these stimuli on reproductive activation. Although reproductive hormones did not explain annual cycles, Gwinner was interested in comparing endocrine mechanisms between seasons. An example was regulation of territorial aggression in spring compared to fall, whereby only in spring circulating testosterone appears to play a major role (Canoine & Gwinner, 2002).

A second candidate pathway for modulating annual timing implicates metabolism and the hypothalamic-pituitary-adrenal-liver axis. In concert with seasonal behavior, birds may change diet and undergo extensive hyper- and hypophagia (Bairlein & Gwinner, 1994). The resulting energetic states were suggested to affect seasonal transitions, especially of migration, which involves shifts between anabolic and catabolic states. Similarly, reproductive decisions were known to be sensitive to nutritional and metabolic state. Thus, energetics, metabolism, and stress were all potentially important for annual cycle timing (e.g., Wikelski et al. (2008), King and Farner (1963)). Gwinner investigated these relationships by studying effects of food availability on seasonal and daily processes (e.g., Scheuerlein and Gwinner (2002), Gwinner and Biebach (1985)) and by studying corticosterone, the main avian glucocorticoid. Research on corticosterone was initially centered on migration, where it was suspected to mediate transitions between fueling and flight phases. In trademark Gwinner style, the research involved both wild and captive birds (Gwinner et al., 1992; Schwabl et al., 1991). The studies suggested associations between migration and daily and seasonal corticosterone profiles but provided no clear answers. Effects of stress were also studied in reproductive context, for example, in African stonechats, where natural predator presence was associated with elevated corticosterone and delayed renesting (Scheuerlein et al., 2001).

Of further relevance to behavioral neuroendocrinology, Gwinner also examined rhythmic processes in cognition. Thus, he demonstrated that hippocampal volume increased with age and migration experience in migratory but not in resident warblers (Healy et al., 1996). Likewise, cognitive testing revealed that migratory warblers had greater spatial memory retention than resident congeners (Mettke-Hofmann & Gwinner, 2003).

Circadian Research and Melatonin

As for circannual rhythms, Gwinner was keenly interested in understanding the physiological mechanisms that constitute circadian processes, focusing in particular on their regulation of organismal functions like behavior, reproduction, and migration.

In mammals, the suprachiasmatic nucleus (SCN) in the brain represents a "master clock," whereas the pineal gland is not a circadian oscillator. In birds, however, in addition to the "SCN," the pineal contains itself an independent oscillator, which communicates with the brain and body via its rhythmic release of melatonin. Gwinner's instrumental work documented that in species like house sparrows, where melatonin and behavioral rhythms are abolished by pinealectomy (Janik et al., 1992), an experimental substitution of rhythmic melatonin levels via injections or addition to the drinking water can restore behavioral rhythmicity (Gwinner & Benzinger, 1978; Heigl & Gwinner, 1994). However, as a keen naturalist, Gwinner expected large differences among bird species in the organization of the circadian system. Research from his and other groups indicated that the role of the pineal and its rhythmic melatonin output contributed to these differences. To reconcile divergent findings across birds, Gwinner proposed the "internal resonance" model, suggesting that the pineal gland, the avian SCN, and in some species the eyes contain self-sustained oscillators that synchronize and amplify each other through resonance (Gwinner, 1989a; Gwinner et al., 1997a). He hypothesized that the importance as well as the strength of the coupling among these three oscillators varies across species, implying that interspecific variation in avian circadian organization is primarily explained by quantitative rather than qualitative differences (Gwinner, 1989a).

In Gwinner's "internal resonance" model, the melatonin rhythm is one major signal that couples and amplifies avian circadian oscillators (Gwinner et al., 1997a). He experimentally supported this prediction by providing birds with a constant melatonin signal via hormone implants, thus abolishing the natural melatonin rhythm. According to the "internal resonance" model, a reduction in the amplitude of one of the oscillators should lead to a damping of the oscillation of the circadian system – and more damped oscillators should more easily entrain to weak *Zeitgeber*. Indeed, house sparrows carrying melatonin implants were more likely and faster to synchronize to a weak *Zeitgeber* compared to sham-treated conspecifics (Hau & Gwinner, 1994); similar effects also occurred in pinealectomized house sparrows (Kumar & Gwinner, 2005).

Gwinner considered these findings exciting not only from a theoretical standpoint but also from an ecological perspective, arguing that a damped melatonin rhythm may in fact be advantageous in certain ecological conditions, seasons, and life history stages (Gwinner et al., 1997a). For example, melatonin rhythms are damped or even absent in winter and summer in some species inhabiting polar regions, while other species maintain melatonin rhythms in high-latitude habitats but with a much lower amplitude than populations from lower latitudes (Silverin et al., 2009). This reduction in melatonin amplitude may allow these species to more easily entrain with existent, but weak, daily changes in their natural environment. Gwinner also considered this phenomenon in the context of migration, where birds migrate across longitudes and latitudes. Returning to his old models, migratory warblers, together with colleagues, he showed that these birds indeed displayed a reduced melatonin amplitude during the migratory season (Gwinner et al., 1993) while resident populations did not (Fusani & Gwinner, 2001). One suggestive aspect of Gwinner's finding was that this reduction in the melatonin amplitude was linked with the display of migratory restlessness in these species (Fusani & Gwinner, 2004). This finding also fits with evidence that birds with a nocturnal lifestyle like owls, which have clear circadian rhythms (Van't Hof et al., 1998), and tropical gulls (Wikelski et al., 2006) hardly show a melatonin rhythm.

Gwinner's important contributions to the circadian field are too extensive to cover in detail. They also included evidence that not only light cycles can serve as *Zeitgeber* for avian circadian rhythms but also rhythmic food availability (Hau & Gwinner, 1992; Heigl & Gwinner, 1999; Abraham et al., 2000). One prescient research line elaborated how early in ontogeny circadian rhythms were expressed. Primarily with Michal Zeman, Gwinner compared circadian development between precocial and altricial birds and demonstrated that embryonic rhythms were sensitive to external conditions (Van't Hof & Gwinner, 1996; Gwinner et al., 1997b; Zeman & Gwinner, 1992, 1993; Zeman et al., 1992). When molecular advances allowed the genetic study of wild species, Gwinner and his group were again at the forefront. Thus, they embarked on a search for the avian SCN by assaying the rhythmic expression of circadian genes in different parts of the brain (Abraham et al., 2002). In his trademark naturalist fashion (Fig. 23.2), Gwinner also immediately applied the nascent information on clock genes to avian ecology and evolution (Fidler & Gwinner, 2003).

Fig. 23.2 Eberhard Gwinner flirting with a Kea while visiting the International Ornithological Congresses in New Zealand, in 1990. (Courtesy Helga Gwinner)

A Visionary of Inclusive Leadership

Gwinner fostered an inclusive approach to research and leadership that was highly unusual in his time and his research institution, with a goal of facilitating diversity. For example, he explicitly supported women in science. By this approach, he created an atmosphere where everybody, from cleaner, animal caretaker, worker, and secretary through school pupil and doctoral student to postdoc and senior researcher felt appreciated, valued, and embedded in the department. This fostered an exceptionally open and scientifically stimulating atmosphere which significantly contributed to the lasting scientific legacy of Gwinner and his team. For sure, the breathtaking views of the department across rolling hills right into the Bavarian Alps aided at widening the scientific horizon. So did vivid scientific discussions regarding planned projects in formal and informal gatherings, such as the daily coffee break (where many ideas for scientific projects were born), visits to the local monastic brewery, evening parties, or afterwork volleyball tournaments. These events generated a familiar atmosphere in which almost everybody identified themselves with, and felt responsibility for, what was – affectionately – called "The Insti". Ebo Gwinner's scientific legacy includes his talent for "leading without leading." Very much supported by his colleague and wife Helga, the two thereby enabled people to unfold their scientific skills and live up to their potentials. Gwinner thus promoted a very contemporary and modern view of scientific excellence that outlived his untimely death in 2004.

A Lasting Legacy: Clocks in a Changing World

This overview of Gwinner's work has focused on fundamental contributions to behavioral neuroendocrinology. Much would need to be added. Gwinner was a visionary researcher who pin-pointed questions that despite rapidly developing methods still await answers. He used his rich background to devise and shape distinct research approaches. Thus, throughout his career, he fostered and exemplified "wild clock" research that is now embraced by chronobiological meetings worldwide (Kronfeld-Schor et al., 2013; Schwartz et al., 2017). He was also visionary in applying his expertise to a changing world, for example, with respect to urbanization and light pollution. During his work on ontogeny of rhythms, he examined effects of realistic light exposure on starlings (Gwinner et al., 1997b). Soon after, with Jesko Partecke, he launched a paired field and captivity project on European blackbirds (*Turdus merula*), whereby reproductive timing, stress, genes, and migratory behavior were compared between urban and forest birds (e.g., Partecke et al. (2004), 2006a, b). Gwinner was also interested in effects of climate change on annual cycles. Toward the end of his career, he replicated an old experiment to assess potential changes in flycatchers, although sadly he did not live to see the resulting indication of evolutionary modification (Helm et al., 2019). These and other topics are now firmly established research areas, whereas the search is still on for the long-term rhythms that had fascinated Gwinner throughout his life.

References

Abraham, U., Gwinner, E., & Van't Hof, T. J. (2000). Exogenous melatonin reduces the resynchronization time after phase shifts of a nonphotic zeitgeber in the house sparrow (Passer domesticus). *Journal of Biological Rhythms, 15*, 48–56.

Abraham, U., Albrecht, U., Gwinner, E., & Brandstatter, R. (2002). Spatial and temporal variation of passer Per2 gene expression in two distinct cell groups of the suprachiasmatic hypothalamus in the house sparrow (Passer domesticus). *The European Journal of Neuroscience, 16*, 429–436.

Aschoff, J. (1955). Jahresperiodik der Fortpflanzung bei Warmblütern. *Studium Generale, 8*, 742–776.

Aschoff, J. (1967). *Circadian rhythms in birds. Proc. XIV Int. Ornithol. Congr., Oxford*. Blackwell.

Bairlein, F., & Gwinner, E. (1994). Nutritional mechanisms and temporal control of migratory energy accumulation in birds. *Annual Review of Nutrition, 14*, 187–215.

Berthold, P., Gwinner, E., & Klein, H. (1971). Circannuale Periodik bei Grasmücken (Sylvia). *Experientia, 27*, 399–399.

Berthold, P., Gwinner, E., & Klein, H. (1972). Circannuale Periodik bei Grasmücken. II. Periodik der Gonadengröße bei Sylvia atricapilla und S. borin unter verschiedenen konstanten Bedingungen. *Journal für Ornithologie, 113*, 407–417.

Blake, G. (1958). Diapause and the regulation of development in Anthrenus verbasci (L.)(Col., Dermestidae). *Bulletin of Entomological Research, 49*, 751–775.

Bluhm, C. K., Schwabl, H., Schwabl, I., Perera, A., Follett, B. K., Goldsmith, A. R., & Gwinner, E. (1991). Variation in hypothalamic gonadotrophin-releasing hormone content, plasma and pituitary LH, and in-vitro testosterone release in a long-distance migratory bird, the garden warbler (Sylvia borin), under constant photoperiods. *The Journal of Endocrinology, 128*, 339–345.

Brandstätter, R., Kumar, V., Abraham, U., & Gwinner, E. (2000). Photoperiodic information acquired and stored in vivo is retained in vitro by a circadian oscillator, the avian pineal gland. *Proceedings of the National Academy of Sciences, 97*, 12324–12328.

Canoine, V., & Gwinner, E. (2002). Seasonal differences in the hormonal control of territorial aggression in free-living European stonechats. *Hormones and Behavior, 41*, 1–8.

Dittami, J., & Gwinner, E. (1985). Annual cycles in the African stonechat Saxicola torquata axillaris and their relationship to environmental factors. *Journal of Zoology, 207*, 357–370.

Dittami, J., & Gwinner, E. (1987). Circannual variations in plasma luteinizing hormone levels in castrated male European starlings (Sturnus vulgaris). *Journal of Biological Rhythms, 2*, 221–226.

Dittami, J., & Gwinner, E. (1990). Endocrine correlates of seasonal reproduction and territorial behavior in some tropical passerines. In M. WADA (Ed.), *Endocrinology of birds: Molecular to behavioral*. Springer.

Dittami, J., Gwinner, H., & Gwinner, E. (1987). Effects of nestboxes on LH, testosterone, testicular size, and the reproductive behavior of male European starlings in spring. *Behaviour, 103*, 68–81.

Fidler, A. E., & Gwinner, E. (2003). Comparative analysis of avian BMAL1 and CLOCK protein sequences: A search for features associated with owl nocturnal behaviour. *Comparative Biochemistry and Physiology. Part B, Biochemistry & Molecular Biology, 136*, 861–874.

Fusani, L., & Gwinner, E. (2001). Reduced amplitude of melatonin secretion during migration in the blackcap (Sylvia atricapilla). In H. J. T. Goos, R. K. Rastogi, H. Vaudry, & R. Pierantoni (Eds.), *Perspective in comparative endocrinology: Unity and diversity*. Bologna.

Fusani, L., & Gwinner, E. (2004). Simulation of migratory flight and stopover affects night levels of melatonin in a nocturnal migrant. *Proceedings of the Royal Society of London. Series B: Biological Sciences, 271*, 205–211.

Goymann, W., Geue, D., Schwabl, I., Flinks, H., Schmidl, D., Schwabl, H., & Gwinner, E. (2006). Testosterone and corticosterone during the breeding cycle of equatorial and European stonechats (Saxicola torquata axillaris and S. t. rubicola). *Hormones and Behavior, 50*, 779–785.

Gwinner, E. (1964). Untersuchungen uber das Ausdrucks-und Sozialverhalten des Kolkraben (Corvus corax L.). *Zeitschrift für Tierpsychologie*.

Gwinner, E. (1966a). Entrainment of a circadian rhythm in birds by species-specific song cycles (Aves, Fringillidae: Carduelis spinus, Serinus serinus). *Experientia, 22*, 765.

Gwinner, E. (1966b). Tagesperiodische Schwankungen der Vorzugshelligkeit bei Vögeln. *Zeitschrift für Vergleichende Physiologie, 52*, 370–379.

Gwinner, E. (1967). Circannuale Periodik der Mauser und der Zugunruhe bei einem Vogel. *Naturwissenschaften, 54*, 447–447.

Gwinner, E. (1968a). Artspezifische Muster der Zugunruhe bei Laubsängern und ihre mögliche Bedeutung für die Beendigung des Zuges im Winterquartier. *Zeitschrift für Tierpsychologie, 25*, 843–853.

Gwinner, E. (1968b). Circannuale periodik als grundlage des jahreszeitlichen funktionswandels bei zugvögeln. *Journal für Ornithologie, 109*, 70–95.

Gwinner, E. (1973). Circannual rhythms in birds: Their interaction with circadian rhythms and environmental photoperiod. *Journal of Reproduction and Fertility. Supplement, 19*, 51–65.

Gwinner, E. (1974). Testosterone induces "splitting" of circadian locomotor activity rhythms in birds. *Science, 185*, 72–74.

Gwinner, E. (1975). Effects of season and external testosterone on the freerunning circadian activity rhythm of European starlings (Sturnus vulgaris). *Journal of Comparative Physiology, 103*, 315–328.

Gwinner, E. (1977). Photoperiodic synchronization of circannual rhythms in the European starling (Sturnus vulgaris). *Naturwissenschaften, 64*, 44–45.

Gwinner, E. (1986). Circannual rhythms. In D. S. Farner (Ed.), *Zoophysiology*. Springer.

Gwinner, E. (1987). Annual rhythms of gonadal size, migratory disposition and molt in garden warblers Sylvia borin exposed in winter to an equatorial or a southern hemisphere photoperiod. *Ornis Scandinavica, 18*, 251–256.

Gwinner, E. (1989a). Melatonin in the circadian system of birds: Model of internal resonance. In T. Hiroshige & K. Honma (Eds.), *Circadian clocks and ecology*. Hokkaido University Press.

Gwinner, E. (1989b). Photoperiod as a modifying and limiting factor in the expression of avian circannual rhythms. *Journal of Biological Rhythms, 4*, 125–138.

Gwinner, E. (1996a). Circadian and circannual programmes in avian migration. *The Journal of Experimental Biology, 199*, 39–48.

Gwinner, E. (1996b). Circannual clocks in avian reproduction and migration. *Ibis, 138*, 47–63.

Gwinner, E., & Benzinger, I. (1978). Synchronization of a circadian rhythm in pinealectomized European starlings by daily injections of melatonin. *Journal of Comparative Physiology, 127*, 209–213.

Gwinner, E., & Biebach, H. (1985). Food availability affects migratory restlessness in caged garden warblers (Sylvia borin). *Naturwissenschaften, 72*, 51–52.

Gwinner, E., & Dittami, J. (1982). Pineal influences on circannual cycles in European starlings: Effects through the circadian system? In J. Aschoff, S. Daan, & G. Groos (Eds.), *Vertebrate circadian systems*. Springer.

Gwinner, E., & Dittami, J. (1990). Endogenous reproductive rhythms in a tropical bird. *Science, 249*, 906–908.

Gwinner, E., & Eriksson, L.-O. (1977). Circadiane rhythmik und photoperiodische Zeitmessung beim Star (Sturnus vulgaris). *Journal für Ornithologie, 118*, 60–67.

Gwinner, E., & Scheuerlein, A. (1998). Seasonal changes in day-light intensity as a potential zeitgeber of circannual rhythms in equatorial Stonechats. *Journal für Ornithologie, 139*, 407–412.

Gwinner, E., & Scheuerlein, A. (1999). Photoperiodic responsiveness of equatorial and temperate-zone stonechats. *The Condor, 101*, 347–359.

Gwinner, E., & Schwabl-Benzinger, I. (1982). Adaptive temporal programming of molt and migratory disposition in two closely related long-distance migrants, the pied flycatcher (Ficedula hypoleuca) and the collared flycatcher (Ficedula albicollis). In F. Papi & H. Wallraff (Eds.), *Avian navigation*. Springer.

Gwinner, E., & Turek, F. (1971). Effects of season on circadian activity rhythms of the starling. *Naturwissenschaften, 58*, 627–628.

Gwinner, E., & Wiltschko, W. (1980). Circannual changes in migratory orientation of the garden warbler, Sylvia borin. *Behavioral Ecology and Sociobiology, 7*, 73–78.

Gwinner, E., & Wozniak, J. (1982). Circannual rhythms in European starlings: Why do they stop under long photoperiods? *Journal of Comparative Physiology, 146*, 419–421.

Gwinner, E., Dittami, J. P., & Beldhuis, H. J. (1988). The seasonal development of photoperiodic responsiveness in an equatorial migrant, the garden warbler Sylvia borin. *Journal of Comparative Physiology A, 162*, 389–396.

Gwinner, E., Gänshirt, G., & Dittami, J. (1989). Starling circannual systems: Are they arrested in long photoperiods? *Journal of Comparative Physiology A, 165*, 35–39.

Gwinner, E., Zeman, M., Schwabl-Benzinger, I., Jenni-Eiermann, S., Jenni, L., & Schwabl, H. (1992). Corticosterone levels of passerine birds during migratory flight. *Naturwissenschaften, 79*, 276–278.

Gwinner, E., Schwabl-Benzinger, I., Schwabl, H., & Dittami, J. (1993). Twenty-four hour melatonin profiles in a nocturnally migrating bird during and between migratory seasons. *General and Comparative Endocrinology, 90*, 119–124.

Gwinner, E., Konig, S., & Zeman, M. (1995). Endogenous gonadal, LH and molt rhythms in tropical stonechats: Effect of pair bond on period, amplitude, and pattern of circannual cycles. *Journal of Comparative Physiology. A, 177*, 73–79.

Gwinner, E., Hau, M., & Heigl, S. (1997a). Melatonin: Generation and modulation of avian circadian rhythms. *Brain Research Bulletin, 44*, 439–444.

Gwinner, E., Zeman, M., & Klaassen, M. (1997b). Synchronization by low-amplitude light-dark cycles of 24-hour pineal and plasma melatonin rhythms of hatchling European starlings (Sturnus vulgaris). *Journal of Pineal Research, 23*, 176–181.

Hau, M., & Gwinner, E. (1992). Circadian entrainment by feeding cycles in house sparrows, Passer domesticus. *Journal of Comparative Physiology A, 170*, 403–409.

Hau, M., & Gwinner, E. (1994). Melatonin facilitates synchronization of sparrow circadian rhythms to light. *Journal of Comparative Physiology A, 175*, 343–347.

Hau, M., Wikelski, M., Gwinner, H., & Gwinner, E. (2004). Timing of reproduction in a Darwin's finch: Temporal opportunism under spatial constraints. *Oikos, 106*, 489–500.

Healy, S. D., Gwinner, E., & Krebs, J. R. (1996). Hippocampal volume in migratory and non-migratory warblers: Effects of age and experience. *Behavioural Brain Research, 81*, 61–68.

Heigl, S., & Gwinner, E. (1994). Periodic melatonin in the drinking water synchronizes circadian rhythms in sparrows. *Naturwissenschaften, 81*, 83–85.

Heigl, S., & Gwinner, E. (1999). Periodic food availability synchronizes locomotor and feeding activity in pinealectomized house sparrows. *Zoology-Analysis of Complex Systems, 102*, 1–9.

Helm, B., & Gwinner, E. (1999). Timing of postjuvenal molt in African (Saxicola torquata axillaris) and European (Saxicola torquata rubicola) stonechats: Effects of genetic and environmental factors. *The Auk, 116*, 589–603.

Helm, B., & Gwinner, E. (2006). Migratory restlessness in an equatorial nonmigratory bird. *PLoS Biology, 4*, e110.

Helm, B., Schwabl, I., & Gwinner, E. (2009). Circannual basis of geographically distinct bird schedules. *The Journal of Experimental Biology, 212*, 1259–1269.

Helm, B., Van Doren, B. M., Hoffmann, D., & Hoffmann, U. (2019). Evolutionary response to climate change in migratory pied flycatchers. *Current Biology, 29*, 3714–3719. e4.

Janik, D., Dittami, J., & Gwinner, E. (1992). The effect of pinealectomy on circadian plasma melatonin levels in house sparrows and European starlings. *Journal of Biological Rhythms, 7*, 277–286.

King, J. R., & Farner, D. S. (1963). The relationship of fat deposition to Zugunruhe and migration. *The Condor, 65*, 200–223.

Kronfeld-Schor, N., Bloch, G., & Schwartz, W. J. (2013). *Animal clocks: When science meets nature*. The Royal Society.

Kumar, V., & Gwinner, E. (2005). Pinealectomy shortens resynchronisation times of house sparrow (Passer domesticus) circadian rhythms. *Naturwissenschaften, 92*, 419–422.

Mettke-Hofmann, C., & Gwinner, E. (2003). Long-term memory for a life on the move. *Proceedings of the National Academy of Sciences of the United States of America, 100*, 5863–5866.

Partecke, J., Van't Hof, T., & Gwinner, E. (2004). Differences in the timing of reproduction between urban and forest European blackbirds (Turdus merula): Result of phenotypic flexibility or genetic differences? *Proceedings of the Royal Society of London. Series B: Biological Sciences, 271*, 1995–2001.

Partecke, J., Gwinner, E., & Bensch, S. (2006a). Is urbanisation of European blackbirds (Turdus merula) associated with genetic differentiation? *Journal of Ornithology, 147*, 549–552.

Partecke, J., Schwabl, I., & Gwinner, E. (2006b). Stress and the city: Urbanization and its effects on the stress physiology in European blackbirds. *Ecology, 87*, 1945–1952.

Pengelley, E. T., & Kelly, K. (1966). A "circannian" rhythm in hibernating species of the genus Citellus with observations on their physiological evolution. *Comparative Biochemistry and Physiology, 19*, 603–617.

Raess, M., & Gwinner, E. (2005). Gonadal status upon spring arrival in long-distance and short-distance migrating stonechats (Saxicola torquata). *Journal of Ornithology, 146*, 325–331.

Rowan, W. (1926). On photoperiodism, reproductive periodicity, and the annual migrations of birds and certain fishes. *Proceedings of the Boston Society of Natural History, 38*, 147–189.

Scheuerlein, A., & Gwinner, E. (2002). Is food availability a circannual zeitgeber in tropical birds? A field experiment on stonechats in tropical Africa. *Journal of Biological Rhythms, 17*, 171–180.

Scheuerlein, A., Van't Hof, T., & Gwinner, E. (2001). Predators as stressors? Physiological and reproductive consequences of predation risk in tropical stonechats (Saxicola torquata axillaris). *Proceedings of the Royal Society of London. Series B: Biological Sciences, 268*, 1575–1582.

Schleussner, G., Dittami, J. P., & Gwinner, E. (1985). Testosterone implants affect molt in male European starlings, Sturnus vulgaris. *Physiological Zoology, 58*, 597–604.

Schwabl, H., Bairlein, F., & Gwinner, E. (1991). Basal and stress-induced corticosterone levels of garden warblers, Sylvia borin, during migration. *Journal of Comparative Physiology B, 161*, 576–580.

Schwartz, W. J., & Daan, S. (2017). Origins: A brief account of the ancestry of circadian biology. In *Biological timekeeping: Clocks, rhythms and behaviour*. Springer.

Schwartz, W. J., Helm, B., & Gerkema, M. P. (2017). *Wild clocks: Preface and glossary*. The Royal Society.

Silverin, B., Gwinner, E., Van't Hof, T. J., Schwabl, I., Fusani, L., Hau, M., & Helm, B. (2009). Persistent diel melatonin rhythmicity during the Arctic summer in free-living willow warblers. *Hormones and Behavior, 56*, 163–168.

Turek, F., & Gwinner, E. (1982). Role of hormones in the circadian organization of vertebrates. In J. Aschoff, S. Daan, & G. Groos (Eds.), *Vertebrate circadian systems*. Springer.

Van't Hof, T. J., & Gwinner, E. (1996). Development of post-hatching melatonin rhythm in zebra finches (Poephila guttata). *Experientia, 52*, 249–252.

Van't Hof, T. J., Gwinner, E., & Wagner, H. (1998). A highly rudimentary circadian melatonin profile in a nocturnal bird, the barn owl (Tyto alba). *Naturwissenschaften, 85*, 402–404.

Wikelski, M., Tarlow, E. M., Eising, C. M., Groothuis, T. G., & Gwinner, E. (2006). Do night-active birds lack daily melatonin rhythms? A case study comparing a diurnal and a nocturnal-foraging gull species. *Journal of Ornithology, 147*, 107–111.

Wikelski, M., Martin, L. B., Scheuerlein, A., Robinson, M. T., Robinson, N. D., Helm, B., Hau, M., & Gwinner, E. (2008). Avian circannual clocks: Adaptive significance and possible involvement of energy turnover in their proximate control. *Philosophical Transactions of the Royal Society of London. Series B, Biological Sciences, 363*, 411–423.

Wiltschko, W. (1974). Evidence for an innate magnetic compass in garden warblers. *Naturwissenschaften, 61*, 406.

Zeman, M., & Gwinner, E. (1992). Rhythmic melatonin production in chick and quail embryos. In *XIX world's poultry congress* (pp. 647–651). WPSA.

Zeman, M., & Gwinner, E. (1993). Ontogeny of the rhythmic melatonin production in a precocial and an altricial bird, the Japanese quail and the European starling. *Journal of Comparative Physiology A, 172*, 333–338.

Zeman, M., Gwinner, E., & Somogyiova, E. (1992). Development of melatonin rhythm in the pineal gland and eyes of chick embryo. *Experientia, 48*, 765–768.

Irving Zucker

24

Brian J. Prendergast

Abstract

Irving Zucker is a pioneer in modern behavioral neuroendocrinology. His work established a central role for biological rhythms in the regulation of hormone-behavior processes, including the estrous cycle, sex behavior, and social/affiliative behaviors. His empirical work has emphasized comparative approaches, ecological considerations, strong inference experimental rigor, and sex differences. He has mentored several generations of scientists and made major contributors to behavioral neuroendocrinology and chronobiology.

Keywords

Circadian · Circannual and ultradian rhythms · Melatonin · Sex behavior · Sex bias · Hibernation

B. J. Prendergast (✉)
Department of Psychology, University of Chicago, Chicago, IL, USA
e-mail: prendergast@uchicago.edu

R. J. Nelson, Z. M. Weil (eds.), *Biographical History of Behavioral Neuroendocrinology*, https://doi.org/10.1007/978-3-031-12970-4_24

Irving Zucker (Irv) was born in 1940 in Montreal, Quebec, the first of two children, to parents who had emigrated separately from Poland a decade earlier. His parents were communists—his father ran a shoe repair shop a few blocks off Boulevard St. Laurent, in the heart of Montreal's Jewish ghetto and later owned a shoe store; and his mother was a homemaker. They lived above the store for many years, and Zucker worked there weekends and summers. He spoke Yiddish at home, which was also one of the languages of instruction in the Jewish parochial school that he attended until sixth grade, when, during a spillover of McCarthyism to Canada, his school was inspected by the red squad under suspicions of fomenting sedition. Because of this intervention and the attendant publicity, several parents withdrew their children from the school which was forced to close. Thereafter, he attended schools run by the protestant school board of greater Montreal.

According to Zucker, his early immersion in a "very left wing" school environment combined with the socialist worldview at home did not factor in leading him toward academics or science, but they had an enormous impact on how he would teach and work with students and colleagues throughout his professional career. Specifically, his youthful environment emphasized treating all people equally, not exploiting others' labors, and eschewing rigid hierarchies. These approaches guided how he established and ran what was to become a world-class research laboratory, how he promoted the work of his students and colleagues, and how he mentored those in his orbit.

McGill University (1957–1961)

In the late 1950s, young anglophone Montrealers who sought post-secondary degrees either attended McGill University or Sir George Williams College, which later became Concordia University. McGill was a commuter school, and Zucker, as many students at the time, lived at home and walked to McGill for classes each day. Though he grew up two miles from McGill, the first time he set foot on campus was the day he registered for classes in 1957.

McGill was a premier institution for the study of behavior and brain function, a subdiscipline of psychology called "Physiological Psychology" which eventually morphed into and became Behavioral Neuroscience. Donald Hebb, Peter Milner, and Dalbir Bindra were on the faculty in the Psychology Department. Zucker recalled courses on *Motivation* taught by Bindra and *Physiological Psychology*, taught by Milner, as exciting and resonant with his interests. As a third-year student, he conducted research in Bindra's lab. In his first week at work, a man he had not yet met was walking by his door at the exact moment that a rat he was handling sank its teeth into his thumb. The ensuing expletive caused the visitor to stop and inquire "what happened?" Of course, it was Hebb, who on learning that a rat had bitten Zucker asked whether it was an experimental or control rat. "I'll look sir," he replied. After consulting his notes, he told Hebb, "It was an experimental rat, sir." Not missing a beat, Hebb inquired: "Are you planning on having one of the controls bite you too?" To have this famous scientist pay attention to him was very important to young Zucker.

The psychology department at McGill was a highly interactive, collaborative, and socially connected environment. Twice each day, faculty and students met for tea—undergraduates, graduate students, postdocs, and faculty—where they would discuss research and recent seminars, or simply socialize. Zucker describes it as an enormously stimulating intellectual environment, which felt more like a family than an academic unit and was a transformational experience. Within a year of joining this group, Zucker was given his own lab room and free access to rats, and under the guidance of Milner worked almost as many hours as the Ph.D. students around him did, a self-described "lab rat." He graduated with a double major in Mathematics and Psychology and published four empirical papers based on his undergraduate work, covering topics such as specific salt hungers (Milner & Zucker, 1965; Zucker, 1965a, b), food hoarding (Zucker & Milner, 1964), sensory deprivation, and exploratory behavior (Zucker & Bindra, 1961).

Zucker credits Hebb and Milner with shaping him as a scientist in enduring ways. Milner was most influential on his intellectual development. He credits Milner with an encyclopedic knowledge of physiological psychology. Milner shared with Zucker a variety of new manuscripts that came across his desk. These experiences greatly broadened his intellectual horizons, and he draws a direct line between his interests in the brain and his relationship with Milner. Hebb's mentoring, on the other hand, took the form of fostering his career. Zucker did not work in Hebb's lab per se, but they interacted often during those years. He credits Hebb with providing a template for how to be a serious scientist while at the same time a friendly and generous mentor. Hebb's credo was "ask not what the students can do for you, ask what you can do for them."

University of Chicago (1961–1964)

In 1961, Zucker began studying for his Ph.D. at the University of Chicago in the Section on Biopsychology. Hebb played a pivotal role in steering him toward Chicago, recommending Robert A. McCleary ("Mac") as a suitable mentor. McCleary had recently come to Chicago from the University of Michigan, and Zucker began working in his lab, following up on McCleary's studies on the effects of septal lesions on active and passive avoidance behaviors in cats. Zucker's thesis examined the extent to which these deficits were specific or generalized to other cognitive processes, work that required long hours employing a Wisconsin general testing apparatus, modified for cats. He recalled finding the research questions moderately interesting, but the effort itself was tedious. The work was well received (Zucker & McCleary, 1964; Zucker, 1965a), but he recalls not wanting to pursue that line of research beyond graduate school. Instead, he wanted to finish his degree as quickly as possible, so that he could explore other lines of research. Zucker completed graduate school in 2 years and 9 months, obtaining his Ph.D. in 1964, at the age of 23.

A few additional significant events in Zucker's life took place while in Chicago. Most importantly, here he met Ellen Krantz. They would marry a few years later. Ellen was an undergraduate at the University of Chicago. She would later join Zucker in Oregon, where she received her Ph.D. in Psychology at the University of Oregon Medical School.

Zucker's interests in behavioral neuroendocrinology grew during his time in Chicago. He was engaged by a paper which showed that early life androgen treatment rendered female rats sexually unresponsive to males, a condition which could not be rescued by estrogen and progesterone treatment (Barraclough & Gorski, 1962). A lab mate recommended that Zucker take a course on the Biology of Sex taught in the Zoology Department. The course used W.C. Young's *Sex and Internal Secretions* (Young, 1961) as a textbook and was taught by Dorothy Price, who first laid out the principles now known as "negative feedback" of the gonads on the pituitary. Price took time to interact with Zucker individually, and she spoke highly of William C. Young whom she knew when they both worked in Carl Moore's Chicago lab.

Having decided to pursue training in behavioral neuroendocrinology, Zucker consulted McCleary who advised him that the best training would be found either at Berkeley with Frank Beach or at Kansas with W.C. Young. Having read the work of Phoenix, Goy, Gerall, and Young (Phoenix et al., 1959), which elaborated the activational-organizational hypothesis of sexual differentiation, tipped the scale in favor of Young's lab, which had recently relocated to the Oregon Regional Primate Research Center in Beaverton. Reflecting on how events in Chicago influenced his movement into the field of behavioral endocrinology, Zucker recalls feeling no urgency at the time to pursue studies that would follow up on his Ph.D. work, and he interpreted this as a sign that he would do well to look elsewhere for questions that he found more exciting.

An exchange that occurred at Zucker's thesis defense underscored his intention to convert to endocrinology. Unlike at McGill, the faculty in Biopsychology at Chicago were not a harmonious group. They would frequently behave aggressively toward each other's students, especially at public thesis defenses. A few years earlier, Zucker had taken a course on *Neuroanatomy* from Bob Moore and Ruth Rhines; he received a "D" grade. (On whether he was deserving of the grade, Zucker admits: "The instructors were being kind.") So, he was not surprised when, at his Ph.D. defense, concern was voiced about whether Zucker would be able to pursue a career in brain research. In an effort to reassure his committee that they had nothing to worry about, he stated that he was headed on to a career in endocrinology and had no intentions of doing neural work ever again. Of course, this sincere promise was not kept less than a decade later, when he and Fred Stephan reported the discovery of the mammalian circadian pacemaker, one of the more important empirical neural localizations of function in the history of behavioral neuroscience.

Oregon (1964–1966)

Zucker began his postdoc in Young's lab in August of 1964. He read Young's work from the 1930s and took an interest in studies on hormonal regulation of sex behavior in female guinea pigs. His first experiments examined how a single progesterone treatment could initially stimulate but thereafter persistently inhibit sex behavior in estrogen-primed female guinea pigs. The work elegantly showed that the persistent inhibition of sex behavior by progesterone could not be accounted for by systemic availability of estrogen, leaving viable the hypothesis that progesterone may block the action of estrogen as a competitive inhibitor at target tissues (possibly neural sites; Zucker, 1966). In 2 years, he completed work that culminated in eight empirical reports on steroid regulation of sex behavior in male and female guinea pigs and rats (e.g., Zucker (1967a, b)), initiating a career-long interest in female reproductive biology.

Young passed away 9 months after Zucker arrived at Oregon. Additional details of his relationship with Young appear elsewhere in this volume. Zucker credits Young with guiding his entry into the field. Zucker also met Harvey Feder, who was a graduate student in Young's lab and after Young's death completed his training supervised by Robert Goy, chairman of the department. Feder would go on to a distinguished career at the Institute for Animal Behavior at Rutgers. Zucker and Feder significantly influenced each other's research and began an enduring friendship that included an exchange of postdoctoral trainees (George Wade and Ted Landau to Rutgers, Alison Fleming, and Larry Morin to Berkeley).

Partway through his time at Oregon, Zucker became aware of steroid-induced changes in gene expression in peripheral tissues and convinced Young to support his travel to the University of Wisconsin to learn RNA and DNA quantification techniques, which he hoped to translate into use on neural tissues. Zucker spent 6 months at UW medical school in Madison, where he trained in a lab in the Department of Pathology.

In March of 1965, Zucker attended the West Coast Sex Society meeting in Berkeley. It was a typically sunny, lovely day in the Bay Area, and after months of Oregon rain, he instantly fell in love with the Berkeley climate and the counterculture in full bloom on Telegraph Avenue. Within a few months, there was an opening in the psychology department at Berkeley. Zucker secured letters of support from Milner, McCleary, and Goy. Many years later, Beach would recount to Zucker the favorable impression those letters made on the search committee. Zucker recalls Beach telling him: "We got a bunch of 'Jesus Christ' letters about you, so I called Eckhard Hess at Chicago [who was not one of Zucker's letter writers] to investigate, and he told me, 'According to McCleary *all* of his students walk on water, but if he said nice things about Zucker, it *might* be true."

So, in early 1966, Zucker interviewed for a position at Berkeley, where he presented his research to faculty of the "Group 2, Experimental and Biological" division of the psychology department. The audience included Beach, David Krech, Arnie Leiman, Mark Rosenzweig, and Al Riley. Two things from the job talk stand out in Zucker's memory: (1) Krech challenging him repeatedly about his findings and conclusions and (2) Al Riley, the group Chair, grinning as Zucker rose to the challenge with rebuttal after rebuttal. Zucker returned to Oregon pleased with the interview and cautiously optimistic about his chances for success. Beach had initially told him that the department would reach their decision in 1 week, but 6 weeks passed before he finally called to extend the offer. Zucker accepted a startup package of $1500 (~$13,000 in today's dollars), Robert Goy generously donated some guinea pig cages from the Primate Center, and Zucker opened his lab in December of 1966. Zucker later discovered the reason behind Beach's delay: the Group 2 faculty were not unanimous about his suitability for the position. Zucker first learned about this intradepartmental conflict in a journal article, of all places. In a paper titled "The Perpetuation and Evolution of Biological Science," published in a 1966 issue of *American Psychologist*, Frank Beach (1966) wrote about the challenges scientists must contend with as their research fields evolve. Beach shared an anecdote to underscore the point:

> The department to which I belong decided to add to its staff a young man or woman with special training and interests in physiological psychology. Several candidates were brought to Berkeley to talk to the faculty about the research they had done and hoped to do. One visitor described a series of elegant experiments involving the application of extremely complex biochemical measurements to single nerve cells after the neurons had been dissected out of the brain. I was mightily impressed, and after the meeting adjourned I asked one of my colleagues if he would favor appointment of the speaker to our department. To my surprise the reply was, "Definitely not! The work is first-rate, but it isn't psychology." Now this reaction was totally unexpected because the individual who expressed it has for many years been a leader in research on brain chemistry and learning ability. When I commented on the apparent discrepancy in his attitude my friend's reply was wryly amusing and quite instructive. He said, "For years I have been arguing that the real future of psychology lies in precisely the kind of research that young man has been doing. But when I suddenly meet the future face-to-face, it scares the hell out of me.

The visitor, of course, was Zucker, who had presented data on responses of individual nerve cells to steroid treatments. The colleague was David Krech, a student of Tolman's and Lashley's and a pioneer in the study of neurochemistry and neuroplasticity. Krech eventually relented, and Zucker's appointment was finalized. Upon arrival at Berkeley, Zucker was assigned in the office immediately adjacent to Krech, and over the years, the two became good friends.

Berkeley (1966–2022)

Zucker recalls getting off to a fast start: thanks to his startup funding, he was conducting experiments within a few weeks of arrival. He was also fortunate to find three graduate students, whom Beach had already admitted but could not accommodate, interested in working with him. Ron Chester, who took a master's degree and left the program, conducted studies of sperm transport and pseudopregnancy in rats (Chester & Zucker, 1970). Brad Powers (Ph.D., 1970) picked up on Zucker's research from Oregon, and his early studies focused on effects of progesterone on sex behavior in rats (Powers & Zucker, 1969). George Wade (Ph.D., 1970) was interested in estrogen influences on food intake (Wade & Zucker, 1970); he completed all of his Ph.D. research in just 18 months.

It was during these early years that Zucker's attention first turned to the study of biological rhythms. As he tells it, his interests in chronobiology were the result of several factors. First, because of his earlier work on taste preferences (at McGill and in Berkeley), he was aware of work by Curt P. Richter, who had published an important volume on the clinical relevance of biological rhythms and had also reported extensively on the effects of lesions and endocrine gland removal on circadian rhythms in rats (Richter, 1965). But the triggering event was an animal behavior seminar that Zucker was teaching which used readings from *Mechanisms of Animal Behavior* (Marler & Hamilton, 1966). The book contained a chapter on circadian rhythms, and Zucker recalls a having visceral response to seeing raster plots of free-running circadian rhythms: "I was blown away, they were so aesthetically pleasing."

Zucker also credits his students with motivating his pursuit of chronobiology research. "Given my interest in food intake," he said, "I started doing studies on the light-dark cycle and food intake. George Wade was indirectly responsible for my doing that: we were studying saccharin preferences at the time, and he observed large differences in how much saccharin rats consumed in the dark versus the light phase. Zucker remembers this capturing his interest and asking Wade at the time, "Are you planning on pursuing that?" [George] said 'no', so I pursued the topic."

Finally, Zucker recalls the chain of events that led him to investigate the neural substrates of circadian rhythms. Zucker and graduate student, Fred K. Stephan (Ph.D., 1972), were interested in visual pathways that influenced homeostasis and neuroendocrine function. At the time that he was conducting studies of light-dark cycle influences on food and water intake in rats, a paper on visual projections that regulated light influences on the pineal gland was published (Moore et al., 1968). At the time, there were two known visual projections from the retina: the primary optic tract (POT) and the inferior accessory optic system. Lesions of the lateral geniculate nucleus in the appropriate location eliminated the transmission of any visual information downstream of the LGN, yet the pineal gland could still respond to light (Chase et al., 1969). The inference was that there must exist additional, as yet unknown, visual projections. Studies had suggested direct retina-to-hypothalamus projections, but it was not until 1971 that clear evidence of retinohypothalamic (RHT) projections was reported by Moore and colleagues, who used tritiated amino acids to show that an RHT diverged from the POT around the optic chiasm and terminated in the suprachiasmatic hypothalamus (Moore et al., 1971; Moore & Lenn, 1972). The first experiment that Zucker and Stephan performed was to lesion the POT of rats (Stephan & Zucker, 1972b). They reported that exposure to continuous illumination (LL) suppressed water intake in POTx rats to the same degree as in neurologically intact rats. POTx rats were perceptually "blind"—they were incapable of performing a simple black-white discrimination task—and yet information about the lighting conditions was still gaining access to the CNS. They were "blind" but could still "see." They concluded that "the visual pathways which underlie pituitary responsiveness to illumination are distinguishable from those which mediate visually guided behavior" and proposed "that neither the primary nor the accessory optic tracts mediate the entrainment of drinking to the day-night cycle; this function may be fulfilled by a direct retino-hypothalamic pathway." They submitted this work to *Physiology and Behavior* in July of 1971, and it was published in February of 1972. In the discussion, they alluded to a "subsequent report... in which circadian and nocturnal rhythms in drinking are abolished by selective lesions of the rat hypothalamus."

This "subsequent report" described experiments that succeeded in eliminating circadian behavioral rhythms of rats via lesions of the hypothalamic suprachiasmatic nuclei. "Circadian rhythms in drinking behavior and locomotor activity of rats are eliminated by hypothalamic lesions" (Stephan & Zucker, *PNAS*, 1972a) is now recognized as a foundational work in behavioral neuroscience—the first report to identify neural substrates necessary for the generation of circadian behavioral rhythms—in the suprachiasmatic nucleus (SCN) of the hypothalamus. It laid the groundwork for decades of research into the biological substrates of circadian timing by laboratories around the world. Other historical accounts offer many more details of the circumstances leading up to their discovery. Highlights include their initial skepticism prompting them to name a replication experiment "Fred's Folly," the work first being rejected for publication in *Science*, and fascinating accounts of the near-simultaneous discovery, by Bob Moore, that the SCN were necessary for the generation of group-level circadian rhythms in neuroendocrine function. At a 1997 meeting commemorating the discovery of the SCN, Moore (Zucker's neuroanatomy instructor at Chicago) conceded that Zucker's neuroanatomy skills had improved (Weaver, 1998; SRBR, 2022).

Zucker confessed that at the time he did not expect hypothalamic lesions at the terminus of the RHT to eliminate circadian rhythmicity. His initial hypothesis was that the SCN were a relay that merely provided input about light to the circadian clock. The clock itself, if it were even a single neural substrate, was probably localized elsewhere. If this were the case, he reasoned, then SCN-lesioned rats should just exhibit free-running circadian rhythms in the presence of a light-dark cycle. When Zucker was informed that the lesions did not simply eliminate entrainment, but that the animals appeared arrhythmic, Zucker recalls telling Stephen, "You've just done something very important. You better go see if it's true in more animals."

For some time, Zucker had suspected that the pacemaker was somewhere in the hypothalamus—Richter (1965) had previously reported results from a single arrhythmic rat that had sustained a lesion to the anterior hypothalamus, but he never reported any histological analyses to permit better specification of the lesion site. Zucker was also aware that the elimination of circadian rhythmicity via lesions was not a sufficient evidence to conclude that the SCN were the pacemaker. But laboratories around the world took note of the Berkeley discovery, and numerous critical experiments were soon performed that addressed the necessity and sufficiency of the SCN in the generation and entrainment of circadian rhythms: studies offered detailed insights into the consequences of SCN damage (Rusak, 1977a), and data were marshalled to indicate that the SCN contained an autonomous pacemaker (Schwartz & Gainer, 1977; Inouye & Kawamura, 1979). By the end of the decade, a picture emerged, based on convergent evidence, of the SCN as the neural basis of the circadian pacemaker (reviewed in Rusak and Zucker (1979)).

The work in Zucker's lab took a slightly different tack, exploring the diversity of biological phenomena that were impacted by the circadian system. In his words, "I preferred horizontal over vertical analyses." Indeed, many of the reports issued by his group in the years after the initial SCN study were focused on characterizing novel roles for the circadian system in the regulation of physiological processes and complex behaviors. Highlights among these are presented below, in a list that is selective rather than comprehensive, and reflects my biases.

Several key follow-ups to the 1972 report were conducted in Zucker's lab by Benjamin Rusak (Ph.D., 1975) who was the first to use appropriate time series analyses in Syrian hamsters to investigate the impact of SCN ablation on circadian activity rhythms (Rusak, 1977a). He evaluated the limits of entrainment of the hamster circadian system and confirmed the necessity of the SCN for circadian function in hamsters (Rusak, 1977b).

Another line of research, performed with Kathleen Fitzgerald (MA thesis), identified an intimate temporal relation between the period of the circadian clock and the length of the estrous cycle; under constant free-running conditions, the estrous cycle length was invariably equal to four times the period of the circadian clock (Fitzgerald & Zucker, 1976). Even when circadian period was artificially lengthened to periods in excess of 25 h via treatment with heavy water, the quadruple-multiple relation remained intact. The work squarely implicated the circadian system in the timing of the estrous cycle and supported a model in which the circadian clock created a "temporal gate" during a fraction of each day during which sexual receptivity could occur if other neuroendocrine conditions permitted. Later work with Marie Carmichael (Ph.D., 1983) and Randy Nelson (who received two Ph.D. degrees for work with Zucker) further dissected the role of the circadian clock in the estrous cycle, showing that gradual entrainment of hamsters to very short T-cycles still failed to desynchronize the activity and estrous rhythms (Carmichael et al., 1981). Subsequent studies by Lawrence Morin, a postdoc, and Fitzgerald showed that "scalloping" behavior of female hamsters, often regarded as a nuisance source of variability, was due to effects of ovarian hormones on the circadian system (Morin et al., 1977). This was the first demonstration that female reproductive hormones provided feedback effects on the circadian clock. In vivo, estradiol altered circadian clock function in a predictable manner over the course of the estrous cycle, advancing the clock every fourth day. Related studies also demonstrated that the circadian clock was sexually differentiated by exposure to gonadal hormones early in life (Zucker et al., 1980).

In the mid- to late 1970s, Zucker developed an interest in how the circadian system influenced seasonal rhythms. His research group was among the first to appreciate the importance of reciprocal interactions between the circadian clock and seasonal timekeeping mechanisms. Studies performed with Eric Bittman (Ph.D., 1978) and Bruce Goldman demonstrated that the SCN played a critical role in reproductive seasonality in hamsters but that the SCN were not necessary targets for melatonin to induce autumnal gonadal regression (Bittman, 1978; Bittman et al., 1979). An important study showed that vernal reproductive photorefractoriness was not due to pineal gland depletion of melatonin but rather a result of insensitivity of neuroendocrine target tissues to the inhibitory effects of melatonin (Bittman & Zucker, 1981). In addition, photoperiodism experiments by Morin elegantly showed that the short-day decline in sex behavior of male Syrian hamsters was not solely a result of seasonal gonadal regression (and the associated withdrawal of gonadal hormone secretions), but that steroid-responsive neural substrates that participate in the generation of male sex behavior simultaneously became refractory to gonadal steroids in winter (Morin & Zucker, 1978).

For many years, Zucker also studied circannual rhythms using golden-mantled ground squirrels. Ground squirrels are not iteroparous within a single season, and so instead of incurring the expenses of maintaining year-round breeding pairs which were only capable of issuing one litter each year, Zucker outsourced his breeding colony to Mother Nature. Each spring the Zucker's lab would receive a phone call from a ranger in the Sierra Nevada, informing them that female squirrels had finally emerged from hibernation. Within a few days, most of these females would become pregnant (in the field), after which time they could be trapped, and carefully transported to the Berkeley lab, where they would deliver pups a few weeks later. Ground squirrels exhibit spectacular seasonal rhythms in adiposity, reproduction, and hibernation, and studies conducted with Terri Lee and John Dark (both began as postdocs) established a role for the SCN in the timing of circannual rhythms in body weight and reproductive physiology (Lee & Zucker, 1991). Later work with Dark and Bud Ruby (Ph.D., 1991) elaborated a more complex role for the SCN in the generation of circannual rhythms in hibernation, indicating that hibernation per se was SCN-independent, but in many individuals, the successful timing of the hibernation season required an intact SCN (Ruby et al., 1996).

Zucker achieved remarkable success in maintaining funding for his research. He received a 5-year Career Development Award from the NIH early in his time at Berkeley and from 1967 through 2012 maintained at least one active R01-level grant from the NIH. For 18 years, two R01s were in play, often with one focusing on photoperiodism and the other on circannual rhythms.

From 1984 through 1988, Zucker was on an NIH study section (Biopsychology), and it was during the regular trips to DC for meetings that he forged a lifelong friendship with Bruce Goldman, a professor of Biology at the University of Connecticut. Goldman and Zucker shared interests in melatonin, reproduction, and comparative approaches to biology, and they would engage in friendly debate at study section meetings and at conferences. Zucker would regularly invite Goldman to provide comments and feedback on his grant proposals prior to submission, and he credits Goldman with offering critiques that helped him submit successful proposals. It was partly due to Goldman's influence that Zucker began many years of study of Siberian hamsters, which proved to be tractable models for investigations of circadian and seasonal patterns of food intake, reproduction, and thermoregulation (torpor). Using Siberian hamsters as a model species, Zucker's lab demonstrated that the SCN played a central role in the timing of daily torpor, but that daily torpor could persist in the absence of the SCN as well (Ruby et al., 1989). Later studies with Matthew Paul (Ph.D., 2005) showed that in the absence of an SCN, the timing of entry into torpor was phase locked to the timing of food intake (Paul et al., 2004). Hamsters were also an excellent model for studying how seasonal changes in photoperiod and melatonin affect the neuroendocrine system, a study performed with David Freeman, a postdoc coming from the Goldman lab, who identified a key mechanism in the control of seasonal reproduction by showing that refractoriness to melatonin developed independently at numerous hypothalamic and thalamic sites after prolonged exposure to short days (Freeman & Zucker, 2001). Also, Michael Gorman (Ph.D., 1995) and Matthew Butler (Ph.D., 2007) used Siberian hamsters to develop important new models of how photoperiodic rodents measure seasonal time (Gorman, 1995; Gorman & Zucker, 1995; Butler et al., 2007, 2010).

Given the diversity of his research interests over the years, it is appropriate to mention that Zucker has been an adherent to the Krogh Principle—that there is an animal of choice for which a physiological process is most conveniently studied (Krogh, 1929; Krebs & Krebs, 1980)—and as a result, his canon has a remarkably comparative appearance. Over the years, his lab has maintained colonies of many different rodent species (squirrels, voles, deer mice, white footed mice, grasshopper mice, cotton rats, and bats, not to mention multiple hamster varieties) each uniquely suited to asking a different question in the realm of behavioral neuroscience.

Zucker also found inspiration in serendipity and embodied Pasteur's legendary adage: "chance favors only the prepared mind". In the early days of keeping Siberian hamsters in Tolman Hall, a vivarium door was inadvertently left ajar, bathing a breeding colony room in hallway light overnight (not a desirable event in a photoperiodism lab). A few weeks later, Zucker and Norah Spears (a postdoctoral fellow working on Siberian hamsters at the time) noticed that the pups which were weaned on the days surrounding the "light accident" exhibited more fully developed gonads than hamsters in the same room that had been born just a few days earlier or later. These accidental data launched several empirical reports that characterized an interval of heightened sensitivity to photoperiod in male and female hamsters around the time of weaning (Spears et al., 1990). Nelson and Glickman (2009) shared a quote from an anonymous grant reviewer that also captures Zucker's enviable combination of encyclopedic knowledge and creative insight: "The PI is world renowned for his uncanny ability to sift important new ideas from the throw-away noise in experimental data..."

Over his career, Zucker received many honors and recognitions for his research. His success at securing grants was one example of this, but he was also regularly solicited to speak at conferences. He served on the editorial boards of multiple journals and advisory boards of several societies. In addition to his NIH Career Development Award, he received the Outstanding Scientific Achievement Award from the Sleep Research Society. Zucker was elected to one term as president of the Society for Research on Biological Rhythms and is a fellow in the American Association for Advancement of Science, the American Psychological Society, and the California Academy of Science.

Perhaps the most significant tribute to his career occurred in 2007 when he received the Daniel S. Lehrman Lifetime Achievement Award, from his peers in the Society for Behavioral Neuroendocrinology. The Lehrman Award recognized his outstanding career as a scientist and as a mentor, both in a formal capacity (to 31 Ph.D. students, 21 postdocs, 9 masters students, 5 visiting scientists, and countless undergraduates who have gone on to remarkable careers in science, medicine, and society) and in an informal capacity (to students and colleagues in neighboring labs at Berkeley and beyond). It is a source of great pleasure to Zucker that multiple of his trainees have received the Beach and Lehrman awards from the SBN.

Zucker officially retired in 2009 but kept his research laboratory open until 2012. The relief from teaching responsibilities and grant writing freed him to pursue new lines of work. One of these is the study of ultradian rhythms. Since 2011, I have had the pleasure of working closely with Zucker, along with several of our colleagues, on studies of ultradian rhythms in behavior. Ultradian rhythms have been largely neglected in chronobiology, and over the past decade, we have been examining how sex, gonadal hormones, the estrous cycle, and the circadian clock modulate the expression of ultradian behavioral rhythms (reviewed in Prendergast and Zucker (2016)).

A second line of research has examined sex biases in biomedical research. In a landmark paper, Zucker and Annaliese Beery (Ph.D., 2008) documented the historical underrepresentation of female subjects in basic and preclinical biology research (Beery & Zucker, 2011; Zucker & Beery, 2010). This work was undertaken at the repeated urging of Greg Demas of Indiana University. Later, Zucker and I worked together on a meta-analysis that found no systematic pattern of greater variability in female mice, contradicting the long-held assumption that females on average are more variable than males (Prendergast et al., 2014). Zucker has also worked on elegant quantitative studies, performed in collaboration with Benjamin Smarr (a postdoctoral fellow) and Lance Kriegsfeld at Berkeley, that examined the sources of variability in female and male animals, with a focus on how trait-specific variability changes differently in the sexes over ultradian, circadian, and infradian timescales (Smarr et al., 2017, 2019). Together, Zucker's recent works in the area of sex bias have contributed to progress in female inclusion, and they stand to have a lasting impact on how sex-conscious research is performed going forward (Zucker et al., 2021). In 2015, his scientific achievements and contributions to research practices were recognized with the Outstanding Science Award from the Swiss Laboratory Animal Science Association.

Like many of his graduate students, I was welcomed into his laboratory with disarming friendliness, patiently guided through literature reviews, supported countless applications, and had an enduring impact on my scientific tastes. I enjoyed foraging with him for fresh cheese rolls and ripe cherries throughout Berkeley and celebratory lunches in the gourmet ghetto, experiences that are undoubtedly shared by many of his students and colleagues. Later in my career, he welcomed me and my family to Berkeley for a sabbatical. He remains an important influence in my career and my life.

As a mentor, Zucker challenged his students to read extensively, to patiently design experiments, and to think critically about outcomes. Zucker mentored with grace and empathy and with an uncommon warmth and kindness. He prided himself on not being a "one-size-fits-all" advisor, instead adapting to the unique needs of individual students.

A review of Zucker's career would not be complete without considering the influences of two of his Berkeley colleagues: Frank Beach and Steve Glickman. Beach was already at Berkeley when Zucker arrived in 1966, and Glickman arrived 2 years later. For 10 years, the three team taught a proseminar in biological psychology, required of all first year Ph.D. students, later joined by Paul Sherman and then by Marc Breedlove. Zucker recalls these courses as wonderful professional experiences: Beach and Glickman consistently emphasized viewing behavior and biology from an evolutionary perspective, and Zucker learned a great deal about the importance of incorporating comparative approaches from them (attested to, in part, by the menagerie of rodents that inhabited the Tolman Hall vivarium over the years). Glickman often retold a story of one afternoon in the first year proseminar, featuring Zucker and Beach sparring over Zucker's strong belief in the importance of conducting "strong inference" research: "Strong Inference" [was] a concept drilled into all graduate and postdoctoral trainees moving through the biopsychology program at Berkeley. Once in a graduate proseminar, Zucker was arguing the virtues of avoiding "I wonder what would happen if..." experiments when Frank Beach quipped, "Well, then I must be doing *weak* inference studies because I often wonder what would happen if...". And so it went" (Nelson & Glickman, 2009). Glickman was a master teacher and was a mentor to Zucker. He claims to have learned the most about the craft of teaching from working alongside Glickman. They were departmental neighbors, they co-mentored many students, and they were dear friends.

From his remarkable initiation into the world of science at McGill, through his long and accomplished career at Berkeley, and continuing into his "retirement,"

Zucker has been a scientific thinker of the highest order. His work has made major contributions to scientific knowledge in multiple disciplines, and he has meaningfully shaped the way science is practiced.

This biography is based on a literature review and a rereading of many papers published by Irv Zucker and his collaborators over the years. It also draws heavily on a series of interviews conducted with Irv from his home in Berkeley, California, during the winter of 2021–2022.

References

Barraclough, C. A., & Gorski, R. A. (1962). Studies on mating behaviour in the androgen-sterilized female rat in relation to the hypothalamic regulation of sexual behaviour. *Journal of Endocrinology, 25*(2), 175–182.

Beach, F. A. (1966). The perpetuation and evolution of biological science. *The American Psychologist, 21*(10), 943–949.

Beery, A. K., & Zucker, I. (2011). Sex bias in neuroscience and biomedical research. *Neuroscience & Biobehavioral Reviews, 35*(3), 565–572.

Bittman, E. L. (1978). Hamster refractoriness: The role of insensitivity of pineal target tissues. *Science, 202*(4368), 648–650.

Bittman, E. L., & Zucker, I. (1981). Photoperiodic termination of hamster refractoriness: Participation of the pineal gland. *Biology of Reproduction, 24*(3), 568–572.

Bittman, E. L., Goldman, B. D., & Zucker, I. (1979). Testicular responses to melatonin are altered by lesions of the suprachiasmatic nuclei in golden hamsters. *Biology of Reproduction, 21*(3), 647–656.

Butler, M. P., Trumbull, J. J., Turner, K. W., & Zucker, I. (2007). Timing of puberty and synchronization of seasonal rhythms by simulated natural photoperiods in female Siberian hamsters. *American Journal of Physiology-Regulatory, Integrative and Comparative Physiology, 293*(1), R413–R420.

Butler, M. P., Turner, K. W., Park, J. H., Schoomer, E. E., Zucker, I., & Gorman, M. T. (2010). Seasonal regulation of reproduction: altered role of melatonin under naturalistic conditions in hamsters. *Proceedings of the Biological Science, 277*, 2867–2874.

Carmichael, M. S., Nelson, R. J., & Zucker, I. (1981). Hamster activity and estrous cycles: Control by a single versus multiple circadian oscillator (s). *Proceedings of the National Academy of Sciences, 78*(12), 7830–7834.

Chase, P. A., Seiden, L. S., & Moore, R. Y. (1969). Behavioral and neuroendocrine responses to light mediated by separate visual pathways in the rat. *Physiology & Behavior, 4*(6), 949–952.

Chester, R. V., & Zucker, I. (1970). Influence of male copulatory behavior on sperm transport, pregnancy and pseudopregnancy in female rats. *Physiology & Behavior, 5*(1), 35–43.

Fitzgerald, K. M., & Zucker, I. (1976). Circadian organization of the estrous cycle of the golden hamster. *Proceedings of the National Academy of Sciences, 73*(8), 2923–2927.

Freeman, D. A., & Zucker, I. (2001). Refractoriness to melatonin occurs independently at multiple brain sites in Siberian hamsters. *Proceedings of the National Academy of Sciences, 98*(11), 6447–6452.

Gorman, M. R. (1995). Seasonal adaptations of Siberian hamsters. I. Accelerated gonadal and somatic development in increasing versus static long day lengths. *Biology of Reproduction, 53*(1), 110–115.

Gorman, M. R., & Zucker, I. (1995). Seasonal adaptations of Siberian hamsters. II. Pattern of change in day length controls annual testicular and body weight rhythms. *Biology of Reproduction, 53*(1), 116–125.

Inouye, S. L., & Kawamura, H. (1979). Persistence of circadian rhythmicity in a mammalian hypothalamic "island" containing the suprachiasmatic nucleus. *Proceedings of the National Academy of Sciences, 76*(11), 5962–5966.

Krebs, H. A., & Krebs, J. R. (1980). The "August Krogh principle". *Comparative Biochemistry and Physiology Part B: Comparative Biochemistry, 67*(3), 379–380.
Krogh, A. (1929). The progress of physiology. *American Journal of Physiology-Legacy Content, 90*(2), 243–251.
Lee, T. M., & Zucker, I. (1991). Suprachiasmatic nucleus and photic entrainment of circannual rhythms in ground squirrels. *Journal of Biological Rhythms, 6*(4), 315–330.
Marler, P. R., & Hamilton, W. J. (1966). *Mechanisms of animal behavior*. Wiley.
Milner, P., & Zucker, I. (1965). Specific hunger for potassium in the rat. *Psychonomic Science, 2*(1), 17–18.
Moore, R. Y., & Lenn, N. J. (1972). A retinohypothalamic projection in the rat. *Journal of Comparative Neurology, 146*(1), 1–4.
Moore, R. Y., Heller, A., Bhatnager, R. K., Wurtman, R. J., & Axelrod, J. (1968). Central control of the pineal gland: Visual pathways. *Archives of Neurology, 18*(2), 208–218.
Moore, R. Y., Karapas, F., & Lenn, N. J. (1971). A retinohypothalamic projection in the rat. *The Anatomical Record, 169*, 382.
Morin, L. P., & Zucker, I. (1978). Photoperiodic regulation of copulatory behaviour in the male hamster. *Journal of Endocrinology, 77*(2), 249–258.
Morin, L. P., Fitzgerald, K. M., & Zucker, I. (1977). Estradiol shortens the period of hamster circadian rhythms. *Science, 196*(4287), 305–307.
Nelson, R. J., & Glickman, S. E. (2009). 2007 Daniel S. Lehrman lifetime achievement award contributions from a pioneer in behavioral endocrinology: Irving Zucker. *Hormones and Behavior, 55*(4), 482–483.
Paul, M. J., Kauffman, A. S., & Zucker, I. (2004). Feeding schedule controls circadian timing of daily torpor in SCN-ablated Siberian hamsters. *Journal of Biological Rhythms, 19*(3), 226–237.
Phoenix, C. H., Goy, R. W., Gerall, A. A., & Young, W. C. (1959). Organizing action of prenatally administered testosterone propionate on the tissues mediating mating behavior in the female Guinea pig. *Endocrinology, 65*(3), 369–382.
Powers, J. B., & Zucker, I. (1969). Sexual receptivity in pregnant and pseudopregnant rats. *Endocrinology, 84*(4), 820–827.
Prendergast, B. J., & Zucker, I. (2016). Ultradian rhythms in mammalian physiology and behavior. *Current Opinion in Neurobiology*, (40), 150–154.
Prendergast, B. J., Onishi, K. G., & Zucker, I. (2014). Female mice liberated for inclusion in neuroscience and biomedical research. *Neuroscience & Biobehavioral Reviews*, (40), 1–5.
Richter, C. P. (1965). *Biological clocks in medicine and psychiatry*. Charles C. Thomas, Publisher.
Ruby, N. F., Ibuka, N. O., Barnes, B. M., & Zucker, I. R. (1989). Suprachiasmatic nuclei influence torpor and circadian temperature rhythms in hamsters. *American Journal of Physiology-Regulatory, Integrative and Comparative Physiology, 257*(1), R210–R215.
Ruby, N. F., Dark, J., Heller, H. C., & Zucker, I. (1996). Ablation of suprachiasmatic nucleus alters timing of hibernation in ground squirrels. *Proceedings of the National Academy of Sciences, 93*(18), 9864–9868.
Rusak, B. (1977a). Involvement of the primary optic tracts in mediation of light effects on hamster circadian rhythms. *Journal of Comparative Physiology, 118*(2), 165–172.
Rusak, B. (1977b). The role of the suprachiasmatic nuclei in the generation of circadian rhythms in the golden hamster, Mesocricetus auratus. *Journal of Comparative Physiology, 118*(2), 145–164.
Rusak, B., & Zucker, I. R. (1979). Neural regulation of circadian rhythms. *Physiological Reviews, 59*(3), 449–526.
Schwartz, W. J., & Gainer, H. (1977). Suprachiasmatic nucleus: Use of 14C-labeled deoxyglucose uptake as a functional marker. *Science, 197*(4308), 1089–1091.
Smarr, B. L., Grant, A. D., Zucker, I., Prendergast, B. J., & Kriegsfeld, L. J. (2017). Sex differences in variability across timescales in BALB/c mice. *Biology of Sex Differences, 8*, 7.
Smarr, B. L., Rowland, N. E., & Zucker, I. (2019). Male and female mice show equal variability in food intake across 4-day spans that encompass estrous cycles. *PLoS One, 14*(7), e0218935.

Society for Research on Biological Rhythms. (2022). Retrieved from https://srbr.org/about-chronobiology/chronohistory on 5 January 2022.

Spears, N., Finley, C. M., Whaling, C. S., Tuthill, C. R., & Zucker, I. (1990). Sustained reproductive responses in Djungarian hamsters (Phodopus sungorus) exposed to a single long day. *Reproduction, 88*(2), 635–643.

Stephan, F. K., & Zucker, I. (1972a). Circadian rhythms in drinking behavior and locomotor activity of rats are eliminated by hypothalamic lesions. *Proceedings of the National Academy of Sciences, 69*(6), 1583–1586.

Stephan, F. K., & Zucker, I. (1972b). Rat drinking rhythms: Central visual pathways and endocrine factors mediating responsiveness to environmental illumination. *Physiology & Behavior, 8*(2), 315–326.

Wade, G. N., & Zucker, I. (1970). Modulation of food intake and locomotor activity in female rats by diencephalic hormone implants. *Journal of Comparative and Physiological Psychology, 72*(2), 328.

Weaver, D. R. (1998). The suprachiasmatic nucleus: A 25-year retrospective. *Journal of Biological Rhythms, 13*(2), 100–112.

Young, W. C. (1961). *Sex and internal secretions* (2nd ed.). William & Wilkins.

Zucker, I. (1965a). Effect of lesions of the septal-limbic area on the behavior of cats. *Journal of Comparative and Physiological Psychology, 60*(3), 344.

Zucker, I. (1965b). Short-term salt preference of potassium-deprived rats. *American Journal of Physiology-Legacy Content, 208*(6), 1071–1074.

Zucker, I. (1966). Facilitatory and inhibitory effects of progesterone on sexual responses of spayed guinea pigs. *Journal of Comparative and Physiological Psychology, 62*(3), 376.

Zucker, I. (1967a). Suppression of oestrous behaviour in the immature male rat. *Nature, 216*(5110), 88–89.

Zucker, I. (1967b). Progesterone in the experimental control of the behavioural sex cycle in the female rat. *The Journal of Endocrinology, 38*(3), 269–277.

Zucker, I., & Beery, A. K. (2010). Males still dominate animal studies. *Nature, 465*(7299), 690.

Zucker, I., & Bindra, D. (1961). Peripheral sensory loss and exploratory behaviour. *Canadian Journal of Psychology/Revue Canadienne de Psychologie, 15*(4), 237.

Zucker, I., & McCleary, R. A. (1964). Perseveration in septal cats. *Psychonomic Science, 1*(1), 387–388.

Zucker, I., & Milner, P. (1964). The effect of amphetamine on established hoarding in the rat. *Psychonomic Science, 1*, 367.

Zucker, I., Fitzgerald, K. M., & Morin, L. P. (1980). Sex differentiation of the circadian system in the golden hamster. *American Journal of Physiology-Regulatory, Integrative and Comparative Physiology, 238*(1), R97–R101.

Zucker, I., Prendergast, B. J., & Beery, A. K. (2021). Pervasive neglect of sex differences in biomedical research. *Cold Spring Harbor Perspectives in Biology, 14*, a039156.

Bruce D. Goldman

25

H. Elliott Albers and Irving Zucker

Abstract

Bruce D. Goldman (b.1940) trained as an endocrinologist at the University of Wisconsin, the Medical College of Georgia, the University of Texas Medical School and the University of California, Los Angeles (UCLA). His early research focused on the regulation of pituitary gonadotropic hormones, with major contributions to our understanding of the actions of these hormones on reproductive physiology and behavior throughout development. He is known for utilizing comparative approaches to study mammalian adaptations to seasonal environmental changes. One of his most important contributions was defining the mechanisms responsible for the transfer of photoperiodic information from the circadian to the reproductive system. His laboratory provided the first strong evidence that the pineal hormone melatonin mediates photoperiodic control of reproduction on a seasonal basis by varying the duration of nightly melatonin secretion in relation to changes in day length. They subsequently documented how seasonal melatonin signals, steroid, and gonadotrophin hormones control daily torpor, hibernation, and pelage changes.

Keywords

Seasonal rhythms · Photoperiod · Melatonin · Circadian rhythms

H. E. Albers (✉)
Georgia State University, Atlanta, GA, USA
e-mail: biohea@gsu.edu

I. Zucker
University of California, Berkeley, CA, USA
e-mail: irvzuck@berkeley.edu

R. J. Nelson, Z. M. Weil (eds.), *Biographical History of Behavioral Neuroendocrinology*, https://doi.org/10.1007/978-3-031-12970-4_25

Bruce Goldman was born on 11 December 1940, in Gary Indiana. He received a BS in Zoology from the University of Michigan in 1962 where he was strongly influenced by the evolutionary biologist Richard Alexander and a MS in Zoology from the University of Wisconsin in 1966 in the laboratory of R.K. Meyer. Goldman's dissertation work was with Virendra Mahesh at the Medical College of Georgia where he received training in steroid biology and was provided an excellent opportunity to pursue his interests in reproductive hormones. His postdoctoral training began with John Porter at the University of Texas Medical School where he learned the techniques of radioimmunoassay that became so important to his career. He finished postdoctoral training with Roger Gorski at UCLA. He joined the faculty of the Department of Biobehavioral Sciences at the University of Connecticut in 1970 and rose to the rank of professor in 1979. For 6 years, he was a senior scientist at the Worcester Foundation of Experimental Biology but ultimately returned to the University of Connecticut, where he remained until he retired as a Professor Emeritus in 2003.

Goldman's research career is distinguished by its remarkable breadth and depth. His work made major contributions to endocrinology, neuroendocrinology, social behavior, biological rhythms, photoperiodism, and hibernation. He wrote influential reviews on a wide range of topics including puberty, the structure of protein and peptide hormones, histology of the pituitary, the physiology of progestins, sexual differentiation, patterns of sexual and maternal behavior, the physiology of melatonin, photic influences on the developing mammal, and effects of photoperiod and steroid hormones on hibernation and daily torpor.

His work was technically and theoretically innovative, emphasizing a comparative approach to great effect. Goldman was motivated by his true love of the process of science, discussion, debate, data collection, and theoretical interpretations. He was often the quietest voice in the room but the one with the biggest ideas. Although his research was aimed at elucidating basic mechanisms, it influenced clinical investigation and practice, notably on the mechanisms underlying seasonality which had an impact on psychiatrists at the National Institute of Mental Health in investigations of seasonal affective disorder.

As a graduate student and a postdoctoral fellow, Goldman worked on the regulation of the pituitary gonadotropic hormones, follicle-stimulating hormone (FSH) and luteinizing hormone (LH), and the role these hormones play in reproductive physiology and behavior. His early studies increased understanding of the induction of ovulation and ovarian maturation resulting from the actions of FSH and LH. Goldman examined the role of FSH and LH during development and was the first to demonstrate that these hormones are actively secreted during the neonatal period. He showed that secretion of FSH and LH in neonates was controlled by negative feedback effects of gonadal hormones and that blocking the action of these hormones in neonates can at least partially block the organizational effects of androgens on adult sexual behavior, presumably by inhibiting the production of testicular androgen (McCullough et al., 1974). During his foundational years at Connecticut, Goldman also began new research initiatives including studies of maternal behavior with graduate student Robert Bridges and investigation of the sexual differentiation of aggressive behavior with graduate student Marylynn Barkley and another student in his department, Michael Selmanoff.

In 1974, Goldman initiated his first work related to photoperiodism and the pineal gland. It was an intrepid move. The pineal gland had a checkered history since the days of Descartes. In the 1960s and 1970s, reviews of the pineal gland reinforced the mystery of its functions: one review noting that "the real physiological role of the gland is simply that of a brake that helps to maintain the homeostatic equilibrium of the organism" (Collu & Fraschini, 1972); another review noted that it "functions as a "regulator of regulators" and is not absolutely indispensable for the normal function of the organism" (Kappers 1969). Nevertheless, by the 1970s, it was becoming clear that there was a nexus between the pineal and reproductive physiology, but even this connection "is plagued with some apparent contradictions, inconsistencies, and misconceptions" (Reiter & Sorrentino Jr., 1970). As such, it is not surprising that pineal research was considered by some to be a career graveyard for young endocrinologists.

By the time Goldman began his studies on the pineal gland, some considered it to be a "transducer," possibly adjusting the level of reproductive activity to seasonal changes in environmental conditions. Many concluded that the pineal hormone had "anti-gonadal effects" although there was some evidence that it could also have "pro-gonadal" actions. Some progress had been made in demonstrating that the circadian clock in the suprachiasmatic nucleus was responsible for photoperiodic time measurement. The major challenge that remained was determining the mechanisms responsible for the transfer of photoperiodic information from the circadian system to the reproductive axis. Some of the first clues came from Goldman's studies with his graduate student, Lawrence Tamarkin, who reported that single injections of melatonin into *pineal-intact* Syrian hamsters (*Mesocricetus auratus*) housed in long summer-like photoperiods produced winter-like gonadal atrophy if the injections were given at the beginning of the dark phase but not if they were administered in the morning or during the middle of the night. In pinealectomized hamsters, three daily injections were required to evoke gonadal atrophy (Tamarkin et al., 1976, 1977). This left open a number of possibilities on how melatonin communicates "time of year" information. Goldman realized they would need to develop a new approach, where dose, time of day, and duration of melatonin could be manipulated and where the response to melatonin could be determined much more rapidly than the 6–8 weeks required in Syrian hamsters. The solution involved the use of a time-programmed subcutaneous infusion system, which Goldman first encountered during his graduate work in Wisconsin, for melatonin administration and a species switch to prepubertal Siberian hamsters (*Phodopus sungorus sungorus*) that display much more rapid responses to photoperiod than adults or individuals of other species. With this approach, the Goldman lab was able to solve much of the pineal conundrum. It was the duration of melatonin secretion that communicated the time of year; melatonin had pro-gonadal effects when its duration was short as in spring-summer-like days and anti-gonadal effects when its duration was long as in autumn-winter-like days (Carter & Goldman, 1983a, b; Bartness & Goldman, 1989). Siberian hamsters selectively infused with melatonin into the suprachiasmatic nucleus, paraventricular nucleus of the thalamus, or nucleus reuniens regions, showed a marked inhibition of gonadal growth (Badura & Goldman, 1992).

In other studies, Goldman and co-workers discovered that photoperiodic information could be communicated from mothers to their fetuses via the maternal rhythm of pineal melatonin secretion (Elliott & Goldman, 1989). Melatonin rhythms of 2-week-old hamsters were differentially affected by the photoperiods experienced by their dams during gestation. Surprisingly, this influence of the mother was observed in male but not female pups.

To investigate temperature regulation, the Goldman's lab studied daily torpor and hibernation in several hamster species. Siberian hamsters display energy savings during daily torpor in winter conditions, and Turkish (*Mesocricetus brandti*) and European hamsters (*Cricetus cricetus*) are true hibernators with multiday bouts, during which they maintain low body temperatures. Seasonal timing mechanisms were not disrupted during hibernation, and increases in testosterone secretion were implicated in terminating hibernation. Using Siberian hamsters, Goldman and Marylyn Duncan (Duncan & Goldman, 1984) showed that dramatic seasonal changes in pelage color and insulation capacity are mediated by seasonal changes in circulating prolactin.

Goldman became enthralled and perhaps a little obsessed with naked mole rats. As eusocial mammals, naked mole rats live in large colonies with a small number of breeders and a large number of nonbreeding subordinates who participate in various tasks that keep the colony viable. In addition to their unusual social structure, naked mole rats also are extremely long-lived, with individuals surviving for up to 30 years in captivity. Given their complex social organization and other challenges, maintaining breeding colonies of naked mole rats is no simple task but one that Goldman relished. In the 1990s, his laboratory was one of the few in the United States to maintain mole rat colonies. Goldman and other researchers noted that the nonbreeding mole rats were remarkably monomorphic in behaviors and physical appearance, even including the external genitalia . These observations were particularly interesting to Goldman in relation to his earlier work in sexual differentiation in rats and his training with Roger Gorski. Goldman was further motivated by his idea that the relative suppression of sex differences might have evolved in naked mole rats in relation to the lack of sexual functions by most members in each colony, with all their activities directed to a common "goal": supporting the reproductive efforts of the few breeding animals. The functions carried out by these nonbreeders might not benefit from sexual differentiation. This eventually led to a decade-long study of sexual differentiation in naked mole rats in collaborative studies with Nancy Forger, Marianne Seney, Sharry Goldman, and Melissa Holmes. They discovered that social status, and not sex, controls the morphology of neural regions that are sexually dimorphic in other mammals (Holmes et al., 2007, 2009). When Goldman retired, he bequeathed the colonies to Melissa Holmes, who had just started her own laboratory in Toronto, and to Daniel McCloskey (CUNY). Those animals – and their descendants – are in active use in many labs to this day. Goldman also extended his studies on eusocial mammals to the Damaraland mole rat (*Fukomys damarensis*) with his student David Freeman, discovering that sexual behavior is unusually independent from the gonads and instead is affected by social cues, in this highly social animal.

Goldman was a prolific mentor throughout his career. Many of his trainees went on to establish their own successful labs and to contribute substantially to the field of behavioral neuroendocrinology. When one first met Goldman, he seemed extremely reserved; however, once the conversation turned to science, he was extremely engaging. He loved in-depth discussions with his trainees. Goldman and his trainees would have lunch together where they discussed papers that had just been published and the latest data from the lab. He had a quiet but effective way of challenging students to tackle difficult problems. He was a hands-on mentor and spent hours at the lab bench.

Goldman contributed significantly to the profession in important ways. He served on influential editorial boards for journals such as *American Journal of Physiology*, *Neuroendocrinology*, *Hormones and Behavior* and the *Journal of Pineal Research*. He was on the editorial board and an associate editor of the *Journal of Biological Rhythms* for more than a decade and an influential voice in the development of the field of behavioral neuroendocrinology through service on the NIH Endocrinology Study Section and the NIH Biopsychology Study Section.

As one of his many trainees, I (HEA) can testify to the huge impact Bruce has had on many lives and careers. I learned more about how to address scientific questions from Bruce than from anyone else in my career. I will always treasure the time I spent in his lab for the science and the fun. IZ has enjoyed nearly five decades of interactions with Bruce. He had a greater impact on work in my lab than any other scientist and was my severest critic. I value our friendship and the privilege of seeing a side of him not frequently on view, namely, his wicked sense of humor.

References

Badura, L. L., & Goldman, B. D. (1992). Central sites mediating reproductive responses to melatonin in juvenile male Siberian hamsters. *Brain Research, 598*(1–2), 98–106.

Bartness, T. J., & Goldman, B. D. (1989). Mammalian pineal melatonin: A clock for all seasons. *Experientia, 45*, 939–945.

Carter, D. S., & Goldman, B. D. (1983a). Antigonadal effects of timed melatonin infusion in pinealectomized male Djungarian hamsters (*Phodopus sungorus sungorus*): Duration is the critical parameter. *Endocrinology, 113*(4), 1261–1267.

Carter, D. S., & Goldman, B. D. (1983b). Progonadal role of the pineal in the Djungarian hamster (*Phodopus sungorus sungorus*): Mediation by melatonin. *Endocrinology, 113*(4), 1268–1273.

Collu, R., & Fraschini, F. (1972). The pineal gland--a neuroendocrine transducer. *Advances in Metabolic Disorders, 6*, 161–175.

Duncan, M. J., & Goldman, B. D. (1984). Hormonal regulation of the annual pelage color cycle in the Djungarian hamster, Phodopus sungorus. II. Role of prolactin. *The Journal of Experimental Zoology, 230*(1), 97–103.

Elliott, J. A., & Goldman, B. D. (1989). Reception of photoperiodic information by fetal Siberian hamsters: Role of the mother's pineal gland. *The Journal of Experimental Zoology, 252*(3), 237–244.

Holmes, M. M., Rosen, G. J., Jordan, C. L., de Vries, G. J., Goldman, B. D., & Forger, N. G. (2007). Social control of brain morphology in a eusocial mammal. *Proceedings of the National Academy of Sciences of the United States of America, 104*(25), 10548–10552.

Holmes, M. M., Goldman, B. D., Goldman, S. L., Seney, M. L., & Forger, N. G. (2009). Neuroendocrinology and sexual differentiation in eusocial mammals. *Frontiers in Neuroendocrinology, 30*(4), 519–533.

Kappers, J. A. (1969). The mammalian pineal organ. *Neurohormones and Neurohumors,* 140–184.

McCullough, J., Quadagno, D. M., & Goldman, B. D. (1974). Neonatal gonadal hormones: Effect on maternal and sexual behavior in the male rats. *Physiology and Behavior, 12*(2), 183–188.

Reiter, R. J., & Sorrentino, S., Jr. (1970). Reproductive effects of the mammalian pineal. *American Zoologist, 10*(2), 247–258.

Tamarkin, L., Westrom, W. K., Hamill, A. I., & Goldman, B. D. (1976). Effect of melatonin on the reproductive systems of male and female Syrian hamsters: A diurnal rhythm in sensitivity to melatonin. *Endocrinology, 99*(6), 1534–1541.

Tamarkin, L., Lefebvre, N. G., Hollister, C. W., & Goldman, B. D. (1977). Effect of melatonin administered during the night on reproductive function in the Syrian hamster. *Endocrinology, 101*(2), 631–634.

Norman Adler

26

Elizabeth Adkins-Regan

Abstract

Norman Tenner Adler was a scientific leader in the second generation of behavioral neuroendocrinologists, the students of the original founders of the field. In his studies of copulation and fertilization in rats—elements of reproduction known to be hormonally regulated—he showed the critical role of males' behavioral mating pattern in generating the female hormonal responses required for fertilization and implantation. This work was instrumental in the development of the idea that the relationship between hormones and behavior is bidirectional, and it showed that behavioral influences on reproductive physiology are functionally adaptive. His research on circadian rhythms further contributed to awareness of the significance of external influences on internal physiology. In addition to his research, Adler also made major contributions as an editor and educator to the development and promotion of the field of behavioral neuroendocrinology and of biological psychology more generally.

Keywords

Norman Tenner Adler · Behavioral neuroendocrinology · Rats · Sexual behavior · Fertilization · Implantation · Progesterone · Sperm transport · Estrous cycles · Circadian rhythms

E. Adkins-Regan (✉)
Department of Psychology, Department of Neurobiology and Behavior, Cornell University, Ithaca, NY, USA
e-mail: er12@cornell.edu

R. J. Nelson, Z. M. Weil (eds.), *Biographical History of Behavioral Neuroendocrinology*, https://doi.org/10.1007/978-3-031-12970-4_26

Fig. 26.1 Norman Adler as a professor and administrator at Yeshiva University. (Reproduced with permission of Yeshiva University)

Norman Tenner Adler (Fig. 26.1) was born on 7 June 1941, in Chicago, Illinois, and graduated from South Shore High School, where his interests in behavior and biology first developed (Adler, 1975). He received his BA degree from Harvard University in 1962, where he was a Woodrow Wilson Fellow, a Harvard College Scholar, and was elected to Phi Beta Kappa. There, a course in physiological psychology taught by Paul Rozin was an important influence, and there he conducted an experiment on classical conditioning of aggressive display by Siamese fighting fish that resulted in his first scientific publication (Adler & Hogan, 1963). He received his MA degree in Endocrinology from the University of California, Berkeley, in 1967, where his advisor was Howard Bern, a leading comparative endocrinologist, and where he also studied animal behavior with Peter Marler. He received his PhD in Psychology from the University of California, Berkeley, in 1967, where Frank Beach was his advisor, one of the founders of the field of behavioral neuroendocrinology. After graduate school, during the 1967–1968 year, he was a postdoctoral trainee in neuroendocrinology at the Brain Research Institute, UCLA (University of California, Los Angeles), in the laboratory of Charles "Tom" Sawyer, a pioneering neuroendocrinologist. In 1968, he joined the faculty of the Department of Psychology at the University of Pennsylvania, where he remained until 1993 and carried out his scientific research program (Adler, curriculum vitae).

Adler's research career began as an important conceptual framework was beginning to take shape in the science of hormones and behavior, a framework to which his own work eventually made a major contribution. The conceptual framework was one in which the relationship between hormones and behavior was bidirectional, a "two-way street." Not only could hormones cause behavioral change, but behavior could cause hormonal change. This important conceptual advance followed upon key discoveries foundational to neuroendocrinology and endocrinology. During the late 1950s, the first method to measure hormones sensitively and accurately (radioimmunoassay) was developed by Rosalyn Sussman Yalow and Solomon Berson (Balthazart, 2020). In another set of breakthroughs, the hypothalamo-pituitary-gonadal (HPG) and hypothalamo-pituitary-adrenal (HPA) axes were discovered. Although it had been known that the posterior pituitary and adrenal medulla were parts of the nervous system, the anterior pituitary was glandular tissue, and its regulation had been a mystery. The work of Geoffrey Harris and Nobel laureates Roger Guillemin and Andrew Schally identified hypothalamic peptides and showed how they were conveyed via the hypothalamo-pituitary portal system to regulate the anterior pituitary hormones that in turn regulated the steroid hormones of the gonads and adrenal cortex (Raisman, 1997; Plant, 2015). Adler's postdoctoral advisor, Charles Sawyer, had demonstrated that ovulation was controlled by the hypothalamus. Taken together, these discoveries confirmed control of the endocrine system's peripheral organs and their hormonal secretions by the brain and therefore potentially by external stimuli or internal brain states. There had long been skepticism in medical endocrinology about whether psychological or behavioral factors could affect reproduction or health. Now there were substantiated mechanisms to make such effects plausible. These advances also helped explain how normal reproductive phenomena such as the induced ovulation of cats or regulation of gonadal state by day length could occur, phenomena then termed neuroendocrine reflexes.

The full implications for understanding the functions and consequences of reproductive behavior itself were largely unexplored territory, however. It is here that Adler made a majority of his most creative and important empirical contributions. In his studies of copulation and fertilization—elements of reproduction already known to be hormonally regulated—he showed the critical role of the male's behavioral mating pattern in generating the female hormonal responses required for fertilization and implantation. This constituted important evidence for the bidirectional relationship between hormones and behavior as well as evidence that behavioral influences on hormones were adaptive by enhancing reproduction.

His most significant line of research on the impact of behavior on physiology revealed in exquisite detail exactly how the seemingly odd copulatory behavior of male rats affects female rats' physiology in ways that promote a successful pregnancy. Male rats, like a number of other rodents, have a mating pattern consisting of a series of mounts with and without penile intromission that are spaced apart in time and that eventually culminate in ejaculation. It was already known that females of some species of mammals are induced ovulators, with ovulation requiring the sensory stimulation of mating, especially vagino-cervical stimulation from penile intromission, a classic neuroendocrine reflex. Rats, like humans, are spontaneous ovulators, however, and do not require such stimulation to ovulate. It seemed as if the male's mating pattern might have some other functions. Beach had suggested the possibility that multiple intromissions might be necessary for the female to become pregnant. Adler was part of the team that did the first experiment providing evidence in support of that hypothesis (Wilson et al., 1965).

In 1969, Adler published a definitive series of eight experiments replicating the importance for female pregnancy of receiving a sufficient number of intromissions and discovering two pathways for the effect (Adler, 1969). The critical experimental manipulation was the creation of low intromission females by mating them with males that had intromitted several times with other females prior to continuing the sequence to ejaculation with the target females, providing the targets with only three or fewer intromissions. Outcomes were then compared to those of high intromission females receiving the entire uninterrupted mating sequence culminating in ejaculation and thus receiving more intromissions (an average of 10). The low intromission females were significantly less likely to become pregnant (20%) compared to the high intromission females (84%). There was a similarly robust difference in the percentages of females that ceased being sexually receptive following mating (18% of low intromission females vs. 100% of high intromission females). Why is this difference in receptivity important? Female rats, unlike humans, have very short ovulatory cycles (4–5 days long) that do not produce fully functional corpora lutea capable of elevating progesterone sufficient to prepare the uterine endometrium for implantation. Cessation of receptivity had been shown previously to occur when females entered the progestational state of pregnancy. Evidently the high intromission females were entering this state but not the low intromission females. Subsequent experiments in the series showed that it was the intromissions specifically that were responsible, rather than, for example, other components of the males' mating behavior or the time spent by the females with the males. The final two experiments targeted another pathway for the greater pregnancy rates of the high intromission females, namely, greater egg fertilization from better sperm transport. All females that were examined after receiving two or more intromissions plus ejaculation had sperm in the uterus and had developing (fertilized) eggs, whereas no females receiving zero or one intromissions plus ejaculation did. The results of the eight experiments were summarized in Fig. 26.2 of the article.

Together with his students and colleagues, Adler went on to further clarify the mechanisms at work in this interesting phenomenon of the essential role of male copulatory behavior in female pregnancy "besides the obvious," as he used to say. He showed that intromissions after another male's ejaculation inhibited pregnancy

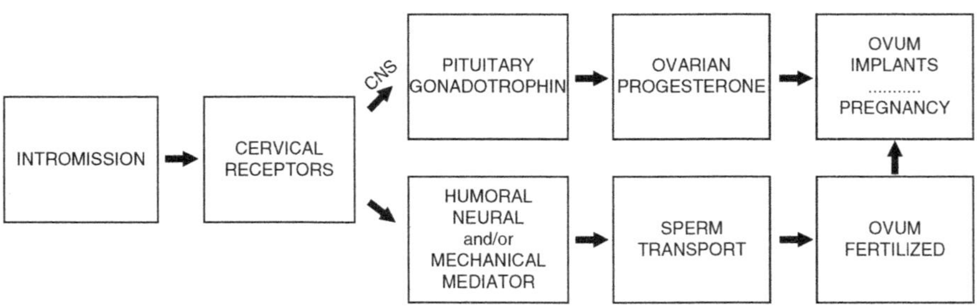

Fig. 26.2 The original summary diagram from Adler (1969) showing the two pathways responsible for the effects of intromissions on pregnancy. (Reprinted with permission of the American Psychological Association from Adler (1969))

produced by the first male (Adler & Zoloth, 1970). He confirmed that females receiving high intromission numbers did indeed have higher circulating progesterone levels following mating than females receiving low intromission numbers, a new neuroendocrine reflex (Adler et al., 1970). In one experiment, females were assigned to receive different numbers of intromissions spaced apart by a wide range of time intervals before receiving an ejaculation. The pregnancy results showed that females' nervous systems could store the information that they had received multiple intromissions over surprisingly long periods of time (up to four hours), a phenomenon consistent with naturally occurring sexual behavior in free-ranging rat groups (Edmonds et al., 1972). Another experiment elucidated the mechanism for a yet longer-term "memory" for the cervical stimulation of intromissions in the form of daily prolactin surges producing a progestational state that could enable sperm to produce a pregnancy as much as 3 days after the intromissions were received (Terkel et al., 1990).

Several of his publications explored the genital sensory field of female rats and effects on it of estrogens or estrous cycle stage (e.g., Adler et al. (1977)). A set of six experiments elucidated the mechanical and temporal parameters of sperm transport facilitation or inhibition by male behavior, with the interesting conclusion that the normal male post-ejaculatory interval was sufficient to avoid disruption of the male's own sperm if he mated again with the female (Matthews & Adler, 1977). Subsequent work showed that vagino-cervical stimulation increased uterine contraction rates, which would facilitate sperm transport, and that greater numbers of intromissions also enhanced the ejaculating male's sperm output (Toner & Adler, 1986a, b).

Adler's research also produced numerous intriguing insights into sociosexual behavior and estrous cycles of rats as well as into additional aspects of male sexual behavior. For example, observations of social interactions in groups of rats, the naturalistic social environment, showed both males and females "taking turns" during the multi-intromission copulatory sequence, with dominant males achieving more ejaculations and females competing for those (McClintock et al., 1982). A study of estrous cycles under different light cycles showed that a light cycle consisting of alternating dim and bright light with a 24-hour period delayed the onset of persistent estrus, a state normally produced by constant light (Weber & Adler, 1979). Observations of female sexual behavior in a seminatural environment during postpartum estrus showed that females time-shared between pup care and gaps in the male's copulatory sequence, but only if they were maternally experienced, as if they had learned how to time-share (Gilbert et al., 1984). Research on ultrasonic vocalizations emitted by males during sexual encounters showed vocalization during the postcopulatory refractory period suggestive of a function in communicating submission to potentially aggressive other males (Adler & Anisko, 1979). Another study showed that sexual rest (a period without any opportunities to mate with females) caused a decline in several indicators of reproductive effectiveness in previously experienced males, with implications for reduced male fecundity at the onset of the breeding season in seasonally breeding species (Weizenbaum et al., 1981). Adler and his students and colleagues were also in the forefront of going inside the brain in an era when such approaches were still relatively novel in hormones and behavior. For example, an electrophysiological study recording from the hippocampus and cortex during male copulation produced results consistent with models positing two different stages, arousal and inhibition (Kurtz & Adler, 1973).

A second major part of Adler's research program, with roots in his postdoctoral work with Sawyer, was the study of biological clocks and especially circadian rhythms. His work on this topic contributed importantly to what was becoming a core area in the field of behavioral neuroendocrinology. As his work on fertilization and pregnancy in rats, his research on rhythms provided further evidence of hormone-behavior bidirectionality, because circadian rhythms of hormone levels and of hormone-regulated behavior are readily entrained by external stimuli, thanks to the brain's control of the endocrine system. Adler's early publications on this topic included work on entrainment of circadian running activity rhythms of rats by food and led to the realization that there must be multiple oscillators rather than just one driving those rhythms (Edmonds & Adler, 1977a, b). As with his work on pregnancy and fertilization in rats, he again went deeper into neural mechanisms. For example, a study using a 2-deoxyglucose uptake approach showed that two other brain regions, the supraoptic and median raphe nuclei, exhibited circadian rhythms of metabolic activity in addition to the suprachiasmatic nuclei (SCN) (Rosenwasser et al., 1985).

The conceptual framework guiding his work on biological clocks became the subject of an important co-authored review (Rosenwasser & Adler, 1986). This masterful discussion addressed how formal analysis of entrained and free-running circadian rhythms (e.g., of wheel running by rats) and study of physiological mechanisms both suggested that there are multiple oscillators that can be experimentally dissociated (desynchronized). Some of the oscillators work together in a coordinated manner and others have hierarchical relationships, with the top level serving as the internal coordinators ("masters" or "pacemakers"). The review laid out quite clearly the evidence for both kinds of oscillator relationships. It brought to the reader's attention that although formal analysis pointed to more than one neural "pacemaker," the SCN was the only one that had been identified. Food-anticipatory entrainment of activity rhythms, for example, still occurred without a functioning SCN and so must be occurring through an unknown locus. The review offered some suggestions for what those other pathways might be. The review also discussed the influence of gonadal hormones on circadian rhythms. One of the Adler group's own studies showed alterations in the free-running rhythms of female rats that were pregnant or had given birth, in support of such an influence of hormones (Rosenwasser et al., 1987). Another set of experiments elucidated the relationship between circadian running rhythms and behavioral depression induced by exposure to inescapable foot shock. Rats whose circadian period lengthened along with behavioral depression performed better (escaped faster) in a new environment where escape from shock was possible (Stewart et al., 1990).

Adler was an excellent communicator and disseminator of advances in behavioral neuroendocrinology. His significant impact on the field came from multiple kinds of contributions to the scientific literature. In addition to his empirical research and reviews in journals, he contributed more than ten major reviews in edited volumes of comparative psychology, reproductive behavior, hormones and behavior, social behavior, and neuroethology. Many emphasized the impact of behavior on reproductive physiology and were instrumental in promoting the concept of hormones and behavior as a bidirectional relationship that should be viewed from an evolutionary perspective. He edited important multiauthored books, including *Neuroendocrinology of Reproduction: Physiology and Behavior* (Adler, 1981) and (with Donald Pfaff and Robert Goy) *Handbook of Neurobiology, vol. 7: Reproduction* (Adler et al., 1985). Together these volumes consolidated and brought to wider attention new developments in behavioral neuroendocrinology and placed them in a broadly integrative framework and an evolutionary context.

Adler was a brilliant intellect, with wide-ranging interests and knowledge and a significant impact on biological psychology generally through his talent for integration, communication and dissemination, and education. He served as the series editor for the 14 volumes of *Handbook of Behavioral Neurobiology* and included volumes on sexual differentiation (Gerall et al., 1992) and on food and fluid intake (Stricker & Woods, 2005) in addition to volumes on reproduction (Adler et al., 1985) and circadian clocks (Takahashi et al., 2001). He also served as a co-editor of the Princeton University Press *Monographs in Neuroethology*. He founded and chaired the Program in the Biological Basis of Behavior at the University of Pennsylvania, which became a model for others around the country. His course offerings included not only behavioral neuroendocrinology and all the fields related to it but also Psychology and Religion, Bioethics of Human Reproduction, Cognitive Neuropsychology, Reproduction and Social Values, and The Human as Animal (an intellectual history of biological psychology) (Adler, curriculum vitae).

Adler was the recipient of numerous awards, including an American Psychological Association Early Career Award (in 1974, when he was the first awardee), a Sigma Xi National Lecturership (1974), the Lindback Award for Distinguished Teaching (1976), two Guggenheim Fellowships (1985), and the Charles A. Dana Foundation Prize for Pioneering Achievement in Higher Education (1988). He was a fellow of the Center for Advanced Study in the Behavioral Sciences, the International Academy of Sex Research, the American Psychological Association, and the Endocrine Society.

Adler was an exceptional mentor. His creativity, intellectual prowess, and wit were inspiring. He gave his students the freedom to develop their own creativity. He supported and encouraged women and men equally in an era when that was rare. Among his 12 PhD students were several who continued to do research in behavioral neuroendocrinology or related fields. They, and their research interests, include James P. Toner (Department of Gynecology and Obstetrics, Emory University School of Medicine), physiology of human reproduction; Avery Gilbert, author of *What the Nose Knows: The Science of Smell in Everyday Life*; Martha McClintock (Department of Psychology, University of Chicago), behavioral and chemosensory influences on reproductive physiology; Stephen Zoloth (Health Sciences, Northeastern University), animal communication; Frank Zemlan (President of P2D Bioscience), drug development for central nervous system disorders; and Elizabeth

Adkins-Regan (Department of Psychology and Department of Neurobiology and Behavior, Cornell University), hormones and avian reproductive behavior.

Beginning in the late 1980s, Adler became increasingly interested in pursuing opportunities to have broader intellectual impact on higher education through administrative service. In 1989, he became an associate dean for the College, School of Arts and Sciences, at his institution, the University of Pennsylvania. He went on to become a vice provost for Research and Graduate Education at Northeastern University in 1993, dean of Yeshiva College in 1995, and special assistant for the Curriculum Development and Research Initiatives in the Office of the Provost at Yeshiva University in 2005 (Adler, curriculum vitae). Eventually, such service precluded continuing an active research career or attending scientific conferences, but it allowed him to influence larger numbers of both graduate and undergraduate students, enabling them to achieve the highest level of liberal arts and intellectual commitment embedded in an ethical framework and worldview.

Adler began his search for a career in high school by trying to decide between becoming a rabbi and a psychoanalyst (Adler 1975). He decided instead to become a biologically oriented scientist, to the lasting benefit of the field of behavioral neuroendocrinology and his many academic descendants.

References

Adler, N. T. (1969). Effects of the male's copulatory behavior on successful pregnancy of the female rat. *Journal of Comparative and Physiological Psychology, 69*, 613–622.

Adler, T. N. (1975). *American Psychologist, 30*, 72–74.

Adler, N. T. (Ed.). (1981). *Neuroendocrinology of reproduction: Physiology and behavior*. Plenum.

Adler, N., & Anisko, J. (1979). The behavior of communicating: An analysis of the 22 kHz vocalization of rats. *American Zoologist, 19*, 493–508.

Adler, N., & Hogan, J. A. (1963). Classical conditioning and punishment of an instinctive response in Betta splendens. *Animal Behaviour, 11*, 351–354.

Adler, N. T., & Zoloth, S. R. (1970). Copulatory behavior can inhibit pregnancy in female rats. *Science, 168*, 1480–1482.

Adler, N. T., Resko, J. A., & Goy, R. W. (1970). The effect of copulatory behavior on hormonal change in the female rat prior to implantation. *Physiology & Behavior, 5*, 1003–1007.

Adler, N. T., Davis, P. G., & Komisaruk, B. K. (1977). Variation in the size and sensitivity of a genital sensory field in relation to the estrous cycle in rats. *Hormones and Behavior, 9*, 334–345.

Adler, N. T., Pfaff, D., & Goy, R. (Eds.). (1985). *Handbook of behavioral neurobiology, vol. 7. Reproduction*. Plenum.

Balthazart, J. (2020). How technical progress reshaped behavioral neuroendocrinology during the last 50 years...and some methodological remarks. *Hormones and Behavior, 118*, 104682.

Edmonds, S. C., & Adler, N. T. (1977a). Food and light as entrainers of circadian running activity in the rat. *Physiology & Behavior, 18*, 915–919.

Edmonds, S. C., & Adler, N. T. (1977b). The multiplicity of biological oscillators in the control of circadian running activity in the rat. *Physiology & Behavior, 18*, 921–930.

Edmonds, S., Zoloth, S. R., & Adler, N. T. (1972). Storage of copulatory stimulation in the female rat. *Physiology & Behavior, 8*, 161–164.

Gerall, A., Moltz, H., & Ward, I. L. (Eds.). (1992). *Handbook of behavioral neurobiology vol. 11. Sexual differentiation*. Plenum.

Gilbert, A. N., Pelchat, R. J., & Adler, N. T. (1984). Sexual and maternal behaviour at the postpartum oestrus: The role of experience in time-sharing. *Animal Behaviour, 32*, 1045–1053.

Kurtz, R., & Adler, N. T. (1973). Electrophysiological correlates of sexual behavior in the male rat. *Journal of Comparative and Physiological Psychology, 84*, 225–239.

Matthews, M., & Adler, N. T. (1977). Facilitative and inhibitory influences of reproductive behavior on sperm transport in rats. *Journal of Comparative and Physiological Psychology, 91*, 727–741.

McClintock, M., Anisko, J., & Adler, N. (1982). Group mating among Norway rats: II. Social dynamics of copulation: Competition, cooperation and mate choice. *Animal Behaviour, 30*, 410–425.

Plant, T. M. (2015). The hypothalamo-pituitary–gonadal axis. *The Journal of Endocrinology, 226*, T41–T54.

Raisman, G. (1997). An urge to explain the incomprehensible: Geoffrey Harris and the discovery of the neural control of the pituitary gland. *Annual Review of Neuroscience (Palo Alto, CA), 20*, 533–566.

Rosenwasser, A. M., & Adler, N. T. (1986). Structure and function in circadian timing systems: Evidence for multiple coupled circadian oscillators. *Neuroscience & Biobehavioral Reviews, 10*, 431–448.

Rosenwasser, A. M., Trubowitsch, G., & Adler, N. T. (1985). Circadian rhythm in metabolic activity of suprachiasmatic, supraoptic, and raphe nuclei. *Neuroscience Letters, 56*, 183–187.

Rosenwasser, A. M., Hollander, S. J., & Adler, N. T. (1987). Effects of pregnancy and parturition on free-running circadian activity rhythms in the rat. *Chronobiology International, 4*, 183–187.

Stewart, K. T., Rosenwasser, A. M., Hauser, H., Volpicelli, J. R., & Adler, N. T. (1990). Circadian rhythmicity and behavioral depression: I. Effects of stress. *Physiology & Behavior, 38*, 149–155.

Stricker, E. M., & Woods, S. (Eds.). (2005). *Handbook of behavioral neurobiology. Vol. 14. Neurobiology of food and fluid intake*. Plenum.

Takahashi, J. S., Turek, F. W., & Moore, R. Y. (Eds.). (2001). *Handbook of behavioral neurobiology vol. 12. Circadian clocks*. Plenum.

Terkel, J., Witcher, J. A., & Adler, N. T. (1990). Evidence for "memory" of cervical stimulation for the promotion of pregnancy in rats. *Hormones and Behavior, 24*, 40–49.

Toner, J. P., & Adler, N. T. (1986a). The pre-ejaculatory behavior of male and female rats affects the number of sperm in the vagina and uterus. *Physiology & Behavior, 36*, 363–367.

Toner, J. P., & Adler, N. T. (1986b). Influence of mating and vaginocervical stimulation on rat uterine activity. *Journal of Reproduction and Fertility, 78*, 239–249.

Weber, A. L., & Adler, N. T. (1979). Delay of light-induced persistent estrus in rats by 24-hour periodic cues. *Science, 204*, 323–325.

Weizenbaum, F., Matthews, M., Whitehouse, J., & Adler, N. (1981). Depressant effects of sexual rest on reproductive behavior and physiology in male rats. *Biology of Reproduction, 25*, 744–751.

Wilson, J. R., Adler, N., & LeBoeuf, B. (1965). The effects of intromission frequency on successful pregnancy in the female rat. *Proceedings of the National Academy of Sciences, 53*, 1392–1395.

Mei-Fang Cheng

27

Gregory F. Ball and Jacques Balthazart

Abstract

Mei-Fang Cheng was a behavioral neuroendocrinologist who was born in Taiwan but pursued graduate and postdoctoral studies in the United States. She then joined the laboratory of Daniel Lehrman at the Institute of Animal Behavior, Rutgers University-Newark, where she spent her entire career first as a research associate and then in 1969 as an assistant professor becoming a full professor in 1979. She also served as an acting director and then director of the Institute of Animal Behavior from 1989 to 1994. Her initial work at Rutgers involving classic experiments of gonadectomy and hormone replacement established links between ovarian steroids and female reproductive behaviors. Then via a combination of experimental approaches, she established the notion of self-stimulation by which the behavior expressed by a female dove retroactively affects her own behavior and physiology. Cheng also demonstrated the presence of an active neurogenesis in the hypothalamus of doves in response to local lesions and the contribution of the new neurons to the behavioral recovery after lesion.

Keywords

Neuroendocrinology · Behavior · Ovarian steroids · Mating behavior · Ring doves

G. F. Ball (✉)
Department of Psychology, University of Maryland, College Park, MD, USA
e-mail: gball@umd.edu

J. Balthazart
University of Liege, GIGA Neurosciences, Liège, Belgium
e-mail: jbalthazart@uliege.be

R. J. Nelson, Z. M. Weil (eds.), *Biographical History of Behavioral Neuroendocrinology*, https://doi.org/10.1007/978-3-031-12970-4_27

An important concept in the history of behavioral neuroendocrinology is that social stimuli such as courtship displays can have profound effects on the endocrine physiology of the receiver to whom these signals are directed. These endocrine changes in the receiver can in turn affect the probability and intensity of a behavioral response to these signals. This notion of the reciprocal interactions between behavioral signals and endocrine state was first championed by Daniel Lehrman at Rutgers University based on his studies of ring doves (*Streptopelia risoria*) (Lehrman, 1965). Mei-Fang Cheng (Fig. 27.1) came to work with Lehrman in 1969 and expanded on these ideas in significant ways. In particular, she established that one effect of social interactions is that a receiver of courtship displays will not only experience endocrine changes but will also change her own behavior. Through a process of self-stimulation via her own behavior, she can adjust her own endocrine physiology to optimize the timing of reproduction. This is a critical concept expanding our views of how behavior and neuroendocrine physiology are intertwined.

Mei-Fang Cheng was born in Taiwan and raised there during the Second World War. She told one of us (GFB) that she still counts in Japanese because she learned her basic arithmetic during the Japanese occupation when they changed the language used in elementary schools to Japanese from Chinese. After the Second World War, Taiwan was in conflict with the People's Republic of China on the mainland, so her early formative years were in a country facing significant existential challenges. However, she loved observing nature and came to the conclusion that she wanted to be an academic if possible. She therefore enrolled in the National Taiwan University and pursued a course of study in psychology. She graduated at the top of her class and wanted to pursue graduate study that was not available in Taiwan at the time.

Based on the advice of one of her professors at Taiwan National, she looked to the United States for additional training. She started in Oregon but soon desired to move east, so she entered a graduate program in experimental psychology at Bryn Mawr College outside of Philadelphia. She worked on human perception questions related to the measurement of length (Cheng, 1968) with her advisor Professor

Fig. 27.1 Mei-Fang Cheng in the early 1980s when she developed the notion of behavioral self-stimulation. (Photo courtesy Ron Barfield)

R. S. Davidson. After completing her PhD in 1965, Cheng joined the lab of Philip Teitelbaum at the University of Pennsylvania deciding to stay in the area in part because she had just married a fellow Taiwanese national who was finishing his PhD in physics at Penn. Teitelbaum was already well known then for his work on how lesions in different regions of the hypothalamus had specific effects on food and water intake in rats. When Cheng arrived, they started developmental studies on food and water intake and compared the behavioral progression after lesion recovery with the natural development of these behaviors (Teitelbaum et al., 1969a, b). Based on this work, Cheng had become hooked on physiological studies in general and found that she was good at small animal surgery and enjoyed doing it.

Cheng then moved to New Jersey to work as a research associate with Daniel Lehrman at Rutgers-Newark. Lehrman came from a tradition of comparative psychology centered at the American Museum of Natural History in New York that studied animals in their natural context rather than as animal models of human processes in artificial situations. He was therefore very ethological in his thinking though he famously critiqued views of behavioral development articulated by Konrad Lorenz (Lehrman, 1953). Lehrman had founded in the 1960s the Institute of Animal Behavior at Rutgers Newark that was dedicated to studies in a field then called psychobiology that integrated ideas from ethology and experimental/comparative psychology. Lehrman had set up a research program on ring doves that established how courtship behavior represented an interplay between hormonal physiology and a sequence of reproductive behaviors. He studied both males and females together as they engaged in the stages of courtship, nest building, copulation, oviposition, incubation, and squab feeding. In males, he was able to employ classic behavioral neuroendocrine methods to explore the role of testicular testosterone. He found, for example, that castrated male ring doves when presented with a female did not engage in the bow-coo display and consequently were not able to stimulate the female's neuroendocrine axis and facilitate oviposition. If he administered exogenous testosterone to the castrated males, then they produced these behaviors and would cause physiological changes in the females as measured by oviduct mass, follicle size, and oviposition (Lehrman, 1963; Lehrman, 1965).

However, when Cheng arrived in his laboratory, he had been stymied for several years in his attempts to ovariectomize female doves. The avian ovary consists of a collection of follicles of various sizes reminiscent of a bunch of grapes of unequal sizes. It is enveloped by a membrane and is near a venous sinus and sensitive organs such as the kidney and adrenals. It therefore requires a very precise surgical technique to extract the entire ovary including each follicle without causing internal bleeding that can lead to death. If any ovarian fragment is left behind, it may implant on the wall of the body cavity and become vascularized and at least partially functional as the ovary is regulated in part by blood-borne hormones such as the gonadotropins. The endocrine part of the ovarian tissue will continue secreting sex steroids into the general circulation, while ova will usually be laid in the abdominal cavity since the connection with the oviduct has been disrupted. In the vast majority of avian species, only the left ovary is developed. In ring doves, if the left ovary is removed, the right ovary will grow, a response called compensatory hypertrophy, that will often result in a new functional ovary. This usually takes 4–6 weeks. So, obtaining a full ovariectomy in doves is a challenging two-step process; the left ovary has first to be removed and one must wait for the right ovary to grow and then remove that. Cheng was able to develop and master a reliable technique that resulted in work by herself and in collaboration with a student of Lehrman, Rae Silver, to demonstrate the importance of estradiol or estradiol plus progesterone for the activation of courtship, copulation, nest building, and the initiation of incubation in female ring doves (Cheng, 1973b; Cheng, 1973a; Cheng & Silver, 1975). These now classic studies laid the foundation for our understanding of the role of ovarian steroid hormones in the regulation of sexual and other reproductive behaviors in birds. With her student Marie Gibson, she later identified by a combination of ovariectomy with stereotaxic lesions or implantation of steroids the neural sites where estradiol is acting to activate many of these behaviors (Gibson & Cheng, 1979).

In the late 1970s, Cheng started to focus more on the broad role played by the female in the social interplay between males and females that was so fundamental to successful ovulation and fitness. In a study with her graduate student Jeff Cohen, they lesioned a mid-brain structure, nucleus intercollicularis (ICo), to assess its role in the activation of the various calls produced by female doves such as the nest coo. One finding from this study was surprising, namely, that the females with lesions to ICo (which did greatly attenuate their ability to engage in the nest coo) had reduced ovarian growth and egg laying compared to females that received sham lesions (Cohen & Cheng, 1981). Similarly, if females received hypoglossal nerve sectioning that reduced their call rate, again those females had reduced reproductive growth compared to controls (Cohen & Cheng, 1979). Both treatment groups in these two studies received a similar amount of courtship stimulation from males, so if one viewed the male courtship behavior as the primary driver of female reproductive physiology and oviposition, then lesions to ICo or nerve cuts would not be expected to inhibit reproductive development. It should be noted that ICo is not directly involved in neuroendocrine function in birds.

These observations led Cheng to establish an entirely new line of research focused on the "self-stimulation" hypothesis (Cheng, 1992) (see Fig. 27.2). She systematically tested in doves how producing and hearing your own nest-coos could stimulate ovarian growth, estradiol secretion, and oviposition (Cheng, 1986; Cheng et al., 1988). She was able to develop converging lines of evidence indicating that females producing and hearing their own nest-coos would experience enhanced follicular growth and oviposition as compared to females who only experienced male bow-coos and nest-coos (Cheng et al., 1988). Cheng completed mechanistic studies illustrating how this process might work. For example, she played back female nest-coos or control sounds to females while recording electrophysiological activity in the hypothalamus and measuring luteinizing hormone (LH) secretion from the pituitary gland. Remarkably she found units in the hypothalamus that were specifically tuned to the female nest-coos and she found that birds hearing the female nest-coos had greatly enhanced LH release as compared to birds only hearing the control sounds (Cheng et al., 1988; Cheng et al., 1998). In a study performed in collaboration with one of us (JB) while he was a postdoctoral researcher in her lab, she

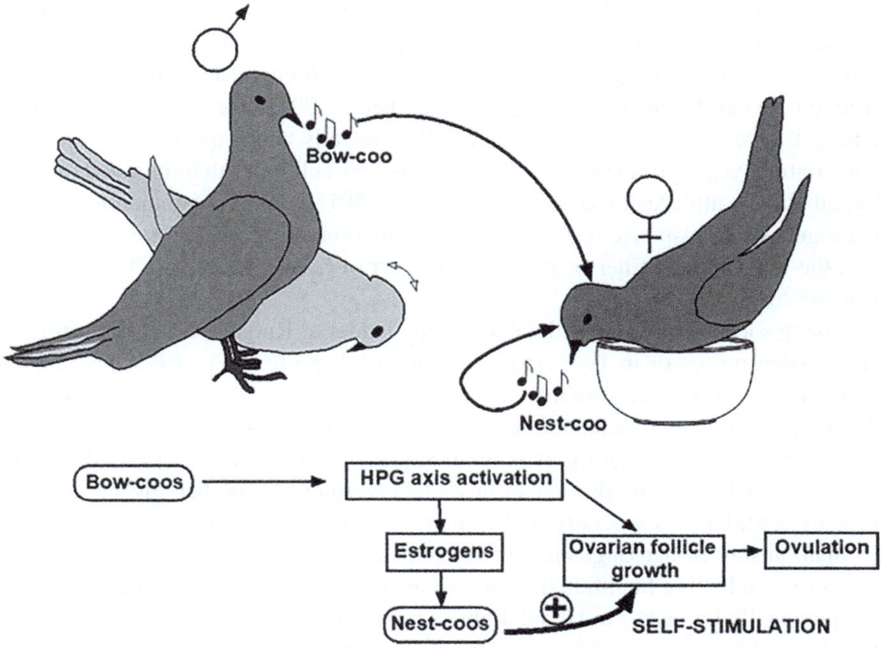

Fig. 27.2 Schematic illustration of the notion of behavioral self-stimulation. Displays of the male (bow-coos) activate the hypothalamo-pituitary-gonadal (HPG) axis of females thus promoting ovarian follicle growth and ovulation. Additionally, the female responds to the coos by producing coos herself and these also contribute to the follicle maturation

extended the notion of self-stimulation by showing that the activity of building the nest by itself similarly affected plasma concentrations of follicle-stimulating hormone near ovulation. Nest building activity also tended to affect egg-laying latency and the final breeding success as measured by the number of squabs raised to the fledging stage (Cheng & Balthazart, 1982). Her work on self-stimulation raised the scenario that females in a volitional manner might regulate much more precisely than previously suspected how they physiologically respond to male sexual stimuli (Cheng, 2008). She argued strongly that this phenomenon of self-stimulation was more generally applicable to other behavioral and neural systems.

Around this time that she was establishing the self-stimulation phenomenon, she returned to investigating a question that she had grappled with during her postdoctoral studies with Philip Teitelbaum, namely, the recovery of function from brain lesions. However, now she investigated how behavioral experience could influence this process (Bernstein et al., 1993). She reported that male ring doves with lesions to the ventromedial region of the hypothalamus exhibited deficits in the bow-coo display and that they would recover more readily if housed with females that they would eventually court and stimulate to lay eggs (Bernstein et al., 1993). Cheng then started to investigate how neurogenesis in adulthood might mediate this recovery of function. She focused on the hypothalamus, an area not known to have a high rate of spontaneous adult neurogenesis, but that she discovered would exhibit a marked increase in neurogenic activity in response to lesion (Cao et al., 2002). She discovered that inhibiting neurogenesis significantly attenuated behavioral recovery from the lesion (Chen et al., 2006; Chen & Cheng, 2007). She also found that specific cell types such as coo-responsive cells and cells that project to the midbrain that controls vocal production were among the new cell types that migrated into the hypothalamus after lesion (reviewed in Cheng (2013)). This phenomenon of adult neurogenesis in response to lesions and the interplay with behavioral interactions was the main focus of her research for the rest of her career (e.g., Cheng (2013), Cheng (2017)).

Cheng successfully climbed the academic ladder at Rutgers–Newark starting as an assistant professor in 1969 and becoming a full professor in 1979. She moved onto leadership positions notably serving as an acting director and then director of the Institute of Animal Behavior, her home unit, from 1989 to 1994. In 2015, Cheng was interviewed as part of the Women in Science program at Rutgers. In this interview, she spoke very frankly about how she had to overcome her shyness and her lack of confidence, especially early in her career because she was a diminutive woman, unsure of her English, and was reluctant to speak at international conferences to avoid being humiliated. Those of us who knew Cheng and interacted with her personally know her to be a delightful person who cared deeply about her colleagues and her trainees. Although she lived with the challenge of working in an Anglophone environment while being a native Chinese speaker, she was determined to rise to the occasion and communicate assertively her data and ideas. That she did with great success. Cheng's legacy is apparent in the enduring impact of her published work and in her graduate and postdoctoral trainees (e.g., John Buntin, Marie Gibson, Jeffrey Cohen, M. Eleanor Sims, Jacques Balthazart, Michael Havens, Sarah Durand, Martha Leah Chaiken, Thomas Akesson, Sharon Barclay, Thorsten Klint, Ming-Xue Zuo, and Jingpian Peng) who have followed in her footsteps.

References

Bernstein, P. L., Zuo, M., & Cheng, M.-F. (1993). Social condition affects the courtship behavior of male ring doves with posterior medial hypothalamic lesions. *Behavioral and Neural Biology, 59*, 120–125.

Cao, J., Wenberg, K., & Cheng, M. F. (2002). Lesion induced new neuron incorporation in the adult hypothalamus of the avian brain. *Brain Research, 943*, 80–92.

Chen, G., & Cheng, M. F. (2007). Inhibition of lesion-induced neurogenesis impaired behavioral recovery in adult ring doves. *Behavioural Brain Research, 177*, 358–363.

Chen, G., Bonder, E. M., & Cheng, M. F. (2006). Lesion-induced neurogenesis in the hypothalamus is involved in behavioral recovery in adult ring doves. *Journal of Neurobiology, 66*, 537–551.

Cheng, M. F. (1968). Tactile-kinesthetic perception of length. *The American Journal of Psychology, 81*, 74–82.

Cheng, M. F. (1973a). Effect of estrogen on behavior of ovariectomized ring doves (Streptopelia risoria). *Journal of Comparative and Physiological Psychology, 83*, 234–239.

Cheng, M. F. (1973b). Effect of ovariectomy on the reproductive behavior of female ring doves (Streptopelia risoria). *Journal of Comparative and Physiological Psychology, 83*, 221–233.

Cheng, M. F. (1986). Female cooing promotes ovarian development in ring doves. *Physiology & Behavior, 37*, 371–374.

Cheng, M. F. (1992). For whom does the female dove coo? A case for the role of vocal self-stimulation. *Animal Behaviour, 43*, 1035–1044.

Cheng, M. F. (2008). The role of vocal self-stimulation in female responses to males: Implications for state-reading. *Hormones and Behavior, 53*, 1–10.

Cheng, M. F. (2013). Hypothalamic neurogenesis in the adult brain. *Frontiers in Neuroendocrinology, 34*, 167–178.

Cheng, M. F. (2017). Adult neurogenesis in injury-induced self-repair: Use it or lose it. *Brain Plast, 2*, 115–126.

Cheng, M. F., & Balthazart, J. (1982). The role of nest-building activity in gonadotropin secretions and the reproductive success of ring doves (Streptopelia risoria). *Journal of Comparative and Physiological Psychology, 96*, 307–324.

Cheng, M. F., & Silver, R. (1975). Estrogen-progesterone regulation of nest-building and incubation behavior in ovariectomized ring doves (Streptopelia risoria). *Journal of Comparative and Physiological Psychology, 88*, 256–263.

Cheng, M. F., Desiderio, C., Havens, M., & Johnson, A. (1988). Behavioral stimulation of ovarian growth. *Hormones and Behavior, 22*, 388–401.

Cheng, M. F., Peng, J. P., & Johnson, P. (1998). Hypothalamic neurons preferentially respond to female nest coo stimulation: Demonstration of direct acoustic stimulation of luteinizing hormone release. *The Journal of Neuroscience, 18*, 5477–5489.

Cohen, J., & Cheng, M. F. (1979). Role of vocalization in the reproductive cycle of ring doves (Streptopelia risoria): Effects of hypoglossal nerve section on the reproductive behavior and physiology of the female. *Hormones and Behavior, 13*, 113–127.

Cohen, J., & Cheng, M. F. (1981). The role of the midbrain in courtship behavior of the female ring dove (Streptopelia risoria): Evidence from radiofrequency lesion and hormone implant studies. *Brain Research, 207*, 279–301.

Gibson, M. J., & Cheng, M. F. (1979). Neural mediation of estrogen-dependent courtship behavior in female ring doves. *Journal of Comparative and Physiological Psychology, 93*, 855–867.

Lehrman, D. S. (1953). A critique of Konrad Lorenz's theory of instinctive behavior. *The Quarterly Review of Biology, 28*, 337–363.

Lehrman, D. S. (1963). The reproductive behavior of ring doves. *Scientific American, 11*, 433–438.

Lehrman, D. S. (1965). Interaction between internal and external environments in the regulation of the reproductive cycle of the ring dove. In F. A. Beach (Ed.), *Sex and behavior* (0th ed.). Wiley.

Teitelbaum, P., Cheng, M. F., & Rozin, P. (1969a). Development of feeding parallels its recovery after hypothalamic damage. *Journal of Comparative and Physiological Psychology, 67*, 430–441.

Teitelbaum, P., Cheng, M. F., & Rozin, P. (1969b). Stages of recovery and development of lateral hypothalamic control of food and water intake. *Annals of the New York Academy of Sciences, 157*, 849–860.

Ingeborg L. Ward

28

Heather N. Richardson and Russell D. Romeo

Abstract

Ingeborg ("Inge") L. Ward made fundamental contributions in our understanding of the sexual differentiation of hormone-dependent behaviors. Among her many discoveries was insight into how maternal stress affected the perinatal endocrine environment and the lasting influences these hormonal changes had on the development of sex-specific behaviors and their neurobiological underpinnings. She also examined how prenatal exposure to both stress and alcohol shaped the process of sexual differentiation and the neurobehavioral consequences of these early developmental experiences. Ward broke through many barriers throughout her career and set the bar high for her colleagues and graduate student trainees in the Experimental Psychology Master of Science graduate program at Villanova University. Through her scientific discoveries and strong commitment to academic mentoring of scientists following her path, Ward's significant contributions to the field of behavioral neuroendocrinology are still evident today.

Keywords

Maternal stress · Sex differentiation · Ontogeny · Hormone-dependent behavioral development

H. N. Richardson
Psychological and Brain Sciences Department, University of Massachusetts Amherst, Amherst, MA, USA
e-mail: hrichardson@cns.umass.edu

R. D. Romeo (✉)
Departments of Psychology and Neuroscience and Behavior, Barnard College of Columbia University, New York, NY, USA
e-mail: rromeo@barnard.edu

R. J. Nelson, Z. M. Weil (eds.), *Biographical History of Behavioral Neuroendocrinology*, https://doi.org/10.1007/978-3-031-12970-4_28

Introduction

Ingeborg L. Ward, or "Inge" as she asked her students to call her, was born in Rotha, Germany, in 1940. After immigrating to the United States, she obtained her BS in Chemistry from Westhampton College in Richmond, VA, in 1960 and then her MS and PhD degrees in 1965 and 1967, respectively, from Tulane University in New Orleans, LA. During her time at Tulane University, she worked with Arnold Gerall, investigating the role that early exposure to steroid hormones played in organizing behavioral potentials in adulthood. This work led to her first two papers on the role of perinatal exogenous testosterone exposure on the reproductive behaviors of female rats (Gerall & Ward, 1966; Ward, 1969). It also ignited her profound interest in studying how hormones modulate behavior and the lasting trace that hormones can have on neurobehavioral and reproductive functions. Her work, spanning over four decades, continues to shape our understanding of behavioral neuroendocrinology. The sections that follow briefly outline some of her major research achievements, including how environmental experiences during gestation can affect the hormonal milieu of the developing fetus, leading to long-term changes in the reproductive behavior of these animals in adulthood.

Perinatal Exposure to Steroid Hormones Modifying Later Behavioral Potential

While launching her independent research program at Villanova University, where she spent her entire academic career, Ward conducted a series of experiments examining the effects of perinatal testosterone exposure on the ability of adult female rats to display female- and male-typical mating behaviors, such as lordosis and mounting, respectively. These early studies, which set the stage for many of her later research projects, indicated that prenatal and/or neonatal administration of testosterone resulted in decreased lordotic behavior and increased mounting and ejaculatory responses, in hormone-primed female rats later in adulthood (Ward, 1969; Ward & Renz, 1972; Hoepfner & Ward, 1988). These behavioral responses, particularly the ejaculatory-like motor patterns, were typically the most robust in adult females when these animals were treated both prenatally and neonatally with testosterone and ovariectomized in adulthood, suggesting additive influences of early hormone exposure on later mating behaviors (Ward, 1969). These initial studies contributed to a growing body of evidence that exposure to hormones early in development could significantly shape an individual's responsiveness to hormones later in life, leading to changes in the quantity and quality of a variety of reproductive behaviors in adulthood.

Ward also showed that blocking androgen production in females during prenatal development reduced levels of mounting and ejaculatory-like behaviors to levels even lower than what is typically observed in unmanipulated females treated with testosterone in later adulthood (Ward & Renz, 1972). These data suggested that females were sensitive to androgenic hormones during fetal maturation, which could have lasting influences on their behavioral potentials in adulthood. As blocking androgen activity early in development had little impact on the internal or external sex organs of treated females, it was suggested by Ward and Renz (1972) that the effects on the behavior were "presumably due to central effects." At the time, some of Ward's contemporaries suggested that the effects of perinatal androgen exposure on later adult behaviors were mediated by the structural organization of peripheral sex organs, such as the phallus (Nadler, 1969). However, given the countless experimental demonstrations that were to come of perinatal hormone exposure organizing the central nervous system (Arnold & Gorski, 1984), it appears that Ward's original suggestion was not only both prescient and insightful but also correct.

In parallel to this groundbreaking research in females, Ward also engaged in a line of research testing how prenatal androgen influences sexual behavior in adulthood in male rats. It had been fairly well established that adult males castrated neonatally and treated with estradiol in adulthood would exhibit lordosis behavior when mounted by a male (Gerall et al., 1967). These data suggested that exposure to testicular hormones, including testosterone, during early postnatal life limited the capacity for adult males to exhibit female-typical mating postures, such as lordosis. However, given that fetal surgical castration was not a viable experimental manipulation, it was unclear what role, if any, prenatal hormones played in organizing these behavioral potentials in adult males. Taking a novel approach, by pharmacologically blocking androgen production prenatally with cyproterone acetate, she was able to demonstrate that inhibiting androgen synthesis prenatally resulted in greater amounts of lordosis behavior in hormone-primed adult males compared to those treated with cyproterone acetate postnatally (Ward, 1972a). These data indicated that prenatal exposure to androgens contributed to the sexual differentiation of reproductive behaviors, reducing the potential for males to display lordosis behavior later in adulthood.

Many of the experiments parsing out the role of early hormone exposure on organizing later behavioral repertoires were executed by experimentally enhancing or reducing perinatal hormone levels. However, the Ward laboratory was able to demonstrate through an elegant study investigating the intrauterine position of developing rat fetuses that the prenatal hormonal milieu played a sublet, yet significant, role on sexual differentiation. Her graduate student and she demonstrated that in the context of uterine blood flow, female fetuses downstream from male fetus showed an increased anogenital distance at birth and greater levels of mounting behavior after ovariectomy and testosterone treatment in adulthood than female fetuses upstream of male fetuses (Meisel & Ward, 1981). These results indicated that a blood-borne factor(s) from male fetuses could alter the somatic and behavioral differentiation of female fetuses. Moreover, their studies also refuted the idea that proximity of male fetuses was a critical determining factor influencing this pattern of differentiation in females (Clemens et al., 1978), as a male fetus adjacent to a female fetus had no significant effect on these somatic and behavioral measures if the male fetus was downstream of the female (Meisel & Ward, 1981).

Altogether, Ward's early work using hormonal manipulations and natural experiments made seminal contributions to the role that hormones played in later adult behaviors. These studies provided a clearer picture regarding which hormones were involved and importantly the developmental stage at which these hormones were most influential. In a related body of research, to which we turn to next, Ward went on to show just how important the timing component of hormone exposure was to later adult behaviors and how experience could shape this aspect of differentiation.

Prenatal Stress and Hormonal Surges

In addition to establishing the impact of exogenous and endogenous perinatal hormone exposure on adult reproductive behaviors, Ward also investigated whether environmental experiences could further modify sexual differentiation and what endocrine mechanisms early in development might be responsible for shaping this process. In one such line of research, Ward determined that prenatal exposure to stress reduced mounting and ejaculatory responses in adult male rats (Ward, 1972b). Further, these prenatally stressed males showed increased lordotic behavior in adulthood when castrated and supplemented with estradiol and progesterone (Ward, 1972b). This was a dramatic demonstration of how an environmental experience could shape sexual differentiation and later reproductive behaviors.

These experiments appear to have been initiated to investigate the relative role of testicular androgens, namely, testosterone, compared to adrenal androgens, namely, androstenedione, in the context of sexual differentiation. Specifically, given that testosterone is a more potent androgen compared to androstenedione, it was reasoned that exposing pregnant dams to restraint stress would induce a significant response from the prenatal adrenal cortex, due to increased maternal adrenocorticotropic hormone (ACTH), resulting in elevated levels of adrenal androstenedione, and possibility reduced production of testicular testosterone. This shift in the ratio of more androstenedione to less testosterone would result in less overall androgenic stimulation of tissues, including the nervous system, during the sexual differentiation process presumably unfolding during prenatal development. Whereas the behavioral results obtained from these prenatally stress males (e.g., exhibiting less mounting behavior and greater lordotic behavior) certainly support such assertation, the stronger evidence came a few years later.

Working with her collaborator, Judith Weisz, at the Milton S. Hershey Medical Center of the Pennsylvania State University, Ward was able to describe the fluctuations in steroid hormone concentrations in male and female fetal rats. Specifically, they showed that plasma testosterone levels were significantly higher in male compared to female fetuses on the 18th and 19th days of gestation (Weisz & Ward, 1980; Ward & Weisz, 1984). This surge was unique to testosterone, as plasma progesterone levels were not different between male and female fetuses, though corticosterone levels were elevated (Weisz & Ward, 1980, Ward & Weisz, 1984). It was concluded that "day 18 and possibly day 19 post conception represents a critical period during which the central nervous system of the male is primed by high levels of testosterone" (pg. 306; Weisz & Ward, 1980).

Ward and her collaborators then went on to demonstrate that exposure to prenatal stress could perturb this pattern of testosterone secretion, such that prenatally stress male fetuses did not show the surge in testosterone on the 18th and 19th days of gestation, as the unstressed, control male fetuses had shown (Ward & Weisz, 1980; Ward & Weisz, 1984). In fact, the data indicated that this surge in testosterone might be shifted by a day or two earlier in development, suggesting that the timing of the surge of testosterone might be more important than just the level of testosterone achieved by the surge (Ward & Weisz, 1980, Ward & Weisz, 1984). They were able to go on to show that stress-induced change in Δ^5-3β-hydroxysteroid dehydrogenase (3β-HSD), a major enzyme important in the synthesis of testosterone, was a likely contributor to these altered fetal plasma testosterone levels. Specifically, male fetuses exposed to prenatal stress showed reduced 3β-HSD activity in the Leydig cells of the testes during the 18th and 19th days of gestation but elevated levels on the 16th and 17th day of gestation, compared to unstressed controls (Orth et al., 1983). Collectively, these data provided a hormonal mechanism through which sexual differentiation progressed during early development, namely, a surge in testosterone produced by the fetal testes, specifically during the 18th and 19th days of gestation, shaped the later behavioral potential of adult males.

It is important to note that investigating the patterns of steroid hormone levels in stressed and unstressed rat fetuses was a colossal undertaking. The amount of plasma needed for the radioimmunoassays required "pooling" many samples across many subjects. The logistics of these experiments, and the labor needed to execute them properly, calls attention to Ward's dedication to rigorous and extensive documentation of the specific prenatal hormonal milieu during this stage of sexual differentiation and how environment influences could shape this milieu.

Later studies were designed to establish the role of endogenous opioids in mediating the effects of prenatal stress on male offspring. More specifically, it was reported that naltrexone, an opioid antagonist, administered to pregnant dams blocked the later behavioral effects induced by prenatal stress in the male offspring (Ward et al., 1986). This effect appears to be mediated by the ability of naltrexone to normalize the levels of Δ^5-3β-hydroxysteroid dehydrogenase (3βHSD) levels in the Leydig cells in the testes of prenatally stress male fetuses, permitting the surge of testosterone to occur on the 18th and 19th days of gestation (Ward et al., 1990).

Furthermore, Ward and her collaborators were also able to show how exposure to prenatal stress shaped the central nervous system, particularly within regions of the brain and spinal cord that showed structural sex differences, namely, the sexually dimorphic nucleus of the preoptic area (SDN-MPOA) and the spinal nucleus of the bulbocavernosus (SNB) and dorsolateral nucleus of the spinal cord (DLN). The volume of the SDN-MPOA is larger, and the motoneurons in the SNB and DLN are more numerous in adult males compared to adult females (Gorski et al., 1978; Breedlove & Arnold, 1980; Jordan et al., 1982). Male rats exposed to prenatal stress during the last week of gestation, during the time of the prenatal testosterone surge, had a significantly smaller SDN-MPOA volume and fewer SBN and DLN motoneurons than non-stressed, control males (Kerchner & Ward, 1992; Grisham et al., 1991). These studies demonstrated that transient exposure to stress prenatally, which dampened the testosterone surge, could alter the sexually dimorphic structures in both the brain and spinal cord with potential ramifications for sexually differentiated reproductive behaviors exhibited in adulthood.

Toward the later part of Ward's research career, she began to investigate the effects of prenatal stress and alcohol exposure, alone or in combination, on later sexually differentiated behavioral and neurobiological endpoints. Working with her longtime collaborator and husband, O. Byron Ward, they were able to show that stress and alcohol had unique effects on the prenatal testosterone surge. In particular, alcohol exposure augmented the prenatal testosterone surge in males, with the combination of alcohol and stress causing an even more substantial reduction in the testosterone surge than prenatal stress alone (Ward et al., 2003). In parallel with these hormonal results, this research also showed that ejaculatory responses were largely unaffected in adult males exposed to prenatal alcohol alone (Ward et al., 1996; Ward et al., 2002; except see Ward et al., 1994), but caused an even greater reduction in the ejaculatory response in males exposed to both prenatal stress and alcohol compared to males exposed to just stress alone (Ward et al., 1996; Ward et al., 2002; Ward et al., 1994).

Taken together, Ward's body of work detailing the prenatal hormonal milieu, as well as describing the significant ways in which prenatal environmental experiences can shape this milieu, have had a profound impact on sexual differentiation research specifically and behavioral endocrinology more broadly. Her studies not only showed the myriad ways that perinatal hormone exposure could influence later adult behaviors but also provided an endocrinological mechanism through which some of these behavioral potentials were organized early in development.

The Ward Lab

Beyond Ward's substantial research contributions to the scientific literature, her investment in scientific training and mentoring has also had a lasting positive impact on the field of behavioral neuroendocrinology. In her time at Villanova University, Ward trained and mentored over 20 graduate students obtaining their terminal master's degree in Experimental Psychology, as well as undergraduate students seeking research experiences at Villanova University. Through her strong mentoring, she helped master's students move onto excellent PhD programs in behavioral neuroendocrinology. She accomplished this by bringing a leadership style to the scientific community that encompassed both rigorously high standards and an unmatched deep commitment to, and appreciation of, her trainees. Ward set up a unique and sustainable model of continuous research questions that advanced science forward by overlapping cohorts of students enrolled in Villanova's 2-year Experimental Psychology Master of Science graduate program. Every year or two, a new student would join her laboratory and team up with a second year graduate student to assist on their master's thesis, which was typically a continuation of a project that was started a year or two earlier by previous students. This positioned the first year student to quickly obtain hands-on laboratory skills and data analysis that could be used for their own thesis research while helping to progress the second year student's research as well. What is especially impressive about this model is that it allowed all students the opportunity to observe and participate in all stages of her research projects that usually spanned several years (i.e., breeding, pre- and/or postnatal natal treatment, development into adulthood, surgical manipulations, behavioral testing, brain tissue collection and processing, microscopic analyses, and data analysis). For example, some students started their 1st year in the graduate program at the tissue collection stage, while others started at the breeding stage, and others started just prior to the surgical manipulations or behavioral testing stages. As these 1st year students moved from the training stage to their own research project, they had the benefit of help from the more senior student(s), while also deepening their own learning by helping to train the next cohort of student(s). This required a great deal of highly coordinated planning and guidance from Ward and a deep commitment to teamwork. A key theme of the Ward laboratory was collaborative science. Modeling this organizational research style, many of Ward's trainees continued with successful research programs and careers founded on these same principles of promoting a scientific environment that values and encourages mentoring and collaboration.

Ward made this all possible through her extraordinary effort in successfully competing for extensive and prestigious federal funding that supported her own career development, her students' training, and her laboratory's scientific projects and discoveries. In 1969, she was awarded her first large grant, which was funded by the National Science Foundation. Just 1 year later, she launched an impressive track record of funding with 12 years of support from the National Institutes of Health (NIH) National Institute of Child Health and Human Development for her research program. She then landed some of the most prestigious awards showing NIH's commitment to a scientist's career: a Research Scientist Development Award (1975–1985) and a Research Scientist Award (1986–1990) from the National Institute of Mental Health. She also was awarded 3 years of funding from the National Institute of Child Health and Human Development for her research project on prenatal stress and alcohol.

As two of her previous graduate students, we can attest to the invaluable support and mentorship she provided to all her trainees. In this way, Ward's significant contributions to behavioral neuroendocrinology will continue to reverberate for many years to come (Fig. 28.1).

Fig. 28.1 A word cloud summarizing key words of the research projects supported by funding from the National Institutes of Health that was awarded to Inge Ward, Ph.D. Larger font reflects the primary focus of study in her long and successful career as a pioneer scientist in the field of behavioral neuroendocrinology. Generated using software from WordClouds.com

References

Arnold, A. P., & Gorski, R. A. (1984). Gonadal steroid induction of structural sex differences in the central nervous system. *Annual Review of Neuroscience, 7*, 413–442.

Breedlove, S. M., & Arnold, A. P. (1980). Hormone accumulation in a sexually dimorphic motor nucleus of the rat spinal cord. *Science, 210*, 564–566.

Clemens, L. G., Gladue, B. A., & Coniglio, L. P. (1978). Prenatal endogenous androgenic influences on masculine sexual behavior and genital morphology in male and female rats. *Hormones and Behavior, 10*, 40–53.

Gerall, A. A., & Ward, I. L. (1966). Effects of prenatal exogenous androgen on the sexual behavior of the female albino rat. *Journal of Comparative Physiological Psychology, 62*, 370–375.

Gerall, A. A., Hendricks, S. E., Johnson, L. L., & Bounds, T. W. (1967). Effects of early castration in male rats on adult sexual behavior. *Journal of Comparative and Physiological Psychology, 64*, 206–212.

Gorski, R. A., Gordon, J. H., Shryne, J. E., & Southam, A. M. (1978). Evidence for a morphological sex difference within the medial preoptic area of the rat brain. *Brain Research, 148*, 333–346.

Grisham, W. G., Kerchner, M., & Ward, I. L. (1991). Prenatal stress alters sexually dimorphic nuclei in the spinal cord of male rats. *Brain Research, 551*, 126–131.

Hoepfner, B. A., & Ward, I. L. (1988). Prenatal and neonatal androgen exposure interact to affect sexual differentiation in female rats. *Behavioral Neuroscience, 102*, 61–65.

Jordan, C. J., Breedlove, S. M., & Arnold, A. P. (1982). Sexual dimorphism and the influence of neonatal androgen in the dorsolateral motor nucleus of the rat lumbar spinal cord. *Brain Research, 249*, 309–314.

Kerchner, M., & Ward, I. L. (1992). SDN-MPOA volume is male rats is decreased by prenatal stress, but is not related to ejaculatory behavior. *Brain Research, 581*, 244–251.

Meisel, R. L., & Ward, I. L. (1981). Fetal female rats are masculinized by male littermates located caudally in the uterus. *Science, 213*, 239–242.

Nadler, R. D. (1969). Differentiation of the capacity for male sexual behavior in the rat. *Hormones and Behavior, 1*, 53–63.

Orth, J. M., Weisz, J., Ward, O. B., & Ward, I. L. (1983). Environmental stress alters the developmental pattern of d5-3b-hydroxysteroid dehydrogenase activity in Leydig cells of fetal rats: A quantitative cytochemical study. *Biology of Reproduction, 28*, 625–631.

Ward, I. L. (1969). Differential effect of pre- and postnatal androgen on the secual beahvior of intact and spayed female rats. *Hormones and Behavior, 1*, 25–36.

Ward, I. L. (1972a). Female sexual behavior in male rats treated prenatally with an anti-androgen. *Physiology and Behavior, 8*, 53–56.

Ward, I. L. (1972b). Prenatal stress feminizes and demasculinizes the behavior of males. *Science, 175*, 82–84.

Ward, I. L., & Renz, F. J. (1972). Consequences of perinatal hormone manipulation on the adult sexual behavior of female rats. *Journal of Comparative and Physiological Psychology, 78*, 349–355.

Ward, I. L., & Weisz, J. (1980). Maternal stress alters plasma testosterone in fetal males. *Science, 207*, 328–329.

Ward, I. L., & Weisz, J. (1984). Differential effects of maternal stress on circulating levels of corticosterone, progesterone, and testosterone in male and female fetuses and their mothers. *Endocrinology, 114*, 1635–1644.

Ward, O. B., Monaghan, E. P., & Ward, I. L. (1986). Naltrexone blocks the effects of prenatal stress on sexual behavior differentiation in male rats. *Pharmacology, Biochemistry and Behavior, 25*, 573–576.

Ward, I. L., Ward, O. B., Hayden, T., Weisz, J., & Orth, J. M. (1990). Naltrexone normalizes the suppression but not the surge of D5-3b-hydroxysteroid dehydrogenase activity in Leydig cells of stressed rat fetuses. *Endocrinology, 127*, 88–92.

Ward, I. L., Ward, O. B., Winn, R. J., & Bielawski, D. (1994). Male and female sexual behavior potential of male rats prenatally exposed to the influence of alcohol, stress, or both factors. *Behavioral Neuroscience, 108*, 1188–1195.

Ward, I. L., Ward, O. B., French, J. A., Hendricks, S. E., Mehan, D., & Winn, R. J. (1996). Prenatal alcohol and stress interact to attenuate ejaculatory behavior, but not serum testosterone or LH in adult male rats. *Behavioral Neuroscience, 110*, 1469–1477.

Ward, O. B., Ward, I. L., Denning, J. H., Hendricks, S. E., & French, J. A. (2002). Hormonal mechanisms underlying aberrant sexual differentiation in male rats prenatally exposed to alcohol, stress, or both. *Archives of Sexual Behavior, 31*, 9–16.

Ward, I. L., Ward, O. B., Affuso, J. D., Long, W. D., French, J. A., & Hendricks, S. E. (2003). Fetal testosterone surge: Specific modulations induced in male rats by maternal stress and/or alcohol consumption. *Hormones and Behavior, 43*, 531–539.

Weisz, J., & Ward, I. L. (1980). Plasma testosterone and progesterone titers of pregnant rats, their male and female fetuses and neonatal offspring. *Endocrinology, 106*, 309–316.

Rae Silver

29

Colin J. Saldanha, Michael N. Lehman, and Lance J. Kriegsfeld

Abstract

Rae Silver is the Helene L. and Mark N. Kaplan professor of Natural and Physical Sciences at Barnard College and professor with appointments in the Department of Psychology, Columbia University, and Pathology and Cell Biology at Columbia College of Physicians and Surgeons. Her scholarship spans multiple topics in reproductive behavior and chronobiology. She is especially well known for her work establishing a humoral signaling system from the suprachiasmatic nucleus (SCN) and, more recently, a portal system associated with the SCN and adjacent circumventricular areas of the mammalian brain. Silver is also one of the founding members of the Society for Behavioral Neuroendocrinology and a dedicated advocate for women in science.

Keywords

Circadian rhythms · Suprachiasmatic nucleus · Mast cells · Humoral signals

C. J. Saldanha (✉)
Departments of Neuroscience and Center for Behavioral Neuroscience, American University, Washington, DC, USA
e-mail: saldanha@american.edu

M. N. Lehman
Brain Health Research Institute, Kent State University, Kent, OH, USA
e-mail: mlehma18@kent.edu

L. J. Kriegsfeld
Departments of Psychology, Integrative Biology, and The Helen Wills Neuroscience Institute, University of California, Berkeley, CA, USA
e-mail: kriegsfeld@berkeley.edu

R. J. Nelson, Z. M. Weil (eds.), *Biographical History of Behavioral Neuroendocrinology*, https://doi.org/10.1007/978-3-031-12970-4_29

A native of Montreal, Quebec, Rae Silver received her BSc with Honors in Physiological Psychology from McGill University and moved to the United States with her husband who was pursuing a graduate degree. Once ensconced on the Upper West Side of Manhattan, Silver completed a MS in Biopsychology from City University of New York. She received her doctoral degree from the Institute of Animal Behavior at Rutgers University under the mentorship of Daniel S. Lehrman, one of the co-founders of the field of behavioral neuroendocrinology. Following assistant professor appointments at Rutgers and Hunter College, and a research associateship at the Museum of Natural History, New York, she was hired as an assistant professor at Barnard College of Columbia University in 1976. She was tenured 3 years later and soon promoted to professor of Psychology at Barnard College and at Columbia University.

Throughout her academic career, she and her team have made fundamental contributions to the field of behavioral neuroendocrinology. With continuous funding from the National Institutes of Health (NIH), the National Science Foundation (NSF), and other nonprofit institutions, she has built a reputation as a rigorous and gifted scientist, a thoughtful mentor, an introspective and forward-thinking leader, and a furious advocate for science education, particularly for women. In this chapter, we will discuss her contributions in each of these areas. We begin with several highlights in her continuing career as a scientist, collaborator, and mentor.

The first of two major chronobiology projects in the Silver's laboratory used the well-documented parental behavior of the ring dove as the context to understand behaviors associated with interval and circadian timing. As Lehrman had described years prior (Lehrman, 1959), in this species, both partners participate in several parental behaviors including nest building, incubation, and brooding. The timing of shared incubation is exquisitely organized. The male tends to incubate for about 6 h in the middle of the day. The rest of the day (about 18 h), it is the female who occupies the nest. Nest exchanges are usually seamless, and at around 1000 h, the male arrives at the nest with a "gift" – usually a twig. The female gets off the nest and the male incubates the eggs or hatchlings for his bout. The female arrives promptly at around 1600 h to resume her incubation, which lasts until 1000 h the next day. This cycle recurs for the entire time the couple incubates the eggs and hatchlings. Silver and her collaborators did several elegant experiments to reveal two kinds of time-keeping in ring doves. More specifically, when access to the nest was prevented for a period of time, the male delayed his arrival at the nest for the morning exchange. In contrast, the female's arrival time for the afternoon exchange remained unchanged in seeming disregard for when the male began incubation and when she had relinquished the nest on the previous day. The team inferred that the male was using an interval timing system which was initiated when he began his incubation bout. In contrast, the female appeared to be using a circadian timing mechanism which was minimally phase-shifted by delaying access to the nest (Gibbon et al., 1984). This project was one of several other research lines being conducted on timing and reproductive behavior using ring doves.

In parallel, the lab took on an old problem in avian photoperiodism. Previous work had strongly suggested that in birds that breed seasonally, vernal increases in daylength were detected by photoreceptors independent of those in the retina and pineal gland (Wilson, 1991). The Silver laboratory was the first to report the expression of the photoreceptive protein opsin, in neurons of the lateral septum and infundibulum in ring doves (Silver et al., 1988). These so-called deep brain receptors had been described for decades, but their location in the brain and their interactions with the reproductive axis remained unclear. Interestingly, opsin-expressing neurons extend processes seemingly through the ependyma into the lateral and third ventricles. Other photoreceptor proteins including transducin and phosducin also localized to the CSF (cerebrospinal fluid)-contacting neurons in the lateral septum, increasing the likelihood that these neurons were indeed photoreceptive (Silver et al., 1988; Saldanha et al., 1994). This hypothesis was indirectly supported by ultrastructural studies that described monosynaptic projections of opsin-expressing neurons onto GnRH-expressing dendrites in the dove (Saldanha et al., 2001). Although the group stopped short of describing actual light sensitivity in the putative deep brain photoreceptors, this was indeed demonstrated to be the case about a decade later (Nakane et al., 2010).

Silver continued to work on ring doves for several years but changed focus upon the discovery of nonneuronal cells that expressed gonadotropin-releasing hormone (GnRH) and appeared to infiltrate the medial habenula following courtship behavior (Silver et al., 1992). These cells appeared after a short 2-h courtship bout and were observed in both sexes. Because of their cellular characteristics including filamentous processes and irregular surfaces, the research team inferred that these cells were from the macrophage or mast cell lineage and appeared to migrate into the brain from the choroid plexus or the vasculature associated with the pia mater. This phenomenon was studied further to reveal a more widespread distribution of these cells following courtship, aggressive behavior toward squab, and even visual isolation from conspecifics. Notably, courting birds had significantly more GnRH-positive mast cell-like structures than long-term castrates and males in visual isolation (Zhuang et al., 1993). These cells were unequivocally identified as mast cells following ultrastructural examination, staining for sulfated proteoglycans, and the identification of secretory granules and filamentous processes (Silverman et al., 1994). Importantly, when an iodinated GnRH analog was administered into the lateral ventricle, no radioactive cells were observed in the brain following courtship. These results strongly suggested that the epitope recognized by the antibody against GnRH was endogenously synthesized by these mast cells and did not reflect endocytosis from the vasculature or the cerebrospinal fluid (Silverman et al., 1994).

The next chapter of this story established the mechanisms whereby mast cells infiltrated the parenchyma in rodents and the role of these cells in the regulation of behavioral states. Capitalizing on the observation that postpartum rats had higher numbers of mast cells in specific brain areas relative to virgin controls, Silver and colleagues harvested mast cells from the periphery, labeled them with vital dyes and administered the labeled cells into a host animal via the carotid artery. Scanning confocal microscopy on serial section revealed that mast cells rapidly took up residence in the basal lamina of endothelial cells, particularly in the thalamus (Silverman et al., 2000). This was the first demonstration of rapid infiltration of the brain by mature mast cells and established that these cells were capable of crossing the blood-brain barrier in a time frame consistent with the idea that parenchymal mast cell number may be acutely regulated by social and other behavioral cues.

Indeed, several years later, evidence suggested that mast cells are associated with anxiety-like behavior (Nautiyal et al., 2008). Mast cell-deficient KitW-sh/W-sh (sash−/−) mice showed increases in anxiety-related behaviors relative to their heterozygote littermates and wild-type controls. Then, using pharmacological agents that block mast cell degranulation but not their ability to cross the blood-brain barrier, the researchers established a role for central, but not peripheral, mast cells in behavioral state. Specifically, when administered intracerebroventricularly (icv) but not peripherally, the mast cell-degranulating agent cromolyn increased anxiety behaviors in both the open field and elevated plus-maze tests (Nautiyal et al., 2008). This work, which began in ring doves and morphed into a story about the mammalian brain, was accompanied, all along by the second major chronobiology project in the Silver's lab. This project involved the neural mechanisms associated with circadian behavior using Syrian hamsters (*Mesocricetus auratus*) as the animal model.

Silver and her lab over the years have made several major contributions to our understanding of the circadian timekeeping system and the role of the suprachiasmatic nucleus (SCN) as the central clock responsible for circadian rhythms in behavior, physiology, and other functions. Indeed, one of the key initial pieces of evidence cementing the role of the SCN as the master oscillator came from groundbreaking work of Silver and her colleagues using neural transplants to demonstrate recovery of circadian rhythmicity after grafting (Hakim et al., 1991; Lehman et al., 1987). Inspired by the work of her colleague and friend, Marie Gibson, who used grafts to restore reproductive function to hypogonadal mice (Gibson et al., 1984), Silver became interested in pursuing transplant models of the SCN to confirm the intrinsic nature of its oscillator and explore potential output signals and targets. Collaborating with Mike Lehman and Eric Bittman, she showed that fetal SCN grafts implanted into the ventricular system of SCN-lesioned hamsters were able to restore free-running rhythms of locomotor behavior; importantly, recovery of function was closely correlated with the presence of the SCN in the graft evidenced by its unique clusters of neuropeptide cells (Lehman et al., 1987). The work was later extended by Martin Ralph who showed, using circadian mutant hamsters, that functional characteristics of the clock, namely, its period, were also transferred to the host by the donor grafts (Ralph et al., 1990). The SCN transplant model was remarkable among other examples of neural transplantation at that time for its clarity and robustness of recovery, the correlation of recovery with anatomical features of the transplant, and the fact that functional recovery could be unambiguously attributed to the donor tissue.

Although it was initially assumed that neural outgrowth from the graft to the host brain was responsible for recovery of rhythms, the initial transplant work showed that while some types of rhythms were restored, others, including neuroendocrine ones such as the reproductive response to photoperiod, were not (Lehman et al., 1987). Complementing this was the work from the lab showing that when the projections of the SCN were severed using a Halász knife, circadian locomotor behavior persisted despite an impairment in photoperiodic responsiveness (Hakim et al., 1991). In addition, the variable location of SCN grafts that restored function, as well as the ability of dissociated SCN cell grafts to do so (Silver et al., 1990), suggested that nonneural, humoral signals might be involved. Silver initiated an innovative collaboration with biomaterial scientists who were exploring the use of semipermeable polymers as vehicles for transplantation and using that approach began experiments encapsulating SCN tissue prior to grafting as a way of preventing neural connections with the host (see Fig. 29.1). After much labor-intensive validation of the technique, which included careful confirmation of the survival of SCN tissue within the polymer and its lack of outgrowth, Silver and colleagues showed the remarkable finding that encapsulated SCN grafts implanted in the ventricular system were able to restore circadian activity rhythms to arrhythmic SCN-lesioned hosts (Silver et al., 1996a). The experimental design included the use of circadian mutant hamsters as either donors or hosts, providing strong evidence that when restored rhythms were seen, they were due to the presence of the grafted clock. Silver and colleagues sometimes jokingly referred to this study as the "bouquet garni"experiment, given the resemblance of the encapsulated graft to the French cooking technique, but nonetheless its results convincingly demonstrated that diffusible signals derived from the SCN were sufficient to restore rhythmicity. The work led to Silver and her collaborators to consider the other roles of CSF-borne

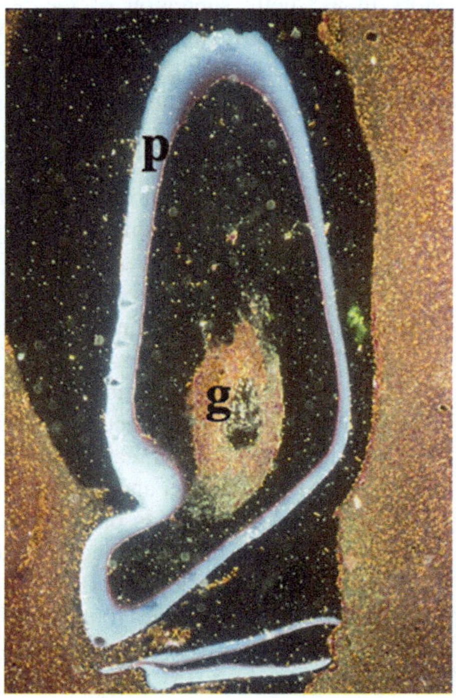

Fig. 29.1 Polymer-encapsulated SCN grafts were able to restore free-running activity rhythms to SCN-lesioned hamsters despite the lack of neural connections with the host brain. P = semi-permeable polymer capsule; g = SCN graft. (From Silver et al., 1996a)

signals in behavior and physiology (Lehman & Silver, 2000) as well as laid the foundation for the lab's recent work characterizing SCN vasculature (see below).

Whereas this work unequivocally established the SCN as the locus of the master clock in the brain and its ability to support behavioral rhythmicity via a diffusible signal, how the SCN maintained cohesive clock function at the tissue level was not well understood. Up until the late 1990s, it was generally believed that the rodent SCN comprised about 20,000 cells, each capable of independent oscillation and, through their intercellular coupling, formed a tissue-level clock. It was assumed that tissue-level rhythms simply required a threshold number of coupled SCN cells to support circadian functioning, although characterization of the SCN into "core" and "shell" regions by Robert Moore suggested potential tissue-level compartmentalization of function. Silver and her group questioned prevailing assumptions by systematically investigating the neuropeptidergic topography of the SCN, inevitably identifying a novel subregion expressing the calcium binding protein, calbindin-D28K, that received the majority of retinal input in Syrian hamsters (Silver et al., 1996b; Bryant et al., 2000). Silver hypothesized that this subregion of the SCN might be uniquely important given its direct access to environmental time.

Indeed, over a number of years, Silver's group established that this small, "core" subregion of the SCN was distinctive in that it was required for rhythmic function (and light entrainment), despite constituting only a small fraction of the entire nucleus (Lesauter & Silver, 1999; Kriegsfeld et al., 2004). She and her group also, surprisingly, established that this core hamster SCN subregion was not endogenously rhythmic but, instead, coordinated the timing of independent cellular oscillators in a "shell" subregion of the SCN to maintain tissue-level rhythmicity (Hamada et al., 2001; Antle et al., 2003). These findings were later extended to mice by Silver and her group, establishing that compartmentalized functional organization was not unique to hamsters (Karatsoreos et al., 2004; Yan & Silver, 2004; Kriegsfeld et al., 2008). To determine the specific pattern of rhythmic emergence across subdivisions of the SCN, in collaboration with David Welsh, Silver and colleagues applied immunohistochemical staining and cluster analysis of live slice imaging to examine spatiotemporal activation patterns of clock gene expression (Foley et al., 2011; Yan et al., 2007) . Through this work, they found that circadian oscillations are manifested as orderly serial activation across specific SCN subregions, further establishing compartmentalized SCN functional organization implied by her earlier work.

Establishing a unique functional partitioning of the SCN led to Silver's interest in how gonadal steroids may act on SCN subregions to influence sex differences in circadian rhythms. Although early work by Serge Daan and Colin Pittendrigh, and later work by Irving Zucker and Lawrence Morin, reported marked effects of gonadectomy and gonadal steroids on circadian functioning, the locus at which these effects occurred was undetermined. By examining androgen receptor expression in male mouse SCN, and the impact of testosterone implants in the SCN of castrated animals, Silver and colleagues found a population of androgen receptors in the core-mediating circadian rhythms and entrainment in male mice (Karatsoreos et al., 2007; Model et al., 2015). These findings revealed that not only does the SCN control rhythmic hormone secretion but also that sex steroid hormones feed back to the SCN to modulate its function and synchronization to environmental time.

Most recently, and quite remarkably, Silver and colleagues discovered a new vascular portal system in the brain that likely contributes to the wide-ranging downstream impact of SCN signaling (Silver et al., 1996b) . Prior to this finding, there was only one known central portal system (i.e., the hypophyseal portal system) that permits communication from hypothalamic neuroendocrine cells to the anterior pituitary, originally characterized by Popa and Fielding in the early 1930s (Gibbon et al., 1984) . At the time, several other researchers reported on the presence of this vasculature, but its significance had gone unrecognized. Based on the pattern of colloid labeling, Popa and Fielding presumed that "the direction of blood flow was certainly hypophysio-hypothalamic" (Bryant et al., 2000). It was not until close to 20 years later, however, that observations by Green and Harris in live rats confirmed directionality of blood flow to be from the median eminence to the anterior pituitary (Kriegsfeld et al., 2004).

Based on knowledge of this neuroendocrine communication conduit, and the earlier work by her group establishing that the SCN can support rhythmic behavior via a diffusible output signal (Silver et al., 1996a), Silver and colleagues became interested in characterizing SCN vasculature and its path. The initial goal was to provide insight into how diffusible clock signals may be communicated through the central circulation to reach target effector loci. Capitalizing on recent developments in tissue clearing techniques and light-sheet microscopy, Silver and her group were able to examine vasculature in the intact brain (see Fig. 29.2). By staining for neuropeptides that demarcate cells of the SCN and collagen to label blood vessels, Silver found an extensive network of capillaries in the SCN connecting with the organum vasculosum of the lamina terminalis (OVLT), a circumventricular organ at the ventral aspect of the third ventricle (Fig. 29.3) (Yao et al., 2021). Circumventricular

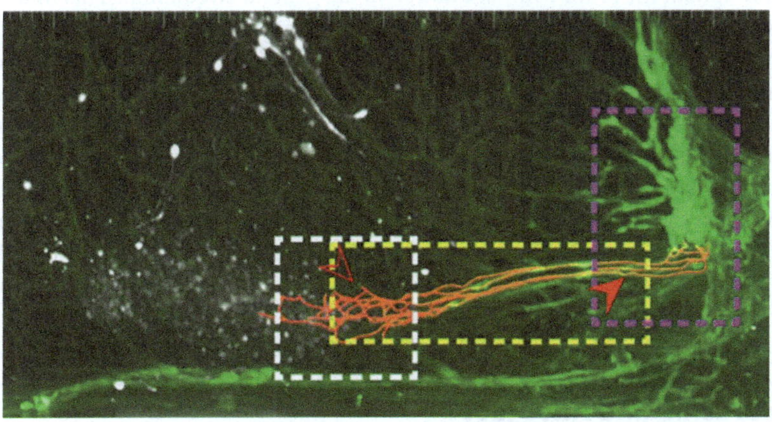

Fig. 29.2 Sagittal view of portal vasculature between the SCN and OVLT in three-dimensional and maximum intensity projection images. Maximum intensity projection with open and closed arrows pointing to a specific vessel lying at the rostral SCN (open arrow) or near the ventral OVLT (closed arrow), respectively. (From Yao et al., 2021)

Fig. 29.3 The Silver family (2021). Standing (L–R) are Geoffrey Silver, Alina Duncan Silver, Rae Silver holding 2-week-old Irving Livingston Silver, Leonard Silver, and Stephen Foley. Seated (L–R) are Molly Sue Silver (3 years), Amanda Rae Silver (5 years), and Darrell Jon Silver

organs are blood-brain barrier-free loci that allow for physiological monitoring (e.g., osmolality in the case of the OVLT) and permit neurochemical communication throughout the brain without dilution in the general circulation. This finding not only established that portal systems may be more common than previously appreciated but also helped to solve the mystery of how neuropeptides can have far-reaching actions at low concentrations. Relevant to Silver's work, the discovery of this novel portal system provided the missing link between SCN diffusible signaling and its targets. Given the broad role of the OVLT in regulating numerous aspects of behavior and physiology that are under circadian control, it is attractive to consider that this portal system allows humoral communication from the SCN to this circumventricular organ. Although more work is needed to establish the direction of blood flow, these findings set the stage for a diverse array of studies in neuroendocrinology and circadian biology.

Stewardship and Advocacy

In addition to her scholarly work, Silver is a passionate steward of the field of behavioral neuroendocrinology. Perhaps her greatest contribution on this front was her role, with Emilie Rissman and Kim Wallen, in the establishment of the Society for Behavioral Neuroendocrinology (SBN) in the mid-1990s. Prior to the SBN, the field appeared somewhat mired in organizational structure, and limitations in terms of animal models, experimental approaches, and indeed behavioral readouts. This was particularly surprising given notable developments in neuroendocrinology including the study of reptiles, amphibians, and birds. Arguably, these new models were pushing boundaries and expanding what we knew about the interactions among hormones, brain, and behavior. Silver and her colleagues were pivotal in changing several traditions in the field and establishing a society that was more diverse in science and scientists. It is fair to say that the vivid diversity across sex, gender, ethnicity, career stage, and indeed scientific questions, approaches, and techniques now seen at SBN is based, in large part, on the push to morph the field into a more welcoming, broad, and inclusive group. Silver has continued to be a force in the progress of the SBN having served as president (2017–2019) and has participated in several initiatives that have built this society.

Her stewardship of the field extends further into science and science education. Among her many contributions toward the advancement of science, she has served in an advisory role at NASA (National Aeronautics and Space Administration), the Human Frontiers Program, and the Office of the Director, NSF. Silver has served as a receiving editor or on advisory boards for several journals including *Journal of Comparative Psychology*, *Endocrinology*, *Journal for Research in Biological Rhythms*, *eNeuro*, and the *European Journal of Neuroscience*.

It would be an oversight to ignore the considerable work Silver has put into developing opportunities specifically for women in STEM (science, technology, engineering, and mathematics) fields. Her position at Barnard College, an undergraduate institution that only admits women, perhaps gave her the right stage on which she could broaden the participation of women in science and mathematics. She has piloted several initiatives on this front including the areas of education in mathematics and experience in computer science, and coding, among others. Her TED (technology, entertainment, and design) talk on increasing opportunities for women in science is widely disseminated and is used by many researchers in their own work broadening participation of under-represented groups in science and engineering.

It is not surprising at all to see that Silver has received many accolades for her involvement in science, education, and mentorship. She was elected as fellow to the American Association for the Advancement of Science, the American Academy of Sciences, and the Helene L. and Mark N. Kaplan Professorship of Natural and Physical Sciences. She received a Teaching Award in recognition of her exceptional contributions to undergraduate education and, in 2015, received the Daniel S. Lehrman Lifetime Contribution Award, named for her graduate mentor, from the Society for Behavioral Neuroendocrinology.

In 2021, Rae and Lenny celebrated 55 years of marriage. They have raised two sons, Geoffrey and Darrell, and are the grandparents of Molly Sue, Amanda Rae, and Irving Livingston (Fig. 29.3). They remain ensconced in a lovely apartment on the Upper West Side of Manhattan, and enjoy spending time at their country house in upstate New York.

Acknowledgments We thank the numerous students, postdoctoral researchers, and collaborators of the Silver laboratory and the editors for spearheading this effort. Our gratitude also extends to our mentor, colleague, and friend, Rae Silver.

References

Antle, M. C., Foley, D. K., Foley, N. C., & Silver, R. (2003). Gates and oscillators: A network model of the brain clock. *Journal of Biological Rhythms, 18*, 339–350.

Bryant, D. N., Lesauter, J., Silver, R., & Romero, M. T. (2000). Retinal innervation of calbindin-D28K cells in the hamster suprachiasmatic nucleus: Ultrastructural characterization. *Journal of Biological Rhythms, 15*, 103–111.

Foley, N. C., Tong, T. Y., Foley, D., Lesauter, J., Welsh, D. K., & Silver, R. (2011). Characterization of orderly spatiotemporal patterns of clock gene activation in mammalian suprachiasmatic nucleus. *The European Journal of Neuroscience, 33*, 1851–1865.

Gibbon, J., Morrell, M., & Silver, R. (1984). Two kinds of timing in circadian incubation rhythm of ring doves. *The American Journal of Physiology, 247*, R1083–R1087.

Gibson, M. J., Krieger, D. T., Charlton, H. M., Zimmerman, E. A., Silverman, A. J., & Perlow, M. J. (1984). Mating and pregnancy can occur in genetically hypogonadal mice with preoptic area brain grafts. *Science, 225*, 949–951.

Hakim, H., Debernardo, A. P., & Silver, R. (1991). Circadian locomotor rhythms, but not photoperiodic responses, survive surgical isolation of the SCN in hamsters. *Journal of Biological Rhythms, 6*, 97–113.

Hamada, T., Lesauter, J., Venuti, J. M., & Silver, R. (2001). Expression of period genes: Rhythmic and nonrhythmic compartments of the suprachiasmatic nucleus pacemaker. *The Journal of Neuroscience, 21*, 7742–7750.

Karatsoreos, I. N., Yan, L., Lesauter, J., & Silver, R. (2004). Phenotype matters: Identification of light-responsive cells in the mouse suprachiasmatic nucleus. *The Journal of Neuroscience, 24*, 68–75.

Karatsoreos, I. N., Wang, A., Sasanian, J., & Silver, R. (2007). A role for androgens in regulating circadian behavior and the suprachiasmatic nucleus. *Endocrinology, 148*, 5487–5495.

Kriegsfeld, L. J., Lesauter, J., & Silver, R. (2004). Targeted microlesions reveal novel organization of the hamster suprachiasmatic nucleus. *The Journal of Neuroscience, 24*, 2449–2457.

Kriegsfeld, L. J., Mei, D. F., Yan, L., Witkovsky, P., Lesauter, J., Hamada, T., & Silver, R. (2008). Targeted mutation of the calbindin D28K gene disrupts circadian rhythmicity and entrainment. *The European Journal of Neuroscience, 27*, 2907–2921.

Lehman, M., & Silver, R. (2000). CSF signaling in physiology and behavior. In L. F. Agnati, J. Fuxe, C. Nicholson, & E. Sykova (Eds.), *Progress in brain research: Volume transmission revisited*. Elsevier.

Lehman, M. N., Silver, R., Gladstone, W. R., Kahn, R. M., Gibson, M., & Bittman, E. L. (1987). Circadian rhythmicity restored by neural transplant. Immunocytochemical characterization of the graft and its integration with the host brain. *Journal of Neuroscience, 7*, 1626–1638.

Lehrman, D. S. (1959). On the origin of the reproductive behavior cycle in doves. *Transactions of the New York Academy of Sciences, 21*, 682–688.

Lesauter, J., & Silver, R. (1999). Localization of a suprachiasmatic nucleus subregion regulating locomotor rhythmicity. *The Journal of Neuroscience, 19*, 5574–5585.

Model, Z., Butler, M. P., Lesauter, J., & Silver, R. (2015). Suprachiasmatic nucleus as the site of androgen action on circadian rhythms. *Hormones and Behavior, 73*, 1–7.

Nakane, Y., Ikegami, K., Ono, H., Yamamoto, N., Yoshida, S., Hirunagi, K., Ebihara, S., Kubo, Y., & Yoshimura, T. (2010). A mammalian neural tissue opsin (Opsin 5) is a deep brain photoreceptor in birds. *Proceedings of the National Academy of Sciences of the United States of America, 107*, 15264–15268.

Nautiyal, K. M., Ribeiro, A. C., Pfaff, D. W., & Silver, R. (2008). Brain mast cells link the immune system to anxiety-like behavior. *Proceedings of the National Academy of Sciences of the United States of America, 105*, 18053–18057.

Ralph, M. R., Foster, R. G., Davis, F. C., & Menaker, M. (1990). Transplanted suprachiasmatic nucleus determines circadian period. *Science, 247*, 975–978.

Saldanha, C. J., Leak, R. K., & Silver, R. (1994). Detection and transduction of daylength in birds. *Psychoneuroendocrinology, 19*, 641–656.

Saldanha, C. J., Silverman, A. J., & Silver, R. (2001). Direct innervation of GnRH neurons by encephalic photoreceptors in birds. *Journal of Biological Rhythms, 16*, 39–49.

Silver, R., Witkovsky, P., Horvath, P., Alones, V., Barnstable, C. J., & Lehman, M. N. (1988). Coexpression of opsin- and VIP-like-immunoreactivity in CSF-contacting neurons of the avian brain. *Cell and Tissue Research, 253*, 189–198.

Silver, R., Lehman, M. N., Gibson, M., Gladstone, W. R., & Bittman, E. L. (1990). Dispersed cell suspensions of fetal SCN restore circadian rhythmicity in SCN-lesioned adult hamsters. *Brain Research, 525*, 45–58.

Silver, R., Ramos, C. L., & Silverman, A. J. (1992). Sexual behavior triggers the appearance of non-neuronal cells containing gonadotropin-releasing hormone-like immunoreactivity. *Journal of Neuroendocrinology, 4*, 207–210.

Silver, R., Lesauter, J., Tresco, P. A., & Lehman, M. N. (1996a). A diffusible coupling signal from the transplanted suprachiasmatic nucleus controlling circadian locomotor rhythms. *Nature, 382*, 810–813.

Silver, R., Romero, M. T., Besmer, H. R., Leak, R., Nunez, J. M., & Lesauter, J. (1996b). Calbindin-D28K cells in the hamster SCN express light-induced Fos. *Neuroreport, 7*, 1224–1228.

Silverman, A. J., Millar, R. P., King, J. A., Zhuang, X., & Silver, R. (1994). Mast cells with gonadotropin-releasing hormone-like immunoreactivity in the brain of doves. *Proceedings of the National Academy of Sciences of the United States of America, 91*, 3695–3699.

Silverman, A. J., Sutherland, A. K., Wilhelm, M., & Silver, R. (2000). Mast cells migrate from blood to brain. *The Journal of Neuroscience, 20*, 401–408.

Wilson, F. E. (1991). Neither retinal nor pineal photoreceptors mediate photoperiodic control of seasonal reproduction in American Tree Sparrows (Spizella-Arborea). *Journal of Experimental Zoology, 259*, 117–127.

Yan, L., & Silver, R. (2004). Resetting the brain clock: Time course and localization of mPER1 and mPER2 protein expression in suprachiasmatic nuclei during phase shifts. *The European Journal of Neuroscience, 19*, 1105–1109.

Yan, L., Karatsoreos, I., Lesauter, J., Welsh, D. K., Kay, S., Foley, D., & Silver, R. (2007). Exploring spatiotemporal organization of SCN circuits. *Cold Spring Harbor Symposia on Quantitative Biology, 72*, 527–541.

Yao, Y. F., Taub, A. B., Lesauter, J., & Silver, R. (2021). Identification of the suprachiasmatic nucleus venous portal system in the mammalian brain. *Nature Communications, 12*.

Zhuang, X., Silverman, A. J., & Silver, R. (1993). Reproductive behavior, endocrine state, and the distribution of GnRH-like immunoreactive mast cells in dove brain. *Hormones and Behavior, 27*, 283–295.

Michael J. Baum

30

Julie Bakker

Abstract

Michael J. Baum is one of the leading scientists studying the neuroendocrine mechanisms underlying the sexual differentiation of the brain and behavior. His research group produced many papers on the perinatal and adult actions of testosterone and its neural metabolite, estradiol, on the expression of sexual behavior in rodent models, including rats and mice, as well as in higher mammalian models including ferrets and nonhuman primates. In particular, his studies using ferrets have made an important contribution to the field by demonstrating that caution is needed in extrapolating findings obtained in rodent models to other mammalian species—thereby emphasizing the value of comparative research. In his role as editor-in-chief, Baum also revitalized *Hormones and Behavior* and established it as the official journal of the Society for Behavioral Neuroendocrinology (SBN). Finally, Baum was a great mentor to many students and postdoctoral fellows, including the author of this biographical sketch.

Keywords

Sexual differentiation · Testosterone · Estradiol · Sexual behavior · Ferrets · Primates

Early Years

Michael J. Baum was born in 1943 in Waterloo, Iowa, USA, where he grew up in a musical family. His father was a professor of piano at Iowa State Teachers College, now known as the University of Northern Iowa, in Cedar Falls. His mother was a

J. Bakker (✉)
GIGA Neurosciences, University of Liège, Liège, Belgium
e-mail: jbakker@uliege.be

R. J. Nelson, Z. M. Weil (eds.), *Biographical History of Behavioral Neuroendocrinology*, https://doi.org/10.1007/978-3-031-12970-4_30

chorus master who taught in a middle school. She was a very strong-willed woman, who wanted to escape from her rural farming community of Czech immigrants. They had three sons with Michael being the oldest. In 1961, Baum moved to Northfield, Minnesota, to start his bachelor's degree in Psychology at Carleton College where he obtained his degree in 1965. During the summer of 1963, he obtained his first research experience working as a research assistant in the Psychology Department laboratory of John Bare, who studied circadian feeding rhythms in rats. He helped collect data on food intake using automated Skinner boxes. This summer internship sparked Baum's interest in scientific research even though a significant part of his workday consisted of cleaning soiled rat cages.

Graduate Years (1965–1970)

As he was completing his bachelor's degree at Carleton, Baum had to choose between the University of Chicago and McGill University for graduate training. He visited the University of Chicago but found the faculty whom he met to be arrogant and unwelcoming. Further, a close Carleton College friend who was 1 year ahead of him had pursued graduate studies in the McGill University psychology department. He convinced Baum to choose McGill (in Montreal) over the University of Chicago. It also seemed like a good idea to be out of the USA in those days, because the country was becoming heavily involved in the Vietnam war—drafting many young men into combat. Baum began his studies at McGill in September 1965. At that time, the only required graduate class was in statistics. The main requirement in the first year of graduate study was to formulate and then conduct a research project and write a master's thesis. Inspired by his previous Carleton summer internship and by reading a paper in *Scientific American* by Richard Wurtman and Julius Axelrod (who won a Nobel Prize for his catecholamine research in 1970), Baum wrote his thesis on the possible role of the pineal gland in mediating the synchronization of rats' circadian feeding rhythms by photoperiod. He observed only minor effects of pinealectomy on this process; he published this paper in *Physiology and Behavior* in 1970 (Baum, 1970).

By contrast, pinealectomy of 3-day-old male rats raised in constant darkness advanced the first expression of male copulatory behaviors (published in *Science*, Baum, 1968). Baum started to study male rat sexual behavior completely on his own initiative. His McGill mentor, Robert Malmo, provided him with all the necessary resources while giving him total freedom in choosing his PhD thesis topic. Baum had to learn through reading papers how to study sexual behavior in rats. No one was around to show him the ropes. Based on his results of pinealectomy on the development of male sexual behavior, Baum got further interested in studying the hormonal mechanisms underlying puberty. In a first study, he attempted to accelerate puberty by administering testosterone to prepubertal male rats. He also administered estradiol to other prepubertal males as control condition, but surprisingly (rather annoyingly at the time), estradiol-treated male rats first mated at an even more precocious age than testosterone-treated males. This work comprised his PhD thesis which was completed in September 1969 and subsequently published in the *Journal of Comparative and Physiological Psychology* (Baum, 1972).

The Dutch Years (1970–1977)

The book, *The Physiology of Puberty* (1965), written by Bernard Donovan and J.J. (Koos) van der Werff ten Bosch, containing both clinical and experimental data, further inspired Baum to pursue his research on the hormonal regulation of puberty. He wrote Donovan to ask him for a postdoctoral position, but because he did not have funding for a postdoc at the time, Donovan recommended that Baum contact Koos van der Werff ten Bosch at the Erasmus University in Rotterdam, the Netherlands. Koos was more than happy to offer Baum a postdoc position since he was setting up a new department, "Endocrinology, Growth, and Reproduction," at the Erasmus University. He was rather urgently looking for people to hire. In the same week that Baum arrived in Rotterdam, another postdoctoral fellow/US citizen, David Goldfoot, joined the research department of Van der Werff ten Bosch. Goldfoot had recently completed his PhD under the supervision of Robert Goy of the Oregon primate center. Over the next 2 years, Goldfoot and Baum worked together on a couple of projects. Importantly, Goldfoot introduced Baum to several "establishment figures" in the American domain of behavioral neuroendocrinology including Robert Goy, Irving Zucker, and Harvey Feder.

Van der Werff ten Bosch provided Baum with full access to his ferret colony which had previously been used to study the effects of hypothalamic lesions on pubertal development in females. Baum restricted his research to studying pubertal development in the male ferret. For that purpose, he had to develop a method to monitor pubertal development by measuring testicular size/volume in vivo in addition to assessing spermatogenesis in testicular biopsies taken repeatedly from the same male ferrets. Fortunately for him, the Medical Faculty of the Erasmus University is next to a major hospital (Dijkzigt Ziekenhuis), and he was invited to come and see how testicular biopsies were performed in men. Witnessing this procedure in an anesthetized man (surely not the highpoint of his career) caused Baum to faint, whereupon he had to be carried out of the surgery room. Upon awakening, Baum proceeded to perfect a method for repeatedly taking biopsies of the testes of the same ferrets over many months and for assigning a numerical rating to stages of spermatogenic development. Together with Goldfoot, Baum went on to publish a paper (*Journal of Endocrinology*, Baum & Goldfoot, 1974) showing that mediobasal hypothalamic lesions accelerated the onset of pubertal development in male ferrets without eliminating their capacity to show an annual cycle of testicular regression and regrowth.

In addition to studying physiological/hormonal aspects of pubertal development, Baum continued to be interested in studying male sexual behavior but now in both rats and ferrets. Probably his paper in the *Journal of Comparative and Physiological Psychology* (Baum, 1976) is one of his most remarkable contributions to the field. In contrast to rats, male ferrets castrated in adulthood and treated with estradiol show full female-typical receptive mating behaviors. Likewise, prenatally androgenized female ferrets do not lose the capacity to show female receptive behaviors in adulthood. These results show that there is no behavioral defeminization in the ferret, as had been observed in essentially all nonhuman primate species studied. These results point to the importance of conducting comparative studies and to not limiting research only to rodent models. During those early years in Rotterdam, Baum also collaborated with his biochemist colleague, Jan Vreeburg, to show that high-affinity estradiol-binding receptors are present in the male rat's hypothalamus (Vreeburg et al., 1975). Together with Vreeburg, he also provided some of the strongest evidence (Baum & Vreeburg, 1973) that the central aromatization of testosterone into estradiol and the reduction of testosterone into dihydrotestosterone are both required for the full adult expression of male sexual behavior in rats. This work provided an important insight into the interpretation of Baum's thesis research in which prepubertal administration of estradiol was even more effective than testosterone in provoking the precocious onset of mating in male rats.

One of several turning points in Baum's career was his sabbatical in 1975 in the laboratory of Joe Herbert at Cambridge University, UK, where Barry Everitt and Barry Keverne were also working. This sabbatical led to long-standing collaborations with both Barry's: With Barry Everitt using c-Fos as a marker for neuronal activity, he published a pioneer and highly cited paper (Baum & Everitt, 1992) showing that both pheromonal and genital somatosensory stimuli activate a neural circuit controlling mating in male rats. With Barry Keverne, he established a role of olfaction for mate recognition in male and female mice—launching an extensive series of studies on this topic that involved both mice (Baum & Keverne, 2002) and ferrets (Kelliher et al., 1998).

At Cambridge, Baum was introduced to primate research. He studied the role of progesterone in modulating the sexual behavior of female rhesus monkeys leading to a *Nature* paper (Baum et al., 1976). It was found that progesterone reduced females' sexual attractivity for a male through a local action in the vagina, not by acting centrally. Surprisingly, he also found that either giving progesterone to an ovariectomized, estradiol-primed female or withdrawing estradiol from an ovariectomized female failed to diminish the expression of receptive mating behavior on those occasions when a male was still motivated to mount and intromit with the female (Baum et al., 1977).

After returning to the Netherlands, Baum collaborated with Koos Slob who was in the same Erasmus University department to study the hormonal regulation of sexual receptivity in female stump-tail macaques using a colony of animals kept at the Dutch Primate Research Center in Rijswijk, the Netherlands. In contrast to rodents and other non-primate mammalian species, surgical removal of both the ovaries and adrenal glands failed to diminish females' sexual receptivity (Baum et al., 1978), further emphasizing the importance of conducting comparative research to understand the range of behavioral responses to sex steroids.

It is worth noting that Baum obtained a permanent research position after completing his Canadian postdoctoral fellowship at the Erasmus University (in 1972).

This meant that he could have stayed for the rest of his career in the Netherlands. So why did he leave a permanent research position to return to the USA in 1978?

Although he learned to speak Dutch fluently and was very productive during his years in the Netherlands, he never really got accustomed to several typical Dutch academic traditions. For one, Baum worked in a medical faculty with a strong hierarchy where in order to climb the "career ladder," you either had to have Dutch roots or an important professorial mentor who would push your career forward. He had neither. Second, Baum never really completely integrated into daily Dutch life. Thus, fed up with Dutch academia, Baum started applying for jobs everywhere, eventually ending up at the Massachusetts Institute of Technology (MIT) in Cambridge, MA, USA.

Back to the USA

Baum was appointed as an associate professor without tenure upon his arrival at MIT where he quickly competed for his first NIH grant to study the neuroendocrine regulation of sexual behavior in the male rat. Mary Erskine (1978–1985) was his first postdoctoral fellow and Stuart Tobet (1979–1985), his first graduate student. Both turned out to be foundational figures in the development of Baum's independent (American) research career. They both continued to collaborate with Baum for long periods after establishing their own laboratories. He continued being successful in obtaining NIH grants for both his rat and ferret work, studying the organizational and activational effects of gonadal hormones on sexual behavior in males and females. Mary Erskine was mostly interested in studying hormonal regulation of female sexual behavior and was a pioneer in studying sexual motivation in the female rat by developing the paced mating paradigm (Erskine & Baum, 1982). Tobet made an important contribution by describing a sexually dimorphic nucleus in the ferret's preoptic/anterior hypothalamic area (Tobet et al., 1986). This discovery of the ferret's "male nucleus" led to a series of important findings including the study by Raul Paredes, a postdoc in his lab from 1992 to 1994. He showed that excitotoxic lesions of the preoptic area/anterior hypothalamus (POA/AH) including the male nucleus altered the partner preference of male ferrets from estrous females to males, providing the first evidence for a neural substrate in the expression of sexual partner preference (Paredes & Baum, 1995). This result was replicated in a later study by Baum's PhD student Heather Kindon (now known as Halem) (Kindon et al., 1996). These studies also set the basis for changing his focus more to the neural basis of sexual motivation since it is probably the most profound sexually dimorphic feature of courtship. Kevin Kelliher, PhD student (1997–2001), was the first to describe the presence of an accessory olfactory bulb in the ferret (Kelliher et al., 2001). Even so, it turned out that the detection of pheromonal signals by the main olfactory system is required for ferrets to identify an opposite-sex mating partner (Kelliher & Baum, 2001; Woodley & Baum, 2004). Another remarkable finding was the identification of a large sex difference in the number of galanin-immunoreactive neurons in the POA/AH, with males having three times as many galanin neurons as females (Park et al., 1997). In addition, these galanin neurons were specifically activated in male but not female ferrets upon mating but not in response to olfactory cues (Bakker et al., 2002).

Although Baum was very productive at MIT, he did not receive tenure. An additional complication was that Baum's MIT Department of Nutrition and Food Sciences was disbanded so that he had to look for positions elsewhere. He preferred to stay in Boston because his wife, Catherine Snow, was at the verge of getting tenure at Harvard University. Baum moved to Boston University (BU) in 1985 and was appointed as associate professor of Biology (again without tenure). Mary Erskine moved with him to BU but as an independent scientist because she had already been able to obtain her own NIH grant support. Baum spent a considerable portion of his career (1978–1993) supported mainly by ("soft") NIH funding including a Career Scientist Award from the NIMH. He eventually received tenure at Boston University in 1993.

Reflex Ovulation: Ancient History?

An interesting aspect of using the ferret as model for studying sexual behavior is that the female undergoes seasonal periods of behavioral (and vulval) estrus when ovulation is induced by sensory stimulation received from the male during penile intromission. Practically speaking, it is very easy to determine when the female is in estrus by just looking to see whether her vulva is swollen (see the appended photo of Baum inspecting a female ferret's vulva). Rona Carroll, a PhD student (1983–1987), made a rather astonishing observation: a single intromission induced a prolonged preovulatory surge in luteinizing hormone (LH; Carroll et al., 1985). This elevation in plasma LH levels began around 1.5 h after the onset of intromission, reached a maximum approximately 6 h later, and was sustained for at least 12 h. Even though intromission duration can vary between 1 and 94 (!) minutes, male ferrets can take their time when mating,[1] the ovaries of every ferret receiving an intromission had corpora lutea 1 week later. Thus, the preovulatory LH surge in the ferret seems to be independent of the amount of mating stimulation and is therefore an all-or-none phenomenon. This observation led to an additional research line in the Baum's lab to elucidate the neuroendocrine mechanisms of reflex ovulation (carried out by PhD students Rona Carroll and Geralyn Lambert-Messerlian and later postdoctoral fellow Julie Bakker). It was taken to the molecular level by analyzing levels of GnRH mRNA of the female ferret's mediobasal hypothalamus in close collaboration with Beverly Rubin at Tufts University. This turned out to be a

[1] When the male mounts the female, he will grab the nape of the female's neck with his teeth and will grip her body by wrapping his forelegs around her rib cage. Periods of pelvic thrusts will be alternated with periods of rest during which the male simply lies over the female but continues to immobilize her by the neck grip. Intromission is characterized by the male arching his back, causing his foreleg grip to slip behind the female's torse. Variable intromission durations have been reported, although they can last up to 3 h. It is unclear when the male ejaculates and how often during the mating.

very laborious study, taking brain samples at many different time points during the LH surge to monitor GnRH gene expression, spending many hours in the dark room developing the radioactive S^{35} signal (Bakker et al., 1999). This study quite often brought tears to the eyes of the first author since technical problems frequently appeared (such as the emulsion not sticking to the slides), but as Baum would always say, "you do the best you can"!

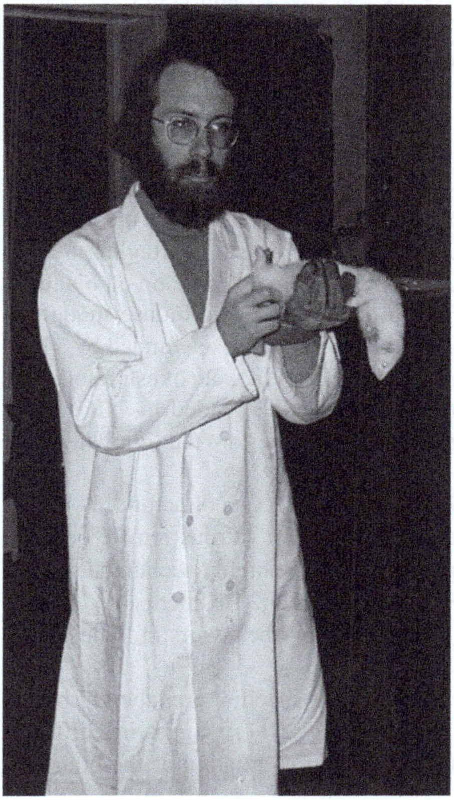

Baum inspecting a female ferret's vulva. (Photo taken in 1973)

Reflex ovulation has been considered as a primitive trait evolutionary speaking, based on the finding that induced ovulation can occasionally occur in species (e.g., rats and humans) that are spontaneous ovulators. It has even been argued recently (Pavlicev & Wagner, 2016) that the neuroendocrine mechanisms underlying female orgasm in women originated in mammals in which ovulation is induced by copulatory stimulation. Even so, there is no scientific evidence of orgasm in induced ovulating species. The evolution of reflex ovulation remains an interesting topic of discussion, and the review (Bakker & Baum, 2000) by Baum and his former postdoc Julie Bakker is still often cited.

Abandoning the Ferrets and Hopping on the Mouse Bandwagon

The development of transgenic mice in the 1990s introduced an important new tool in the field of behavioral neuroendocrinology. Finally, one was able to "knock out" a gene controlling the expression of a neuropeptide or enzyme in order to determine its effects on behavior. Indeed, Baum and his colleagues (Baum et al., 1994) were among the earliest to use this technology to study the effect of a null mutation of the c-fos gene on the expression of mating behavior in male mice. Surprisingly, despite being greatly stunted in size and lacking teeth, homozygous c-fos null mutant mice mated quite normally with females that were twice their size. Eventually, it was discovered that knock-out mice had some important disadvantages, such as the presence of compensatory mechanisms or not being able to distinguish between developmental and adult effects of the inactivated gene. Even so, after the introduction of these mouse models, it became increasingly difficult to get funding for any nontraditional model systems such as the ferret. Baum finally gave up on his ferret research in 2007 and completely switched to using transgenic mouse models to study (1) the functioning of the vomeronasal system in both sexes funded by a NIH grant together with his former postdoc Jim Cherry and (2) the role of estradiol in the behavioral feminization of the mouse brain and behavior, funded by NIH grants together with Julie Bakker. Some highlights of these studies are the following:

1. Removal of the vomeronasal organ (VNO) had no effect on olfactory sex pheromone discrimination but did reduce the preference for estrous female odors in male mice (Pankevich et al., 2004) arguing against the highly acclaimed role for the VNO in sex discrimination (e.g., Stowers et al. (2002), Kimchi et al. (2007)) and in suppressing the expression of male copulatory behaviors in female mice (Martel & Baum, 2009).
2. Estradiol feminizes lordosis behavior during a specific prepubertal period (between postnatal days 15 and 25; Brock et al. (2011)) suggesting different critical windows for male versus female sexual differentiation of the brain and behavior.
3. State-of-the-art techniques such as optogenetics and DREADDs (designer receptors exclusively activated by designer drugs) revealed the relative roles of the main and the accessory olfactory systems in mediating partner preference and mating performance in mice of both sexes (Kang et al., 2009; DiBenedictis et al., 2015; McCarthy et al., 2017; Kunkhyen et al., 2017).

Other Important Accomplishments

Baum's work as the editor of *Hormones and Behavior* for 8 years (1996–2004) is another major accomplishment in his career. He raised the journal's reputation and impact factor to a higher level, thereby establishing *Hormones and Behavior* as the official journal of the newly formed Society for Behavioral Neuroendocrinology

(SBN) and the top journal in the field of behavioral neuroendocrinology. He also served as the president of the Society for Behavioral Neuroendocrinology (SBN) between 2001 and 2003, and he received the Daniel Lehrman Lifetime Achievement Award in 2014. Beginning in 1984, Baum became a member of the International Academy of Sex Research where over the years he strove to link his animal work to theoretical and practical issues in the domain of human sexuality. In this vein, he published an article on the relevance of animal work on the brain and behavioral sexual differentiation to the human situation (Baum, 2006). He also served as the president of the IASR for 2 years (2006–2007). Baum also served stints on federal grant review committees including the National Institute of Drug Abuse study section (1979–1982), the psychobiology panel at NSF (1988–1989), and the NIH Neuroendocrinology, Neuroimmunology, Rhythms, and Sleep Study section (2008–2012).

It should be mentioned that Baum is an excellent writer and that he has written many highly cited review papers, including his review from 1979 on a comparative analysis of the differentiation of coital behavior in mammals which is his most cited contribution.

Conclusion

Baum started his scientific career by studying circadian rhythms in feeding behavior in rats and then passed through his "pubertal period" in the Netherlands where he got acquainted with ferrets, and finally ended up in Boston, first at MIT followed by Boston University, studying the neuroendocrine mechanisms underlying the sexual differentiation of the brain and behavior, with a special focus on the role of olfactory cues in mate recognition and sexual motivation. His work with ferrets made an especially important contribution to the field, demonstrating that one should be very cautious in extrapolating findings obtained in rodent models to other mammalian species—thereby emphasizing the value of comparative research. Baum has always had a very critical approach to his own work as well as that of others. His research has been primarily hypothesis-driven; his students and postdocs had to come up with very strong scientific arguments to convince him why it would be interesting to conduct a particular experiment ("why do you want to do that"?), also often not realizing that he had already done that experiment in the past. His great enthusiasm for science in general and his capability to convey his scientific rigor to his many students and postdocs made him a great mentor. It is thus not surprising that many of his former students and postdocs stayed in close contact and continued to collaborate with him for many years.

References

Bakker, J., & Baum, M. J. (2000). Neuroendocrine regulation of GnRH release in induced ovulators. *Frontiers in Neuroendocrinology, 21*, 220–262.

Bakker, J., Rubin, B. S., & Baum, M. J. (1999). Changes in mediobasal hypothalamic gonadotropin releasing hormone mRNA levels induced by mating or ovariectomy in a reflex ovulatory, the ferret. *Endocrinology, 140*, 595–602.

Bakker, J., Woodley, S. K., Kelliher, K. R., & Baum, M. J. (2002). Sexually dimorphic activation of galanin neurons in the ferret's dorsomedial preoptic area/anterior hypothalamus after mating. *Journal of Neuroendocrinology, 14*, 116–125.

Baum, M. J. (1968). Pineal gland: Influence on the development of copulation in male rats. *Science, 162*, 586–587.

Baum, M. J. (1970). Light-synchronization of rat feeding rhythms following sympathectomy or pinealectomy. *Physiology & Behavior, 5*, 325–329.

Baum, M. J. (1972). Precocious mating in male rats following treatment with androgen or estrogen. *Journal of Comparative and Physiological Psychology, 78*, 356–367.

Baum, M. J. (1976). Effects of testosterone propionate administered perinatally on sexual behavior of female ferrets. *Journal of Comparative and Physiological Psychology, 90*, 399–410.

Baum, M. J. (1979). Differentiation of coital behavior in mammals: A comparative analysis. *Neuroscience and Biobehavioral Reviews, 3*, 265–284.

Baum, M. J. (2006). Mammalian animal models of psychosexual differentiation: When is "translation" to the human situation possible? *Hormones and Behavior, 50*, 579–588.

Baum, M. J., & Everitt, B. J. (1992). Increased expression of c-fos in the medial preoptic area after mating in the male rat: Role of afferent input from the medial amygdala and midbrain central segmental field. *Neuroscience, 50*, 627–646.

Baum, M. J., & Goldfoot, D. A. (1974). Effect of hypothalamic lesions on maturation and annual cyclicity of the ferret testis. *The Journal of Endocrinology, 62*, 59–73.

Baum, M. J., & Keverne, E. B. (2002). Sex difference in attraction thresholds for volatile odors from male and estrous female mouse urine. *Hormones and Behavior, 41*, 213–219.

Baum, M. J., & Vreeburg, J. T. M. (1973). Copulation in castrated male rats following combined treatment with estradiol and dihydrotestosterone. *Science, 182*, 283–285.

Baum, M. J., Everitt, B. J., Herbert, J., Keverne, E. B., & de Greef, W. J. (1976). Reduction of sexual interaction in rhesus monkeys by a vaginal action of progesterone. *Nature, 263*, 606–608.

Baum, M. J., Keverne, E. B., Everitt, B. J., Herbert, J., & de Greef, W. J. (1977). Effects of progesterone and estradiol on sexual attractivity of female rhesus monkeys. *Physiology & Behavior, 18*, 659–670.

Baum, M. J., Slob, A. K., de Jong, F. H., & Westbroek, D. L. (1978). Persistence of sexual behavior in ovariectomized stumptail macaques following dexamethasone treatment or adrenalectomy. *Hormones and Behavior, 11*, 323–347.

Baum, M. J., Brown, J. J. G., Kica, E., Rubin, B. S., Johnson, R. S., & Papaioannou, V. E. (1994). Effect of a null mutation of the c-fos proto-oncogene on sexual behavior of male mice. *Biology of Reproduction, 50*, 1040–1048.

Brock, O., Baum, M. J., & Bakker, J. (2011). The development of female sexual behavior requires prepubertal estradiol. *The Journal of Neuroscience, 31*, 5574–5578.

Carroll, R. S., Erskine, M. S., Doherty, P. C., Lundall, L. A., & Baum, M. J. (1985). Coital stimuli controlling luteinizing hormone secretion and ovulation in the female ferret. *Biology of Reproduction, 32*, 925–933.

DiBenedictis, B. T., Olugbemi, A. O., Baum, M. J., & Cherry, J. A. (2015). DREADD-induced silencing of the medial olfactory tubercle disrupts the preference of female mice for opposite-sex chemosignals. *eNeuro, 2*: e0078-15.2015, 1–16.

Erskine, M. S., & Baum, M. J. (1982). Effects of paced coital stimulation on termination of estrus and brain indoleamine levels in female rats. *Pharmacology Biochemistry and Behavior, 17*, 857–861.

Kang, N., et al. (2009). A direct main olfactory bulb projection to the 'vomeronasal' amygdala in female mice selectively responds to volatile pheromones from males. *The European Journal of Neuroscience, 29*, 624–634.

Kelliher, K. R., & Baum, M. J. (2001). Nares occlusion eliminates heterosexual partner selection without disrupting coitus in ferrets of both sexes. *The Journal of Neuroscience, 21*, 5832–5840.

Kelliher, K. R., Chang, Y.-M., Wersinger, S. R., & Baum, M. J. (1998). Sex difference and testosterone modulation of pheromone-induced neuronal Fos in the ferret's main olfactory bulb and hypothalamus. *Biology of Reproduction, 59*, 1454–1463.

Kelliher, K. R., Baum, M. J., & Meredith, M. (2001). The ferret's vomeronasal organ and accessory olfactory bulb: Effect of hormone manipulation in adult males and females. *The Anatomical Record, 263*, 280–288.

Kimchi, T., Ju, X., & Dulac, C. (2007). A functional circuit underlying male sexual behavior in the female mouse brain. *Nature, 448*, 1009–1014.

Kindon, H. A., Paredes, R. G., & Baum, M. J. (1996). Medial preoptic/anterior hypothalamic lesions induce a female-typical profile of sexual partner preference in male ferrets. *Hormones and Behavior, 30*, 514–527.

Kunkhyen, T., McCarthy, E. A., Korzan, W., Doctor, D., Han, X., Baum, M. J., & Cherry, J. A. (2017). Optogenetic activation of accessory olfactory bulb input to the forebrain differentially modulates investigation of opposite vs same-sex urinary chemosignals and stimulates mating in male mice. *eNeuro, 4*: e0010-17.2017, 1–16.

Martel, K. L., & Baum, M. J. (2009). Adult testosterone treatment but not surgical disruption of vomeronasal function augments male-typical sexual behavior in female mice. *The Journal of Neuroscience, 29*, 7658–7666.

McCarthy, E. A., Kunkhyen, T., Korzan, W. J., Naik, A., Maqsudlu, A., Cherry, J. A., & Baum, M. J. (2017). A comparison of the effects of male pheromone priming and optogenetic inhibition of accessory olfactory bulb forebrain inputs on the sexual behavior of estrous female mice. *Hormones and Behavior, 89*, 104–112.

Pankevich, D. E., Baum, M. J., & Cherry, J. A. (2004). Olfactory sex discrimination persists whereas the preference for urinary odorants from estrous females disappears in male mice after vomeronasal organ removal. *The Journal of Neuroscience, 24*, 9451–9457.

Paredes, R. G., & Baum, M. J. (1995). Altered sexual partner preference in male ferrets given excitotoxic lesions of the preoptic area/anterior hypothalamus. *The Journal of Neuroscience, 15*, 6619–6630.

Park, J.-J., Baum, M. J., & Tobet, S. A. (1997). Sex difference and steroidal stimulation of galanin immunoreactivity in the ferret's doral preoptic area/anterior hypothalamus. *The Journal of Comparative Neurology, 389*, 277–288.

Pavlicev, M., & Wagner, G. (2016). The evolutionary origin of orgasm. *Journal of Experimental Zoology. Part B, Molecular and Developmental Evolution, 326*, 326–377.

Stowers, L., Holy, T. E., Meister, M., Dulac, C., & Koentges, G. (2002). Loss of sex discrimination and male-male aggression in mice deficient for Trp2. *Science, 295*, 1493–1500.

Tobet, S. A., Zahniser, D. J., & Baum, M. J. (1986). Sexual dimorphism in the preoptic/anterior hypothalamic area of ferrets: Effects of adult exposure to sex steroids. *Brain Research, 364*, 249–257.

Vreeburg, J. T. M., Schretlen, P. J. M., & Baum, M. J. (1975). Specific, high-affinity binding of estradiol in cytosols from several brain regions and pituitary of intact and castrated adult male rats. *Endocrinol, 97*, 969–977.

Woodley, S. K., & Baum, M. J. (2004). Differential activation of glomeruli in the ferret's main olfactory bulb by anal scent gland odours from males and females: An early step in mate identification. *The European Journal of Neuroscience, 20*, 1025–1032.

C. Sue Carter

31

A. Courtney DeVries

Abstract

C. Sue Carter was born on 25 December 1944 and grew up in rural Missouri. She earned a BA in Biology from Drury College, which was located a few miles from her family home, and then a PhD in Zoology from the University of Arkansas. It was during her postdoctoral fellowship at Michigan State University that she first became engaged in behavioral neuroendocrinology research. Over the subsequent five decades, Carter held professorships at six universities and served as Director of the Kinsey Institute; each change in institution expanded her scientific interests and added depth and complexity to her research program. She is perhaps best known within behavioral neuroendocrinology for (1) introducing prairie voles as a model organism for studying the physiological regulation of social bonds and biparental behavior and (2) for catalyzing the study of oxytocin and vasopressin in behavior. However, she also has published extensively on the importance of oxytocin in promoting human social behavior, and the behavioral consequences of aberrant oxytocin signaling. Lastly, Carter was an exceptional mentor to trainees and colleagues, and the strongest possible advocate for women and individuals from groups underrepresented in science and academia.

Keywords

Monogamy · Social bonding · Vasopressin · Oxytocin · Prairie voles · Kinsey Institute

A. C. DeVries (✉)
Department of Medicine and Neuroscience, West Virginia University, Morgantown, WV, USA
e-mail: courtney.devries@hsc.wvu.edu

R. J. Nelson, Z. M. Weil (eds.), *Biographical History of Behavioral Neuroendocrinology*, https://doi.org/10.1007/978-3-031-12970-4_31

Early Life, Education, and Training

Carol Sue Carter, professionally known as C. Sue Carter, was born on 25 December 1944 in San Francisco, California. Her father was working in the Kaiser shipyards at that time. She claims that without consulting her, the family migrated back to their original home in the Missouri Ozarks when she was 3. There she grew up immersed in rural America, in turn developing the work ethic and fierce determination that would propel her through the inevitable ups and downs of academia. Carter was a middle child of eight offspring spread over a 27 year period, so she claims the advantage of being the recipient of a great deal of alloparenting. Throughout middle school and high school, she raised cattle to save money for her college education.

Carter attended Willard High School in Willard, Missouri, and then earned her Bachelor of Arts degree at Drury College (now Drury University) which was located approximately 10 miles away in Springfield, Missouri. She graduated *summa cum laude*, with a major in Biology. In recognition of her extraordinary career, Drury University awarded Carter the 2016 Distinguished Alumni Award for Lifetime Achievement.

After leaving Drury, Carter pursued a PhD in Zoology at the University of Arkansas. Upon matriculation, her goal was to complete a 2 year Master of Science degree and then teach high school. However, the course of her future serendipitously changed when she was awarded a 3 year NSF predoctoral scholarship, with the understanding that she would complete a PhD in that time period. She accomplished this under the joint mentorship of Douglas A. James, a conservationist with expertise in avian ecology and ethology, and Jack Marr, a behavioral biologist, whose interests lead her to do a thesis entitled "Early olfactory experience in the guinea pig, *Cavia porcellus L.*". This graduate work formed the basis of Carter's first two publications in *Animal Behavior* (Carter & Marr, 1970; Carter, 1971) on developmental aspects of olfactory imprinting preference; the work cemented her love of research and desire for an academic career. A shift toward studying hormonal mechanisms of behavior began as she launched her postdoctoral training at Michigan State in the laboratory of Lynwood G. Clemens. Clemens had trained with Frank A. Beach, Robert W. Goy, and Roger A. Gorski, and this training brought her into the field of behavioral neuroendocrinology. When Carter joined his lab in 1969, Clemens was studying the effects of early life exposure to gonadal steroids on adult reproductive behaviors in rats. Carter brought hamsters to that laboratory as well as a comparative and developmental perspective on sexual behavior that she retained throughout her career (Carter et al., 1972). In 1970, Carter married Stephen W. Porges (a psychologist) and moved with him to West Virginia University (WVU). At WVU, she was sponsored by Martin Schein for an independent NIH Postdoctoral Fellowship. There she also expanded her research to include hormonal regulation of postcopulatory sexual receptivity (Carter & Schein, 1971), quickly establishing herself as an expert in hormones and sex behavior and initiating her transition to research independence.

Institutions, Promotions, and Leadership Positions

Carter and Porges relocated to the University of Illinois, Urbana-Champaign, in 1972, where she was offered a research position in a newly formed laboratory for psychopharmacology sponsored by the State of Illinois Department of Mental Health. During this period, Carter developed and taught a popular elective seminar in Human Sexuality; it was the first time this subject had been offered at that university and it attracted more than 200 undergraduate students. In 1974, largely through the efforts of Ed Banks, a senior ethologist, the University of Illinois created a faculty position for Carter which comprised appointments in three separate units. She began her faculty career as an Assistant Professor in the Departments of Psychology, Ecology Ethology, and Evolution and the School of Basic Medical Sciences. She rapidly rose through the ranks to attain the rank of Professor in 1984. It was during this period that three crucial shifts in Carter's research occurred that would shape the next several decades of her career: (1) she began to direct her work to the emerging field of behavioral neuroscience and to studies of sex differences in the brain (Greenough et al., 1977); (2) she incorporated increasingly broad measures of social behavior in a multitude of species, including lemmings (Huck et al., 1979); and (3) she began to collaborate with Lowell Getz, a mammologist working with prairie voles (*Microtus ochrogaster;* Dluzen et al., 1981).

At that point in time, prairie voles had rarely been studied in the laboratory and their mating system (social monogamy) was unknown. Carter's first several prairie vole studies examined basic reproductive biology and confirmed that females responded to ovarian hormones in a way that was similar to other induced ovulators (Dluzen & Carter, 1979). However, this species did not exhibit an estrous cycle and was uniquely dependent on social cues for reproductive activation, which was initiated by male pheromones (Carter et al., 1980). Social cues in turn precipitated a cascade of endocrine events in the olfactory bulb that implicated the vomeronasal organ in the facilitation of reproduction in this species (Dluzen et al., 1981). However, it was Carters' interactions with Getz, a behavioral ecologist who conducted population studies on several species of *Microtus*, that prompted her to begin thinking about reproductive behavior in the context of a monogamous social system. The early papers by Carter and Getz describing the social organization (Getz & Carter, 1980) and mating system (Getz et al., 1981) of prairie voles laid the groundwork for Carter's research developing prairie voles as a model for studying the physiological mechanisms underlying the formation of adult social bonds and parental care. Carter and Getz published several additional experimental papers and highly cited review papers on monogamy and pair-bonding over the subsequent 20 years (for example, Carter et al., 1995). Getz remained an engaging colleague, good friend, and reliable source for outbreeding Carter's prairie vole colony for decades.

Toward the end of her time at Illinois, and despite being the mother of a 1 year-old baby, Carter took a sabbatical in the Physiology Department at Stanford University (1981). She worked primarily with Julian Davidson, a renowned behavioral neuroendocrinologist who was one of the founding editors of the journal

Hormones and Behavior. Davidson was remarkably effective in translating fundamental rodent research into the study of hormones and sexual behavior. Carter's first forays into clinical research were studies with Davidson on the effects of age and menstrual cycle on the sexual arousal of women (Morrell et al., 1984) and the effects of testosterone replacement on tactile sensitivity in hypogonadal men (Segraves et al., 1985). The latter human studies extended into a collaboration with Davidson that continued into the early 1990s (Burris et al., 1991), a time during which Davidson took a 1 year sabbatical to NIH.

Shortly after returning from Stanford, and while still on the faculty at the University of Illinois, Carter served a 1 year rotation (1982–1983) as a Program Officer in Psychobiology at the National Science Foundation, in Washington, DC. She used her position to advocate for increased funding for research on the biological bases of behavior and greater support of interdisciplinary research. She also used that period to gestate a second child, returning to Illinois in time for his birth. However, Carter's work inside the beltway had precipitated a case of what she refers to as "Potomac fever," a common, but noninfectious, disorder characterized by the realization of the importance of federal funding for the growth of science. It was an increased awareness of opportunities to collaborate with government laboratories and a personal desire to raise their family in a less extreme climate than offered by central Illinois that prompted Carter and Porges to begin considering faculty positions in the Washington area.

Thus, soon after being promoted to Professor at Illinois (1984), Carter left to join the Department of Zoology at the University of Maryland, College Park (1985). Diane Witt, who had worked with Carter as an undergraduate honors student in Illinois, moved with her to Maryland to pursue a PhD. Witt's thesis research centered around behavioral and hormonal factors influencing estrus induction in prairie voles. In 1990, Witt and Carter published their first paper examining the effects of oxytocin on social interactions and sexual behavior in prairie voles (Witt et al., 1990). It was the first of more than 125 papers Carter has published on oxytocin to date and was a harbinger of her extensive exploration of the neuroendocrine mechanisms underlying pair-bonding (described in more detail in section "Scientific Impact and Key Contributions to the Field of Behavioral Neuroendocrinology"). Carter was appointment Distinguished University Professor at the University of Maryland in 1997 in recognition of her scholastic achievements, including her role in mentoring trainees and other scientists, often beyond her own field. She was the first woman at the institution to be awarded this distinction.

During her tenure at the University of Maryland, Carter also was a Guest Researcher within the Developmental Endocrinology Branch of the National Institute of Child Health and Human Development at NIH. Through collaborations with George Chrousos, Philip Gold, and Margaret Altemus, intramural NIH clinician-researchers with expertise in neuroendocrine and behavioral sequelae of stress, Carter began studying the effects of stress and immune function on women's mental health. She also collaborated with Chrousos on studies of reproductive success in common marmosets (*Callithrix jacchus*), a highly social primate species, and the effects of lactation on stress reactivity and immune function in postpartum women. These interactions also added a new layer to Carter's prairie vole research as she explored the hypothesis that improved stress regulation was a consequential benefit of social bonding and mediated by oxytocin. It also brought to her attention the physiology "peculiarities" shared by prairie voles (Taymans et al., 1997) and socially monogamous species of New World monkeys (Johnson et al., 1996), in turn prompting her to speculate that there were likely common physiological mechanisms underlying social monogamy across mammalian species (Carter & Perkeybile, 2018).

After 16 years at Maryland, Carter and Porges were recruited back to the University of Illinois, this time to the Chicago campus, to serve as Co-Directors of the Brain-Body Center and Professors of Psychiatry. Carter's research during that period was well-funded by a series of NIH Program Project grants and R01s. In addition, in collaboration with Porges, she was awarded a large NIH renovation grant that they used to create a 20,000 sq foot laboratory, facilitating the integration of basic and clinical research. Carter also developed novel assays for the noninvasive measurement of oxytocin in human saliva, which are now widely accepted and used. During this period, Carter continued to collaborate with and to mentor clinicians focused on various aspects of psychiatry and child development, including Suma Jacob. Jacob was a child psychiatrist, originally trained by Martha McClintock, who had an interest in studying oxytocin in the context of autism. Together, Jacob and Carter conducted one of the first studies of genetic variation in the oxytocin receptor among individuals diagnosed with autism, published in 2007 (Jacob et al., 2007) and widely replicated.

While oxytocin effects on behavior remained central to her research, the emphasis of Carter's research at UI Chicago shifted to understanding the importance of developmental exposure to oxytocin and vasopressin on physiological responses, brain development, attachment, and bonding. She conceptualized oxytocin signaling, and its modulation of stress responses, as an important mechanism through which disruption of parent-offspring and adult-adult bonds could promote pathology. Carter published several papers supporting a link between oxytocin and vasopressin and symptom severity in patients with schizophrenia (e.g., Rubin et al., 2010) and proposed that dysregulation of oxytocin also could underlie the skewed social behaviors that are characteristic of autism spectrum disorders (e.g., Jacob et al., 2007) and Williams syndrome (e.g., Dai et al., 2012). She also became involved with projects on the role of oxytocin and birth experiences in postpartum depression through collaborations with a clinical research group at McGill University in Montreal (e.g., Zelkowitz et al., 2014).

From 2013 to 2014, Carter spent a brief period as a Professor of Psychiatry at the University of North Carolina, Chapel Hill, before being recruited to Indiana University as the Executive Director of the Kinsey Institute and the Rudy Professor of Biology. The world-renowned Kinsey Institute was established at Indiana University in 1947 to promote understanding of human sexuality, gender, and reproduction through historical preservation of documents and art, research, education, and outreach. As director of the Kinsey Institute from 2014 to 2019, Carter made the bold move to expand, for the first time, the institute's emphasis beyond sex, gender, and reproduction to include the context of relationships, specifically the psychological and emotional functions and contributions to well-being. Carter argued that sex can have very different meaning and elicit very different psychological and physiological responses when it occurs consensually within a relationship, consensually outside of a relationship, or without consent. She also proposed to increase research at the interface of medicine and sexuality and to build further on the study of how physical and psychological trauma can alter sexuality.

She was the first biologist to be appointed director since Alfred Kinsey, who likewise trained as a zoologist. Some critics interpreted this pivot toward sex in the context of relationships by a director who studied monogamous rodents as an attempt to sanitize the Kinsey Institute and bow to political pressure; nothing could have been further from the truth. Those leveling this criticism do not appear to have read or truly understood Carter's work, which emphasized that prairie voles were socially, but *not* sexually, monogamous. Furthermore, Carter had built her career on studying the overlapping and diverging physiological mechanisms involved in sexual behavior, social bonding, and parental behavior in prairie voles and had applied her knowledge to understand disorders characterized or exacerbated by disrupted social behaviors. Alas, for an institute that is very comfortable when mired in controversy, expanding its mission to understand love may have been a step too far. In 2019, Carter completed her term as Director of the Kinsey Institute and returned to full-time research, as a Distinguished University Professor. The current director of the Kinsey Institute replaced "love" with "sexuality" on the Kinsey Institute website but retained the emphasis on relationships and well-being.

In 2021 Carter was offered a position as a Professor of Psychology at the University of Virginia. At UVA Carter has merged her research program with that of her epigeneticist collaborator, Jessica Connelly (Bell et al., 2015). Carter and Connelly have worked together for more than a decade, with current R01 funding from NICHD. Their program continues to study both prairie vole and human behavior. It also focuses on the developmental effects of early social experience, a consistent theme throughout Carter's career. In addition, Carter and Connelly are refining noninvasive methods for measuring peptides and peptide receptors, which can be used in behavioral research, including many translational studies, and are becoming accepted as "biomarkers" for wellness and longevity. Thus, this most recent phase of her academic journey studying longevity began when Carter was in her mid-70s, yet somehow seems fitting for someone whose first published research paper appeared in 1970. (Table 31.1).

Scientific Impact and Key Contributions to the Field of Behavioral Neuroendocrinology

Quantitative metrics of scientific impact demonstrate Carter's broad influence. She has published over 400 articles and chapters and edited five books. Her papers have been cited nearly 33,000 times and her h-index is 90 (Google Scholar, January

Table 31.1 Academic positions held by Carter

Years	Position	Department	University
1974–1977	Assistant professor	Ecology ethology, and evolution and psychology	University of Illinois, Champaign
1977–1984	Associate professor	Ecology ethology, and evolution and psychology	University of Illinois, Champaign
1984–1985	Professor	Ecology ethology, and evolution and psychology	University of Illinois, Champaign
1985–1997	Professor	Zoology	University of Maryland, College Park
1997–2001	Distinguished university professor	Biology	University of Maryland, College Park
2001–2012	Professor	Psychiatry	University of Illinois, Chicago
2013–2014	Professor	Psychiatry	University of North Carolina, Chapel Hill
2014–2019	Executive director Rudy professor	Kinsey institute Biology	Indiana University, Bloomington
2019–present	Distinguished university research scientist	Kinsey institute	Indiana University, Bloomington
2021–present	Professor	Psychology	University of Virginia Charlottesville

2022). Although these metrics place Carter in the highest echelon of scientists, they do not adequately capture the three areas in which she is likely to have the greatest lasting impact on the field of behavioral neuroendocrinology: (1) establishing prairie voles as models to study hormonal regulation of social bonding; (2) catalyzing the study of oxytocin as a behaviorally active hormone in a wide array of species, including humans; and (3) enticing trainees, clinicians, and scientists in other fields to incorporate hormonal measures into their behavioral studies.

Carter popularized prairie voles as a wild rodent that could be bred and studied easily in the laboratory. Prior to Carter's first paper on the effects of ovarian hormones on sexual and social behaviors in prairie voles published in 1979 (Dluzen & Carter, 1979), there were only four papers identified by PubMed using the keywords "prairie vole behavior" (compared to 653 papers in January 2022). The second crucial step in establishing prairie voles as a model of social monogamy was developing a reliable index of social bonding; in 1992 Carter, Jessie Williams (Carter's postdoctoral fellow), and undergraduate student Kenneth Catania (now a neurobiologist at Vanderbilt University who received a MacArthur Award in 2006 and was named a Guggenheim Fellow in 2014) designed a three-chamber testing paradigm that allowed rapid assessment of preference for the experimental animal's partner versus an unfamiliar vole (Williams et al., 1992a, b). The first paper using this behavioral test established that female prairie voles developed partner preferences more rapidly if they had mated with their partner, but that sex was not required. The behavioral task remains widely used and has been indispensable in identifying the hormonal mediators and neural circuitry underlying social bond formation. To date more than 225 papers have been published on pair-bonding in prairie voles.

Carter can also be credited with catalyzing the study of oxytocin as a behaviorally active hormone. In the late 1980s when Carter became interested in oxytocin's potential role in monogamy, it had been extensively studied in the context of uterine contractions and milk letdown. Barry Keverne's group at this time was studying mother-infant interactions in sheep and published a series of studies indicating that stimulation of the cervix during birth caused the release of oxytocin which in turn was crucial for the onset of maternal behavior, including the formation of a mother-infant bond (summarized in Keverne and Kendrick (1994)). Witt et al. (1990) demonstrated for the first time that administration of oxytocin altered social behavior between adult prairie voles (Witt et al., 1990). More than 200 papers have subsequently been published on oxytocin and prairie voles, and approximately 25% of these papers list Carter as an author. Furthermore, more than 2600 papers have been published on the broad topic of oxytocin and social behavior.

Based on the very similar structures of oxytocin and vasopressin and their intriguing physiological relationship, Carter began studying the role of vasopressin in social bonding in prairie voles, as well. A collaboration between Carter's lab and Thomas Insel's lab at NIMH, resulted in a *Nature* paper demonstrating that ICV vasopressin administration facilitates partner preferences in prairie voles (Winslow et al., 1993). Following publication of this report, there was a sharp increase in papers on vasopressin and social behavior; to date there are more than 1100 papers on this topic.

Carter's influence on the field of behavioral neuroendocrinology is further amplified through the training of dozens of individuals across five decades, ranging from undergraduate research assistants to colleagues in other fields. Carter was an exemplary mentor; her passion for scientific discovery was infectious and being part of her laboratory was exhilarating. She equipped her trainees with the skills needed to develop independence by providing support and guidance without micromanaging projects. She encouraged trainees to think broadly and to take risks on projects that had the potential to challenge entrenched scientific ideas. Furthermore, she was thrilled when trainees chose to continue studying oxytocin or vasopressin, pair-bonding, or prairie voles after leaving her lab; there was no mentor-trainee competitiveness. Carter also was a positive role model for aspiring women faculty; she contended, and perhaps more importantly demonstrated, that it was possible to integrate a healthy family life and successful scientific career. Indeed, Carter was celebrating and finding ways to accommodate her trainees' growing families long before universities started talking about policies to support work-life integration. Carter trained more than 30 graduate students and postdoctoral fellows who have remained in academia or joined NIH or NSF.

A Family of Accomplished Academicians and Professionals

Carter met her husband, Stephen W. Porges, at Michigan State where he was completing a PhD in Psychology and she was beginning a postdoctoral fellowship. Porges went on to become the leading expert in the physiological control of heart rate variability and associated behavioral consequences. Of note, he proposed the polyvagal theory, helping to redefine the relationship between the evolution of the autonomic nervous system and social behavior in mammalian species, and how physiological perturbations could elicit behavioral changes and psychiatric disorders (Porges, 1995).

Although Carter and Porges shared scientific interests and often discussed each other's data, they maintained separate careers, publishing together only a handful of times and mostly in the past two decades. Porges is a Professor Emeritus at the University of Maryland and University of Illinois at Chicago, and a Distinguished University Scientist at Indiana University and Professor of Psychiatry at the University of North Carolina. Carter and Porges are a rare example of a two-career academic couple, who were fortunate to land independent faculty positions at the same institutions throughout their careers.

Carter and Porges raised two sons who have likewise established themselves in professional careers. Eric Porges earned a PhD in Psychology at the University of Chicago and is an Assistant Professor in the Department of Clinical and Health Psychology at the University of Florida. He studies the neurobiology of cognitive development across the life span, with an emphasis on the consequences of neural damage caused by HIV and alcohol use disorder. His younger brother, Seth Porges, received his BS and MS in Journalism from Northwestern University and is a freelance scientific writer who also directs and produces documentaries. He is widely published and his documentaries, including Class Action Park on HBOmax, have won awards at several film festivals.

Conclusion

C. Sue Carter has led an extraordinary life and career as a scientist and mentor. She grew up in a large family in rural Missouri and raised cattle to pay her college tuition. She was broadly trained, earning a BS in Biology from Drury College and a PhD in Zoology from the University of Arkansas, and then undertook postdoctoral training in behavioral endocrinology with Clemens at Michigan State University. Over the subsequent five decades, Carter made several key discoveries in behavioral neuroendocrinology, including delineating the roles of oxytocin, vasopressin, and corticosteroids in the regulation of social bonds, biparental behavior, and alloparental behavior in prairie voles, demonstrating the stress-buffering effects of lactation and pro-social behavior, and providing evidence of the importance of oxytocin in promoting human social behavior. Carter held professorships at six universities across her career, including the University of Illinois (Champaign), University of Maryland (College Park), University of Illinois (Chicago), University of North Carolina (Chapel Hill), Indiana University (Bloomington), and University of Virginia (Charlottesville). She also served 5 years as Director of the Kinsey Institute at Indiana University. This wide range of experiences provided Carter with unique perspectives that enriched her research program. Lastly, one of Carter's most enduring legacies is likely to be that she was an exceptional mentor to dozens of trainees and colleagues, many of whom were women and individuals from groups under-represented in science and academia.

Acknowledgments I thank Sue for sharing 50 years' worth of stories and insights with me. It was wonderful to be reminded of the many personal stories behind the scientific discoveries. Unfortunately, for the reader, I only had sufficient space to share a small fraction of them here.

References

Bell, A. F., Carter, C. S., Steer, C. D., Golding, J., Davis, J. M., Steffen, A. D., Rubin, L. H., Lillard, T. S., Gregory, S. P., Harris, J. C., & Connelly, J. J. (2015). Interaction between oxytocin receptor DNA methylation and genotype is associated with risk of postpartum depression in women without depression in pregnancy. *Frontiers in Genetics*. https://doi.org/10.3389/fgene.2015.00234

Burris, A. S., Gracely, R. H., Carter, C. S., Sherins, R. J., & Davidson, J. M. (1991). Testosterone therapy is associated with reduced tactile sensitivity in human males. *Hormones and Behavior, 25*, 195–205.

Carter, C. S. (1971). Effects of olfactory experience on the behaviour of the guinea pig, *Cavia porcellus*. *Animal Behaviour, 21*, 241–256.

Carter, C. S., & Marr, J. N. (1970). Olfactory imprinting and age variables in the guinea pig, *Cavia porcellus*. *Animal Behaviour, 18*, 238–244.

Carter, C. S., & Perkeybile, A. M. (2018). The monogamy paradox: What do love and sex have to do with it? *Frontiers in Ecology and Evolution, 6*, 202.

Carter, C. S., & Schein, M. W. (1971). Sexual receptivity and exhaustion in the female golden hamster. *Hormones and Behavior, 2*, 191–200.

Carter, C. S., Clemens, L. G., & Hoekema, D. A. (1972). Neonatal androgen and adult sexual behavior in the golden hamster. *Physiology and Behavior, 9*, 89–95.

Carter, C. S., Getz, L. L., Gavish, L., McDermott, J. L., & Arnold, P. (1980). Male-related pheromones and the activation of female reproduction in the prairie vole (Microtus ochrogaster). *The Biology of Reproduction, 23*, 1038–1045.

Carter, C. S., DeVries, A. C., & Getz, L. L. (1995). Physiological substrates of monogamy: The prairie vole model. *Neuroscience and Biobehavioral Reviews, 19*, 303–314.

Dai, L., Carter, C. S., Ying, J., Bellugi, U., Pournajafi-Nazarloo, H., & Korenberg, J. R. (2012). Oxytocin and vasopressin are dysregulated in Williams Syndrome, a genetic disorder affecting social behavior. *PLoS One, 7*(6), e38513.

Dluzen, D. E., & Carter, C. S. (1979). Ovarian hormones regulating sexual and social behaviors in female prairie voles, Microtus ochrogaster. *Physiology and Behavior, 23*, 597–600.

Dluzen, D. E., Ramirez, V. D., Carter, C. S., & Getz, L. L. (1981). Male urine stimulates localized and opposite changes in luteinizing hormone-releasing hormones and norepinephrine within the olfactory bulb of female prairie voles. *Science, 212*, 573–575.

Getz, L. L., & Carter, C. S. (1980). Social organization in *Microtus ochrogaster* populations. *The Biologist, 62*, 56–69.

Getz, L. L., Carter, C. S., & Gavish, L. (1981). The mating system of the prairie vole, *Microtus ochrogaster*: Field and laboratory evidence for pair-bonding. *Behavioral Ecology and Sociobiology, 8*, 189–194.

Greenough, W. T., Carter, C. S., Steerman, C., & DeVoogd, T. J. (1977). Sex differences in dendritic patterns in hamster preoptic area. *Brain Research, 126*, 63–72.

Huck, U. W., Carter, C. S., & Banks, E. M. (1979). Estrogen and progesterone interactions influencing sexual and social behavior in the brown lemming, *Lemmus trimucronatus*. *Hormones and Behavior, 12*, 40–49.

Jacob, S., Brune, C. W., Carter, C. S., Leventhal, B., Lord, C., & Cook, E. H., Jr. (2007). Association of the oxytocin receptor gene (OXTR) in Caucasian children and adolescents with autism. *Neuroscience Letters, 417*, 6–9.

Johnson, E. O., Kamilaris, T. C., Carter, C. S., Calogero, A. E., Gold, P. W., & Chrousos, C. P. (1996). The biobehavioral consequences of psychogenic stress in a small, social primate (*Callithrix jacchus jacchus*). *Biological Psychiatry, 40*, 317–337.

Keverne, E. B., & Kendrick, K. M. (1994). Maternal behaviour in sheep and its neuroendocrine regulation. *Acta Paediatrica. Supplement, 397*, 47–56.

Morrell, M. J., Dixen, J. M., Carter, C. S., & Davidson, J. M. (1984). The influence of age and cycling status on sexual arousability in women. *American Journal of Obstetrics & Gynecology, 148*, 66–71.

Porges, S. W. (1995). Orienting in a defensive world: Mammalian modifications of our evolutionary heritage. A Polyvagal Theory. *Psychophysiology, 32*(4), 301–318.

Rubin, L. H., Carter, C. S., Drogos, L., Pournajafi-Nazarloo, H., Sweeney, J. A., & Maki, P. M. (2010). Peripheral oxytocin is associated with reduced symptom severity in schizophrenia. *Schizophrenia Research, 124*, 13–21.

Segraves, R. T., Madsen, R., Carter, C. S., & Davis, J. M. (1985). Erectile dysfunction associated with pharmacological agents. In R. T. Segraves & H. W. Schoenberg (Eds.), *Diagnosis and treatment of erectile disturbance* (pp. 23–63). Plenum Press.

Taymans, S. E., DeVries, A. C., DeVries, M. B., Nelson, R. J., Friedman, T. C., Castro, M., Detera Wadleigh, S., Carter, C. S., & Chrousos, G. P. (1997). The hypothalamic-pituitary-adrenal axisof prairie voles (Microtus ochrogaster): Evidence for target tissue glucocorticoid resistance. *General and Comparative Endocrinology, 106*, 48–61.

Williams, J. R., Catania, K., & Carter, C. S. (1992a). Development of partner preferences in female prairie voles (Microtus ochrogaster): The role of social and sexual experience. *Hormones and Behavior, 26*, 339–349.

Williams, J. R., Carter, C. S., & Insel, T. R. (1992b). Partner preference development in female prairie voles (Microtus ochrogaster) is facilitated by mating or the central infusion of oxytocin. *Annals of the New York Academy of Sciences, Oxytocin in Maternal, Sexual and Social Behavior., 652*, 487–489.

Winslow, J. T., Hastings, N., Carter, C. S., Harbaugh, C. R., & Insel, T. R. (1993). A role for vasopressin in pair bonding in monogamous prairie voles. *Nature, 365*, 545–548.

Witt, D. M., Carter, C. S., & Walton, D. (1990). Central and peripheral effects of oxytocin administration in prairie voles (*Microtus ochrogaster*). *Pharmacology, Biochemistry and Behavior, 37*, 63–69.

Zelkowitz, P., Gold, I., Feeley, N., Hayton, B., Carter, C. S., Tulandi, T., Abenhain, H. A., & Levin, P. (2014). Psychosocial stress moderates the relationships between oxytocin, perinatal depression, and maternal behavior. *Hormones and Behavior, 66*, 351–360.

Elizabeth Adkins-Regan

32

Richmond Thompson

Abstract

Elizabeth K. Adkins-Regan is an American behavioral neuroendocrinologist. Born in Washington, DC, in 1945, she was educated at the Universities of Maryland and Pennsylvania where she earned a PhD in Norman Adler's laboratory. She later joined the faculty at Cornell University where she remained until her retirement in 2018. Her work focused on explorations of the roles of sex steroid hormones and neuropeptides in the regulation of sexual differentiation and social behaviors in birds. Her many and important contributions and focus on both proximate and ultimate determinants of behavior have enriched our understanding of the regulation of social behavior.

Keywords

Organization/activation hypothesis · Testosterone · Birds · Nucleus taeniae · Bed nucleus of the stria terminals · Septum

Elizabeth K. Adkins-Regan was born in July 1945 in Washington, DC, and raised in nearby Maryland, where she attended college at the University of Maryland. Although initially interested in chemistry, she became a psychology major, presumably reflecting a deeper fascination with behavior; she graduated with a BS in 1967. However, her dual interests in chemistry and behavior began to coalesce in graduate school at the University of Pennsylvania. A course in comparative psychology

R. Thompson (✉)
Oxford College of Emory University, Oxford, GA, USA
e-mail: rick.thompson@emory.edu

R. J. Nelson, Z. M. Weil (eds.), *Biographical History of Behavioral Neuroendocrinology*, https://doi.org/10.1007/978-3-031-12970-4_32

during her first year focused her behavioral interests on the proximate and ultimate causes of animal behavior, and the work of a new faculty hire at Penn, Norman Adler, who would become her PhD advisor, helped her see how she could study the chemical mechanisms of animal behavior. The rest, as they say, is history; she had an incredibly productive career, primarily spent at Cornell University, examining how hormones shape the development and expression of reproductive behaviors in vertebrate animals. Although her research included a variety of species, from anole lizards to pigs to king quail, most focused on birds, reflecting her lifelong love of a vertebrate group that likewise sparked the imagination of many other great scientists whose intellectual roots trace back to the most notable of all, Charles Darwin.

To understand the importance of Adkins-Regan's work, particularly the early experiments that launched her career, it is critical to understand the intellectual context in which they were conducted. The organization/activation model of hormone actions during sexual differentiation, developed in guinea pigs (Phoenix et al., 1959), was on the ascendence when Adkins-Regan was in graduate school, and with it the tendency to presume its mechanistic foundation – that testosterone secreted by the testes early in development masculinizes bodies and brains – was the key to understanding differences between males and females. Whether because of Adkins-Regan's early recognition that hormone mechanisms must be understood within evolutionary context or simply her desire to study the birds she loved, not the rodents that most psychologists, including her own advisor, were using at the time, she began to branch out to see what generalities associated with the organization/activation model might apply within the avian clade. Whatever the initial motivation, her bold decision to begin work in Japanese quail, for in which little endocrine work had been reported and virtually nothing on the hormonal control of sexual differentiation and behavior, paid off richly. What followed was a series of papers, the first of which was co-authored with her graduate advisor establishing Japanese quail as a viable species for studying the hormonal basis of reproductive behavior (Adkins & Adler, 1972), but the rest of which were primarily solo authorships or driven largely by her intellectual vision. In the experiments described in them, she was able to demonstrate clear effects of circulating sex steroids on reproductive behavior in adult Japanese quail, as well as long-term influences of brief hormone exposure early in development, both generally consistent with the organization/activation model. The work demonstrated her clear desire to integrate chemistry into the study of behavior, as she was able to identify the biochemical pathways through which testosterone differentially affects unique components of reproductive behavior in adult quail, stimulating copulation itself through aromatization into estradiol, whereas crowing and strutting, two male-typical sexual displays, depend directly on androgens. Her work thus began to dissociate the hormone mechanisms that affect different components of reproductive behavior, thereby illustrating how diverse reproductive phenotypes can be generated through, for example, genetic differences within specific biochemical pathways.

Working with one of her early graduate students, James Watson, she went on to identify where the steroid receptors that mediate those behavioral effects are produced in quail brains and to demonstrate that, as in mammals, the preoptic area is the critical region where testosterone's conversion to estradiol enhances male copulatory behavior. Projects with subsequent graduate students, typically developed independently per their own individual interests but always reflecting her guiding principle that mechanisms should be explored in evolutionary contexts, went on to identify the role that numerous hormone-sensitive regions of the avian brain play in a variety of reproductive behaviors. Those included the nucleus taeniae (medial nucleus homologue, sexual motivation), the bed nucleus of the stria terminals (brooding), and the septum (courtship song, territorial aggression). It is worth noting that her work on the role of the septum in avian social regulation, initiated by the late Jim Goodson when he was just beginning in her lab, helped establish how peptides in the vasopressin/oxytocin family evolved complex, context-dependent social functions in vertebrate animals. More broadly, the anatomical work done with her students helped establish evolutionary conservation across species and vertebrate clades between brain structures that we now recognize play key roles in social regulation in most, if not all, living vertebrates.

However, Adkins-Regan was also one of the first scientists to illustrate some limitations, at least of the organization part of the organization/activation model, in relation to phylogeny. In her seminal paper, "Hormonal Basis of Sexual Differentiation of Japanese Quail" (Adkins, 1975), she demonstrated that short-term exposure to testosterone, delivered to the eggs prior to hatching, did not create males as it does in mammals. Instead, early testosterone treatment led to the development of animals that acted more like females as adults, as did early exposure to estradiol, a metabolite of testosterone. She and her students and colleagues were ultimately able to establish that estrogens produced by developing ovaries prior to hatching typically produce those effects in genetic females, and they determined the critical periods and dose dependencies for many of estardiol's effects on early sexual differentiation. Further, even though her own work demonstrated that in birds males appeared the "default sex," an unfortunate term coined in mammals to describe the sex that develops in the absence of an active hormone signal, Adkins-Regan went on to show that during early development androgens, potentially provisioned in the yolk by mothers, affect the behavior of adult male quail, surprisingly decreasing how sexually active they become (Adkins-Regan, 1985a). In her reviews, Adkins-Regan often highlighted how extraordinary examples in nature can generate questions that eventually lead to novel insights into how hormones work; her counterintuitive finding that early androgens in developing males can decrease masculine behaviors when they become adults would have been considered such an example at the time, although she did not typically focus on her own work in those reviews. She should have, because those findings foreshadowed our current recognition that hormone signals play active roles in the differentiation of both sexes, even the so-called "default" sex, and that testosterone is not simply a hormone that promotes all things masculine, but rather a complex modulator that can produce different types of effects in different contexts, in different individuals, or as in the quail she studied, at different times of life.

Together, Adkins-Regan's early work on sexual differentiation in quail, summarized nicely in her reviews in *American Zoologist* (Adkins, 1978) and *Science*

Progress (Adkins-Regan, 1985b), was seminal for establishing what was the most generalizable feature of the organization model – that exposure to sex steroids during early development can produce lasting effects on behaviors related to reproduction in at least two vertebrate clades with genetic sex determination, birds as well as mammals. However, her work also began to illuminate the limitations of that model in relation to phylogeny. That is, Adkins-Regan's work helped establish that the divergent evolutionary paths of birds and mammals led to unique ways of coupling genetic sex determination to the gonadal mechanisms that drive much of the downstream process. In mammals, males are the heterogametic sex, and the genetic switch is the SRY gene on their unique Y sex chromosome, which drives early differentiation of the testes and thus the elevations of testosterone that lead to the development of many male characteristics. In contrast, female birds are the heterogametic sex, having ZW sex chromosomes instead of the male ZZ pattern. Although the sex-determining switch(es) had not yet been identified in birds, because of Adkins-Regan's work in Japanese quail, ring doves, and zebra finches, we learned that it drives the early ovarian development crucial for estradiol production and the subsequent development of many, but not all, female behavioral tendencies.

Although it was her early work on the role that sex steroids play in the differentiation and activation of reproductive behavior in Japanese quail that set her career in motion, it was her work in zebra finches that highlighted Adkins-Regan's fascination with how complex social behaviors are hormonally regulated, as well as with how complex those mechanisms can be. She and her students were among the first to highlight the paradox of sexual differentiation in zebra finches, in which they discovered that, in common with Japanese quail, estrogens promote female-typical patterns of copulatory behavior, but can also promote singing, a male-typical behavior. They also observed that decreasing early estrogen synthesis in females reduced tendencies to form pair bonds with males and increased tendencies to pair bond with other females, a pattern consistent with a role of endogenous estrogens in promoting behaviors typical of most females. Again paradoxically, treatment with *extra* estrogens also increased tendencies to bond with other females, at least if the treated females were raised exclusively with other females. Thus, Adkins-Regan had demonstrated that the same hormone could promote some behaviors typical of females and some more typical of males. Embracing the complexity, she highlighted what was fundamentally important about her findings – that there are dissociations in how early hormones can induce lasting effects on different components of behaviors typically thought to be part of a uniform suite of sexuality, and that those mechanisms can interact with unique social experiences to produce a range of behavioral phenotypes, including tendencies to pair with same – or other-sex individuals.

Ultimately, Adkins-Regan was able to dissociate lasting hormone effects on discrete behaviors as a function of which biochemical pathway they work through, where within the brain they act, or when they are secreted during development, which she demonstrated can even extend into puberty in zebra finches. Studies conducted with one of her last graduate students, Nicole Baran, also showed that it is not just steroid exposure early in development that can produce lasting effects on social behavior, but also peptides in the vasopressin/oxytocin family. Together,

Adkins-Regan's body of work on sexual differentiation helped us appreciate how diverse social phenotypes can be generated. Further, based upon the complex patterns of effects that estrogens have upon zebra finch sexual differentiation discussed above and a detailed description that she and her colleagues provided of hormone profiles in developing finches, in 1990 Adkins-Regan made what proved a prescient proposition in light of our current recognition of the importance of brain-derived steroids, which was that peripheral steroids produced by the gonads may not always be what is critical, and that steroids produced in different parts of the body and at different times may have unique, sometimes contrasting, effects on brains and behavior (Adkins-Regan et al., 1990). Together, her work highlighting how unique steroid sources, time-courses of action, biochemical pathways, and interactions with social experience shape complex reproductive characteristics not only helps us understand sexual differentiation in birds, but even how the sexual mosaicism recently discovered within human brains might be generated (Joel et al., 2015).

Of course, Adkins-Regan's research was not limited to experiments probing how the organization/activation model did and did not generalize to birds. Indeed, her broad evolutionary thinking, coupled with the unique interests brought to the lab by her students, enabled her to establish an incredibly broad research program. That included experiments that challenged stereotypical ideas of sexual dimorphism, for example, showing that female zebra finches, under conditions of mate competition, can be as aggressive as males, and that even male Japanese quail, which do not form pair bonds or care for young and should thus be exemplars of ardent, less-than-choosy males, exhibit mate preferences based on assessments of females' tendencies to incubate eggs. She explored novel sperm competition mechanisms in Japanese quail associated with the production of foam, deposited along with sperm, which disrupts the ability of rivals' sperm to fertilize eggs. She and her students demonstrated that maternal diet, learning, and hormones such as prolactin can affect offspring sex ratios and the subsequent attractiveness of offspring, and that female birds can even adjust their own hormone levels, and perhaps consequently the fertilization rates of individual pairings and the sex ratios of their clutches, as a function of variation in the male's mating behavior. To show how broadly comparative she was, she even studied sexual differentiation and chemical communication in pigs! This catalogue is not exhaustive, but nonetheless illustrates how her broad thinking, honed by an encyclopedic knowledge of endocrinology and evolution, was integrated with the unique interests brought to the lab by her students and colleagues. The overarching emphasis was not to narrowly drill down to bottom-line mechanisms, but rather to expand our ideas of how hormones evolved complex regulatory functions in relation to the life histories of animals in the world.

Her broad thinking was also evident in her reviews, which typically highlighted "natural wonders" that made us consider how hormone regulatory mechanisms evolved in a diverse world, and which most famously championed comparative approaches when she revisited Frank Beach's original Snark Hunting metaphor when she asked, "Is the Snark Still a Boojum?" (Adkins-Regan, 1990). Her willingness to mix mechanistic and evolutionary levels of analysis to understand hormone regulatory mechanisms, and her advocation that we all do the same, was also evident in the textbook she wrote, *Hormones and Animal Social Behavior* (2005). By the time she retired from Cornell in 2018, her publications, including those describing original experiments, reviews, commentaries, and her textbook, totaled at least 153, all written while serving the field of behavioral neuroendocrinology in many ways, from being an ever-ready reviewer of manuscripts and grants to serving as president of the Society for Behavioral Neuroendocrinology. The quality of her work and the value of her service have been recognized by multiple awards during her distinguished career, from being an NSF graduate fellow at the beginning to receiving the Danny Lehrman Lifetime Achievement Award from the Society for Behavioral Neuroendocrinology in 2019. And all of this was achieved while maintaining a balanced lifestyle, primarily in Ithaca, New York, with her husband Dennis Regan, one that included time in her garden battling wood chucks, cross-country skiing and hiking, playing her beloved piano, enjoying great food and wine, and taking adventurous trips, often to see the birds that she has taught us so much about living in the wild. Ultimately, Elizabeth Adkins-Regan demonstrated how to be an excellent scientist with balance and grace.

References

Adkins, E. K. (1975). Hormonal basis of sexual differentiation in the Japanese quail. *Journal of Comparative and Physiological Psychology, 89*, 61–71.

Adkins, E. K. (1978). Sex steroids and the differentiation of avian reproductive behavior. *American Zoologist, 18*, 501–509.

Adkins, E. K., & Adler, N. T. (1972). Hormonal control of behavior in the Japanese quail. *Journal of Comparative and Physiological Psychology, 81*, 27–36.

Adkins-Regan, E. (1985a). Embryonic exposure to an aromatization inhibitor increases copulatory behavior of male quail. *Behavioural Processes, 11*, 153–158.

Adkins-Regan, E. (1985b). Mechanisms of sex determination in vertebrates. *Science Progress, 69*, 553–568.

Adkins-Regan, E. (1990). Is the snark still a boojum? The comparative approach to reproductive behavior. *Neuroscience and Biobehavioral Reviews, 14*, 243–252.

Adkins-Regan, E., Abdelnabi, M., Mobarak, M., & Ottinger, M. A. (1990). Sex steroid levels in developing and adult male and female zebra finches (Poephila guttata). *General and Comparative Endocrinology, 78*, 93–109.

Joel, D., Berman, Z., Tavor, I., Wexler, N., Gaber, O., Stein, Y., Shefi, N., Pool, J., Urchs, S., Margulies, D. S., Liem, F., Hänggi, J., Jäncke, L., & Assaf, Y. (2015). Sex beyond the genitalia: The human brain mosaic. *Proceedings of the National Academy of Sciences of the United States of America, 112*(50), 15468–15473.

Phoenix, C. H., Goy, R. W., Gerall, A. A., & Young, W. C. (1959). Organizing action of prenatally administered testosterone propionate on the tissues mediating mating behavior in the female guinea pig. *Endocrinology, 65*, 369–382.

Arthur P. Arnold

33

Barney A. Schlinger

Abstract

Art Arnold is a leading figure in behavioral neuroendocrinology and the genetics of vertebrate sexual differentiation. This review provides a personal perspective on Arnold's life and career, a lifetime filled with breakthrough biomedical discovery and dedicated scientific training and mentoring. His integrity, his devotion to family, and his deep commitment to education are evident throughout.

Keywords

Sexual differentiation · Sex dimorphisms · Brain · Spinal cord · CNS · Aromatase · Testosterone · Sex chromosomes · Estradiol · Neuroendocrine · Zebra finch · Four-core genotype

This book of special recognition is reserved for individuals who have not only performed career-long exemplary neuroendocrine research, but their lifetime of scientific exploration also has enduringly transformed concepts that have steered current and future study. In many cases, these individuals have worked tirelessly within and outside of academia to promote neuroendocrine science conveying its importance to society and medicine. In the lab, in the classroom, and in administration, Arthur Palmer Arnold (Art) is, and has been for decades, a leader in the fields of behavioral neuroscience and endocrinology as well as in the molecular genetics of sex, all

B. A. Schlinger (✉)
Department of Integrative Biology and Physiology and Department of Ecology and Evolutionary Biology, University of California, Los Angeles, Los Angeles, CA, USA
e-mail: schlinge@lifesci.ucla.edu

R. J. Nelson, Z. M. Weil (eds.), *Biographical History of Behavioral Neuroendocrinology*, https://doi.org/10.1007/978-3-031-12970-4_33

foundational concerns of behavioral neuroendocrinology. As a consequence, Arnold is an exceptionally appropriate person for inclusion in this book of neuroendocrine pioneers.

Of course, much of Arnold's scientific contributions can be assessed by examining his CV, or his lab website, or by performing a PubMed or Google Scholar search of his many notable publications (as of 21 December 2021, Arnold had 292 published papers with 32,568 citations and an h index of 93). Some information is also available in a previous published biography describing accomplishments that led to his receipt of the Daniel Lehrman Lifetime Achievement Award from the Society for Behavioral Neuroendocrinology (SBN) (Schlinger et al., 2011). These metrics, however, do not fully showcase the history of Arnold's scientific accomplishments, his talents as a teacher and mentor, his skills in writing or his decades-long wise council for a vast number of academic committees and programs. These measures also do not describe the beautiful personal life that Arnold has built. Altogether, this forms an intriguing and heartwarming biography of one scientist's life which I hope to accurately present in what follows.

It goes without saying that science is often a group effort, and Arnold has had many excellent students, postdocs, and colleagues with whom, as a team, he has made many scientific advances. It is impossible to name them all or to appropriately acknowledge their various contributions. As the theme here is neuroendocrinology, this chapter will focus on those facets of his research that most strongly impinge on neuroendocrine thinking and spotlight those students and postdoctoral fellows who remained in academia and retained a research focus on neuroendocrinology.

Early Years

Art Arnold was born on 16 March 1946 at Temple University Hospital in Philadelphia, Pennsylvania. His parents, Wiley E. Arnold and Leona (Lee) Barth Arnold, were of German/European decent whose grandparents had settled in the Midwest after arriving in America by the mid-1800s. Art's parents first settled in Chicago, where his older brother and sister, Rick and Carol, were born, but his father's career necessitated a family move to the Philadelphia suburbs where Arnold was born and raised. Home life was guided by a strong Christian ethic. Having grown up during the depression, Arnold's dad aspired for better; he was thrifty and financially cautious and provided a stable middle-class life for the Arnolds. Arnold was exposed to a loving, secure, and happy environment while also endowed with a spirit of ambition for something better.

A transformative experience occurred when he was selected to be an exchange student with the American Friends Service Committee program which sent Arnold to Berlin, for a year (in 1961–1962) of high school at the fresh age of fifteen and a half. He had a year of German language study prior to the trip, and, although his German family spoke some English, they decided to only speak to Arnold in Deutsch. Not surprisingly, he quickly grasped the new language. Notably, his German "parents" were both physicians and booklovers. This provided a new perspective on science and learning that differed from his home environment in Pennsylvania. Importantly, Arnold carried to Germany a strong commitment to represent his country and the American people in the best possible light. The perceived successes and failures in this regard were of significant maturational value to this aspiring adolescent young man.

Education

Nothing in particular seems to stand out in Arnold's early life that would suggest he would become a scientist, or even a biologist. Indeed, although he was a very good student, his high school biology and chemistry classes were unstimulating experiences. This perspective all changed upon taking his first Introduction to Psychology class at Grinnell College. His teacher was a Skinnerian, and learning theory was all the rage. He was taught that behavior could be quantified. At home for the Christmas holidays, Arnold told his father that he'd been taught "people don't think, they just behave." While admittedly this was an inaccurate portrayal of learning theory, the quantitative observational approach to the study of behavior obviously captured Arnold's intellect and steered the future course of his academic life. Behavior could be seen as predictable, governed by laws of nature.

Arnold gained additional appreciation for a more tangible understanding of behavior after taking a course in Physiological Psychology (at the time neuroscience per se was not an independent discipline but was woven within courses in anatomy, physiology, and psychology). He gained some experimental laboratory research experience with live rats before graduating Phi Beta Kappa from Grinnell in 1967 with a major in Psychology.

One day at a campus function in the dining hall, a student server caught Arnold's eye. Not the bashful type, he asked her out. Caroline consented and ultimately, they married and became lifelong partners. Caroline became an extraordinarily successful children's book author and artist. She deserves her own biography. As this piece is about Art, however, Caroline Arnold will remain in the background. There is no doubt, however, that much of Arnold's academic, scientific, and, of course, personal life derives from Caroline's sacrifices as well as her love, talents, personality, character, and patience.

The true watershed moment in Arnold's career development was his choice to attend Rockefeller University for graduate school. In a new graduate program in behavior and neurophysiology, Arnold was exposed to a relatively small cohort of outstanding scientists with interests in behavior and physiology. He joined the lab of Fernando Nottebohm. As an assistant professor at the time, Fernando was a junior faculty member in the lab of Peter Marler. Marler's position as Head of Laboratory gave Arnold significant exposure to both top scientists. More generally, the Rockefeller culture allowed Arnold to forge friendships via his interactions with other notable neuroendocrinologists (including Bruce McEwen and Don Pfaff), as well as a cohort of outstanding graduate students. Altogether, this environment was the springboard that vaulted Art into his career as a behavioral neuroscientist.

One key experience for Arnold as a Rockefeller student involved a 4-month field course in Africa with Peter Marler and several other extraordinary professors and fellow students. This not only gave Arnold a new perspective on the advantages (and difficulties) of studying animals in nature, but it also led to one of Arnold's first publications, a paper about mating behavior of an African antelope (Floody & Arnold, 1975).

The focus of Arnold's graduate work involved detailed behavioral, endocrine, and neurobiological studies of singing behavior in zebra finches. It goes without saying that this was a fortuitous choice of species for study. The zebra finch is now one of the most popular animal models for studies in vertebrate biology. Following Arnold's work establishing a gonadal hormone basis for song and song development in these finches (Arnold, 1975a, b), Arnold and Nottebohm were then the first to demonstrate large structural sex differences in the brain of a vertebrate (Nottebohm & Arnold, 1976). These were the circuits that guided the production and learning of passerine birdsong, and which exhibited significant accumulation of steroid hormones (Arnold et al., 1976). Arnold would continue work on the zebra finch model after leaving Rockefeller, but for postdoctoral work, he remained at Rockefeller to learn neurophysiological techniques in the lab of Hiroshi Asanuma. There, Arnold solidified his expertise in neuroscience by performing studies of motor neural circuits of cats (e.g., Asanuma et al., 1976; Shinoda et al., 1978). Arnold was hired as an assistant professor at the University of California, Los Angeles (UCLA), and in 1976 he moved with Caroline, their young daughter Jennifer, and their son Matthew (Matt) to Los Angeles where he has remained ever since (Fig. 33.1).

Fig. 33.1 Art, Jennifer, and Caroline Arnold in the fall of 1970, on the streets of New York, when Art was a PhD student at the Rockefeller University in Manhattan. (Photo by Harold W. Fuson, Jr.)

Career

Arnold set up his lab at UCLA two floors below ground across from a vivarium in the basement of one wing of Franz Hall, the building that housed the Psychology Department. His office was quite far off on the eighth floor of the Franz Hall Tower. By some standards, the lab space was no "laboratory" at all, but a collection of small rooms with assorted tables, not benches: a microscope here, slide holders, and staining trays there. Given the limitations imposed by this space, Arnold's productivity as well as that of his students and postdocs was quite remarkable. He quickly established a colony of zebra finches and set to expand work he had started as a PhD student exploring hormonal and neural mechanisms as well as brain sex differences underlying the control of zebra finch song.

At the outset, Arnold worked to develop some techniques he would need to move forward (such as Arnold, 1980a, b, 1981). He also published a highly important paper describing sex differences in hormone accumulation in the zebra finch brain (Arnold & Saltiel, 1979). Arnold had established himself as a world leader in the identification and study of structural brain sex differences.

Notably, one of Arnold's first seminars at UCLA included Roger Gorski in the audience. Gorski had established himself as a premier scientist studying hormonal control of rodent reproductive physiology and behavior. Upon seeing Arnold's photomicrographs of the impressive sex differences in the zebra finch brain, Gorski decided to re-examine the brains of male and female rats. He was excited to see a very large sex difference present in the hypothalamus of these rats visualized by just examining Nissl-stained sections and looking under the light microscope (Gorski et al., 1978). Suddenly, large structural brain sex differences were not just restricted to birds but were found in a center of neuroendocrine control in the dominant mammalian neuroendocrine model organism of the day. Other morphological sex differences were soon reported in other neural systems (e.g., De Vries et al., 1981; DeLacoste-Utamsing & Holloway, 1982; Simerly et al., 1985). These discoveries of quantifiable sex differences in the structure of brain regions had a major impact because the process of sexual differentiation of the brain could now be studied at the level of cells, rather than just at the level of behavioral or physiological output of the brain. The study of a few salient sexual dimorphisms in the rodent brain thus gave rise to general concepts of brain sexual differentiation that were assumed to apply broadly to most or all sex differences in brain function. A new era for accurate quantitative study of neural sex differences was born, with Arnold key to its parturition.

Arnold was also beginning to attract talented students and postdocs to his lab. Marc Breedlove was Arnold's first doctoral student, and what a team they made. Recognizing that motor systems outside of the brain per se were instrumental in activating sexually dimorphic behavior, they established the spinal nucleus of the bulbocavernosus (SNB) as a major model system for studying mammalian sexual differentiation as well as steroid effects on peripheral motoneurons (Breedlove & Arnold, 1980, 1981, 1983a, b, c; Breedlove et al., 1982, 1983). Around the same time, Cindy Jordan, Arnold's lab technician and later doctoral student, also began working on spinal neuromuscular circuits. Several years later, this grew into a powerful model system for understanding hormonal control of the neuromuscular junction (e.g., Jordan et al., 1988, 1989a, b; Jordon et al., 1990). Sex steroid hormones directly regulated developmental synapse elimination.

Sarah Bottjer joined Arnold's lab as a postdoctoral fellow and helped expand his work on the zebra finch song system with a number of significant discoveries. Of particular importance was their work that established one nucleus called MAN (then called the magnocellular nucleus of the neostriatum) as crucial for song learning (a core feature of the entire oscine birdsong clade) (Bottjer et al., 1984). Another postdoctoral fellow, Eliot Brenowitz, added a comparative perspective to Arnold's songbird research program including work on duetting songbirds in which, unlike zebra finches, both males and females contribute to the species' song (Brenowitz et al., 1985; Brenowitz & Arnold, 1985, 1986, 1989).

A key question arising from the identification of sexually dimorphic neural circuits asks how they arise. Investigation of the sexually dimorphic SNB system of rats by Arnold and his postdoctoral fellows (Ernie Nordeen, Kathy Nordeen, and Dale Sengelaub) reported that, developmentally, androgens prevented neuronal death. In this way, androgens secreted by the testes of male embryos rescued motoneurons of the SNB which normally die in females. This firmly established cell death and hormonal rescue as crucial mechanisms in the ontogeny of sexually dimorphic neural circuits in vertebrates (Nordeen et al., 1985). The Nordeens then successfully extended this line of work in Arnold's lab to studies of the zebra finch song system (Nordeen et al., 1986, 1987a, b).

Dale Sengelaub continued working in Arnold's lab developing the rat spinal cord as an important model system for exploring steroid effects on spinal cord neuronal structure and function. Together with another postdoctoral fellow Elizabeth Kurz, they showed that sex steroids caused significant dendritic growth and reorganization in fully differentiated steroid-sensitive neurons (Kurz et al., 1986). This was followed by a series of significant studies with Arnold and other postdocs demonstrating clear steroid hormone effects on spinal neuron number and anatomy (Sengelaub & Arnold, 1986, 1989; Sengelaub et al. 1989a, b).

Around this same time, Arnold was fortunate to have visiting scholars from Japan join his laboratory. Akira Matsumoto, an expert quantitative electron microscopist, working with Arnold and Paul Micevych (Arnold's close collaborator at UCLA), showed for the first time that androgens could regulate synaptic organization and gap junctions in the adult mammalian CNS, in the SNB (Matsumoto et al., 1988a, b). Thus, numerous studies by members of the Arnold lab of the 1980s exploited the advantages of the SNB and songbird systems to determine which cellular processes (cell death, synaptogenesis, size, and organization of dendritic trees and cell bodies) were controlled by the process of sexual differentiation, and in particular by neonatal or adult variations in the effects of sex steroid hormones,

Arnold entertained a somewhat new line of inquiry after I joined his laboratory in early 1988. Knowing that there was extensive evidence that estradiol established sex differences in the neural song system (a concept radically changed later based on studies by Arnold), interest then developed in understanding more about the estrogen synthetic enzyme aromatase in songbirds. We embarked on a series of

Fig. 33.2 From left, Barney Schlinger, Tony Campagnoni, and Art Arnold at their poster at the Society for Neuroscience meeting approximately in 1994. Tony was a major mentor of Arnold's, helping him learn methods and concepts of molecular biology in the 1990s

studies showing elevated presence of aromatase in the male songbird brain, with an absence of obvious activity outside of the brain (Schlinger & Arnold, 1991). We showed that the brain of the zebra finch could be the actual source of circulating estradiol in males (Schlinger & Arnold, 1992, 1993). Arnold then established a collaboration with Anthony Campagnoni at UCLA, who helped us clone the aromatase gene and develop cell culture techniques and molecular approaches to further evaluate this novel neural expression of aromatase in songbirds (Shen et al., 1994, 1995). We found evidence for aromatase expression in astrocytes (Schlinger et al., 1994). We developed an antibody that was specific for zebra finch aromatase which worked exceptionally well for immunocytochemical studies (Saldanha et al., 2000) and is still in use to this day (Fig. 33.2).

A brief personal note. While my postdoctoral position with Arnold had me associated with the Department of Psychology, I was hired as an Assistant Professor at UCLA in the Department of Physiological Science in 1993. Three years later, when our department made a push to expand our neuroscience faculty, Arnold agreed to switch departments. He went from being my mentor to being my colleague. When he became Department Chair, a position he occupied for 8 years, I became his Vice-Chair, and when I became Chair of the Department (that was renamed Integrative Biology and Physiology during Arnold's term as Chair), he became one of my faculty. All the while our friendship grew, a relationship that I cherish, and which made me honored to write this biography.

Mid-Career Transformation

At UCLA, Arnold rapidly advanced to Full Professor and by the year 2000, he was already recognized as Distinguished Professor. Despite having devoted much of this career to investigations into CNS sex dimorphisms starting with zebra finches that he began studying as a PhD student, Art Arnold made a dramatic turn mid-career (what for many would be considered late-career), to understand the molecular genetic basis for vertebrate sexual differentiation. For some, these appear to be related fields and a simple a natural step forward. Far from that, however, this was a transformation of Arnold from neuroscientist, endocrinologist, and behavioral biologist into, I dare say, a hard-core molecular biologist/geneticist (though due to Arnold's humility he would disagree). Due to this remarkable shift, Arnold has had notable impact on disparate fields of animal biology that guide the thinking of all behavioral neuroendocrinologists, as well as biomedical scientists of all ilk with interest in sex differences.

This shift occurred in the mid-1990s as Arnold began to appreciate that gonadal hormones could not fully explain the appearance of sex differences in the zebra finch brain. Working as a postdoc in Arnold's lab, Juli Wade found that genetic females experimentally treated to induce growth of testicular rather than ovarian tissue possessed a song system with female characteristics, with little or no signs of masculinization (Wade & Arnold, 1996; Wade et al., 1996). The logical conclusion was that sex chromosome genes, not gonadal hormones, had a role in sex-specific song system development (Arnold, 1996, 1997). This was generally confirmed (in a mammalian model) somewhat later when Arnold's postdoctoral fellow Laura Carruth and their collaborator Ingrid Reisert demonstrated a sex chromosome effect on neuronal sexual differentiation in vitro (Carruth et al., 2002). Sabbaticals during the 1990s that Arnold took in the labs of Anthony Campagnoni at UCLA and Andrew Sinclair in Melbourne, Australia, gave him the laboratory tools to begin asking questions concerning the molecular genetics of avian sex (Fig. 33.3).

Around the same time, a remarkable gyandromorphic zebra finch was discovered in the lab of Fernando Nottebohm, who generously donated the bird for study in the Arnold lab. Gyandromorphic birds are phenotypically sexually laterally asymmetrical, with male plumage on one side, and female on the other. With doctoral student Bob Agate and former postdoc Bill Grisham, Arnold set out to assess whether this asymmetry was present in the bird's song system neuroanatomy and how those features might relate to gonadal sex steroid synthesis and sex-specific gene expression. Recall that male birds are homogametic (ZZ) and females heterogametic (ZW). The results of this analysis provided strong evidence that the gynandromorph overwhelmingly expressed two copies of a Z-linked gene on their right side and only one on their left. The birds also expressed one W-linked gene on the left side with virtually none on the right (Agate et al., 2003). Moreover, the right side of the bird had a testis and what appeared to be a fully masculine neural song system, whereas the left side of the bird had an ovary and an only partially masculine song system. Clearly, the sexual genotype of the right and left sides corresponded with morphology of the song system, an observation impossible to explain by gonadal

Fig. 33.3 Art Arnold has a close encounter with an Australian Magpie (Gymnorhina tibicen), during his sabbatical research in Andrew Sinclair's lab in Melbourne in 1998. (Photo by Caroline S. Arnold)

hormonal control alone. The dogma that gonadal hormones controlled all sexual differentiation of brain and behavior was now turned on its head. Equipped with all his new experimental tools, his intellectual appreciation for molecular genetics, and his intense focus, Arnold was positioned to transform our thinking of sexual differentiation producing a twenty-first-century model for this fascinating and profoundly important biological process (Arnold, 2012, 2017).

One project resulting from this work led Arnold to a discovery that, perhaps surprisingly at first glance, he asserts as one of his favorite lab papers. Organisms with sex chromosomes have a potential problem: the homogametic sex (XX in female mammals) could have double the expression of X-linked genes than in the heterogametic (XY) males. Via dosage-compensation mechanisms, the genes on one X chromosome in mammalian XX cells are largely and effectively silenced, making expression of most X genes comparable in XX and XY cells. Work by Yuichiro Itoh, a postdoc, and MD-PhD student Esther Melamed, showed that in contrast to mammals, birds did a very poor job of equalizing expression of Z genes in ZZ (male) and ZW (female) cells (Itoh et al., 2007). Thus, according to Arnold, "Male birds are flying around with most of their Z chromosome expressed higher than in females…and so I think that more than a few of those genes are likely to be… biasing development in the male direction." Certainly, a foundational discovery that applies to an entire vertebrate class is something of which to be especially proud. The study also proved that effective dosage compensation of sex chromosomes was not a requirement of genomes, as had been previously assumed.

A key new piece of Arnold's toolbox was a genetically engineered "four core genotypes" (FCG) mouse model created by Paul Burgoyne and Robin Lovell-Badge, from the National Institute for Medical Research in London, now the Crick

Institute. Arnold formed a strong friendship with Burgoyne, who became his mentor and collaborator (Turner et al., 2020). The mouse model involves removal of the testis-determining *Sry* gene from the Y chromosome, and insertion of an *Sry* transgene onto an autosome. The model produces four types of progeny: XX and XY mice lacking *Sry*, which develop ovaries, and XX and XY mice with the *Sry* transgene, which develop testes. The availability of FCG mice meant that any sex difference in a mouse phenotype could be attributed to either effects of gonadal hormones or sex chromosome complement (De Vries et al., 2002). A wide range of investigators adopted the model. In collaboration with Arnold at UCLA, several groups used the FCG model to show that sex differences in diverse mouse phenotypes were caused in part by XX vs. XY differences, not just by gonadal hormones. These include sex differences in autoimmune disease (Smith-Bouvier et al., 2008; Itoh et al., 2019), metabolism and adiposity (Chen et al., 2012; Link et al., 2020), cardiac ischemia/reperfusion (Li et al., 2014), neural tube closure defect (Chen et al., 2008), pulmonary hypertension (Umar et al., 2018), pain and analgesia (Taylor et al., 2020; Gioiosa et al., 2008), and learning (Seu et al., 2014). Other research programs discovered sex chromosome effects on various cardiovascular diseases and development (Arnold et al., 2017; Shi et al., 2021), Alzheimer's and longevity (Davis et al., 2020), reproductive and social behaviors (Cox et al., 2014), neonatal lung injury (Grimm et al., 2021), and neuroanatomy (Arnold, 2020). In a few cases, the X or Y genes responsible for sex chromosome effects have now been identified, validating the results using FCG mice and reflecting cascading influences of these discoveries (Itoh et al., 2019; Link et al., 2020; Davis et al., 2020; Cunningham et al., 2020). Several of Arnold's collaborators at UCLA provided the expertise that led to some of these discoveries. These include Rhonda Voskuhl, Karen Reue, and Mansoureh Eghbali.

A large body of work like this requires considerable financial support. Arnold proved himself to be an expert at acquiring those funds via his masterfully written grant proposals. His first major grant was funded in 1977 and at the writing of this chapter, he is a PI on a grant set to expire in 2026. During this 49-year period of continuous funding, he usually held multiple grants simultaneously. Notably, he also authored a series of NIH T32 proposals entitled "Neuroendocrinology, Sex Differences, and Reproduction." This award funded the education of trainees within the Laboratory of Neuroendocrinology, an affinity group of ~12 faculty initiated by the great neuroendocrinologists Charles Sawyer and Roger Gorski under the umbrella of the UCLA Brain Research Institute. Under Arnold's stewardship, this faculty and trainee collective thrives to this day.

Teaching and Mentoring

In more ways than just acquiring funds, Arnold has made significant contributions to teaching as a faculty member at UCLA, and he did so both as an admired instructor and by significant administrative responsibilities. A few of these duties stand out. For example, between 1989 and 2001, Arnold was the Associate Director for

Education for the UCLA Brain Research Institute. Quite remarkably, during this same period, he chaired the UCLA Interdepartmental PhD Program for Neuroscience (1995–2001) and he chaired the UCLA Interdepartmental Undergraduate Program for Neuroscience (1996–1998).

As an instructor, Arnold's courses covered the gamut from graduate to undergraduate, including general courses in Biology and Psychology to more focused courses on animal behavior, developmental neuroscience, and neuroendocrinology, to a very popular and creative team-taught undergraduate analytic course entitled "Sex: From Biology to Gendered Society." Even in retirement from UCLA (not from research), Arnold created and teaches a new course for undergraduate and graduate students on "Sex differences in physiology and disease." Art's lecture style forces students not to memorize but to think, often describing concepts by asking students to design experiments to answer particular questions. Many students really enjoyed Art's style. One student stated: "I liked how Arnold's lectures always had the 'main point' clearly highlighted, and that he asked challenging questions about how you would go about finding the effects of so-and-so hormone on the body in an experiment. While challenging, I really think I benefitted a lot from this kind of thinking." Another student said: "Dr. Arnold has grown on me. He helped me with my paper one time and validated my arguments, he keeps the class awake with his subtle dry humor, and he has genuine interest in his field of study. PS. He asks loaded questions, but for good reasons."

Honors

Art has received numerous honors over his career including a large number of Keynote lectures. Most relevant to this biography include Art's election as the Inaugural President (from 1997 to 1999) of the Society for Behavioral Neuroendocrinology (SBN) and, as described at the outset, his receipt of the Lehrman Lifetime Achievement award of the SBN. Art was asked to serve as the Founding Editor-in-Chief of the journal *Biology of Sex Differences*, the official journal of the Organization for the Study of Sex Differences. That same society established the Arthur Arnold Distinguished Lecture in 2019, and the Laboratory of Neuroendocrinology described previously created the UCLA Arthur Arnold Innovator Lecture also in 2019. In 1988, Art was named a Fellow of the John Simon Guggenheim Memorial Foundation and a Fellow of the American Association for the Advancement of Science (Fig. 33.4).

The Present and Future

This piece has no conclusion because Art's life and work continues. He retains his laboratory at UCLA with a highly skilled long-term associate Dr. Xuqi Chen and, at least at present, one master's student. He sustains vital collaborations that continue to publish novel and important discoveries at a high rate and in top-tier journals. As

Fig. 33.4 UCLA Integrative Biology and Physiology Departmental website portrait, 2019

pointed out above, his funding will last for several more years, and there is no doubt he will continue to be consulted by colleagues at UCLA as well as scientists worldwide on the role of sex chromosomes in vertebrate sexual differentiation and much more. His recent successful collaborative effort to create an FCG rat model will confirm and expand experimental and conceptual approaches to the study of sex differences. These animals will soon be available, and potentially there will be many years of advances in the study of sex differences based on Art's creation of this new resource. We can all look forward to many more years of science, of wisdom, of laughter, and of discussions about sex chromosomes with Art Arnold. We are all the better for his continued presence in our lives and in neuroendocrine science.

Acknowledgments Thanks to Art for checking this piece for accuracy and for providing the figures.

References

Agate, R. J., Grisham, W., Wade, J., Mann, S., Wingfield, J., Schanen, C., Palotie, A., & Arnold, A. P. (2003). Neural not gonadal origin of brain sex differences in a gynandromorphic finch. *Proceedings of the National Academy of Sciences, 100*(8), 4873–4878. https://doi.org/10.1073/pnas.0636925100

Arnold, A. P. (1975a). The effects of castration on song development in zebra finches (Poephila guttata). *The Journal of Experimental Zoology, 191*(261), 278.

Arnold, A. P. (1975b). The effects of castration and androgen replacement on song, courtship, and aggression in zebra finches (Poephila guttata). *The Journal of Experimental Zoology, 191*(309), 326.

Arnold, A. P. (1980a). Quantitative analysis of sex differences in hormone accumulation in the zebra finch brain: Methodological and theoretical issues. *The Journal of Comparative Neurology, 189*(421), 436.

Arnold, A. P. (1980b). A technique for simultaneous steroid autoradiography and retrograde labelling of neurons. *Brain Research, 192*(210), 212.

Arnold, A. P. (1981). Quantitative analysis of steroid autoradiograms. *The Journal of Histochemistry and Cytochemistry, 29*(207), 212.

Arnold, A. P. (1996). Genetically triggered sexual differentiation of brain and behavior. *Hormones and Behavior, 30*, 495–505.

Arnold, A. P. (1997). Sexual differentiation of the zebra finch song system: Positive evidence, negative evidence, null hypotheses, and a paradigm shift. *Journal of Neurobiology, 33*, 572–584.

Arnold, A. P. (2012). The end of gonad-centric sex determination in mammals. *Trend in Genetics, 28*, 55–61.

Arnold, A. P. (2017). A general theory of sexual differentiation. *Journal of Neuroscience Research, 95*, 291–300.

Arnold, A. P. (2020). Four Core Genotypes and XY* mouse models: Update on impact on SABV research. *Neuroscience & Biobehavioral Reviews, 119*, 1–8.

Arnold, A. P., & Saltiel, A. (1979). Sexual difference in pattern of hormone accumulation in the brain of a song bird. *Science, 205*, 702–705.

Arnold, A. P., Nottebohm, F., & Pfaff, D. W. (1976). Hormone concentrating cells in vocal control and other areas of the brain of the zebra finch (Poephila guttata). *The Journal of Comparative Neurology, 165*(487), 512.

Arnold, A. P., Cassis, L. A., Eghbali, M., Reue, K., & Sandberg, K. (2017). Sex hormones and sex chromosomes cause sex differences in the development of cardiovascular diseases. *Arteriosclerosis, Thrombosis, and Vascular Biology, 37*, 746–756.

Asanuma, H., Shinoda, Y., Arnold, A. P., & Zarzecki, P. (1976). Reexamination of functional arrangements of pyramidal tract neurons in the motor cortex of the cat. *Experimental Brain Research, Supp I*, 440–444.

Bottjer, S. W., Miesner, E. A., & Arnold, A. P. (1984). Forebrain lesions disrupt development but not maintenance of song in passerine birds. *Science, 224*, 901–903.

Breedlove, S. M., & Arnold, A. P. (1980). Hormone accumulation in a sexually dimorphic motor nucleus of the rat spinal cord. *Science, 210*(564), 566.

Breedlove, S. M., & Arnold, A. P. (1981). Sexually dimorphic motor nucleus in the rat spinal cord: Response to adult hormone manipulation, absence in androgen insensitive rats. *Brain Research, 225*, 297–307.

Breedlove, S. M., & Arnold, A. P. (1983a). Hormonal control of a developing neuromuscular system: I. Complete demasculinization of the male rat spinal nucleus of the bulbocavernosus using the antiandrogen flutamide. *The Journal of Neuroscience, 3*(417), 423.

Breedlove, S. M., & Arnold, A. P. (1983b). Hormonal control of a developing neuromuscular system: II. Sensitive periods for the androgen induced masculinization of the rat spinal nucleus of the bulbocavernosus. *The Journal of Neuroscience, 3*(424), 432.

Breedlove, S. M., & Arnold, A. P. (1983c). Sex differences in the pattern of steroid accumulation of motoneurons of the rat lumbar spinal cord. *The Journal of Comparative Neurology, 215*(211), 216.

Breedlove, S. M., Jacobson, C. D., Gorski, R. A., & Arnold, A. P. (1982). Masculinization of the female rat spinal cord following a single neonatal injection of testosterone propionate but not estradiol benzoate. *Brain Research, 237*(173), 181.

Breedlove, S. M., Jordan, C. L., & Arnold, A. P. (1983). Neurogenesis in the sexually dimorphic spinal nucleus of the bulbocavernosus in rats. *Developmental Brain Research, 9*(39), 43.

Brenowitz, E. A., & Arnold, A. P. (1985). Lack of sexual dimorphism in steroid accumulation in vocal control brain regions of duetting song birds. *Brain Research, 344*, 172–175.

Brenowitz, E. A., & Arnold, A. P. (1986). Interspecific comparisons of the size of neural song control regions and song complexity in duetting birds: Evolutionary implications. *The Journal of Neuroscience, 6*(2875), 2879.

Brenowitz, E. A., & Arnold, A. P. (1989). Accumulation of estrogen in a vocal control brain region of a duetting song bird. *Brain Research, 480*, 119–125.

Brenowitz, E. A., Arnold, A. P., & Levin, R. N. (1985). Neural correlates of female song in tropical duetting birds. *Brain Research, 343*, 104–112.

Carruth, L. C., Reisert, I., & Arnold, A. P. (2002). Direct effect of sex chromosome genes on brain sexual differentiation. *Nature Neuroscience, 5*, 933–934.

Chen, X., Watkins, R., Delot, E., Reliene, R., Schiestl, R. H., Burgoyne, P. S., & Arnold, A. P. (2008). Sex difference in neural tube defects in p53 null mice is caused by differences in the complement of X not Y genes. *Developmental Neurobiology, 68*, 265–273.

Chen, X., McClusky, R., Chen, J., Beaven, S. W., Tontonoz, P., Arnold, A. P., & Reue, K. (2012). The number of X chromosomes causes sex differences in adiposity and metabolism in mice. *PLoS Genetics, 8*(5), e1002709.

Cox, K. H., Bonthuis, P. J., & Rissman, E. F. (2014). Mouse model systems to study sex chromosome genes and behavior: Relevance to humans. *Frontiers in Neuroendocrinology, 35*, 405–419.

Cunningham, C. M., Umar, S., Doshi, M., Li, M., Ruffenach, G., Hrncir, H., Arnold, A. P., & Eghbali, M. (2020). Y-chromosome gene, Uty, protects against pulmonary hypertension by reducing lung pro-inflammatory cytokines. *The FASEB Journal*. https://doi.org/10.1096/fasebj.2020.34.s1.02378

Davis, E. J., Broestl, L., Abdulai-Saiku, S., Worden, K., Bonham, L. W., Miñones-Moyano, E., et al. (2020). The second X chromosome confers resilience against Alzheimer's disease-related deficits in male and female mice. *Science Translational Medicine, 12*, eaaz5677. https://doi.org/10.1126/scitranslmed.aaz5677

de Vries, G. J., Buijs, R. M., & Swaab, D. F. (1981). Ontogeny of the vasopressinergic neurons of the suprachiasmatic nucleus and their extrahypothalamic projections in the rat brain – presence of a sex difference in the lateral septum. *Brain Research, 218*, 67–78.

De Vries, G. J., Rissman, E. F., Simerly, R. B., Yang, L.-Y., Scordalakes, E. M., Auger, C., Swain, A., Lovell-Badge, R., Burgoyne, P. S., & Arnold, A. P. (2002). A model system for study of sex chromosome effects on sexually dimorphic neural and behavioral traits. *The Journal of Neuroscience, 22*, 9005–9014.

DeLacoste-Utamsing, C., & Holloway, R. L. (1982). Sexual dimorphism in the human corpus callosum. *Science, 216*, 1431–1432.

Floody, O. R., & Arnold, A. P. (1975). Uganda kob (Adenota kob thomasi): Territoriality and the spatial distributions of sexual and agonistic behaviors at a territorial ground. *Zeitschrift für Tierpsychologie, 37*(192), 212.

Gioiosa, L., Chen, X., Watkins, R., Klanfer, N., Bryant, C. D., Evans, C. J., & Arnold, A. P. (2008). Sex chromosome complement affects nociception in tests of acute and chronic exposure to morphine in mice. *Hormones and Behavior, 54*, 124–130.

Gorski, R. A., Gordon, J. H., Shryne, J. E., & Southam, A. M. (1978). Evidence for a morphological sex difference within the medial preoptic area of the rat brain. *Brain Research, 148*, 333–346.

Grimm, S. L., Dong, X., Zhang, Y., Carisey, A. F., Arnold, A. P., Moorthy, B., Coarfa, C., & Lingappan, K. (2021). Effect of sex chromosomes versus hormones in neonatal lung injury. *JCI Insight*, 146863. https://doi.org/10.1172/jci.insight.146863

Itoh, Y., Melamed, E., Yang, X., Kampf, K., Wang, S., Yehya, N., Van Nas, A., Replogle, K., Band, M. R., Clayton, D. F., Schadt, E. E., Lusis, A. J., & Arnold, A. P. (2007). Dosage compensation is less effective in birds than in mammals. *Journal of Biology, 6*, 2. PMCID: PMC2373894.

Itoh, Y., Golden, L. C., Itoh, N., Matsukawa, M. A., Ren, E., Tse, V., Arnold, A. P., & Voskuhl, R. R. (2019). The X-linked histone demethylase Kdm6a in CD4+ T lymphocytes modulates autoimmunity. *The Journal of Clinical Investigation, 130*, 3852–3863.

Jordan, C. L., Letinsky, M. S., & Arnold, A. P. (1988). Synapse elimination occurs late in the hormone sensitive levator ani muscle of the rat. *Journal of Neurobiology, 19*(335), 356.

Jordan, C. L., Letinsky, M. S., & Arnold, A. P. (1989a). The role of gonadal hormones in neuromuscular synapse elimination in rats. I. Androgen delays the loss of multiple innervation in the levator ani muscle. *The Journal of Neuroscience, 9*, 229–238.

Jordan, C. L., Letinsky, M. S., & Arnold, A. P. (1989b). The role of gonadal hormones in neuromuscular synapse elimination in rats. II. Multiple innervation persists in the adult levator ani muscle after juvenile androgen treatment. *The Journal of Neuroscience, 9*, 239–247.

Jordon, C. L., Letinsky, M. S., & Arnold, A. P. (1990). Critical period for the androgenic block of neuromuscular synapse elimination. *Journal of Neurobiology, 21*, 760–767.

Kurz, E. M., Sengelaub, D. R., & Arnold, A. P. (1986). Androgens regulate dendritic length of sexually dimorphic mammalian motoneurons in adulthood. *Science, 232*(395), 398.

Li, J., Chen, X., McClusky, R., Ruiz-Sundstrom, M., Itoh, Y., Arnold, A. P., & Eghbali, M. (2014). The number of X chromosomes influences protection from ischemia/reperfusion injury in mice: One X is better than two. *Cardiovascular Research, 102*, 375–384.

Link, J., Wiese, C. B., Chen, X., Avetisyan, R., Ronquillo, E., Ma, F., Guo, X., Yao, J., Allison, M., Chen, Y.-D., Rotter, J., Moustafa, J., Small, K., Iwase, S., Pellegrini, M., Vergnes, L., Arnold, A. P., & Reue, K. (2020). X chromosome dosage of histone demethylase KDM5C determines sex differences in adiposity. *The Journal of Clinical Investigation*, 140223. https://doi.org/10.1172/JCI140223. PMCID: PMC7598065.

Matsumoto, A., Micevych, P. E., & Arnold, A. P. (1988a). Androgen regulates synaptic input to motoneurons of adult rat spinal cord. *The Journal of Neuroscience, 8*, 4168–4176.

Matsumoto, A., Arnold, A. P., Zampighi, G. A., & Micevych, P. E. (1988b). Androgenic regulation of gap junctions between motoneurons in rat spinal cord. *The Journal of Neuroscience, 8*, 4177–4183.

Nordeen, E. J., Nordeen, K. W., Sengelaub, D. R., & Arnold, A. P. (1985). Androgens prevent normally occurring cell death in a sexually dimorphic spinal nucleus. *Science, 229*(671), 673.

Nordeen, K. W., Nordeen, E. J., & Arnold, A. P. (1986). Estrogen establishes sex differences in androgen accumulation in zebra finch brain. *The Journal of Neuroscience, 6*(734), 738.

Nordeen, E. J., Nordeen, K. W., & Arnold, A. P. (1987a). Sexual differentiation of androgen accumulation within the zebra finch brain through selective cell loss and addition. *The Journal of Comparative Neurology, 259*, 393–399.

Nordeen, K. W., Nordeen, E. J., & Arnold, A. P. (1987b). Estrogen accumulation in zebra finch song control nuclei: Implications for sexual differentiation and adult activation of song behavior. *Journal of Neurobiology, 18*(569), 582.

Nottebohm, F., & Arnold, A. P. (1976). Sexual dimorphism in vocal control areas of the song bird brain. *Science, 194*, 211–213.

Saldanha, C. J., Tuerk, M. J., Kim, Y.-H., Fernandes, A. O., Arnold, A. P., & Schlinger, B. A. (2000). The distribution and regulation of telencephalic aromatase expression in the zebra finch revealed with a specific antibody. *The Journal of Comparative Neurology, 423*(619), 630.

Schlinger, B. A., & Arnold, A. P. (1991). Brain is the major site of estrogen synthesis in a male songbird. *Proceedings of the National Academy of Sciences of the United States of America, 88*, 4191–4194.

Schlinger, B. A., & Arnold, A. P. (1992). Circulating estrogens in a male songbird originate in the brain. *Proceedings of the National Academy of Sciences of the United States of America, 89*, 7650–7653.

Schlinger, B. A., & Arnold, A. P. (1993). Estrogen synthesis in vivo in the adult zebra finch: Additional evidence that circulating estrogens can originate in brain. *Endocrinology, 133*, 2610–2616.

Schlinger, B. A., Amur-Umarjee, S., Campagnoni, A. T., & Arnold, A. P. (1994). Neuronal and non-neuronal cells express high levels of aromatase in primary cultures of developing zebra finch telencephalon. *The Journal of Neuroscience, 14*, 7541–7552.

Schlinger, B. A., Carruth, L., Xu, J., & deVries, G. J. (2011). State-of-the Art (Arnold) Behavioral Neuroendocrinology (A Tribute to Art Arnold upon his receipt of the Daniel S. Lehrman award for Lifetime Achievement in Behavioral Endocrinology). *Hormones and Behavior, 60*, 1–3.

Sengelaub, D. R., & Arnold, A. P. (1986). Development and loss of early projections in a sexually dimorphic rat spinal nucleus. *The Journal of Neuroscience, 6*, 1613–1620.

Sengelaub, D. R., & Arnold, A. P. (1989). Hormonal control of neuron number in sexually dimorphic spinal nuclei of the rat. I. Testosterone regulated death in the dorsolateral nucleus. *The Journal of Comparative Neurology, 280*, 622–629.

Sengelaub, D. R., Jordan, C. L., Kurz, E. M., & Arnold, A. P. (1989a). Hormonal control of neuron number in sexually dimorphic spinal nuclei of the rat. II. Development of the spinal nucleus of the bulbocavernosus in androgen insensitive (Tfm) rats. *The Journal of Comparative Neurology, 280*, 630–636.

Sengelaub, D. R., Nordeen, E. J., Nordeen, K. W., & Arnold, A. (1989b). Hormonal control of neuron number in sexually dimorphic spinal nuclei of the rat. III. Differential effects of dihydrotestosterone on the spinal nucleus of the bulbocavernosus and dorsolateral nucleus. *The Journal of Comparative Neurology, 280*, 637–644.

Seu, E., Groman, S., Arnold, A. P., & Jentsch, J. D. (2014). Sex chromosome complement influences operant responding for a palatable food in mice. *Genes, Brain, and Behavior*;epub May 23. PMCID: PMC4079740.

Shen, P., Campagnoni, C. W., Kampf, K., Schlinger, B. A., Arnold, A. P., & Campagnoni, A. T. (1994). Isolation and characterization of a zebra finch aromatase cDNA: In situ hybridization reveals high aromatase expression in brain. *Molecular Brain Research, 24*, 227–237.

Shen, P., Schlinger, B. A., Campagnoni, A. T., & Arnold, A. P. (1995). An atlas of aromatase mRNA expression in the zebra finch brain. *The Journal of Comparative Neurology, 360*, 172–184.

Shi, W., Sheng, X., Dorr, K. M., Hutton, J. E., Emerson, J. I., Davies, H. A., Andrade, T. D., Wasson, L. K., Greco, T. M., Hashimoto, Y., Federspiel, J. D., Robbe, Z. L., Chen, X., Arnold, A. P., Cristea, I. M., & Conlon, F. L. (2021). Cardiac proteomics reveals sex chromosome-dependent differences between males and females that arise prior to gonad formation. *Developmental Cell*, S1534-5807. Online ahead of print. PMID: 34655525.

Shinoda, Y., Ghez, C., & Arnold, A. P. (1978). Spinal branching of rubrospinal axons in the cat. *Experimental Brain Research, 30*, 203–218.

Simerly, R. B., Swanson, L. W., & Gorski, R. A. (1985). The distribution of monoaminergic cells and fibers in a periventricular preoptic nucleus involved in the control of gonadotropin release: Immunohistochemical evidence for a dopaminergic sexual dimorphism. *Brain Research, 330*, 55–64.

Smith-Bouvier, D. L., Sasihar, M., Du, S., Tiwari-Woodruff, S. K., King, J. K., Arnold, A. P., Singh, R. R., & Voskuhl, R. R. (2008). A role for sex chromosome complement in the female bias in autoimmune disease. *The Journal of Experimental Medicine, 205*, 1099–1108.

Taylor, A. M. W., Chadwick, C. I., Mehrabani, S., Hrncir, H., Arnold, A. P., & Evans, C. J. (2020). Sex differences in kappa opioid anti-nociception is influenced by the number of X chromosomes in mouse. *Journal of Neuroscience Research*. https://doi.org/10.1002/jnr.24704

Turner, J. M. A., Mahadevaiah, S. K., Arnold, A. P., & Lovell-Badge, R. (2020). Paul S. Burgoyne (1946-2020). *Development, 147*, dev197467. https://doi.org/10.1242/dev.197467

Umar, S., Cunningham, C. M., Itoh, Y., Moazeni, S., Vaillancourt, M., Sarji, S., Centala, A., Arnold, A. P., & Eghbali, M. (2018). The Y chromosome plays a protective role in experimental hypoxic pulmonary hypertension. *American Journal of Respiratory and Critical Care Medicine, 197*, 952–955.

Wade, J., & Arnold, A. P. (1996). Functional testicular tissue does not masculinize development of the zebra finch song system. *Proceedings of the National Academy of Sciences of the United States of America, 93*, 5264–5268.

Wade, J., Springer, M. L., Wingfield, J. C., & Arnold, A. P. (1996). Neither testicular androgens nor embryonic aromatase activity alter morphology of the neural song system in zebra finches. *Biology of Reproduction, 55*, 1126–1132.

Alison Sarah Fleming

34

Andrea Gonzalez

Abstract

Alison Sarah Fleming is a Canadian neuroscientist (or behavioral neuroendocrinologist) whose research program focused on maternal motivation. Spanning 40 years, Fleming studied multiple species, most notably rats and humans, with work ranging from sensory and experiential factors influencing maternal behavior, to the underlying physiology and neuroanatomy, to the impact of early life adversity on subsequent parenting and later generations. She is a true *tour de force*, not only shaping the field through the depth and breadth of her work, but also, if you have ever met her, through her vibrant, engaging personality and infectious laugh.

Keywords

Maternal behavior · Sensory factors · Hypothalamic-pituitary-adrenal axis · Early life adversity

Spanning a 40 year career, Alison Fleming was focused on the study of various aspects of maternal behavior, in particular maternal motivation, using a multi-method, cross-species approach. The most influential sources on her research program were her mentors, and other leaders in the field, Jay Rosenblatt, Daniel Lehrman, and T.C. Schneirla. In addition, Nikolaas Tinbergen's concept of the four

A. Gonzalez (✉)
McMaster University, Department of Psychiatry & Behavioural Neurosciences, Offord Centre for Child Studies, Hamilton, ON, Canada
e-mail: gonzal@mcmaster.ca

R. J. Nelson, Z. M. Weil (eds.), *Biographical History of Behavioral Neuroendocrinology*, https://doi.org/10.1007/978-3-031-12970-4_34

different levels for answering questions related to causality, function, evolution, causation, and development provided the framework to how she approached research, although Fleming argued that only the integration of the four levels would provide the most complete picture. She balanced a demanding academic career – teaching, conducting research, and "mothering" countless students and trainees while raising her own three daughters. She is a pioneer in the field of behavioral neuroendocrinology, not only laying the groundwork for studying sensory and experiential aspects of maternal behavior but also exemplifying the successes and struggles of professional mothers. Her passion and engaging nature led to numerous international collaborations, was a draw to students, and was celebrated during her retirement party in 2013, where over 30 former trainees and colleagues from around the world attended the event (Fig. 34.1).

Born in Great Britain, Fleming grew up in New York City, was educated in New Jersey (where she obtained her PhD in 1973 with Jay Rosenblatt) and Berkeley, California (during her postdoctoral fellowship with Irving Zucker from 1972 to 1974), and settled in Toronto, Canada, where she started and finished her successful career at the University of Toronto in the Department of Psychology (1975–2013). She is currently Professor *Emeritus* at the University of Toronto and splits her time among Toronto, Tennessee (with her two granddaughters), and Mexico (at her retirement residence). In retirement, she continues to balance research collaborations with spending time with her family and friends, walking, reading, and

Fig. 34.1 Fleming lab

attending plays and movies and has manifested her creativity on a new canvas, with her art (http://alisonfleming.ca/), which is as varied and prolific as her research!

It was during a course and work with Burt Slotnick at Columbia University in her sophomore year that Fleming marks as the start of her academic journey, when she was struck with the thought "what makes mothers want to mother" on a walk to Low Memorial Library. However, her path was laid long before that and began with her own mother. Fleming's mother was a professional economist with the United Nations during a time when the only females pictured at international conferences were administrative staff. She held a high-powered, influential position which came at a cost to her children who were uprooted with a move from the United Kingdom to the United States and who were primarily raised by nannies. The seed of Fleming's research program was born of a need to understand why her mother opted to be a professional and have children – and how mothering was not necessarily a natural process for her. Although Fleming did not know her trajectory was set to be a professional academic in her own right, she knew that she wanted to be a mother someday and was determined to do things differently.

It would be challenging to provide a detailed discussion of each of Fleming's published works given her productivity and range – over 180 manuscripts and chapters over the course of 40 years. Instead, below I highlight two main areas of her work on maternal behavior, which include (1) experiential and hormonal changes in the transition to motherhood (or induction using virgin female rats) and (2) the influence of early life experiences. As highlighted earlier, one of the strengths of Fleming's work was her cross-species approach to highlight similarities in function and mechanism. Most of work cited below will focus on her studies using rat models of maternal behavior with some references to her human work when it is aligned with the relevant themes.

Experiential and Hormonal Changes

The phenomenology of rat maternal behavior is well characterized (Weisner & Sheard, 1933) and appears in a very stereotypical pattern. At parturition, the dam is immediately responsive, retrieving all pups to a nest site, mouthing, and licking them, and adopting a nursing posture over the litter within the first 30 min after birth and without any prior experience interacting with pups (Fleming & Rosenblatt, 1974). This contrasts with nulliparous rats that do not exhibit any maternal responsive behaviors when initially presented with pups (Weisner & Sheard, 1933) and either actively avoids or attacks them (Fleming & Luebke, 1981). However, after a "pup-induction" or "pup-sensitization" process, which involves continually exposing the virgin to foster pups over a period of time, the virgin rat becomes habituated to the pups, laying close to them after 1–2 days (Fleming & Luebke, 1981; Fleming & Rosenblatt, 1974), and will begin to show a pattern of maternal responsiveness similar to that of a new mother after 5–10 days of repeated exposure (Fleming & Rosenblatt, 1974). These differences in maternal responsiveness are more widespread, affecting an overall difference in emotionality, with dams exhibiting an overall increased willingness to approach an unfamiliar intruder, or explore a new environment (Fleming & Luebke, 1981). These differences in emotionality and responsiveness to pups between dams and virgin females are hormonally mediated, with a regimen of progesterone and estradiol that facilitates the onset of maternal behavior in the virgin rat, also reducing pup-avoidance and measures of timidity (Fleming et al., 1989) and via hormonal effects on olfactory-mediated responses (Fleming & Rosenblatt, 1974).

Hormonal regulation underpinning maternal responsiveness was initially pioneered by Fleming's mentor Jay Rosenblatt (Terkel & Rosenblatt, 1972) and further characterized by many leaders in the field (Bridges, 2016; Insel, 1990; Rosenblatt, 1990), illustrating the diverse array of hormones at play, including steroid hormones (estrogen and progesterone), protein hormones (oxytocin, prolactin, and vasopressin), glucocorticoids, as well as neurotransmitters (e.g., dopamine, norepinephrine), which collectively underpin important aspects of pregnancy, parturition, lactation, and maternal responsiveness (Fleming, 1987), including increasing the likelihood that the dams will respond to offspring by enhancing maternal attractiveness to infant cues while reducing fearfulness and neophobia (A. Fleming et al., 1999). Against this backdrop of hormones commonly associated with maternal behavior, Fleming and her lab largely focused on dopamine (Olazábal et al., 2013a, b) and glucocorticoids, the latter of which will be highlighted in some detail below.

Hormones associated with our main stress system, the hypothalamic-pituitary-adrenal (HPA) axis, change during gestation, birth, and lactation. In a series of elegant studies, the Fleming lab was the first to investigate whether changes in glucocorticoids were associated with differences in the qualitative expression of maternal behaviors. They examined the role of corticosterone on maternal behaviors through adrenalectomies and corticosterone replacement in primiparous and virgin rats. Adrenalectomies were performed on day 17 of pregnancy (or sham surgeries), and corticosterone was administered in varying, increasing dosages (0, 25, 100, 300, or 500 μg/ml) through drinking water after parturition, or through implantation of corticosterone pellets (Rees et al., 2004). Blood levels of corticosterone were measured to ensure the experimental protocol was achieving the intended effects. Maternal behaviors were recorded across ten consecutive days following parturition and corticosterone replacement. Compared to the sham surgery group, adrenalectomies decreased licking of pups, with higher concentrations of corticosterone replacement increasing hovering and licking and spending more time in the nest across compared to groups receiving the lower (or no) dosages. Interestingly, behaviors related to maternal motivation, such as retrieval and nursing responses, were not affected, suggesting that glucocorticoids had more of an impact on the quality of maternal responsiveness (Rees et al., 2004). Subsequent studies indicated that dams exhibited better retention of their postpartum experiences interacting with pups when exposed to additional corticosterone, indicating that corticosterone likely also plays a role in enhancing memory in the maternal context (Graham et al., 2006). A final set of experiments with sensitized virgin female rats showed that replacement of corticosterone was not sufficient to promote maternal responding and caused an inhibition of maternal behavior in this group. These findings highlighted that the backdrop of pregnancy hormones plays an important role in how glucocorticoids influence later maternal behavior (Rees et al., 2006).

Many of the same hormones implicated in nonhuman mammalian parenting are also involved in early expression of mothering in humans (Lonstein et al., 2015). As with rats, these pregnancy and parturition hormones constitute the background against which glucocorticoid in humans, cortisol, has its effects. As mentioned above, one of the many strengths of Fleming's research program was her attempt to bridge questions of mothering across species. She was one of the leaders, highlighting the importance of cortisol in human mothering showing that immediately after birth, mothers with higher levels of baseline plasma cortisol tended to also show higher levels of contact and affectionate approach responses toward their infants (Fleming, 1987). Later studies highlighted the role of parity with higher cortisol levels associated with affectionate behaviors (e.g., affectionate burping, stroking, hugging) in primiparous mothers, and higher cortisol positively related to greater caretaking behaviors (e.g., instrumental burping, wiping face, adjusting blanket) in multiparous mothers (Fleming et al., 1997a, b). In addition to examining the central role of cortisol and observed behaviors, the Fleming lab also explored its role in attribution to infant cues in new mothers. In an ingenious set of experiments, new mothers (1 day postpartum) were presented with various odors in Baskin-Robbins's ice cream containers! Odors included the mother's own infant's T-shirt (body odor) and urine sample (cotton swatch); another, same-sex infant's T-shirt and urine samples; and a non-infant, control odor (the spice, marjoram). Salivary cortisol was collected prior to any presentation of odors or interaction with the infants. Fleming and her team found that parity played an important role in the findings. In primiparous, but not multiparous, mothers, higher cortisol concentrations were associated with greater ratings of attraction to infant T-shirt body odors and urine of both own infant and other infant. This attraction was specifically related to infant odors and not generalized to the non-infant odor (marjoram). Interestingly, an opposite parity effect was found when testing discriminatory abilities in new mothers. Multiparous mothers with higher cortisol levels had higher accuracy scores when asked to identify their own infant's body odor compared to others' body odor across multiple trials. There was no association between cortisol and recognition in primiparous mothers (Fleming et al., 1997b).

Given the emerging role of the HPA axis in mothering in both rats and humans, Fleming was keenly interested in how multiple sensory systems contributed to maternal motivation and responses. Using a new experimental paradigm, conducted almost immediately after birth (days 1 or 2 postpartum), Fleming and her colleagues tested the role of cortisol in discriminating a different sensory system, infant cries. In this study, mixed parity females were presented with either audiotapes of two infant pain cries and two infant hunger cries, or control stimuli (blank tape). Each of the four cries was included on one tape; each cry was 45 s long with a 2 min interstimulus interval. Following the presentation of each stimulus segment (pain or hunger cry, or control), participants completed a visual analogue scale (ranging from not at all to extremely) to describe several affective states, such as alertness and sympathy. Salivary cortisol was measured at three time points, baseline, before stimulus presentation, and then 20 and 40 min after the stimulus presentation ended (cries vs. control) (Stallings et al., 2001). Overall, primiparous women had higher cortisol levels compared to multiparous mothers, but neither group exhibited a differential response to the stimuli until affect ratings were considered. Mothers of both parity who rated themselves as more sympathetic to the cry stimuli had significantly higher baseline cortisol levels, compared to mothers showing lower sympathy. These mothers also showed a greater decline in cortisol concentrations over time. Primiparous mothers also exhibited overall higher sympathy ratings to cries compared to multiparous mothers; however, multiparous mothers had greater discrimination in their sympathetic responses to infant pain versus hunger cries (Stallings et al., 2001). In a subsequent study with adolescent and adult mothers, only the older group of mothers exhibited a cortisol response to infant cries, which was positively correlated with sympathy ratings and affectionate behaviors. There was no such association in young mothers (Giardino et al., 2008). Fleming's initial work on the relevance and importance of infant stimuli on parenting behaviors laid an important foundation for current research examining brain activation in response to infant cries and other stimuli (Provenzi et al., 2021; Swain et al., 2011; Witteman et al., 2019).

Early Life Experiences

Fleming had a long-standing interest in how early life experiences influenced later mothering and what mechanisms may underlie those behavioral differences. Following work by Seymour Levine and Myron Hofer, some Fleming's first studies involved maternal deprivation models, where pups were removed from the nest for varying lengths of time (Lovic et al., 2001). It was on a site visit to the University of Waterloo in 1997 that Fleming was introduced to the "pup-in-a-cup" paradigm, where researchers Glen Ward and Patricia Wainwright were conducting nutrition studies in rats. Fleming returned to UTM, excited about the idea of artificially rearing rats, and a model of complete maternal deprivation where one could "reconstruct" mom within an experimental paradigm – a new arm of research for the lab was born. In Fleming's artificial rearing paradigm, on postnatal day (PND) 4, pups were removed from the dam and a cheek cannula is inserted for feeding. Pups were placed in a cup with corncob bedding, with all cups floating in a warm water bath (34–37 °C) for temperature regulation (see Fig. 34.2) until weaning on PND 18 when their feeding tube was removed, and they were placed in pairs until adulthood.

Pups were fed via infusion pumps (PHD 22/2000 syringe, Harvard Apparatus, Holliston, MA) to which they were connected by a cheek cannula tubing. Pups were fed via the infusion pump connected to the feeding tubes which delivered mil (Messer diet) for 10 min every hour, 24 h per day. The amount of milk infused was calculated based on mean pup body weight. Beginning on PND 4, pups received a volume of milk equal to 33% of the mean body weight, and this amount increased by 1% daily. Each morning, the pups were disconnected from the pumps weighed, and all tubing was flushed with double-distilled water. New infusion rates were programmed based on daily body weights.

Fig. 34.2 Pup-in-a-cup, artificial rearing paradigm

This artificial rearing regimen allowed for a range of manipulations, including feeding schedules, odors, and most importantly to the lab's paradigm, artificial "licking"-like stimulation (Lomanowska & Melo, 2016). To accomplish the latter, soft paintbrushes were used to stroke the pups' general body and anogenital region. Swabbing the anogenital region was necessary for physiological functions, including the induction of defecation and urination, whereas the overall general body stroking mimicked maternal licking, crucial for development (Gonzalez et al., 2000). The number of stroking stimulations was varied in experimental groups to examine whether mimicking maternal licking mitigated the effects of maternal deprivation. In most of the AR studies, two female offspring were removed and artificially reared, one received minimal anogenital stimulation (two swabs per day; AR-MIN), and those receiving maximal stimulation (two anogenital, and six daily bouts of body stroking distributed throughout daytime hours; AR-MAX), two control female sibling (mother reared; MR)) were given a sham surgery (cheek puncture) and left with the dam until weaning. The litter was culled to eight pups (four females, four males). Studies using the artificial rearing paradigm demonstrated that AR rats exhibited impairments in maternal behavior toward own offspring later in life, which carried across generations (Gonzalez et al., 2000; Melo et al., 2006; Palombo et al., 2010). Underlying systems were affected, including with changes in the dopaminergic system (Afonso et al., 2011; Lovic et al., 2006, 2010) increased sensitivity to natural and drug rewards (Lomanowska et al., 2006) and changes to executive function, including compromised cognitive flexibility (ability to shift strategies in the face of new information) and increased impulsivity (Lovic & Fleming, 2004). At the cellular level, changes were seen in *c-fos* expression in the medial preoptic area and parietal and piriform cortices in juvenile AR rats, compared to MR rats (Gonzalez & Fleming, 2002)). Juvenile AR rats also had increased numbers of neurons and astroglial cells, but reduced expression of neural proteins involved in synaptic and neural plasticity, suggesting a state of "reduced cortex functionality" (Chatterjee et al., 2007). In most studies, providing additional licking-like stimulation for the AR-MAX rats, partially mitigated effects (Gonzalez et al., 2000; Chatterjee et al., 2007; Lovic & Fleming, 2004; Lovic et al., 2010; Palombo et al., 2010). Overall, this work showed the importance of early life adversity in later expression of maternal behaviors and how adversity impacted underlying systems and brain structures.

As with her other studies, Fleming wanted to bridge her work between rodents and humans. However, one could draw parallels between the artificial rearing model and the Romanian orphanage work by the Bucharest Early Intervention Project (Smyke et al., 2009). Fleming was interested in how retrospective reports of early life experiences influenced mothering and the mechanisms underlying any behavioral differences. To capture early life adversity, we used both the Childhood Trauma Questionnaire to assess severity of five sub-types of maltreatment; physical, sexual, and emotional abuse; and emotional and physical neglect and a Life History Calendar to capture whether mothers lived with both biological parents continuously since birth, and whom they lived with over the course of the first 18 years of life (number of transitions). Building on Fleming's nonhuman animal work and the literature on the importance of the HPA axis and executive function (Drury et al., 2016) in early adversity, we began examining the association of maternal ELA on cortisol dysregulation, executive dysfunction, and maternal behavior. A series of studies showed that ELA, as measured by consistency of care and/or history of childhood maltreatment, was associated with dysregulated diurnal cortisol patterns, specifically higher cortisol awakening response, greater blunting across the course of the day, and greater variability in levels across two consecutive days (Gonzalez et al., 2009a, b). In adolescent mothers, ELA was also associated with higher cortisol levels after a stressor (Krpan et al., 2005). Although executive function was not studied as an important factor to maternal behavior in humans at the time, based on the AR work (Lovic & Fleming, 2004), we included measures of working memory and cognitive flexibility in a study of mothers 4–6 months postpartum. Using the Cambridge Automated Neuropsychological Test Assessment Battery (CANTAB), we found that executive dysfunction, greater number of working memory errors, and poorer cognitive flexibility were associated with increased maternal insensitivity (Gonzalez et al., 2012). However, executive function was a significant mediator between maternal ELA and parenting only when considering cortisol levels such that ELA was associated with higher levels of cortisol, which were associated with greater executive function difficulties which in turn were associated with greater insensitivity. The executive dysfunction findings were further replicated in samples of teenage mothers (Chico et al., 2014; Almanza-Sepulveda et al., 2018) and extended to altered amygdala activation in brain imaging studies (Barrett et al., 2012).

Summary

What began as a seed of a question about "what makes mothers, mother?" while walking to the library blossomed into a lifelong research journey and discoveries from molecular insights to how experiences shape behavior. The conceptualization of maternal motivation is complex, involving both approaches, hedonic reaction to infant cues, such as odors, and the absence of avoidance to aversive stimuli such as infant cries (Gonzalez et al., 2009a, b). Fleming's work uncovered the physiology and neuroanatomy of underlying maternal motivation, as well as the role of current (parity status) and past (early life adversity) experiential factors. Although not addressed in this chapter, she has also examined genetic and epigenetic factors related to mothering. Her work was built upon the foundational research of her mentors, but she has left her own legacy for future generations.

References

Afonso, V. M., King, S. J., Novakov, M., Burton, C. L., & Fleming, A. S. (2011). Accumbal dopamine function in postpartum rats that were raised without their mothers. *Hormones and Behavior, 60*(5), 632–643. https://doi.org/10.1016/j.yhbeh.2011.08.016

Almanza-Sepulveda, M. L., Chico, E., Gonzalez, A., Hall, G. B., Steiner, M., & Fleming, A. S. (2018). Executive function in teen and adult women: Association with maternal status and early adversity. *Developmental Psychobiology, 60*(7), 849–861. https://doi.org/10.1002/dev.21766

Barrett, J., Wonch, K. E., Gonzalez, A., Ali, N., Steiner, M., Hall, G. B., & Fleming, A. S. (2012). Maternal affect and quality of parenting experiences are related to amygdala response to infant faces. *Social Neuroscience, 7*(3), 252–268. https://doi.org/10.1080/17470919.2011.609907

Bridges, R. S. (2016). Long-term alterations in neural and endocrine processes induced by motherhood in mammals. *Hormones and Behavior, 77*, 193–203. https://doi.org/10.1016/j.yhbeh.2015.09.001

Chatterjee, D., Chatterjee-Chakraborty, M., Rees, S., Cauchi, J., de Medeiros, C. B., & Fleming, A. S. (2007). Maternal isolation alters the expression of neural proteins during development: 'Stroking' stimulation reverses these effects. *Brain Research, 1158*, 11–27. https://doi.org/10.1016/j.brainres.2007.04.069

Chico, E., Gonzalez, A., Ali, N., Steiner, M., & Fleming, A. S. (2014). Executive function and mothering: Challenges faced by teenage mothers. *Developmental Psychobiology, 56*(5), 1027–1035. https://doi.org/10.1002/dev.21185

Drury, S. S., Sánchez, M. M., & Gonzalez, A. (2016). When mothering goes awry: Challenges and opportunities for utilizing evidence across rodent, nonhuman primate and human studies to better define the biological consequences of negative early caregiving. *Hormones and Behavior, 77*, 182–192. https://doi.org/10.1016/j.yhbeh.2015.10.007

Fleming, A. S. (1987). Psychobiology of rat maternal behavior: How and where hormones act to promote maternal behavior at parturition. In B. R. Komisaruk, H. I. Siegel, M. F. Cheng, & H. H. Feder (Eds.), *Reproduction: A behavioral and neuroendocrine perspective* (Vol. 474, pp. 234–251). New York Academy of Sciences.

Fleming, A. S., & Luebke, C. (1981). Timidity prevents the virgin female rat from being a good mother: Emotionality differences between nulliparous and parturient females. *Physiology & Behavior, 27*(5), 863–868. https://doi.org/10.1016/0031-9384(81)90054-8

Fleming, A. S., & Rosenblatt, J. S. (1974). Maternal behavior in the virgin and lactating rat. *Journal of Comparative and Physiological Psychology, 86*(5), 957–972. https://doi.org/10.1037/h0036414

Fleming, A. S., Cheung, U., Myhal, N., & Kessler, Z. (1989). Effects of maternal hormones on 'timidity' and attraction to pup-related odors in female rats. *Physiology & Behavior, 46*(3), 449–453. https://doi.org/10.1016/0031-9384(89)90019-x

Fleming, A. S., Ruble, D., Krieger, H., & Wong, P. (1997a). Hormonal and experiential correlates of maternal responsiveness during pregnancy and the puerperium in human mothers. *Hormones and Behavior, 31*(2), 145–158. https://doi.org/10.1006/hbeh.1997.1376

Fleming, A. S., Steiner, M., & Corter, C. (1997b). Cortisol, hedonics, and maternal responsiveness in human mothers. *Hormones and Behavior, 32*(2), 85–98. https://doi.org/10.1006/hbeh.1997.1407

Fleming, A., O'Day, D., & Kraemer, G. (1999). Neurobiology of mother–infant interactions: Experience and central nervous system plasticity across development and generations. *Neuroscience & Biobehavioral Reviews, 23*(5), 673–685. https://doi.org/10.1016/s0149-7634(99)00011-1

Giardino, J., Gonzalez, A., Steiner, M., & Fleming, A. S. (2008). Effects of motherhood on physiological and subjective responses to infant cries in teenage mothers: A comparison with non-mothers and adult mothers. *Hormones and Behavior, 53*(1), 149–158. https://doi.org/10.1016/j.yhbeh.2007.09.010

Gonzalez, A., & Fleming, A. S. (2002). Artificial rearing causes changes in maternal behavior and c-fos expression in juvenile female rats. *Behavioral Neuroscience, 116*(6), 999–1013. https://doi.org/10.1037/0735-7044.116.6.999

Gonzalez, A., Lovic, V., Ward, G. R., Wainwright, P. E., & Fleming, A. S. (2000). Intergenerational effects of complete maternal deprivation and replacement stimulation on maternal behavior and emotionality in female rats. *Developmental Psychobiology, 38*(1), 11–32.

Gonzalez, A., Atkinson, L., & Fleming, A. S. (2009a). Attachment and the comparative psychobiology of mothering. In M. de Haan & M. Gunnar (Eds.), *Handbook of developmental social neuroscience* (pp. 225–245). Guilford Press.

Gonzalez, A., Jenkins, J. M., Steiner, M., & Fleming, A. S. (2009b). The relation between early life adversity, cortisol awakening response and diurnal salivary cortisol levels in postpartum women. *Psychoneuroendocrinology, 34*(1), 76–86. https://doi.org/10.1016/j.psyneuen.2008.08.012

Gonzalez, A., Jenkins, J. M., Steiner, M., & Fleming, A. S. (2012). Maternal early life experiences and parenting: The mediating role of cortisol and executive function. *Journal of the American Academy of Child & Adolescent Psychiatry, 51*(7), 673–682. https://doi.org/10.1016/j.jaac.2012.04.003

Graham, M. D., Rees, S. L., Steiner, M., & Fleming, A. S. (2006). The effects of adrenalectomy and corticosterone replacement on maternal memory in postpartum rats. *Hormones and Behavior, 49*(3), 353–361. https://doi.org/10.1016/j.yhbeh.2005.08.014

Insel, T. R. (1990). Oxytocin and maternal behavior. In N. A. Krasnegor & R. S. Bridges (Eds.), *Mammalian parenting: Biochemical, neurobiological and behavioural determinants* (pp. 260–280). Oxford University Press.

Krpan, K. M., Coombs, R., Zinga, D., Steiner, M., & Fleming, A. S. (2005). Experiential and hormonal correlates of maternal behavior in teen and adult mothers. *Hormones and Behavior, 47*(1), 112–122. https://doi.org/10.1016/j.yhbeh.2004.08.006

Lomanowska, A. M., & Melo, A. I. (2016). Deconstructing the function of maternal stimulation in offspring development: Insights from the artificial rearing model in rats. *Hormones and Behavior, 77*, 224–236. https://doi.org/10.1016/j.yhbeh.2015.05.017

Lomanowska, A. M., Rana, S. A., McCutcheon, D., Parker, L. A., & Wainwright, P. E. (2006). Artificial rearing alters the response of rats to natural and drug-mediated rewards. *Developmental Psychobiology, 48*(4), 301–314. https://doi.org/10.1002/dev.20139

Lonstein, J. S., Lévy, F., & Fleming, A. S. (2015). Common and divergent psychobiological mechanisms underlying maternal behaviors in non-human and human mammals. *Hormones and Behavior, 73*, 156–185. https://doi.org/10.1016/j.yhbeh.2015.06.011

Lovic, V., & Fleming, A. S. (2004). Artificially-reared female rats show reduced prepulse inhibition and deficits in the attentional set shifting task—reversal of effects with maternal-like licking stimulation. *Behavioural Brain Research, 148*(1–2), 209–219. https://doi.org/10.1016/s0166-4328(03)00206-7

Lovic, V., Gonzalez, A., & Fleming, A. S. (2001). Maternally separated rats show deficits in maternal care in adulthood. *Developmental Psychobiology, 39*(1), 19–33. https://doi.org/10.1002/dev.1024

Lovic, V., Fleming, A. S., & Fletcher, P. (2006). Early life tactile stimulation changes adult rat responsiveness to amphetamine. *Pharmacology Biochemistry and Behavior, 84*(3), 497–503. https://doi.org/10.1016/j.pbb.2006.06.013

Lovic, V., Palombo, D. J., & Fleming, A. S. (2010). Impulsive rats are less maternal. *Developmental Psychobiology, 53*(1), 13–22. https://doi.org/10.1002/dev.20481

Melo, A. I., Lovic, V., Gonzalez, A., Madden, M., Sinopoli, K., & Fleming, A. S. (2006). Maternal and littermate deprivation disrupts maternal behavior and social-learning of food preference in adulthood: Tactile stimulation, nest odor, and social rearing prevent these effects. *Developmental Psychobiology, 48*(3), 209–219. https://doi.org/10.1002/dev.20130

Olazábal, D. E., Pereira, M., Agrati, D., Ferreira, A., Fleming, A. S., González-Mariscal, G., Lévy, F., Lucion, A. B., Morrell, J. I., Numan, M., & Uriarte, N. (2013a). Flexibility and adaptation of the neural substrate that supports maternal behavior in mammals. *Neuroscience & Biobehavioral Reviews, 37*(8), 1875–1892. https://doi.org/10.1016/j.neubiorev.2013.04.004

Olazábal, D. E., Pereira, M., Agrati, D., Ferreira, A., Fleming, A. S., González-Mariscal, G., Lévy, F., Lucion, A. B., Morrell, J. I., Numan, M., & Uriarte, N. (2013b). New theoretical and experimental approaches on maternal motivation in mammals. *Neuroscience & Biobehavioral Reviews, 37*(8), 1860–1874. https://doi.org/10.1016/j.neubiorev.2013.04.003

Palombo, D. J., Nowoslawski, M., & Fleming, A. S. (2010). Motherless rats show deficits in maternal behavior towards fostered pups. *Developmental Psychobiology*. https://doi.org/10.1002/dev.20422

Provenzi, L., Lindstedt, J., de Coen, K., Gasparini, L., Peruzzo, D., Grumi, S., Arrigoni, F., & Ahlqvist-Björkroth, S. (2021). The paternal brain in action: A review of human fathers' fMRI brain responses to child-related stimuli. *Brain Sciences, 11*(6), 816. https://doi.org/10.3390/brainsci11060816

Rees, S. L., Panesar, S., Steiner, M., & Fleming, A. S. (2004). The effects of adrenalectomy and corticosterone replacement on maternal behavior in the postpartum rat. *Hormones and Behavior, 46*(4), 411–419. https://doi.org/10.1016/j.yhbeh.2004.03.010

Rees, S. L., Panesar, S., Steiner, M., & Fleming, A. S. (2006). The effects of adrenalectomy and corticosterone replacement on induction of maternal behavior in the virgin female rat. *Hormones and Behavior, 49*(3), 337–345. https://doi.org/10.1016/j.yhbeh.2005.08.012

Rosenblatt, J. S. (1990). Landmarks in the physiological study of maternal behavior with special reference to the rat. In N. A. Krasnegor & R. S. Bridges (Eds.), *Mammalian parenting: Biochemical, neurobiological and behavioural determinants* (pp. 40–60). Oxford University Press.

Smyke, A. T., Zeanah, C. H., Fox, N. A., & Nelson, C. A. (2009). A new model of foster care for young children: The bucharest early intervention project. *Child and Adolescent Psychiatric Clinics of North America, 18*(3), 721–734. https://doi.org/10.1016/j.chc.2009.03.003

Stallings, J., Fleming, A., Corter, C., Worthman, C., & Steiner, M. (2001). The effects of infant cries and odors on sympathy, cortisol, and autonomic responses in new mothers and nonpostpartum women. *Parenting: Science and Practice, 1*(1), 71–100.

Swain, J. E., Kim, P., & Ho, S. S. (2011). Neuroendocrinology of parental response to baby-cry. *Journal of Neuroendocrinology, 23*(11), 1036–1041. https://doi.org/10.1111/j.1365-2826.2011.02212.x

Terkel, J., & Rosenblatt, J. S. (1972). Humoral factors underlying maternal behavior at parturition: Cross transfusion between freely moving rats. *Journal of Comparative and Physiological Psychology, 80*(3), 365–371. https://doi.org/10.1037/h0032965

Weisner, B. P., & Sheard, N. M. (1933). *Maternal behavior in the rat*. Oliver & Boyd.

Witteman, J., van IJzendoorn, M., Rilling, J., Bos, P., Schiller, N., & Bakermans-Kranenburg, M. (2019). Towards a neural model of infant cry perception. *Neuroscience & Biobehavioral Reviews, 99*, 23–32. https://doi.org/10.1016/j.neubiorev.2019.01.026

David Crews

35

Isaac Miller-Crews, Randy J. Nelson, and Andrea C. Gore

Abstract

David Crews' 50-year career was remarkable for the diversity of disciplines in which he made seminal discoveries: evolutionary biology, reproductive physiology, sex determination, brain sexual differentiation, behavioral neuroendocrinology, and transgenerational epigenetics. He never limited himself to a single or few species; instead, he would identify a critical gap in knowledge and then find the perfect species with which to conduct experimentation to fill that gap. In this article, we will discuss four of the key areas for which Crews is known: that the brain is intrinsically bisexual; that reproductive physiological properties and sex behavior can be disassociated; that sex determination involves active molecular cascades in both males and females (as opposed to the female being a "default" state); and that epigenetic transgenerational inheritance can influence sexual selection.

Keywords

Sex determination · Brain sexual differentiation · Behavioral neuroendocrinology · Endocrine-disrupting chemicals · Transgenerational epigenetics · Reproductive physiology · Evolutionary diversity

---

I. Miller-Crews (✉)
Department of Integrative Biology, The University of Texas at Austin, Austin, TX, USA
e-mail: imillercrews@utexas.edu

R. J. Nelson
Department of Neuroscience, Rockefeller Neuroscience Institute, West Virginia University, Morgantown, WV, USA
e-mail: Randy.Nelson@hsc.wvu.edu

A. C. Gore
Division of Pharmacology & Toxicology, College of Pharmacy, The University of Texas at Austin, Austin, TX, USA

R. J. Nelson, Z. M. Weil (eds.), *Biographical History of Behavioral Neuroendocrinology*, https://doi.org/10.1007/978-3-031-12970-4_35

Early History

David Crews made a career out of observing diversity in reproductive behaviors and physiology, asking questions that challenged the dogma, and taking advantage of unique and unconventional species to do so. However, he did not start out as a biologist, nor did his K-12 education foretell success in academia. As a military dependent, he moved every 1–2 years, living in Panama, Norway, West Germany, Florida, and South Carolina, to name a few. He spent his time reading, or preferably outdoors where he engaged in fishing, chasing and catching lizards and snakes (Fig. 35.1), and skiing. His love of reptiles was born in summer of 1959, when he captured three coral snakes in Gold Head Branch State Park in Florida while vacationing with family. However, his academic performance did not match his passion for nature and the outdoors. Because he had so underperformed in high school, he was unable to get into a US university; instead, he started at the University of Maryland Extension in Munich, Germany. He did well enough his first 2 years to be

Fig. 35.1 David Crews doing what he loves: catching wild reptiles, interacting with Miss Piggy the eastern box turtle, and making friends with a horse

able to transfer to the main campus in College Park, Maryland. Crews graduated with a B.A. in Psychology and a minor in Sociology.

With the goal of becoming a social worker, Crews worked at the Bureau of Social Science Research (BSSR) in Washington, D.C. He also worked for two summers in the field with inner city youths who were brought together in a camp that sought to build cooperation skills. During the second summer (1967), Crews was in the woods camping and canoeing with his group when the Newark riots broke out. Although he was able to maintain the peace within his group, who learned of the riots via transistor radio, Crews decided that social work was not for him.

During his senior year at Maryland, Crews took a course in comparative neuroanatomy under William Hodos. He was captivated and was able to obtain a position as a research assistant in the Department of Experimental Psychology at Walter Reed Army Institute of Research working under Michie Vane, Walle Nauta's technician, learning histology and cutting brains.

Crews' path to graduate school began with Bill Hodos telling him about the Institute of Animal Behavior at Rutgers University and his writing a letter of recommendation. Gene Peterson, who had supervised Crews during his time at the BSSR, assisted Crews with his letter to Daniel Lehrman, his eventual supervisor. Crews was brought to Newark for a memorable interview, in which Crews was bemoaning his low GRE scores, whereupon Lehrman reached into a drawer and pulled out his undergraduate transcript showing his spotty academic record.

Crews did not want to work on ring doves, Lehrman's model species in the lab. Rather, Crews undertook work in anolis lizards (*Anolis carolinensis*; Fig. 35.2),

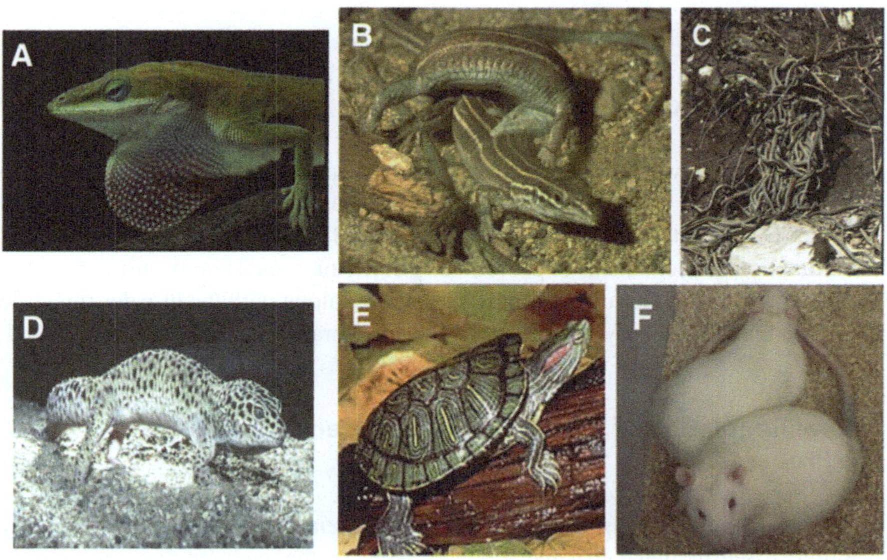

Fig. 35.2 Photographs of some of the species used by David Crews in his research. (**a**) Green anole, *A. carolinensis*. (**b**) The whiptail lizard, *C. uniparens*, shown in pseudocopulation between two females. (**c**) Red-sided garter snake (*T.s. parietalis*) mating ball. (**d**) Leopard gecko, *E. macularius*. (**e**) Red-eared slider turtle, *T. scripta*. (**f**) Laboratory rat, *R. norvegicus*

inspired by work done at the American Museum of Natural History by G. Kingsley Noble. Crews' work demonstrated first, similar to previous studies in the ring dove, that courtship behavior in the male stimulated female ovarian growth in the spring. Second, he discovered that females also influenced the male physiological state. Third, females viewing aggression between males failed to grow their ovaries: this was the first experimental demonstration of this phenomenon.

Tragically, Danny Lehrman died in 1972. Jay Rosenblatt and Colin Beer stepped up to serve as his advisors for the last year in graduate school. Ultimately, Crews received his PhD in 1973. He joined Paul Licht's lab at UC-Berkeley where he was an NSF-supported postdoctoral fellow, working on steroid biochemistry and miniaturizing radioimmunoassays for in vitro studies in lower vertebrates, as well as work on gonadotropins. This was followed by a year of additional postdoctoral training with Ernest William at Harvard, doing fieldwork and behavioral mate selection in anolis lizards.

Crews joined the Psychology and Biology faculty at Harvard as an Assistant Professor, in 1976, was promoted to Associate Professor in 1979, and then moved to the University of Texas at Austin in 1982 until he retired nearly 40 years later in 2019.

This chapter on David Crews may seem unusually personal because it is co-authored by Isaac Miller-Crews (son, and PhD candidate at the University of Texas at Austin), Andrea Gore (spouse, and Professor at the University of Texas at Austin), and Randy Nelson (friend and colleague, and Professor at West Virginia University). We have stuck to the facts as much as possible and ask for indulgence from the reader of any personal perspective.

Research

Nature's Own Experiments

From the start, Crews utilized "nature's own experiments" to expand and challenge our knowledge on how the environment influences reproductive biology. A true comparative and integrative biologist, Crews shined a light across biological levels, from functional mechanisms to evolutionary consequences. Here we bring together a selection of the impactful work Crews produced, along with the broader scientific tenets that guided not only this research but the generations of scientists he inspired.

Neural Regulation of Sexual Behavior: Serendipity Favors the Prepared Mind

Research on the whiptails, *Cnemidophorus uniparens*, started from Crews simply wanting to see firsthand that these lizards were parthenogenetic, i.e., an entirely female species that reproduces asexually. By chance, while observing behavior in anole lizards in the lab, he looked over and saw the whiptails performing behaviors which Crews later coined "pseudosexual behavior." In other words, within the dyad, the first female exhibited the female-typical receptive posture of arched back and raised tail. The second female mounts and then wraps its tail to appose the cloacal areas. Crews coined the term "donut" to refer to this posture (Fig. 35.2). The fact that an all-female parthenogenetic *C. uniparens* continues to display behaviors typical of both sexes from the ancestral, extant whiptail species perfectly illustrated the bipotentiality of feminine and masculine behavior within a single individual. This species thus provided a unique opportunity to gain insight into the evolution of the neuroendocrine substrates that underlie sexual behavior. Although not the first researcher to discover this phenomenon, Crews was primed to realize the potential importance of this observation, which developed into a multi-decade research program on the ontogeny of sexual behavior and the bisexual brain (Crews, 2010, 2012; O'Connell & Crews, 2022).

Crews went on to discover in *C. uniparens* that whether an individual exhibited "male-like" or "female-like" pseudosexual behavior depended upon the reproductive hormonal state and follicle development within the individual (Crews & Fitzgerald, 1980). The lizards engaged in feminine receptive behaviors were in the follicular phase of their cycles, when developing follicles release increasing concentrations of estrogens. Those lizards exhibiting male-like mounting behaviors were postovulatory, when the *corpora lutea* release increasing circulating concentrations of progesterone. These lizards did not have high androgen concentrations, which was surprising since male courtship and mounting in male lizards from the ancestral sexual species are androgen-dependent. Instead, male-like pseudosexual mating behaviors in *C. uniparens* are induced by the high progesterone concentrations, with progesterone subsuming roles normally played by androgens in males (Grassman & Crews, 1986).

Importantly, Crews was able to follow up with studies of how these dual progesterone and estrogen signals are integrated into the brain. This research provided strong support for the concept of the bisexual brain, as the preoptic area was more active during male-typical pseudosexual behavior and the ventromedial hypothalamus more active during female-typical pseudosexual behavior, mirroring what occurs in the separate sexes of sexual species (Rand & Crews, 1994). Additional work showed that steroid hormone receptor genes for estrogens, androgens, and progestins were expressed in both brain regions, providing a direct molecular link from the gonads to the brain (Young et al., 1994). In fact, how estrogen regulates gene expression in the preoptic area (typically considered the hub for masculine behavior) is a major difference between *C. uniparens* and females of the sexual ancestral species (Godwin & Crews, 1995). Crews subsequently targeted multiple neural systems to detail the functional neural networks, across both genes and brain regions, from dopamine (Woolley & Crews, 2004; Dias & Crews, 2008; O'Connell et al., 2012), to nitric oxide (Sanderson et al., 2005; O'Connell et al., 2011), to serotonin (Dias & Crews, 2008).

Crews' comparative perspective challenged dogma that brain sexual differentiation has a female "default" state (Wade & Crews, 1991; Crews, 1993, 2002). Rather, his work in evolution led him to theorize that the female evolved as the fundamental sex, and the evolution of males, androgens, and sexual dimorphisms did not occur until much later (Crews, 2002). This idea was further bolstered by studies showing that developmental exposure to aromatase exposed the genetic ability to produce functional males (Wennstrom & Crews, 1995).

Behavioral Neuroendocrinology: Pursuit of Novelty

For sexual reproduction to occur, animals need to be in the right place, at the right time, and physiologically prepared. Yet, from within these constraints, a diversity of novel regulatory systems has evolved in vertebrates. Crews recognized that the neuroendocrine mechanisms controlling such processes were also subject to evolutionary pressures, leading to his research on the ecology of neuroendocrine mechanisms that control behavior. These ideas arose from his reading an article (Aleksiuk & Lavies, 1975) on the northernmost reptile in North America, the Canadian red-sided garter snake, *Thamnophis sirtalis parietalis* (Fig. 35.2). Increasing temperatures in the spring signal the males to emerge from long hibernation *en masse.* Females emerge singly over the next few weeks, and males will form "mating balls" around a females that can contain up to ~100 snakes. Crews took advantage of this species to discover a dissociation between sexual behavior and gonadal physiology (Crews et al., 1984). In fact, at the times of spring emergence, male red-sided garter snakes have regressed gonads and extremely low testosterone concentrations, yet they initiate vigorous sexual behavior, independent of sex steroid hormones. This was a novel mechanism that challenged traditional endocrine dogma of a functional association between gonadal growth, steroid hormone concentrations, and sexual behavior. Finally, Crews demonstrated that the organizational effects of hormones on sexual behavior were in fact trans-seasonal in this species, with the hormone levels from the previous season necessary to engage in sexual behavior after an intervening hibernation (Crews, 1991). This research also led to the identification, isolation, and synthesis of a novel class of vertebrate pheromones critical to male courtship and mating (Mason et al., 1989).

Temperature-Dependent Sex Determination: Questions, Not Species

Most vertebrates are gonochoric species, having two unique sexes represented in separate individuals that reproduce sexually. The vast majority of experimental work has utilized species with chromosomal sex determination such as mammals with X and Y chromosomes. A portion of the Y chromosome (*Sry*) triggers a cascade of molecular processes that results in the development of testes (males); XX individuals (female) lack *Sry*, yet the same genes exhibit different patterns of activity, causing ovaries to develop. Thus, both male and female sex determination are organized (in contrast to the idea of a female being the "default" sex), shaping processes that ultimately determine the adult sexual phenotype.

By contrast to standard laboratory model systems (e.g., mammals, birds), the factors that trigger sex determination in reptiles, amphibians, and some fish can differ profoundly (Crews, 2002). These species lack sex chromosomes and, instead, depend upon environmental factors to trigger sex determination and sexual differentiation. For example, in temperature-dependent sex determination (TSD), the temperature of the embryo dictates gonadal sex.

In Crews' research, it was ultimately biological questions that matter, and then species were selected that challenge the dogma. For example, Crews was a pioneer in establishing the mechanisms by which TSD occurs and leads to the induction of sexual phenotypes (Ramsey & Crews, 2009; Shoemaker & Crews, 2009; Shoemaker-Daly et al., 2010). Importantly, the pursuit of this question necessitated integration of species new to Crews. Using the red-eared slider turtle, *Trachemys scripta* (Fig. 35.2), Crews uncovered the foundational functional mechanisms by which an external physical factor, in this case temperature, is converted into a physiological signal that ultimately triggers development of one sex or the other. By expanding the comparative framework, Crews realized that the molecular pathways invoked by temperature in reptiles with TSD were homologous to those involved in mammalian sex determination (Crews & Bull, 2009; Shoemaker & Crews, 2009). Taken together, this line of research challenged the then-canonical dogma that there existed a "default" and an "organized" sex, generally considered to be females and males, respectively. Instead, Crews recognized that indeed both sexes are organized, with differences within a sex just as important as the differences between the sexes.

To determine the long-term effects of incubation temperature, Crews selected the leopard gecko, *Eublepharis macularius* (Fig. 35.2), to prove that temperature during incubation determined not only gonadal sex but also the differences observed in phenotypic sexuality – morphology, endocrinology, and reproductive behavior – in adulthood (Gutzke & Crews, 1988; Crews & Groothuis, 2005). This work was followed by several decades of research into how variation in incubation temperature is reflected in different patterns of neural and genetic activity in different brain areas.

Transgenerational Epigenetics: How the Past Can Influence the Future

A major research thread running through Crews' career is that the signals from the external environment are integrated internally and that this has evolutionary implications on the individual and their reproductive success. This philosophy extended to his recent research on endocrine-disrupting chemicals (EDCs). While attending a lecture on how the physiological effects of EDCs can be inherited across multiple generations, Crews recognized the potential long-standing evolutionary ramifications of EDC exposures. This led to his influential contributions to the then burgeoning field of behavioral epigenetics.

Prior research identified reproductive dysfunctions induced by the EDC vinclozolin, a fungicide, in male rats (*Rattus norvegicus;* Fig. 35.2) descended from great-grandmothers exposed to vinclozolin during pregnancy (Anway et al., 2005). F3 generation males descended from vinclozolin-exposed ancestors had transgenerational epigenetic modifications to DNA methylation in the testes. Crews postulated that this transgenerational inheritance of epigenetic traits might alter the attractiveness of modified males to a female conspecific. In a mate preference paradigm, female rats evinced a strong preference for vehicle-descended F3 males over vinclozolin-descended F3 males (Crews et al., 2007). Crucially, this research program was the first to implicate EDC pollution on sexual selection in future generations.

Crews broadened this work to extend into additional behavioral domains, such as sociality, learning, and anxiety-related behavior, to understand how descendants responded to life challenges (Crews et al., 2014; Gillette et al., 2014). Next, Crews targeted the underlying functional mechanisms in genomic activation and neural development that culminated in these changes in behavioral response to stress (Crews et al., 2012; Gillette et al., 2015).

Crews expanded the theoretical framework by which epigenetic modifications are discussed today. He coined the term "germline-dependent" epigenetic modification to describe when epigenetic alterations are incorporated directly into the germline (Crews, 2011). This mechanism of transgenerational inheritance is distinct from "context-dependent" epigenetic which relies on continued exposure each generation, a situation that likely reflects the real-world situation of permanent pollution of the world by EDCs.

Achievements, Awards, and Accomplishments

David Crews received numerous awards, accolades, and honors for his work. He was elected as Fellow to the American Academy of Arts and Sciences, American Association for the Advancement of Science, American Psychological Society, and American Psychological Association. Included in his awards were the Daniel S. Lehrman Lifetime Achievement Award (Society for Behavioral Neuroendocrinology), the University of Texas Research Excellence Award for Best Research Paper, and multiple NIMH Research Scientist and Merit awards, and he held several distinguished lectureships. Toward the end of his career, Crews served as the editor-in-chief of the *Journal of Experimental Zoology-A* for 7 years and retired from that role in January 2021 with a special honorary issue in 2022 (Nelson, 2021, 2022). Crews' lifetime of work was published in over 400 papers and three books.

References

Aleksiuk, M., & Lavies, B. (1975). Manitoba's fantastic snake pits. *National Geographic, 148*, 715–723.

Anway, M. D., Cupp, A. S., Uzumcu, M., & Skinner, M. K. (2005). Epigenetic transgenerational actions of endocrine disruptors and male fertility. *Science, 308*, 1466–1469.

Crews, D. (1991). Trans-seasonal action of androgen in the control of spring courtship behavior in male red-sided garter snakes. *Proceedings of the National Academy of Sciences of the United States of America, 88*(9), 3545–3548.

Crews, D. (1993). The organizational concept and vertebrates without sex chromosomes. *Brain, Behavior and Evolution, 42*(4–5), 202–214.

Crews, D. (2002). Diversity and evolution of neuroendocrine mechanisms underlying reproductive behavior. In J. Becker, S. M. Breedlove, D. Crews, & M. McCarthy (Eds.), *Behavioral endocrinology* (2nd ed., pp. 223–288). MIT Press.

Crews, D. (2010). Neural control of sexual behavior. In M. D. Breed & J. Moore (Eds.), *Encyclopedia of animal behavior* (Vol. 2, pp. 541–548). Academic Press.

Crews, D. (2011). Epigenetic modifications of brain and behavior: Theory and practice. *Hormones and Behavior, 59*(3), 393–398.

Crews, D. (2012). The (bi)sexual brain. *EMBO Reports, 13*, 779–884.

Crews, D., & Fitzgerald, K. T. (1980). "Sexual" behavior in parthenogenetic lizards (Cnemidophorus). *Proceedings of the National Academy of Sciences of the United States of America, 77*(1), 499–502.

Crews, D., & Groothuis, T. (2005). Tinbergen's fourth question, ontogeny: sexual and individual differentiation. Animal biology (Leiden, Netherlands), 55(4), 343–370. https://doi.org/10.1163/157075605774841003

Crews, D., & Bull, J. J. (2009). Mode and tempo in environmental sex determination in vertebrates. Seminars in Cell & Developmental Biology, 20(3), 251–255. https://doi.org/10.1016/j.semcdb.2009.02.004

Crews, D., Camazine, B., Diamond, M., Mason, R., Tokarz, R. R., & Garstka, W. R. (1984). Hormonal independence of courtship behavior in the male garter snake. *Hormones and Behavior, 18*(1), 29–41.

Crews, D., Gore, A. C., Hsu, T. S., Dangleben, N. L., Spinetta, M., Schallert, T., Anway, M. D., & Skinner, M. K. (2007). Transgenerational epigenetic imprints on mate preference. *Proceedings of the National Academy of Sciences of the United States of America, 104*(14), 5942–5946.

Crews, D., Gillette, R., Scarpino, S. V., Manikkam, M., Savenkova, M. I., & Skinner, M. K. (2012). Epigenetic transgenerational inheritance of altered stress responses. *Proceedings of the National Academy of Sciences of the United States of America, 109*(23), 9143–9148.

Crews, D., Gillette, R., Miller-Crews, I., Gore, A. C., & Skinner, M. K. (2014). Nature, nurture and epigenetics. *Molecular and Cellular Endocrinology, 398*(1–2), 42–52.

Dias, B. G., & Crews, D. (2008). Regulation of pseudosexual behavior in the parthenogenetic whiptail lizard, Cnemidophorus uniparens. *Endocrinology, 149*(9), 4622–4631.

Gillette, R., Miller-Crews, I., Nilsson, E. E., Skinner, M. K., Gore, A. C., & Crews, D. (2014). Sexually dimorphic effects of ancestral exposure to vinclozolin on stress reactivity in rats. *Endocrinology, 155*(10), 3853–3866.

Gillette, R., Miller-Crews, I., Skinner, M. K., & Crews, D. (2015). Distinct actions of ancestral vinclozolin and juvenile stress on neural gene expression in the male rat. *Frontiers in Genetics, 6*, 56.

Godwin, J., & Crews, D. (1995). Sex differences in estrogen and progesterone receptor messenger ribonucleic acid regulation in the brain of little striped whiptail lizards. *Neuroendocrinology, 62*(3), 293–300.

Grassman, M., & Crews, D. (1986). Progesterone induction of pseudocopulatory behavior and stimulus-response complementarity in an all-female lizard species. *Hormones and Behavior, 20*(3), 327–335.

Gutzke, W. H., & Crews, D. (1988). Embryonic temperature determines adult sexuality in a reptile. *Nature, 332*(6167), 832–834.

Mason, R. T., Fales, H. M., Jones, T. H., Pannell, L. K., Chinn, J. W., & Crews, D. (1989). Sex pheromones in snakes. *Science, 245*(4915), 290–293.

Nelson, R. J. (2021). Inaugural Editorial. Journal of Experimental Zoology. Part A, Ecological and Integrative Physiology, 335(4), 385. https://doi.org/10.1002/jez.2455

Nelson, R. J. (2022). Introduction to the special issue honoring the career of David Crews. Journal of Experimental Zoology. Part A, Ecological and Integrative Physiology, 337(1), 5–6. https://doi.org/10.1002/jez.2545

O'Connell, L. A., & Crews, D. (2022). Evolutionary insights into sexual behavior from whiptail lizards. *Journal of Experimental Zoology Part A: Ecological and Integrative Physiology, 337*(1), 88–98.

O'Connell, L. A., Matthews, B. J., & Crews, D. (2011). Neuronal nitric oxide synthase as a substrate for the evolution of pseudosexual behaviour in a parthenogenetic whiptail lizard. *Journal of Neuroendocrinology, 23*(3), 244–253.

O'Connell, L. A., Mitchell, M. M., Hofmann, H. A., & Crews, D. (2012). Androgens coordinate neurotransmitter-related gene expression in male whiptail lizards. *Genes, Brain, and Behavior, 11*(7), 813–818.

Ramsey, M., & Crews, D. (2009). Steroid signaling, temperature-dependent sex determination – Reviewing the evidence for early action of estrogen during ovarian determination in turtles. *Seminars in Cell & Developmental Biology, 20*, 283–292.

Rand, M. S., & Crews, D. (1994). The bisexual brain: Sex behavior differences and sex differences in parthenogenetic and sexual lizards. *Brain Research, 663*(1), 163–167.

Sanderson, N. S. R., Weissler, E., & Crews, C. (2005). The nitric oxide synthase inhibitor L-NAME suppresses androgen-induced male-like pseudocopulatory behavior in whiptail lizards. *Brain Research, 1052*(2), 236–239.

Shoemaker, C. M., & Crews, D. (2009). Analyzing the coordinated gene network underlying temperature-dependent sex determination in reptiles. *Seminars in Cell & Developmental Biology, 20*(3), 293–303.

Shoemaker-Daly, C. M., Jackson, K., Yatsu, R., Matsumoto, R., & Crews, D. (2010). The genetic network underlying temperature-dependent sex determination is endogenously regulated by temperature in isolated cultured *Trachemys scripta* gonads. *Developmental Dynamics, 239*, 1061–1075.

Wade, J., & Crews, D. (1991). The relationship between reproductive state and "sexually" dimorphic brain areas in sexually reproducing and parthenogenetic whiptail lizards. *The Journal of Comparative Neurology, 309*(4), 507–514.

Wennstrom, K. L., & Crews, D. (1995). Making males from females: The effects of aromatase inhibitors on a parthenogenetic species of whiptail lizard. *General and Comparative Endocrinology, 99*(3), 316–322.

Woolley, S. C., & Crews, D. (2004). Species differences in the regulation of tyrosine hydroxylase in Cnemidophorus whiptail lizards. *Journal of Neurobiology, 60*(3), 360–368.

Young, L. J., Lopreato, G. F., Horan, K., & Crews, D. (1994). Cloning and in situ hybridization analysis of estrogen receptor, progesterone receptor, and androgen receptor expression in the brain of whiptail lizards (Cnemidophorus uniparens and C. inornatus). *The Journal of Comparative Neurology, 347*(2), 288–300.

Martha K. McClintock

36

Gretchen L. Hermes

Abstract

Martha Kent McClintock was born on 22 February 1947 in Pasadena, California. She obtained her bachelor's degree from Wellesley College in 1969. She produced what is likely the most famous honors thesis in history, which was ultimately published in *Nature* in 1971, describing synchronization and suppression of menstrual cycles of her cohorts living in a Wellesley dormitory. McClintock received her PhD from the University of Pennsylvania in 1976 working with Norman Adler; her doctoral dissertation described the effects of social and behavioral regulation of neuroendocrine function in the mating system of *Rattus norvegicus*. She joined the Department of Psychology at the University of Chicago where she remained for her entire career. Among other elements, her work describes the role of androstenedione in human sweat and how it could influence physiological and psychological responses. McClintock also reported the importance of "paced" mating in reproductive success of female rats; she was committed to studying reproduction in seminatural spaces to fully understand behavior-neuroendocrinology interactions. In 1999, she founded the Institute for Mind and Biology at the University of Chicago and in 2003 helped start the Center for Interdisciplinary Health Disparities Research to understand why Black women have higher mortality from breast cancer than white women. She is a member of the American Academy of Arts and Sciences and the Institute of Medicine. Taken together her work was at the interaction among behavior, reproductive neuroendocrinology, and immunology.

G. L. Hermes (✉)
Department of Psychiatry, Yale School of Medicine, New Haven, CT, USA
e-mail: gretchen.hermes@yale.edu

R. J. Nelson, Z. M. Weil (eds.), *Biographical History of Behavioral Neuroendocrinology*, https://doi.org/10.1007/978-3-031-12970-4_36

Keywords
Menstrual synchrony · Suppression of reproductive cycles · Androstenedione · Breast cancer · Seminatural mating tests · Rats

Driven by curiosity despite my self-consciousness, I mention[ed] that the same thing happens in humans. Didn't they know that? All of them being male, they didn't. In fact, I got the impression that they thought it was ridiculous...– Martha McClintock, Chicago Mesic, (1996). Recalling the first time she proposed the existence of menstrual synchrony

The Roots of Creativity and Curiosity

Rupert Maclaurin, an economics professor at MIT, had a vision. He wanted to build low-cost housing for young families – especially at his school and neighboring Harvard – who were being priced out of the market in the post-VJ Day exhalation economy. Maclaurin was a New Zealand-born economist in what was then a university entirely focused on engineering. In the prior decade, he had begun a movement that would ultimately transform MIT's economics department from an academic backwater focused solely on the needs of the school's engineering students to an economics powerhouse today. His housing project, auspiciously located in Concord, Massachusetts, was called Conantum, derived from an old Iroquois word meaning "morning food," the most important sustenance of the day ahead. It originally consisted of more than one hundred homes, carefully sited to preserve as much of the natural environment as possible, all connected to each other and the development's amenities, including a fully-fledged cooperative, by an extensive trail system. Tragically, as his creation was coming together, Rupert Maclaurin took his life in 1959. But by then a child on the cusp of adolescence, who would go on to create, participate in, and study community after community, many multidisciplinary, dedicated to innovation, and woven together with trails, had spent the better part of her childhood immersed in what he left behind.

Along the way, Martha McClintock would combine animal research, a broad array of tools, unconventional testing methods, sophisticated data analyses, creative syntheses, and courage to forge and reinforce connections between psychology and sociobiology, elucidate the psychosomatics of reproduction, illuminate the dynamics of previously poorly understood forms of human communication – as well as the impact of social neglect on morbidity and mortality, especially during critical periods of development – and explicate the role of downward causation in the sexuality of women, the widespread health disparities experienced by Black Americans, and other phenomena.

Martha K. McClintock was born on 22 February 1947, in Pasadena, California, the first of four children, and only daughter, of Frank and Mary McClintock (Fig. 36.1). Her family's roots were in New England, however, and when her father, a mechanical engineer, secured a position at MIT, they returned, fortuitously at the same time Conantum was springing to life.

Fig. 36.1 Martha McClintock, photo taken in 2021

It has been said of many life stories that they are so incredible, the dream factory fifteen miles down the road from Pasadena would reject a faithful script as not believable. And certainly any tale that ends with a young woman who had (a) been informed by the guidance department at her high school she was ideally suited to take up a career as "a bronco buster or cherry picker," (b) instead gaining acceptance to and enrolling in a prestigious college, conducting research in her college dormitory with her fellow residents as its subjects, (c) resulting in conclusions that shook the foundations of what was understood about the interaction of physiology and behavior, (d) work published in *Nature* when she was *23, 3 years before completing her doctorate*, would seem to fall into this category (McClintock, 1971).

But the screenwriter who understands how paradigm-shifting scientists are made, and make themselves in turn, would find plenty to work with, experiences that helped set her on a course that extended far beyond *Menstrual Synchrony and Suppression*, and uniquely prepared her to fulfill her destiny. In the end, the only miracle involved was the seemingly biological, organic way in which the required inputs came together and built upon one another, and really, that was no miracle, either; it was McClintock, the precociously keen observer and questioner absorbing everything around her – though those qualities had their sources too.

There was her father, who would treat her as a peer in an era when sexism in science, as outrageous as it continues to be, was far worse. Women could not even apply for undergraduate admission to schools like Harvard, Yale, and Princeton until years after McClintock began her studies. Frank McClintock's work focused on mechanical stress and strain, which fundamentally means testing and pressing against boundaries, and he would often conduct related experiments with his children. Together they would explore what makes metal break, for example, by testing how many times they could bend a paper clip before it breaks, whether they could put the pieces back together, and if so, how its properties differed from its original form when reconstructed.

There was her mother, who had no advanced degrees, but crossed credentialing boundaries as a serious amateur naturalist, in the spirit of the original meaning of those words, someone dedicated to the study of plants and animals in their natural environments, by observation rather than experiment, whose expertise is a product of love, not a desire for money. Mary McClintock was always questioning, always wondering of anything and everything "why do we think that is?" She inculcated a constant curiosity and inquisitiveness in her daughter, a profound sense of awe and appreciation of the beauty of nature that went far beyond the typical tropes of sunsets and the like.

With both parents, she spent time hiking woods and mountains without trails, through which she not only further developed her keen sense of observation but a scientist's instincts for "where you are in an investigation, a gut sense of where to go, and how you're going to get there," McClintock recalls. It was an experience that "just made scientific thinking embodied and natural, made exploring, taking risks, very transcendental." One could also see the influence of both parents coming together in the time she spent as a child – and throughout her career– "gently poking at nature in its context in order to learn more."

Most of her childhood trailblazing took place in the mountains of the American West, but the environments of Conantum and Concord were also central to who she became. Under her mother's tutelage and the influence of the nature, architecture, design, and culture of what was officially known as Kalmia Woods, she spent much of her childhood outdoors. As she looks back on it: "Many afternoons and weekends were spent with friends playing in the same woods, fields, and rivers that Thoreau had explored…because [we] shared interests and were at similar life stages, many people in Conantum acted like an extended family, providing a sense of community that has endured."

The influence of Concord was more complex. On the one hand, the reaction of the townspeople, particularly to her Black, Jewish, and other friends they "didn't think belonged," created in her a keen sense of social justice that would manifest repeatedly in her work. On the other hand, there were the Concord public schools which, like Massachusetts public schools in general at the time (and long after), rivaled the best in the world, and no doubt contributed to a desire to become a K-12 teacher that persisted even after she graduated from college. In the end, her undergraduate classes, graduate students, and, more generally, the communities she would go on to create would become the beneficiaries of the calling she felt instead.

One school program in particular made a lasting impression on her. In her middle school years, she and her class were chosen to participate in the Harvard School-University Program for Research and Development project run by education professor Don Oliver and his students. Oliver simultaneously reveled in debate yet revered community, always grounding his work in the interactions of everyday life. The work of the School-University Program ultimately led to the development of a social studies curriculum for middle and high schools designed to stimulate discussion of social issues that had a significant influence on how social studies is taught in K-12 schools.

In the Concord iteration of the developing program, the focus was on critical thinking. Using Supreme Court cases and other primary source materials, Martha and her junior high classmates learned about inductive and deductive reasoning, what constitutes data, and proof, and more. As she described it via an APA editor in 1983, She and her classmates "had the thrill of finding out that supposed facts had been based on value judgments and created by faulty logic. They discovered that the world was not black and white and took great delight, as only adolescents can, in pointing this out to their other teachers and parents. McClintock was quite proud to be singled out by a substitute teacher in a letter to the John Birch Society newspaper as a prime example of the degeneracy of American youth, corrupted by liberal teaching methods."

It was the beginning of what McClintock would call her "oppositional" period when she became a "holy terror." It also helped build in her both the instinct and skills necessary to puncture conventional wisdom, both of which she would need to call on often in the journey ahead. Still, she recalls, she ultimately decided to continue her education at Wellesley at the urging of her mother, and for reasons it would be hard to describe as anything more than conventional at the time.

As Martha recalls, Mary cajoled, "Whether you marry or not, you'll depend on women and should go somewhere you'll learn how to do that. She said that with a good education, I'd have something interesting to think about while folding diapers, which has also proven true. I had no plans when I arrived there, beyond maybe being a first-grade teacher. I was fourth-generation Wellesley, and thought the women who'd gone before me were what I'd grow up to be–a well-educated mother and community volunteer. My grandmother got engaged at Wellesley and knew she would move to western Massachusetts, where my grandfather farmed. She took astronomy because she figured in the country she'd see the stars and could teach her children their names" (Horn, 1999).

Once there, however, her rich childhood experiences, irrepressible brilliance, and the more fundamental, distinctive elements of her nature made her, in some ways, a young woman in a hurry. She first majored in molecular biology, and then philosophy, only to abandon them when she realized they likely "required years of highly structured training before there was an opportunity for independent work or thought." She ultimately settled on psychology as her focus "because it was in the center of a widening spectrum of interests and had the fewest number of required courses" (Fleishman, 1983).

In those days, Wellesley was a relative oasis for aspiring women, a place that gave McClintock and her classmates "the luxury of learning for its own sake and somehow managed to give them a subliminal sense that they could undertake and accomplish what they chose," albeit, in many cases, "a sense that often was not realized until long after graduation." McClintock studied anthropology with Annemarie Shimony, experimental psychology with Laurel Furumoto and Claire Zimmerman, and biology with Elizabeth Conant, Louise Palmer Wilson, and Harriet Creighton. From Nia Janis, with whom she was studying the influence of photography on nineteenth-century painting, she "learned by her teacher's example that a lecture could be an art form in itself" (Fleishman, 1983). Perhaps not coincidentally, during her first two summers, McClintock worked as a photochemistry research technician at Polaroid for the redoubtable Vivian Walworth, a force of nature who brooked no inequities. McClintock found that lab work was "technically satisfying."

But it was an experience the following summer that truly proved pivotal. She had been taken on as an NSF trainee at Jackson Laboratory in Bar Harbor, the renowned institution that first confirmed that cancer is a genetic disorder, discovered stem cells, performed the first bone marrow transplants, did the foundational research making organ transplants possible, has been involved in work leading to 35 Nobel Prizes, and most portentously, literally wrote both the book and the genomic database on the laboratory mouse. For all that, the lab work McClintock was assigned was not particularly exciting – she was tasked with graveyard shift data collection from the animals on the circadian dynamics of their stress response – but she reports, "the atmosphere was exciting" and when not on task, she was free to roam Acadia National Park, a naturalist's paradise, which was literally across the street.

As it happened, she also regularly took lunch with a group of researchers deeply involved with pheromone research. Most of what she recalls from these conversations, she's noted dryly, are "the best places to go trout fishing." One day, however, the discussion was about how these chemical signals were causing groups of their female mice to all ovulate at the same time. What happened next was life changing. As McClintock recounts: "Driven by curiosity despite my self-consciousness, I mention[ed] that the same thing happens in humans. Didn't they know that? All of them being male, they didn't. In fact, I got the impression that they thought it was ridiculous. But they had the courtesy to frame their skepticism as a scientific question: 'What is your proof?' I said it was what happened in my dormitory. And they said unless you address it scientifically, that evidence is worthless" (Mesic, 1996).

And for most young would-be scientists, that would have been that. And it well could have been, but for three things. First, McClintock was McClintock. Second, in her senior year, she studied endocrinology with Virginia Fiske, whose "approach to science and confidence in her students encouraged many [young women] to continue in related fields." Fiske was a pioneering researcher on the pineal gland, a tiny pea-shaped gland in the brain that had itself been scoffed at in a way. Before Fiske discovered in 1961 that exposing laboratory rats to bright light over an extended period of time caused slight decreases in the size of the pineal gland – despite its location in the brain's dark recesses – most scientists thought it had no function at all, that it was vestigial perhaps, like the appendix (which, we are now realizing, may also have an important function after all). Close observation and gentle probing, against conventional wisdom, of something otherwise overlooked, had begun to yield unthought-of secrets (today, we know this gland is the main source of natural melatonin in our bodies); how could this not resonate with, and inspire, the student of nature from Conantum, the daughter of Frank and Mary McClintock? When Fiske invited the class to participate in a new pilot study on the role of the pineal gland in the circadian rhythm of cortisol production, McClintock committed to conduct and write a literature review on the entrainment of endocrine rhythms by light cycles.

Third, her advisor in the Psychology Department, Patricia Sampson, encouraged her to pick up the gauntlet of the Jackson Lab scientists instead of writing that lit review, wisely intuiting it would be a lot more interesting. All 135 fellow students in her dormitory – a healthy N – agreed to participate and dutifully recorded the particulars of their cycles. Not only did the data confirm that the cycles of roommates and friends both were and became more synchronous with each other than with others, but women who had little or no contact with men cycled less frequently, just as the mice studied at the Jackson Lab, in which contact with males precipitates ovulation. McClintock wrote up these results as her senior thesis, and irrespective of graduating with High Honors in her field – despite what she describes as an "undistinguished grade average" – *that, again, could have been that.* Especially when her first instinct on graduation was to persist with her plans to become a K-12 teacher and apply for teacher certification, via various masters-level programs, while teaching marine and field biology to children on Nantucket. How many undergraduate senior theses, after all, ever turn into anything more?

But she also applied to PhD programs because she loved research, and one in particular spoke to her protean nature: an interdisciplinary doctoral degree combining biology, psychology, and anthropology offered by Harvard. She was accepted and took the 20-minute trip from Wellesley to Cambridge that at first seemed more like a journey between continents. When she arrived, she learned that the interdisciplinary program she had applied for had disappeared. Then, upon the departure of key faculty members, so did the facilities she was counting on to do physiological psychology research. Leaving her at loose ends, in what could only be described as an alien landscape, one where she was considered the alien, the "no-woman's land" of Harvard. As she recounts: "Title Nine hadn't happened yet. I wasn't allowed to eat in the faculty club with my chairman or to use the stacks in some libraries or the squash courts. I was explicitly told by faculty that they admitted me only because the field needed people to do parametric studies, which are essentially the housework of the discipline, and that women kept nice, neat lab notebooks. There were no women faculty, which was demoralizing" (Horn, 1999).

It was in this context that she met, in her peregrinations, as any naturalist at heart roaming the Yard would seem destined to, E.O. Wilson, then a Professor of Zoology, and Curator of Entomology at the Museum of Comparative Zoology.

Wilson was famously shy and southern genteel, but fiercely determined and "roused by the amphetamine of ambition." Like McClintock, he moved easily and comfortably between animal and human models, with a naturalist's understanding that authenticity, not anthropomorphism, unites them. Like McClintock, he had a naturalist's strong predilection for holistic rather than atomistic views of the subject at hand – he had just published what's now widely considered one of the most important volumes in ecology, *The Theory of Island Biogeography*. And he was well-known for his work on ants, particularly, at the time, his unlocking and classification, with mathematician William Bossert, of their pheromone-based chemical communication system. His eureka moment in that work was a gentle probe of nature (for all but one ant) that would not have been out of place in the fields or woods of Conantum: "I took these ants, they're just like a grain of salt. And I could tease apart the abdomen of a fire ant and then remove different glands from the rear abdomen…And this was an exciting moment. I came to a little gland, finally teased it loose, took the contents of it, making an artificial trail away from the nest and the ant nest exploded. They came pouring out. That was it" (WBUR News & Wire Services, 2021).

For all of these reasons, he was quite enthusiastic about McClintock's senior thesis when she shared it with him and urged her to submit it for publication to *Nature.* This work, which began as a Wellesley undergraduate, marked the commencement of McClintock's work on the social regulation of reproductive function and, more broadly, the role of downward causation and the impact and influence of higher levels of organization – the organism, the community, the biome, and the sociopolitical and physical environment – on behavior and gene expression, from breast cancer in marginalized communities to processes of aging and the lethal effects of social isolation. Subsequently, she proposed that menstrual synchrony was caused by two opposing pheromones in a coupled oscillator system: one that shortens cycles and one that lengthens them (McClintock, 1971).

Doctorate and Norman Adler at the University of Pennsylvania

The celebrity that accompanied McClintock's precipitous and groundbreaking publication did nothing to alter the reality that the PhD program she had come to Cambridge to complete had collapsed before she even arrived, leaving, in its wake, a Potemkin version of the graduate experience she had hoped for. She cast about for another program and landed on another prestigious opportunity at the University of Pennsylvania's Department of Psychology. For a young female investigator, it might have appeared to be no more than a lateral move, since UPenn, like Harvard and most of the other Ivies, was still not yet co-educational. But its Department of Psychology graduate program welcomed women and it had Norman Adler, who became her PhD advisor.

Norman Adler was remembered for many things upon his untimely passing at the age of 72: his affinity for corny jokes; his quick and sophisticated wit, often peppered with references that never reached ground level; and his unapologetic passion for bad horror films and *Waiting for Godot*. Also his chameleon-like capacity to adapt his way into any conversation on any topic; his boundless, positive energy; his ability to make a lasting impression even on his worst days, as when he lamented to a student that he had "not done one creative thing" that day; and his love and respect for his students, whom he treated as true collaborators and equals, generous with the resources of his lab and credit for work done, which enabled him, like Niels Bohr in physics, to attract a remarkable collection of students who became collaborators with each other as well, all of which made a deep and lasting impression on McClintock and how she would pursue her life's work going forward.

In Adler, she found someone with whom she could share her top-of-mind interest in one of the many interdisciplinary eddies or swirls in the field, the behavioral control of endocrine function, and the discipline of behavioral endocrinology, to which he introduced her. Adler's spirit of collaboration was contagious at Penn, and it led to unusually strong connections with colleagues in other departments for his department in general, and its graduate students in particular, resulting, in McClintock's case, in especially supportive relationships with biologists W. John Smith, a leading researcher in the area of animal communication, and Ingrid Waldron, whose focus on gender differences in health and mortality was/is also invaluable in McClintock's own work.

It was also at Penn where McClintock's naturalism came into full flower. Her doctoral work concerned the behavioral regulation of neuroendocrine function in the mating system of Norway rats. McClintock was concerned – and convinced – that studies using the specially bred rats used by every laboratory in the world would be of limited, if any, validity at the level of organization and existence that interested her: the level of the organism, the whole rat, which was itself a rebellion against the run of play in a scientific environment that was becoming increasingly and foolishly reductionist, infatuated with genes, molecules, and subatomic particles.

So, McClintock decided to use wild rats instead. But how to procure them? She called on the Philadelphia public health patrol to live-trap 12 individuals for her. This was initially tragicomically disappointing – the first batch was delivered to her dead because the public health authorities could not imagine that she would want such creatures alive.

Once this misunderstanding was resolved, McClintock set about creating a suitably natural environment for her charges. As recounted in Miriam Horn's *Rebels in White Gloves*, she… "contrived for her rats a home built of sticks, rocks, and wire mesh, with trails and places to nest and hide, all monitored by cameras so that even while the rats scurried through heaps of litter and nooks and tubes, they could be constantly surveilled."

By the time she took up her position at the University of Chicago, this approach had become signature. A Natalie Angier *The New York Times* profile observed (Angier, 1995): "One of the enduring cliches about scientists is that they remain forever young at heart: full of childlike wonder and curiosity about the natural world, their eyes as round as pies, their energy infinite, their speech, a variant on the worshipful 'Wow!'"

Of course, as Angier notes "McClintock is no child, and her sense of curiosity is too richly alloyed by years of research, skepticism, point, and counterpoint to fit into any child's brain. Yet one thing about McClintock is distinctly kidlike: she loves to make a mess."

"Most researchers who study animal behavior keep their subjects in standard-issue cages and try to minimize the confounding variables, to zero in on one behavior isolated from all distractions," for example. "McClintock does just the opposite. When she puts animals into an enclosure, she deliberately turns the place into a dump. Showing off the room that has been established for her lab's collection of Norway rats, she declares proudly that it is just the sort of environment where the rodents feel most at home. There is litter all over the floor, and plastic tubes and nooks and tight places for the rats to slither through, and burrows for breeding.

"Some of my students think there should be smashed hubcaps and grain silos in here as well," she said."

As it happens, what may sound like an endearing quirk to the mass media has yielded profound results. For example, for decades it had been believed, based on experiments using laboratory rats, that addictions to substances like cocaine were fairly hopeless. Time and again, researchers would report that given a choice between cocaine and food, rats would press the cocaine lever over and over until they died of starvation, or as a result of the impact of the cocaine itself. But in the late 1970s, shortly after McClintock completed her doctorate, psychologist Bruce Alexander wondered whether these well-established results were merely an artifact of the housing conditions imposed, which, in many cases, consisted of little, if anything, more than a sterile, empty glass or transparent plastic cage, in which they were expected to make existential decisions. Like McClintock, Alexander developed a more naturalistic environment, affectionately known as "Rat Park," as context for similar investigations and found no preference for cocaine over food at all; in fact, in most cases, rats with their housing and social needs attended to typically sampled the cocaine no more than once, out of curiosity, if at all.

In McClintock's own acutely sensitive application of the principles she developed, the impact of naturalism was often initially more subtle, yet even more wide-ranging in its ultimate implications. Before her now-classic experimental work in the mid-1970s, for example, scientists had always studied rat sex by putting one male and one female together in a cage, as if they were heteronormative human beings. And as if performing on cue, the animals acted just like little people, or at least the way people act when subjected to strictures imposed by others with thumbs consciously or unconsciously on the scales, with the males initiating all intimate activity and, as such, appearing to be far more interested in their reproductive duties, QED predictably confirming the human male establishment in its beliefs about the natural state of relations between men and women, with often horrific consequences for the latter. But this behavior, McClintock found, was almost entirely a construct of the wholly artificial environment in which they had been placed. As Horn relates: "By examining hundreds of hours of videotape, often a frame at a time, she discovered that female rats were not at all passive or coy. In fact, it was she who initiated sex by entering a male's personal space. Scientists observing rats in small cages had never witnessed this behavior, because the female was already within that space. The male's response was misunderstood as initiative. Still more intriguing was her discovery that sex was not a private matter between two consenting rat adults but a kind of orgy, with females working as a group to maximize each of their chances at conception, enticing the males and then passing them around" (Horn, 1999).

Not surprisingly, this is not a model scientists have been as eager to ascribe to humans, nor has McClintock, because, it is important to note, her naturalism was always bidirectional – if she was trying to show something about animals, then she wanted the experimental environment to be as natural to the animal as possible; if she was using an animal model to say something about humans, then she wanted to create conditions for the animal that were as close, for the animal, on its own terms, to the human condition she was probing, because while she moves between models with an ease that's one of her key strengths, this same facility has made her acutely sensitive to – and adamant about – what's been widespread carelessness in making these transitions.

The power of her later work on social isolation, for example, owed much to the naturalistic character of the separation imposed. Many prior researchers had tested the deleterious impact of isolation by housing isolate animals such that they could not see, hear, or smell other rats, a circumstance more akin to solitary confinement in prison – for which no one needed an animal model to ascribe negatives – than a faithful model of the epidemic of loneliness, the social deserts in seas of plenty, that mark the widespread day-to-day manifestations of isolation we see all around us today. To more accurately, powerfully, and poignantly recreate such conditions in rat populations, McClintock's isolated animals *could* see, hear, and smell each other; all they were denied, like so many in human society today, was the ability to reach out and *touch* the rats in other cages.

From Doctorate to Residency

There are two kinds of trailblazers in the intellectual realm, and McClintock has been both. The first strikes out for *terra incognita*, with no assurance of *terra* or that it is *cognoscibilis* (knowable). The resistance they encounter is, at least at first, loving, parental or maternal, concerned, not hostile, and, in any case, they're soon beyond the reach or hearing of catcalls or derision. The greatest tests they will ever face are those moments when they realize they're surrounded only by themselves. If they fail, then they are forgotten or forgiven, prodigal upon their return. If they succeed, they are well-armed with new weaponry the inevitable academic opposition has never seen before.

The second kind forges new paths, their own paths, within territory that's already well understood, well posted, and marked by convention, with well-organized and respected forces always at the ready to enforce the rules. These pioneers find themselves facing the cognitive equivalent of building a new street (or "shortcut" as their detractors will call it) through downtown Manhattan on their own say-so, sometimes, conventionally, on the ground or, more often, constructed literally out of thin air, leaping across chasms from building to building – rather than taking the elevator down, walking to the corner, waiting for the light, walking across the street, back up the street, and back into an elevator like everyone else.

A small example of McClintock's trailblazing of the second type is what happened after she completed her doctorate in 1974. She wanted to immerse herself in psychiatry, but did not want to have to traverse the usual 4 years of med school (let alone the premed requirements that then applied) to get there, so, armed with no more than a postgraduate fellowship from the National Institute of Mental Health, she convinced UPenn to take her on as a psychiatry resident, even though she had no medical degree.

The postdoctoral fellowship allowed her to return to her earlier interest, the interaction of hormones and behavior in humans; in the residency, she concentrated on outpatient psychiatry and psychosomatics. Although some viewed her as "leaving the field" when she shifted her attention from typical academic research to medicine's professional development environment – always a dangerous accusation in academia, and not the last time such sentiments would be grumbled – McClintock saw the residency as something akin to fieldwork, and eventually most would be compelled to accept "the field" as being wherever she was and whatever intersected there.

She was fortunate enough to have Lester Luborsky as her mentor in this work, arguably the founding father of psychotherapy research, who would ultimately come to author nine books and over 400 articles in the area. Luborsky was particularly interested in uncovering what we would now call "evidence-based best practices," a task to which he brought a rare capacity for applying scientific sensibilities to the personal processes inherent to the domain, a gift McClintock would find much occasion to develop across multiple fields as well, especially since some of its key components/ingredients, such as his "creative, can-do genius for inventing methods for particular purposes, jury-rigging if necessary," were already well ingrained in her, if not native from the start. What was more surprising to her was the discovery that she thoroughly enjoyed the therapeutic process. Indeed, how the human mind works in the presence of others – consciously and unconsciously – became bedrock in the research and faculty appointments that would soon follow.

Faculty Appointment at the University of Chicago

When it came time to choose an environment where she could thrive, McClintock was ready for a fundamental change. She sought and obtained an appointment at the University of Chicago, in part because Chicago had a long tradition of equal regard for women. "It was one of the first research universities to go coed, a hundred years ago. When I got there, Hannah Gray was president and a third of the faculty were women" (Horn, 1999).

Though even now, she would acknowledge the challenges faced by women and other marginalized groups throughout academia. When asked what advice she would give a young female or BIPOC researcher, she is both positive and steely: "There are good guys – find them – and learn to tolerate anxiety. Pushing across boundaries, swimming upstream, viewing a phenomenon from unconventional perspectives is certainly exhilarating if not addictive. But it is also exhausting and at times frightening unless you like debating for the sake of a fight. For me, the balm has been collaboration with like-minded scholars and the camaraderie of a lab filled with wonderfully diverse students."

Chicago was – and is – also renowned for its emphasis on interdisciplinary thinking and collaboration, as exemplified by the 20+ broadly cross-disciplinary committees embedded throughout its academic divisions, headlined by the Committee on Social Thought, as well as myriad other interdisciplinary programs. It was as if an institution of higher learning had been created for her and others of her inclination. When McClintock joined the faculty, it was in multiple disciplines – psychology, human development, evolution and ecology, and consult liaison (CL) psychiatry – and she ultimately came to hold appointments at four of U of C's integrative peaks, the Committee on Human Development, which, like Social Thought, has evolved into a standalone academic department – Comparative Human Development; the Committee on Evolutionary Biology; the Committee on Neurobiology; and the Committee on Biopsychology, for which she ultimately served as chair.

As she had expanded her reach into medicine at Penn, McClintock's remarkable career intersected in important ways with Daniel X. Freedman's chairmanship of Chicago's Psychiatry Department, which at first blush seems as quixotic as a medical resident who had never attended medical school. Freedman, who had made her hire possible, was a deeply trained and passionate psychopharmacologist. McClintock was a biologically trained psychologist and not a prescriber, which meant her presence on the CL Service was less about molecules and medication and more about observing and listening to the subjective experience of patients. Her appointment represented, in microcosm, Freedman's guiding philosophical vision for the service, grounded in *both* psychopharmacology *and* detailed observation. Not surprisingly, the name of the Department during Freedmen's tenure and beyond at Chicago was the Department of Psychiatry and Behavioral Neuroscience. By McClintock's and others' accounts, Freedman's department was "diverse, lively, with great heterogeneity in training backgrounds, significant collaboration between MDs and PhDs, deeply committed to the life of the mind." The CL Service linked her with a vibrant and versatile fertility clinic, led by Richard L. Landau, Chief of the Section of Endocrinology, and Thomas M. Jones, with clinical interests ranging from preoperative transgender support to fertility to anorexia nervosa.

Similarly, once onboard the Chicago Unlimited, with the help and support of Bernice Neugarten, one of the first scholars to deeply explore human aging, along with another gerontological pioneer, Gunhild Hagestad, McClintock came to view "a life span perspective on almost any problem" as both necessary and of value, and expanded her work on the pheromonal origins of menstrual synchrony to cover a variety of points in the reproductive life span, including the birth cycle and reproductive senescence. Or to put it more colloquially, the scope and findings in her work exploded, and not all of what she found required a keen sense of smell to appreciate.

Among her many research projects were studies of how female rats coordinate their reproductive cycles to allow them to give birth to pups *en masse*, allowing for communal nursing pools and healthier offspring (Mennella et al., 1990). As it turns out, the female rats generate pheromones that either enhance or suppress the fertility of their neighbors. This rodent synchrony of fertility offers an animal model for understanding why and how such group cycling occurs. Her lab has also discovered that when female rats are prevented from interacting communally with other females, they go into the rat's version of menopause comparatively early (LeFevre & McClintock, 1991) and are at heightened risk of breast cancer and other cancers (Cavigelli et al., 2006).

This discovery, in combination with what it led to, is as good an illustration as any of how and why McClintock's insights have come to cover such astonishingly broad swaths of territory in the landscape of who we are – and explains how I came to write this chapter. Having observed that denial of the company of their fellows caused female rats to become more vulnerable, earlier, to a variety of cancers, caused her to wonder if she might be at the precipice of key insights into why social isolation among humans increases one's risk for an early death, building on Lisa Berkman's landmark longitudinal Alameda Study, begun in 1965, which found social isolation as deleterious to health as cigarette smoking (Berkman & Syme, 1979). My own initial research interests emerged from my doctoral work with and for McClintock, in which we sought to extend and expand on Berkman's findings, using McClintock's animal model to probe the fundamental biology and biological mechanisms behind the negative impact of neglect. Using female Sprague-Dawley rats, we found that isolation not only did lead to early incidence and greater rates of breast cancer but also significantly shortened life spans, prematurely aging ovaries, dysregulated stress hormone responses, pro-inflammatory status, cognitive impairment, and runaway tumor growth (Hermes, 2003) (Hermes et al., 2005, 2006, 2009; Hermes & McClintock, 2008). We further hypothesized and showed the increased cancer burden, in particular, was linked to glucocorticoid receptor expression in mammary tissues, which in turn affected anti-apoptotic pathways.

Because McClintock's work extended in so many different directions like this, it is well beyond the scope and limitations of this chapter to do it justice, and we've been compelled, instead, to cite examples only in endeavoring to illuminate the underlying philosophy, themes, esthetic, and techniques that inform her oeuvre. Overall, it includes 160 publications and (most definitely still) counting that have been cited by other scholars more than 14,000 times in their own work.

Becoming a Teacher

Because teaching is a calling, and because it called enough to McClintock as a young woman, however diffidently and stereotypically, to have had a real chance to pull her off her current course and into a K-12 classroom after she graduated from college, it seems inevitable teaching and mentoring, and the impact of same, would have to be given space in considering her career overall, and the rich, collegial Chicago environment of interdisciplinary study merely ensured fertile ground, given her nature, for it to be especially remarkable. With her longtime friend and colleague Susan Goldin-Meadow, currently the Beardsley Ruml Distinguished Service Professor in Psychology, and William Wimsatt, professor emeritus in the Department of Philosophy, the Committee on Conceptual and Historical Studies of Science, and the Committee on Evolutionary Biology at the University of Chicago, she developed a celebrated core curriculum course entitled *Mind*, described as follows in the current course catalog: "Mind explores how our mental states and processes shape our individual behavior but are also profoundly shaped by our social contexts and grounded in our biological nature. By introducing a wide range of phenomena that illustrate the constructive nature of our experience of reality (perceptually, conceptually, affectively, socially, culturally), the lectures engage students in considering fundamental issues about the nature of mind."

Thirty years on, this course remains a cornerstone social science sequence for undergraduates at the U of C. Dr. McClintock was also the recipient of the Faculty Award for Excellence in Graduate Teaching and Mentoring in 1994, recognizing her exemplary graduate teaching, and impact on graduate students, in fields ranging from human development to evolution and ecology, to computational neuroscience and medicine. McClintock's cross-disciplinary contributions are legendary on the Chicago campus. As the preeminent cultural anthropologist, Richard Shweder, Harold H. Swift Distinguished Service Professor of Human Development, noted: "McClintock was the ideal colleague in the interdisciplinary research program that is the Department of Comparative Human Development at the University of Chicago. She knew how to think big and research narrowly. She was a holistic, synthetic, and generous thinker and intellectual who could talk the talk and walk the walk of science, with philosophers and theorists and methodologists of various stripes in biology, psychology, and cultural anthropology. She was a biologist who even bio-phobic, anti-reductionist social scientists could love."

Changing the Landscape of Neuroscience Through Collaborations and Architecture

Beginning in 1990, McClintock was involved with a leading group of scientists supported by the MacArthur Foundation. This group, called the Mind-Body Network, (possibly neurobiology's answer to the Algonquin Group), was comprised of individuals from a wide range of fields, including psychiatry, medicine, cognitive neuroscience, psychophysiology, neurophysiology, neuroimmunology, and the history of science. The group examined questions at the interface of mind, the brain, and behavior that were near and dear to McClintock, and included how the person and the social group actively influence health and disease; how psychology could unite with neuroscience to render a unified view of the workings of the mind in the context of a physical brain; how autonomic, endocrine, and immune pathways link the brain and body; and how the social world affects the brain and body to produce short-term and long-term health outcomes.

Among the scientists participating in the Network was John Cacioppo, who was a co-founder of the field of social neuroscience and became the Blake Professor of Social Psychology at the University of Chicago. Cacioppo was at Ohio State when he joined the MacArthur Network. McClintock recruited him to the Psychology Department at University of Chicago and to his role as a founding member of the Institute for Mind and Biology. David Spiegel, Wilson Professor and Associate Chair of Psychiatry at Stanford University School of Medicine, and also a member of the Mind-Body Network, explored the effects of group therapy on quality of life and length of survival of women with metastatic breast cancer. This esteemed group also included Anne Harrington, Ford Professor of the History of Science at Harvard, who coined the core question "How does the social environment get under the skin?" An important interlocutor and investigator, she curated an exhibit entitled *Emotions and Disease* at the National Library of Medicine as part of her Network involvement (Fee et al., 1997).

McClintock's interface with Cacioppo, Spiegel, Harrington, and others was multifaceted, reflecting her capacity for and interest in interdisciplinary work and community-building among scientists and humanists. She responded to the generativity of the group with a set of important and creative experiments that represented translations and reverse translations of their findings while setting the stage for 20 more years of work in the area of behavioral neuroendocrinology (Fig. 36.1).

Institute for Mind and Biology at the University of Chicago

In the early 1990s, while McClintock was still an active member of the MacArthur Mind-Body Network, she despaired that her animal laboratory and those of fellow members of the Committee on Biopsychology did not meet federal guidelines for animal research and were scattered across campus and even off-campus. With the sponsorship of the University of Chicago, the National Science Foundation awarded her two million dollars toward the construction of a Biopsychological Sciences Building (BPSB) to be used by faculty in the Social and Biological Science Divisions. The BPSB was designed as a scientific instrument to facilitate the integration of social and psychological research with the state of the art in endocrinology, molecular genetics, systems neuroscience, and immunology research. During the following decade, interest in the relationship between mind and biology increased dramatically. As a result of this national and institutional support, the Institute for Mind and Biology (IMB) was established in 1999, with McClintock as the founding director. Its mission was "to enable transdisciplinary research answering fundamental 'big questions' about the mind and its dynamic interactions with the biological systems of the body. Our goal is to do revolutionary science essential for the scientific and medical disciplines seeking to understand the relationships between the mind, the brain and the body, both in health and in disease. We shall study these interactions in their social and cultural contexts, further our understanding of psychology and identify the specific cellular and genetic mechanisms that mediate the reciprocal interactions between the mind and biology…."

Furthermore, "The University of Chicago [will be] the only university with a building designed as a scientific instrument to facilitate the integration of social and psychological research with the best of endocrinology, molecular genetics, systems neuroscience, and immunology."

McClintock was deeply involved in designing the Institute and, to the amazement of many – though probably not those who knew her father or her work – in the supervision of the building process (she would become the first woman in the United States to design and supervise the building of a free-standing institute for the study of neuroscience). Meetings with graduate students were organized around daily construction walkthroughs.

McClintock's hard hat had its own protected space in the lab as work proceeded. In a recent conversation, McClintock described how the $12 million building structure, designed by one of Chicago's most prestigious architectural firms, Holabird & Root, was meant to serve as an instrument for the study of links between the environment and biology (Fig. 36.2).

"At the time, there was no facility at Chicago for studying animal behavior along with biology or neuroscience. The biomedical model of animals – as furry test tubes and organ donors living in centrally ventilated racks – held sway, without any way to control the environment in which animals lived or to observe their behavior. On the Chicago campus, biopsychology labs were scattered from the Museum of Science and Industry, to converted row houses, to Quonset huts and did not meet federal regulations, let alone have the behavioral and biological laboratories contiguous. At the time, open labs were all the rage – multiple investigators were assigned lineal feet of benches in a large warehouse-like space, based on the size of their grants – not driven by the science questions being asked." By contrast, she explained, "My plan for IMB was for wet labs to be housed near the animal housing and behavioral testing rooms to facilitate analyses of hormones, nervous system, immune function, and gene expression, a building with laboratory floors and animal floors. To facilitate scientific collaboration, the idea was to have the wet labs along one corridor, the offices along another, then the computer rooms as a pass through, creating different types of contiguous spaces within an investigator group. Shared communal research rooms and infrastructure were also in the center. The offices for dedicated grants administrators, the building manager, IT support and animal care

Fig. 36.2 *Institute for Mind and Biology* at the University of Chicago. (Designed by Holabird & Root and completed in 1999. ©Martha McClintock)

staff were central and all became deeply supportive of the interdisciplinary research" (McClintock, 2022).

Inaugural members of the Institute included Cacioppo; Leslie Kay, a specialist in olfactory and limbic system neurophysiology; Dario Maestripieri, a primatologist whose work focuses on the phylogenetic history of animal behavior with regard to social and mating systems, life history, social behavior, communication, and cognition; Brian J. Prendergast, a psychologist with a research emphasis on how information about diurnal and seasonal time are imparted to the reproductive and immune systems; and Jill Mateo, a field researcher cataloging and analyzing developmental and biological mechanisms that enhance survival in *Belding's ground squirrels*, a highly social species found in the western United States.

The Institute has expanded to nine faculty members with undergraduate, graduate, and postdoctoral students from psychology, neuroscience, evolutionary biology, and the medical school represented by the diverse research projects within the IMB, ranging from critical periods for songbirds, to the perceptual resolution of color, to how focus of attention is maintained and the trade-offs between it and working memory, to the connection between sense of smell and brain function in older adults, and much more. Moreover, from the time she first saw, in her mind's eye, its gothic colonnade-inspired atrium, she always envisioned it as the cornerstone of an even bigger vision, a natural sciences quad on the campus, as it is today.

Center for Interdisciplinary Health Disparities Research (CIHDR)

With construction of the Institute of Mind and Biology complete and the close of meetings of the MacArthur Network, another opportunity emerged for McClintock to examine how environmental effects get beneath the skin to influence health outcomes. As part of a multi-site federal program to develop centers for the study of population health and health disparities, the Center for Interdisciplinary Health Research at the University of Chicago was initiated in order to explore and understand why Black women in the United States and West Africa develop more aggressive and lethal breast cancers at a younger age than those experienced by White women of Northern European ancestry, drawing on the diverse talents of social workers, psychologists, physicians, and molecular geneticists. McClintock's enthusiasm for this project was instantaneous, having developed a keen sense of social responsibility in Conantum decades before.

McClintock's colleague, Sarah Gehlert, the co-Director of the Health Disparities Center, commented on McClintock's initial engagement with the project: "When my colleagues and I were trying to assemble a team of investigators from across the University of Chicago campus to respond to an RFA to construct a transdisciplinary center to address health disparities, it became clear that we needed a behavioral scientist. Funmi Olopade and I planned how to approach McClintock, whom we both knew slightly. We approached her with a well-thought-out plan, but before we got too far, she shouted "I'm in!" It was the start of a great collaboration."

"I later asked McClintock why she and I "got" the new approach when others were struggling," Gehlert continued, "and she said, 'it is because you and I always take the big picture.' Our ability to carry out this new approach to science by creating new intellectual spaces above and beyond single disciplines (and *with* members of South Side Chicago communities) helped us to capture the complex, multilayered phenomenon of Black-White health disparities in breast cancer.

In the course of the initiative, Gehlert, now Larson Professor of Health, Ethnicity, and Poverty at the University of Southern California, championed community-based participatory research, first conducting focus groups in African-American South Side communities around the University of Chicago to shape CIHDR's research questions, ensure a focus on stressors meaningful to their lived experiences, and incorporate the community's views of breast cancer and barriers to treatment (Masi & Gehlert, 2009) (Dookeran et al., 2010).

This yielded strong parallels between her measurements of the built-environment stressors faced by African-American women – necessitating vigilance around their homes – and the rat model of McClintock's project, which explicated how social isolation increased vigilance, dysregulated the stress hormone response, and increased burden of spontaneous breast cancers with the full range of pathologies seen in women (McClintock et al., 2005). Suzanne Conzen, Simmons Chair in Cancer Research at UT Southwestern, then used a genetically homogenous transgenic mouse model designed to rapidly develop breast cancers with stress hormone receptors, finding that social isolation altered gene expression in the mouse breast cancers, specifically the normal fat cells surrounding the enlarging tumors (Williams et al., 2009). Cancer needs energy and fat can provide it. Women with worse prognosis and high stress hormone activation relapsed quickly (Pan et al., 2011), and stress hormone receptors become more prevalent in the breast cancers of older women (Belova et al., 2009).

Critically, Olopade, Palmer Professor and Director of the Center for Clinical Cancer Genetics at the University of Chicago, ruled out the hypothesis that African-American women in Chicago were more vulnerable to breast cancer mortality simply because they had inherited unique genes from their West African ancestors (Olopade, 2021). Together, these four Center projects posited and demonstrated a revolutionary causal pathway from a stressful social and physical world downward through women's lived experience, to stress hormones, to gene expression changes in breast tissue, and then back up to growth of more virulent cancers.

Menstrual Synchrony, Mate Choice, and Vasanas

Menstrual synchrony and suppression have proven to be neither solely menstrual nor limited to synchrony. In rats, a species without menstruation, airborne compounds shared among a female group alter the timing of ovulation (McClintock, 1978, 1984). Likewise, compounds in fertile women's underarm sweat can either advance or delay the time of ovulation, depending on when in the cycle they are produced (Stern & McClintock, 1998). These menstrual cycle compounds have no effect on the timing of menstruation, which is largely driven by when ovulation occurs during the menstrual cycle. Instead, it alters the pulses of ovulatory hormones that trigger ovulation (Shinohara et al., 2001). Likewise, colleagues found that male rat compounds increase ovulatory hormone pulses in female rats (Rajendren et al., 1990). And likewise, men's underarm sweat is sufficient to regularize women's cycles, also by quickening pulses of ovulatory hormones (Preti et al., 2003). Collectively, this work clearly demonstrated that there are human pheromones regulating ovulation.

But McClintock, never one to rest on laurels, soon realized that groups of women with fertile menstrual cycles, repeated year after year, were an artifact of modernity: good nutrition, few environmental stressors, and, above all, reliable birth control. These were rare conditions during human evolution. E. O. Wilson and Jeanne Altmann, then a Professor of Ecology and Evolution at the University of Chicago, had eloquently argued for thinking about the full reproductive life span, and the conditions that shaped its evolution. This meant considering the timing of ovulation in the context of puberty and menopause, as well as in-between birth cycles of conception, pregnancy, lactation, weaning, and, ultimately, the ovulatory cycles beginning the next birth cycle. Collaborating with Julie Mennella, her colleague on the faculty of the Monell Chemical Senses Center and a large team of graduate students, McClintock reported that compounds collected from lactating women and their breastfeeding infants dramatically disrupted the timing of ovulation (Jacob et al., 2004), as they had in Norway rats (McClintock, 1983; Gudermuth et al., 1984). Not only that, it increased the recipient women's sexual motivation, measured in terms of their everyday mental life, sexual thoughts, and fantasies, not the triggered reflexes of releaser pheromones nor even sexual frequency which is constrained by so many other factors (Bullivant et al., 2004; Spencer et al., 2004).

Could chemical signals go so far as to guide the choice of a mate? It was well established that mice use their olfactory systems to recognize genotypes of potential mates, choosing mates based on the number of alleles that matched their own within specific gene regions encoding immune molecules. Suma Jacob, Segal Professor of Psychiatry at the University of Minnesota, who was then a postdoc in McClintock's lab, described the research: "We had the unique opportunity to study how young Hutterites in South Dakota, an isolated, ethnically homogeneous religious group, managed to avoid higher rates of miscarriage by choosing spouses who are not too genetically similar to themselves. Women preferred the scent of some men over that of other men because of specific genes these women inherited from their father. This genetic mechanism detecting differences in self versus other may [in turn] influence social behaviors such as mammalian mate selection."

The women consciously detected the scents, discriminating differences in a single allele, yet did not recognize that they were from men or even humans. Instead, they described them as any number of everyday natural odors. It was not the perceived identity of the scent that drove the women's choices, but it was its pleasantness. The identical scent would be deeply pleasant to some women and repulsive to others, depending on its genetic match with the perceiver (Jacob et al. 2002a, b).

In all cases, however, the active compounds had not been isolated and synthesized. They were vagile, volatile, and transmitted through contact, but were unidentified. Previously, Jacob had tunneled into the labyrinth of olfactory social communication by starting with known compounds, androstadienone and estratetraenol, which had already been isolated from human sweat, saliva, tears, and urine. Her PhD dissertation with McClintock – and a team of students affectionately called the ORG (Olfactory Research Group) – revealed their psychological and social functions. The compounds did modulate the emotional and attentional states of both women and men, and regulated the autonomic nervous system, though did not trigger specific social behaviors or desires as a releaser pheromone might (Jacob et al. 2002a, b). They did so even in minute nanomolar concentrations, and even when masked with strong clove oil, precluding conscious detection as an odor. Brain imaging revealed widely distributed metabolic effects well beyond the olfactory system and consistent with regulation of emotions and attention (Jacob et al., 2001).

Building on this success, McClintock's next graduate student, Tom Hummer, now in the Department of Psychiatry at Indiana University's School of Medicine, showed androstadienone attunes the mind specifically to emotional – and not social – images by reducing the drive from the visual cortex to the amygdala, increasing downstream activity in prefrontal and orbitofrontal cortex, with no effects on the olfactory system (Hummer & McClintock, 2009).

So, what to call these unconscious social chemosignals that clearly do not exert their function as odors? They regulate ovulation, the autonomic nervous system, and the brain, but are not the releaser pheromones that trigger the kinds of stereotyped behaviors E. O. Wilson had so gleefully demonstrated by painting McClintock's name in ant pheromones on his lab table and releasing ants from a vial to dutifully follow the trail, revealing her name with their squirming bodies. Nor were they just the primer pheromones the mouse pheromone group at Jackson Laboratory had debated as McClintock perched on a pile of boxes in the Ham Station conference room. Always pressing against the strictures of any discipline, she thought perhaps a non-Western culture would already know and understand the kind of concept involved and consulted Wendy Doniger, the Mircea Eliade Distinguished Service Professor of History of Religions at Chicago. Remarkably, they were able to trace a path back through centuries to *vasana*, a medieval *Sanskrit* noun derived from the verb "to perfume," and defined as "an impression left involuntarily on the mind" (McClintock et al., 2001).

Olfaction and Aging

Since her "retirement," McClintock has seemed incapable of following the life cycle she has spent much of her career engaged with, a human who keeps developing, developing, and developing, while others come to rest. (During the writing of this chapter, McClintock, at the age of 75, made her first trip to Antarctica. During the review and editing process, she was on the road to the Platte River to observe the epic migration of a half a million Sandhill Cranes.) With Jayant M. Pinto, Professor of Surgery and the Committee on Molecular Medicine at the University of Chicago and other colleagues that form the National Social Life and Aging Project, she has continued a program of research aimed at understanding the effects of aging on olfaction and vice versa.

This work has produced medically relevant findings while identifying potential underlying mechanisms related to the aging process, showing for the first time that olfactory dysfunction predicts development of depression in older adults (Adams et al., 2018). Additional results suggest that olfactory function is a strong predictor of mortality and a potential leading indicator of slowed cellular regeneration and toxic environmental exposures (Pinto et al., 2017). Following on her career-long interest in human sexuality, this more recent research has demonstrated that olfactory dysfunction in older adults is associated with decreased sexual motivation and emotional satisfaction, potentially due to evolutionarily conserved neurological links between olfaction and sexuality (Zhong et al., 2018). In a sleep study on sleep-disordered breathing (SDB), a highly prevalent but underdiagnosed condition in older adults, McClintock has reported an association with impaired odor identification, suggesting that SDB affects pathways in the central nervous system involving chemosensory processing (McSorley et al., 2017). She and Pinto also found, for the first time, in a nationally representative sample, that home-dwelling older adults with normal cognition and difficulty identifying odors face higher odds of being diagnosed with dementia 5 years later, independent of other significant risk factors (Adams et al., 2018).

Epilogue

When asked what she would like to emphasize in this chapter, McClintock instantly responded: "My work has been about *downward causation*, about how social and mental events shape physical and biological ones. This has been at the heart of my interest in reproductive biology, women's health, and the effects of social isolation and health disparities on development and aging." As this chapter comes to a close, it should be clear that McClintock has embodied the very scientific and humanistic principles that she examined and espoused for five decades – making creative breakthroughs out of every extraordinary environment in which she has been a part.

She has taken environments as opportunities to find new paths to the peaks of her field, bringing countless graduate students with her to gain a vision of the role of the environment in shaping biology, behavior, and development. The implication of her work for her graduate students has probably been one of the more unpredicted, surprising outcomes of her work. Implicit in an understanding of the importance of the environment is a moral and ethical responsibility. Students who have developed programs of research have also managed psychiatric services for the Department of Corrections, addiction and HIV/AIDS treatment programs in communities at the margin, educational training and precepting for underrepresented minorities, and more. On behalf of her graduate students and esteemed colleagues, a deep abiding thanks to McClintock's husband Joel Charrow and their children, Ben and Julia, who shared, across these deeply fulfilling years, with many more to come, their remarkable and adored partner and parent.

References

Adams, D. R., Kern, D. W., Wroblewski, K. E., McClintock, M. K., Dale, W., & Pinto, J. M. (2018). Olfactory dysfunction predicts subsequent dementia in older U.S. adults. *Journal of the American Geriatrics Society, 66*(1), 140–144.

Angier, N. (1995). Scientist at work: Martha K. McClintock: How biology affects behavior and vice versa.

Belova, L., Delgado, B., Kocherginsky, M., Melhem, A., Olopade, O. I., & Conzen, S. D. (2009). Glucocorticoid receptor expression in breast cancer associates with older patient age. *Breast Cancer Research and Treatment, 116*(3), 441–447.

Berkman, L. F., & Syme, S. L. (1979). Social networks, host resistance, and mortality: A nine-year follow-up study of Alameda County residents. *American Journal of Epidemiology, 109*(2), 186–204.

Bullivant, S. B., Sellergren, S. A., Stern, K., Spencer, N. A., Jacob, S., Mennella, J. A., & McClintock, M. K. (2004). Women's sexual experience during the menstrual cycle: Identification of the sexual phase by noninvasive measurement of luteinizing hormone. *Journal of Sex Research, 41*(1), 82–93.

Cavigelli, S. A., Yee, J. R., & McClintock, M. K. (2006). Infant temperament predicts life span in female rats that develop spontaneous tumors. *Hormones and Behavior, 50*(3), 454–462.

Dookeran, K. A., Dignam, J. J., Ferrer, K., Sekosan, M., McCaskill-Stevens, W., & Gehlert, S. (2010). p53 as a marker of prognosis in African-American women with breast cancer. *Annals of Surgical Oncology, 17*(5), 1398–1405.

Fee, E., Sternberg, E., Harrington, A., & Brown, T. (1997). *Emotions and disease: An exhibition at the National Libray of Medicine. U. S. N. L. o. Medicine*. Friends of the National Library of Medicine.

Fleishman, E. A. (1983). Distinguished scientific awards for an early career contribution to psychology. *American Psychologist, 59*(5), 57–60.

Gudermuth, D. F., McClintock, M. K., & Moltz, H. (1984). Suppression of postpartum fertility in pairs of female rats sharing the same nesting environment. *Physiology & Behavior, 33*(2), 257–260.

Hermes, G. L. (2003). *The embodiment of social isolation: Bio-behavioral systems, pre-disease pathways, and patterns of age* (Ph.D.). University of Chicago.

Hermes, G. L., & McClintock, M. K. (2008). Isolation and the timing of mammary gland development, gonadarche, and ovarian senescence: Implications for mammary tumor burden. *Developmental Psychobiology, 50*(4), 353–360.

Hermes, G. L., Jacobs, L. F., & McClintock, M. K. (2005). The sectored foraging field: A novel design to quantify spatial strategies, learning, memory, and emotion. *Neurobiology of Learning and Memory, 84*(1), 69–73.

Hermes, G. L., Rosenthal, L., Montag, A., & McClintock, M. K. (2006). Social isolation and the inflammatory response: Sex differences in the enduring effects of a prior stressor. *American Journal of Physiology. Regulatory, Integrative and Comparative Physiology, 290*(2), R273–R282.

Hermes, G. L., Delgado, B., Tretiakova, M., Cavigelli, S. A., Krausz, T., Conzen, S. D., & McClintock, M. K. (2009). Social isolation dysregulates endocrine and behavioral stress while increasing malignant burden of spontaneous mammary tumors. *Proceedings of the National Academy of Sciences of the United States of America, 106*(52), 22393–22398.

Horn, M. (1999). *Rebels in white gloves: Coming of age with the Wellesley Class of '69*. Crown.

Hummer, T. A., & McClintock, M. K. (2009). Putative human pheromone androstadienone attunes the mind specifically to emotional information. *Hormones and Behavior, 55*(4), 548–559.

Jacob, S., Kinnunen, L. H., Metz, J., Cooper, M., & McClintock, M. K. (2001). Sustained human chemosignal unconsciously alters brain function. *Neuroreport, 12*(11), 2391–2394.

Jacob, S., Garcia, S., Hayreh, D., & McClintock, M. K. (2002a). Psychological effects of musky compounds: Comparison of androstadienone with androstenol and muscone. *Hormones and Behavior, 42*(3), 274–283.
Jacob, S., McClintock, M. K., Zelano, B., & Ober, C. (2002b). Paternally inherited HLA alleles are associated with women's choice of male odor. *Nature Genetics, 30*(2), 175–179.
Jacob, S., Spencer, N. A., Bullivant, S. B., Sellergren, S. A., Mennella, J. A., & McClintock, M. K. (2004). Effects of breastfeeding chemosignals on the human menstrual cycle. *Human Reproduction, 19*(2), 422–429.
LeFevre, J., & McClintock, M. K. (1991). Isolation accelerates reproductive senescence and alters its predictors in female rats. *Hormones and Behavior, 25*(2), 258–272.
Masi, C. M., & Gehlert, S. (2009). Perceptions of breast cancer treatment among African-American women and men: Implications for interventions. *Journal of General Internal Medicine, 24*(3), 408–414.
McClintock, M. K. (1971). Menstrual synchorony and suppression. *Nature, 229*(5282), 244–245.
McClintock, M. K. (1978). Estrous synchrony and its mediation by airborne chemical communication (Rattus norvegicus). *Hormones and Behavior, 10*(3), 264–275.
McClintock, M. (1983). Modulation of the estrous cycle by pheromones from pregnant and lactating rats. *Biology of Reproduction, 28*(4), 823–829.
McClintock, M. K. (1984). Estrous synchrony: Modulation of ovarian cycle length by female pheromones. *Physiology & Behavior, 32*(5), 701–705.
McClintock, M. (2022). *Design plans for the institute of mind and biology*. G. L. Hermes.
McClintock, M. K., Jacob, S., Zelano, B., & Hayreh, D. J. (2001). Pheromones and vasanas: The functions of social chemosignals. *Nebraska Symposium on Motivation, 47*, 75–112.
McClintock, M. K., Conzen, S. D., Gehlert, S., Masi, C., & Olopade, F. (2005). Mammary cancer and social interactions: Identifying multiple environments that regulate gene expression throughout the life span. *The Journals of Gerontology. Series B, Psychological Sciences and Social Sciences, 60*(Special_Issue_1), 32–41.
McSorley, V. E., Pinto, J., Schumm, L. P., Wroblewski, K., Kern, D., McClintock, M., & Lauderdale, D. S. (2017). Sleep and olfaction among older adults. *Neuroepidemiology, 48*(3–4), 147–154.
Mennella, J. A., Blumberg, M. S., McClintock, M. K., & Moltz, H. (1990). Inter-litter competition and communal nursing among Norway rats: Advantages of birth synchrony. *Behavioral Ecology and Sociobiology, 27*, 183–190.
Mesic, Penelope (1996). Life Science, *Chicago Magazine*, March 1996
Olopade, O. I. (2021). Doing better and being better in breast cancer care: An interview with Funmi Olopade. *Disease Models & Mechanisms, 14*(9), dmm049198.
Pan, D., Kocherginsky, M., & Conzen, S. D. (2011). Activation of the glucocorticoid receptor is associated with poor prognosis in estrogen receptor-negative breast cancer. *Cancer Research, 71*(20), 6360–6370.
Pinto, J. M., Wroblewski, K. E., Huisingh-Scheetz, M., Correia, C., Lopez, K. J., Chen, R. C., Kern, D. W., Schumm, P. L., Dale, W., & McClintock, M. K. (2017). Global sensory impairment predicts morbidity and mortality in older U.S. adults. *Journal of the American Geriatrics Society, 65*(12), 2587–2595.
Preti, G., Wysocki, C. J., Barnhart, K. T., Sondheimer, S. J., & Leyden, J. J. (2003). Male axillary extracts contain pheromones that affect pulsatile secretion of luteinizing hormone and mood in women recipients. *Biology of Reproduction, 68*(6), 2107–2113.
Rajendren, G., Dudley, C. A., & Moss, R. L. (1990). Role of the vomeronasal organ in the male-induced enhancement of sexual receptivity in female rats. *Neuroendocrinology, 52*(4), 368–372.
Shinohara, K., Morofushi, M., Funabashi, T., & Kimura, F. (2001). Axillary pheromones modulate pulsatile LH secretion in humans. *Neuroreport, 12*(5), 893–895.
Spencer, N. A., McClintock, M. K., Sellergren, S. A., Bullivant, S., Jacob, S., & Mennella, J. A. (2004). Social chemosignals from breastfeeding women increase sexual motivation. *Hormones and Behavior, 46*(3), 362–370.

Stern, K., & McClintock, M. K. (1998). Regulation of ovulation by human pheromones. *Nature, 392*(6672), 177–179.

Williams, J. B., Pang, D., Delgado, B., Kocherginsky, M., Tretiakova, M., Krausz, T., Pan, D., He, J., McClintock, M. K., & Conzen, S. D. (2009). A model of gene-environment interaction reveals altered mammary gland gene expression and increased tumor growth following social isolation. *Cancer Prevention Research (Philadelphia, Pa.), 2*(10), 850–861.

WBUR News & Wire Services, E. O. (2021). Wilson, pioneering Harvard biologist known as "ant man," dies WBUR.org,WBUR Boston, December 21, 2021, https://www.wbur.org/news/2021/12/27/edward-wilson-sociobiology-pioneerdies

Zhong, S., Pinto, J. M., Wroblewski, K. E., & McClintock, M. K. (2018). Sensory dysfunction and sexuality in the U.S. population of older adults. *The Journal of Sexual Medicine, 15*(4), 502–509.

Kim Wallen

37

Michelle Tomaszycki

Abstract

Kim Wallen is an American behavioral neuroendocrinologist known for his studies on sex differences in behavior in rhesus macaques. His work has emphasized the importance of studying animals in naturalistic contexts to understand organizational and activational effects of hormones. Kim Wallen investigated the Organizational Hypothesis in a large study in which animals received flutamide or physiologically relevant doses of testosterone prenatally. He studied the effects of these hormonal manipulations on a wide variety of complex social behaviors. Wallen's work has provided important information about the timing and sensitivity of social behaviors, hormones, the nervous system, and developmental milestones to prenatal androgens. His work in adult monkeys has challenged the notion that female monkeys are passive during sexual encounters, that females are always motivated to mate, and that androgens regulate female sexual desire. Kim Wallen's findings have elucidated the importance of social factors and context in studying hormone-behavior relationships.

Keywords

Organizational Hypothesis · Sex difference · Social context · Testosterone · Flutamide · Estrus · Puberty · Vocalization · Rough play · Mating

M. Tomaszycki (✉)
Program in Neuroscience, University of Illinois, Urbana-Champaign, Urbana, IL, USA
e-mail: mlt10@illinois.edu

R. J. Nelson, Z. M. Weil (eds.), *Biographical History of Behavioral Neuroendocrinology*, https://doi.org/10.1007/978-3-031-12970-4_37

Kim Wallen was born 3 September 1947 in Asheville, NC. His father was a social psychologist at Black Mountain College. At a young age, his father moved the family out to Estacada, Oregon, to start a communal farm. Wallen lived there until he was 8 years old. Farm life shaped his interest in animal behavior and biology. One of his earliest recollections was at 5 years of age when his dad decided to become a goat herder. Wallen observed newly acquired female goats finding their place in the hierarchy and the fierce aggression that bucks display during rut. He also processed the goats with his father, which allowed him to observe their internal physiology.

As a child, Wallen was a self-proclaimed behaviorist. Inspired by Skinner, whom his dad had known at Harvard, Wallen created his own Skinner box and trained a rat. Unfortunately, the rat eventually escaped and was killed by the family dog. Later, when Wallen went out to Antioch College for his undergraduate studies, he avoided the Psychology department because it was largely focused on behaviorism at the time. However, Antioch College afforded Wallen the opportunity to explore various career options, including photography. The college had a co-op program in which students attended classes for 6 months and worked for 6 months. Elliot Valenstein, who was at Antioch at the time, encouraged Wallen to reach out to Robert Goy at the Oregon Primate Center for his first co-op experience. After hesitating a bit, Wallen contacted Goy and spent the next 6 months conducting research. As we will see, this relationship was enduring and had a major impact on Wallen's career.

Most scientists start by publishing in small journals, but Wallen's first publication was in *Science*. Wallen had a work opportunity at the UCLA Neuroscience Institute with Lyn Clemens and Roger Gorski. They found that, after applying potassium chloride (KCl) to the cortex of ovariectomized female rats treated with estradiol, they immediately went into heat, seemingly proving that progesterone was unnecessary to induce estrus (Clemens et al., 1967). Their conclusions were later disproved (the KCl treatment actually induced adrenal activity and resulted in high progesterone levels), but Wallen learned an early lesson that one can never think of every possible alternative explanation.

Wallen earned a B.A. in Biology from Antioch College in 1970. Once he graduated, he was faced with the prospect of being drafted into the Vietnam War – his lottery number was 5. He therefore applied for alternative service. His first application was to work in photography, but it was determined the service project was likely to further his career (he had taken a photography class during college). Strangely enough, they instead agreed to alternative service in the lab of Bob Goy.

Wallen's older brother, Kurt, was also working with Bob Goy at the time. Kim Wallen did 3 years of research with Goy, following Goy from Oregon to the Wisconsin Regional Primate Center. Wallen was not sure that he wanted to do a PhD, but realized it was necessary after talking with a former director of the primate center at Cayo Santiago. Wallen was admitted to the Neuroscience program in 1973 when it was in its infancy (the program was only in its second year). Wallen earned his PhD from the University of Wisconsin-Madison in 1978.

After completing a year as a postdoctoral fellow at the Wisconsin Regional Primate Research Center, Wallen joined the Psychology department at Emory College (now Emory University) in 1979. David Edwards, a professor at Emory, had encouraged him (via David Goldfoot) to apply. Wallen was ultimately Emory's second choice, but the other candidate (Chris Coe) turned them down to do a postdoc elsewhere. Wallen has been at Emory University ever since. He is presently Samuel Candler Dobbs Professor of Psychology and Behavioral Neuroendocrinology at Emory University. Wallen has been a huge contributor to the field of behavioral neuroendocrinology. He has over 135 publications and has graduated 16 PhDs, all of them women. He has served as President of the Society for Behavioral Neuroendocrinology (1999–2001) and was Editor-in-Chief of Hormones and Behavior from 2012 to 2019.

Effects of Social/Environmental Context on Adult Female Sexual Behavior: Activational Effects on Estrogens

When Wallen arrived at Emory, there was little funding for research (he was given a mere $600 to purchase a modem to dial into the college server). One thing he did have, however, was access to the Yerkes Regional Primate Research Center Field Station, which allows monkeys to live outdoors in socially complex groups of approximately 100 individuals. Wallen's first project was to investigate the effects of space on mating throughout the cycle. At the time that Wallen initiated this research, the prevailing belief was that male rhesus monkeys initiated most sexual encounters and that, because they were able to, females mated continuously throughout the estrous cycle. Thus, sexual behavior in female monkeys was deemed to be uncoupled from hormonal influences. The conceptualization of the passive female was largely born out of Harlow's work on socially isolated monkeys (reviewed in Wallen (1996)). Wallen's first project published at Emory demonstrated that the size of the enclosure affects how much animals mate throughout the cycle, mating more frequently in the luteal phase when they were housed in a small cage than when they were housed in a larger enclosure (Wallen, 1982). Next, Wallen et al. (1984) investigated sexual behavior in monkeys housed under the semi-naturalistic conditions of the Yerkes Field Station and determined that over 90% of sexual encounters were initiated by female monkeys. Furthermore, in these complex social groups, mating was tightly coupled with the estrous cycle (Wallen et al., 1984). One important social factor determining when mating occurs is female-female competition. In lower-ranking females, mating was more tightly linked to the estrous cycle than in higher-ranking females (Wallen, 1990). Thus, Wallen's studies of sexual behavior in adult female rhesus monkeys highlighted the female's role in determining when mating occurs and emphasized the need to consider the social and environmental context of the testing environment.

In complex social groups, mating occurs during female estrus, suggesting that ovarian hormones play a role in female sexual motivation. Historically, however, researchers had assumed that androgens, not estrogens, regulate sexual behavior in females, particularly women (reviewed in Cappelletti and Wallen (2016)). When Wallen ovariectomized females and gave them estrogens or aromatizable androgens, female proceptive behaviors increased, indicating that estrogens are effective in increasing female sexual behavior (Wallen & Goy, 1977). Mating by females treated with a GnRH agonist decreased sharply, but was restored when ovarian function was restored, further suggesting that ovarian hormones are key in regulating sexual behavior (Wallen et al., 1986). However, in these studies, male behavior was a potential confound. To eliminate this confound, Wallen and colleagues (Zehr et al., 1998) tested ovariectomized females during the nonbreeding season. Estradiol-treated ovariectomized females initiated sex more frequently than ovariectomized females who did not receive estradiol or nonpregnant controls (Zehr et al., 1998). Thus, research by Wallen and his colleagues conclusively demonstrated that it is estrogens and not androgens that regulate female rhesus monkey sexual behavior.

Effects of Social Context on Sex Differences in Behavior: Organizational Effects of Androgens

Another important contribution to the field of behavioral neuroendocrinology is Wallen's work on the organizational effects of hormones on social behavior. As stated above, Kim trained with Robert Goy during his undergraduate career, as a postbaccalaureate, as a PhD student, and as a postdoctoral fellow. Goy had been a co-author on the well-known 1959 article that set forth the Organizational Hypothesis, that hormones, through their actions early in development, cause sex differences in behavior (Phoenix et al., 1959). This famous study used female guinea pigs as a model, but Goy applied the Organizational Hypothesis to the study of rhesus monkeys. Goy was perfectly situated to do this, as he had learned about rhesus monkey behaviors while visiting Harry Harlow's lab, and had succeeded Harlow as the director of the Wisconsin Regional Primate Center in 1971.

Wallen first studied the effects of social context on the development of male rhesus monkey behavior. Most experimental approaches to studying behavior controlled and restricted the social environment to eliminate potential confounds. Thus, most of the early work in rhesus monkeys used subjects that had been taken from their mothers at an early age and raised with only minimal access to peers. This contrasts markedly with how free-ranging rhesus monkeys live. Rhesus monkeys live in large social groups, where matrilineally related females have tight social relationships and matrilines are ranked in a strict hierarchy. Each group contains only a couple of adult males and males migrate out of their natal group at sexual maturity.

Wallen sought to examine which social experiences during development are necessary for normal adult male mating behavior. Wallen found that male monkeys who had greater exposure to mothers and peers had the highest rates of mounting behavior and were the most successful at engaging in sexual behavior in adulthood (Goy & Wallen, 1979). Juvenile male rhesus monkeys who had only 30 min per day of access to peers were less likely to exhibit foot-clasp mounting (the posture that allows intromission during mating) than were males who had continuous access to peers (Wallen et al., 1981). Wallen and colleagues concluded that "under (our) circumstances, only alterations in the more intimate and cooperative aspects of social behavior are evident, rather than pronounced deficiencies in all aspects of social interaction" (Wallen et al., 1981, p. 308). Thus, rearing by the mother is an important component of behavioral development, as is exposure to same-aged peers, with animals faring better if they had continuous exposure to both. By controlling the rearing environment, therefore, previous researchers were creating behavioral results that had little connection to normal rhesus monkey behavior.

Wallen's early research demonstrated that social context matters for males, but what about females? At a time when most researchers in scientific fields studied only males (a trend that persists to some extent today), Wallen's contribution to studying both sexes cannot be underestimated. Wallen, at the Yerkes Regional Primate Center Field Station, was in a prime position to study the Organizational Hypothesis in socially living monkeys of both sexes.

Although previous researchers had examined the Organizational Hypothesis in developing female rhesus monkeys (Wallen's was the fourth such study), his studies differed from previous ones in four important ways. First, previous studies had studied these effects in females that were raised with varying degrees of social isolation. In contrast, Wallen studied socially reared monkeys living in stable groups in outdoor enclosures of 75–125 individuals from established matrilines. This allowed Wallen to study organizational effects of hormones in animals that could freely socially interact with mothers, peers, siblings, and related adult females, along with exposure to adult males and unrelated adult females (Wallen, 2005). Such social complexity allowed Wallen and his lab to study social behaviors, like interest in infants and vocalizations, that were not possible in simpler social contexts. Second, Wallen used a lower dose of testosterone, which allowed him to explore the sensitivity of various behaviors to testosterone (Wallen, 2005). Although the dose used in his studies was somewhat of an accident, the dose was one that would be more likely to occur naturally. In contrast, previous studies had used supernormal doses of testosterone, essentially blasting the system with hormones. Third, Wallen was the first to study demasculinization in males by exposing them to flutamide, an androgen receptor antagonist. All in all, Wallen's study had ten different treatment groups: control females, control males, androgen-treated females, flutamide-treated males, as well as flutamide treatments in females and androgen treatments in males (it was difficult to determine sex in utero). Within each treatment group, animals were exposed to one of two 30–35-day treatments, one during the second trimester (considered the "early" treatment) and one during the third trimester (considered the "late" treatment). This huge undertaking allowed Wallen and his colleagues to simultaneously study effects of hormonal manipulations in both sexes during two different sensitive periods of prenatal development. Finally, most studies of development had looked solely at sexual behaviors. Although rough play was included in most studies of primate behavior, it was included mostly because it was thought to be preparation for adult sexual behavior (reviewed in Goy and Wallen (1979)). Wallen expanded our understanding of the organizational effect of hormones by studying a variety of nonsexual behaviors, many of which were not possible to study in impoverished rearing conditions.

The lower dose of testosterone resulted in 83% lower testosterone concentrations in mothers compared to previous studies and resulted in no significant effects on female genital morphology (Herman et al., 2000). Early (second trimester) flutamide treatments in males resulted in incomplete masculinization; some early flutamide males had penises posterior to their scrotum (Herman et al., 2000). Late (third trimester) flutamide males developed a smaller penis in comparison to control males (Herman et al., 2000). Through this study, Wallen and colleagues were able to determine that the threshold of androgen sensitivity was higher than the dose provided and that flutamide had the capacity to demasculinize male genital morphology.

One might expect that these minor effects on genital morphology would translate to no effects on behavior. However, Goy and colleagues had already determined that the effects of hormones on behavioral sex differentiation could be independent of the effects on genital sex differentiation (Goy et al., 1988). While genital differentiation primarily occurs during the second trimester, behavioral differentiation can also be impacted by exogenous androgens administered during the third trimester (Goy et al., 1988). The use of a lower dose of testosterone in Wallen's studies allowed him and his research team the ability not only to test the extent to which effects on behavior are separable from the effects on genital morphology but also to determine whether behavior (central effect) is more sensitive to hormones than genital morphology (peripheral effect).

Classically, studies of the effects of prenatal androgens on behavior have focused on foot-clasp mounting and rough play. This is because rough play is one of the largest sex differences in infant and juvenile monkeys, with males engaging in more rough play than females (Wallen, 1996). Foot-clasp mounting is also more common in developing males than in females, as it prepares males to successfully mate later in life (Goy & Wallen, 1979). Previous studies had found that testosterone administered late in gestation resulted in higher rates of rough play in females and that treatments both early and late in gestation increased mounting (Goy et al., 1988). In Wallen's study, the lower dose of testosterone had no statistically significant effects on rough play or mounting in females (Wallen, 2005). In males, flutamide treatments late in gestation paradoxically increased mounting behavior relative to control males (Wallen, 2005). Thus, Wallen and colleagues demonstrated that lower doses of testosterone were insufficient to significantly increase female rough play and mounting and challenged the assumption that flutamide acts centrally to potently block androgen receptors.

In addition to studying classic behaviors such as rough play and mounting, Wallen's use of monkeys living in complex social groups enabled him and his colleagues to study the organizational effects of hormones on behaviors that had not been previously studied. One such category of behavior is interest in infants. In contrast to play and mounting, juvenile females are more likely to engage with infants than are males: they often touch them, play with them, carry them around, and even "kidnap" them (Herman et al., 2003). This sort of behavior can only be studied in large social groups, in which offspring of multiple ages can interact. However, the only prenatal treatment that altered interest in infants was flutamide administered late in gestation to females, in which interest in infants was decreased, a paradoxical result (Herman et al., 2003).

Vocal behavior is another social behavior that had not yet been studied in the context of prenatal androgens. Rhesus monkeys use many vocalizations to communicate with each other and, in adulthood, females use recruitment screams to solicit help from matrilineal relatives during agonistic encounters, whereas males are unlikely to use these vocalizations. These screams convey information about the severity of aggression and the relative rank of the opponent (Gouzoules et al., 1984). I met with Harold Gouzoules (also at Emory) when I was applying to graduate school, because I was interested in studying sex differences in vocal development. While there, I met Wallen, and we discussed the possibility that the sex differences observed in rhesus monkeys (namely, that females are more vocally voluble than males) might be due to prenatal androgens. Wallen readily agreed to be a co-advisor and invited me to join his study. This demonstrates Wallen's commitment to understanding sex differences broadly defined and also demonstrates his openness to exploring new avenues of research. I had no prior experience in the field of behavioral neuroendocrinology, so he was also really taking a chance on me.

We began by studying sex differences in infant vocalizations. One of the primary contexts in which infants vocalize is when they are separated from, or rejected by, the mother. The fact that the monkeys were housed in large outdoor enclosures made this possible. Females and males both vocalized a lot during maternal separation but used different calls. Females used the harmonically rich "coo," which sounds exactly like "oo" (Tomaszycki et al., 2001). Males, in contrast, used geckers, which are short, broadband (i.e., noisy) calls given in rapid succession, typically accompanied by convulsive jerking of the body (envision a child throwing a tantrum in a store) (Tomaszycki et al., 2001). As one might expect based on the types of calls employed by each sex, mothers responded more to calls by males than by females (Tomaszycki et al., 2001).

Prenatal hormones affected these sex differences. Early androgen treatments completely masculinized female calling behavior, and late androgen females were somewhat masculinized (half of the call features were male-typical) (Tomaszycki et al., 2001). Contrary to expected, early flutamide treatments in females also masculinized vocalizations (Tomaszycki et al., 2001). Also surprising was that masculinization of calling behavior did not affect maternal responsiveness to females (Tomaszycki et al., 2001). Flutamide treatments in males (both early and late) demasculinized vocal behavior to some extent, and, in contrast to findings in females, maternal responsiveness to these individuals was decreased (Tomaszycki et al., 2001).

As monkeys reach 1 year of age, they become more independent from their mother. This makes them vulnerable to aggression. Rhesus monkeys are a highly aggressive species, and therefore referential agonistic screams become important, with adult females using these screams more than adult males. A young monkey must learn not only to produce an acoustically correct version of the scream but also to produce it in the appropriate context. We found no sex differences in proper contextual usage of screams but found that juvenile females produced more adultlike screams than did juvenile males (Tomaszycki et al., 2005). Androgen and flutamide treatments administered to females late in gestation resulted in screams that were less adultlike, but no treatment in males resulted in more adultlike screams (Tomaszycki et al., 2005).

Taken together, we demonstrated that prenatal androgens have organizing effects on rhesus monkey vocal behavior. Furthermore, vocal behavior may have a lower threshold of sensitivity than other behaviors, such as mounting and rough play. Finally, mothers may not be paying attention to the calls themselves, but instead to the genital morphology of the infant, as maternal responsiveness was only altered in the treatment groups that had altered genital morphology (early and late flutamide males).

Wallen and his team continued to study the effects of prenatal androgens in adulthood. During early puberty (3.5 years of age), early flutamide males had increased testosterone and luteinizing hormone compared to control males, and both early and late flutamide treatments resulted in increased testes volume (Herman et al., 2006). Thus, Wallen and colleagues concluded that flutamide likely decreases negative feedback on luteinizing hormone and that flutamide treatments cannot compensate for this effect. These findings suggest that hormonal systems are resilient. These effects on hormone levels in males were no longer apparent a year later. There were also no effects of prenatal hormones on timing of puberty or first ovulation in females (Zehr et al., 2005). Treatments also did not eliminate mating behavior. All males engaged in sexual behavior with few differences according to treatment (Herman et al., 2006), and all females conceived at least once during their first three breeding seasons (Wallen, 2005). However, Wallen determined that social rank impacted the timing of puberty in both sexes, such that higher-ranking animals went through puberty earlier than lower-ranking animals (reviewed in Stephens and Wallen (2013)).

Wallen was also the first to consider sex differences in cognition in rhesus monkeys and the first to test them in outdoor enclosures. Herman and Wallen (2007) conducted tests of spatial navigation using spatial cues and local markers; they found that females outperformed males when relying solely on either one of these cues. Flutamide treatments in both sexes resulted in poor performance, such that more flutamide-treated animals were removed from the study due to this compared to any other group (Herman & Wallen, 2007). There were no effects of androgen treatments (Herman & Wallen, 2007).

Although much of Wallen's research interests were based on social behavior, he also collaborated with researchers at other institutions to understand as much as possible about the effects of prenatal androgens on developing monkeys. With Dennis McFadden, Wallen explored the effects of prenatal hormones on click-evoked otoacoustic emissions in adult male and female rhesus monkeys. They were the first to determine that female rhesus monkeys had stronger click-evoked otoacoustic emissions than did males (McFadden et al., 2006). Flutamide treatments in males, regardless of timing, resulted in more female-like otoacoustic emissions, and androgen treatments in females administered late in gestation resulted in more male-like otoacoustic emissions (McFadden et al., 2006). Late androgen-treated males were hyper-masculinized (McFadden et al., 2006). It is interesting that the two sets of studies focused on vocal/auditory work (this study and the studies on infant/juvenile vocalizations) both showed effects of prenatal androgens in the predicted direction. Taken together, Wallen's findings suggest that vocal and auditory processes, an important component of communication in rhesus monkeys, are more sensitive to the effects of prenatal androgens than are other social behaviors.

Wallen also collaborated with Nancy Forger to understand how these treatments affect underlying circuitry. They focused on Onuf's nucleus, a motoneuron region that innervates the muscles involved in erectile function and ejaculation (Forger et al., 2018). Males had more motoneurons in this region and larger soma sizes in comparison to females (Forger et al., 2018). Males treated early in gestation with flutamide had slightly fewer motoneurons in Onuf's nucleus, and androgen treatments in males had no effects (Forger et al., 2018). This study established a sex difference in Onuf's nucleus and suggests that prenatal androgens play some role in the emergence of this effect.

In sum, Kim Wallen's contributions to our understanding of the Organizational Hypothesis are extensive. His research has provided important information about the relative contributions of prenatal androgens and complex social rearing environment to sex differences in behavior. Wallen considered many types of social behaviors, going beyond many studies that only focus on male-biased and sexual behavior, and including studies that considered the timing of developmental milestones (such as puberty). He explored behaviors not studied before in rhesus monkeys, like spatial navigation and otoacoustic emissions, and effects on underlying circuitry. Wallen also further established that different behaviors have different timing and sensitivities to prenatal androgens. Some behaviors, like rough play and mounting, seem to be less sensitive, whereas vocal behavior and otoacoustic emissions seem to be more sensitive. Finally, through using lower doses of testosterone that had few effects on genital morphology, Wallen was able to further establish that the effects of prenatal androgens on behavior are independent of their effects on genital morphology (Fig. 37.1).

This tremendous project was not without its problems. This work required a large crew of individuals to administer injections 7 days per week, morning (7 am) and afternoon (4 pm), even through the holidays. To administer injections, we had to walk into the large outdoor enclosures with 100 monkeys, identify individual females, and get them to run into the indoor enclosure to receive their injection. Once inside, the females were trained to transfer into a box, and then into a cage, where they would present their leg for the injection. Minutes later, they were returned to the outdoor enclosure. We had the females trained very well, but we also had new adult males from Cayo Santiago. During a training session in 1996, one of the adult males jumped on my back, leaving a large gouge. Rhesus monkeys transmit Herpes B. To monkeys, it is much like Herpes A in humans. When it is transmitted to humans, it is almost always fatal. I was lucky – the monkey who attacked me was not shedding the virus at the time and I did not get sick. To this day, I have a scar. I was a bit scared to go back into the enclosures, but Wallen counseled me to go back in as soon as possible. It was good advice, as I got over my fear rather

Fig. 37.1 Kim Wallen

quickly (and the monkey who attacked me had been removed from the enclosure). However, another member of our lab was not so lucky. Elizabeth (Beth) Griffin began her work as a volunteer undergraduate assistant working with me, but eventually transitioned to a paid position as a technician in Wallen's lab. Annually, we assisted the veterinarians in bringing all animals in our study into the indoor enclosure. Once inside we anesthetized them and checked their health. During one such health check at the end of October 1997, Griffin got something from a monkey (urine or feces) in her eye. Within 2 weeks, she was hospitalized. She died on 10 December 1997. The impact on all of us in the Wallen lab cannot be described. Wallen considered leaving academia and we all reconsidered studying monkeys. But we all came together for mutual support and somehow managed to continue.

Recent Research Contributions to Behavioral Neuroendocrinology

Luckily for the field, Wallen chose to remain in academia and continue research. Griffin's death occurred when the animals were between 1 and 2 years of age. Wallen continued studying these animals into puberty and beyond. More recently, he has focused on two different avenues of research that are especially noteworthy in terms of their contributions to behavioral neuroendocrinology. First, in collaboration with Mar Sanchez and Jocelyne Bachevalier, among others, he embarked on another enormous project to study the effects of neonatal amygdala lesions on behavioral development in rhesus monkeys. They found few behavioral effects, although these lesions reduced time females spent with the mother in infancy compared to control females (Raper et al., 2014) and advanced puberty in females (Stephens et al., 2015). The few effects of amygdala lesions challenge our current thinking about the role of the amygdala in social behavior. Second, Wallen has translated his findings in monkeys to humans by studying hormone-behavior relationships, mainly in relation to visual sexual stimuli. Wallen and his colleagues again focused on the amygdala and determined that males have greater activation of this region in response to viewing sexual stimuli than do females (Hamman et al., 2004). Furthermore, 46, XY women with complete androgen insensitivity syndrome are female-typical in response to visual sexual stimuli, with less activation of the amygdala compared to males (Hamann et al., 2014). Thus, Wallen extended his research program to consider central effects and to consider how his research in monkeys translates to humans.

Conclusion

The focus of this chapter has been on Wallen's research on the effects of social context on hormones and behavior. His studies have shaped how we design experiments in the field of behavioral neuroendocrinology. Wallen has emphasized the need to consider social and contextual factors when studying hormone-behavior relationships. Wallen has contributed to our understanding of both organizational and activational effects of hormones on social behavior, through studies that mimic naturalistic conditions by focusing on rhesus monkeys living in complex social groups in outdoor enclosures. Wallen has also examined central effects on behavior, examining the role of the amygdala on social behavior in rhesus monkeys living in semi-naturalistic conditions. He has then translated those findings to the human condition. Wallen has instructed us that hormone-behavior relationships cannot be separated from the social and environmental contexts in which they occur.

References

Cappelletti, M., & Wallen, L. (2016). Increasing women's sexual desire: The comparative effectiveness of estrogens and androgens. *Hormones and Behavior, 78*, 178–193.

Clemens, L. G., Wallen, K., & Gorski, R. A. (1967). Mating behavior: Facilitation in the female rat after cortical application of potassium chloride. *Science, 157*, 1208–1209.

Forger, N. G., Ruszkowski, E., Jacobs, A., & Wallen, K. (2018). Effects of sex and prenatal androgen manipulations on Onuf's nucleus of rhesus macaques. *Hormones and Behavior, 100*, 39–46. https://doi.org/10.1016/j.yhbeh.2018.03.003

Gouzoules, S., Gouzoules, H., & Marler, P. (1984). Rhesus monkey (*Macaca mulatta*) screams: Representational signalling in the recruitment of agonistic aid. *Animal Behaviour, 32*, 182–193. https://doi.org/10.1016/S0003-3472(84)80336-X

Goy, R. W., & Wallen, K. (1979). Experiential variables influencing play, foot-clasp mounting and adult sexual competence in male rhesus monkeys. *Psychoneuroendocrinology, 4*, 1–12. https://doi.org/10.1016/0306-4530(79)90035-0

Goy, R. W., Bercovitch, F. B., & McBrair, M. C. (1988). Behavioral masculinization is independent of genital masculinization in prenatally androgenized female rhesus macaques. *Hormones and Behavior, 22*, 552–571. https://doi.org/10.1016/0018-506X(88)90058-X

Hamann, S., Stevens, J., Hassett Vick, J., Bryk, K., Quigley, C., Berenbaum, S. A., & Wallen, K. (2014). Brain responses to sexual images in 46, XY women with complete androgen insensitivity syndrome are female-typical. *Hormones and Behavior, 66*, 724–730.

Hamman, S., Herman, R. A., Nolan, C. L., & Wallen, K. (2004). Men and women differ in amygdala responses to visual sexual stimuli. *Nature Neuroscience, 7*(4), 411–416.

Herman, R. A., & Wallen, K. (2007). Cognitive performance in rhesus monkeys varies by sex and prenatal androgen exposure. *Hormones and Behavior, 51*, 496–507. https://doi.org/10.1016/j.yhbeh.2007.01.005

Herman, R. A., Jones, B., Mann, D. R., & Wallen, K. (2000). Timing of prenatal androgen exposure: Anatomical and endocrine effects on juvenile male and female rhesus monkeys. *Hormones and Behavior, 38*, 52–66. https://doi.org/10.1006/hbeh.2000.1608

Herman, R. A., Measday, M. A., & Wallen, K. (2003). Sex differences in interest in infants in juvenile rhesus monkeys: Relationship to prenatal androgen. *Hormones and Behavior, 43*, 573–583. https://doi.org/10.1016/S0018-506X(03)00067-9

Herman, R. A., Zehr, J. L., & Wallen, K. (2006). Prenatal androgen blockade accelerates pubertal development in male rhesus monkeys. *Psychoneuroendocrinology, 31*, 118–130. https://doi.org/10.1016/j.psyneuen.2005.06.004

McFadden, D., Pasanen, E. G., Raper, J., et al. (2006). Sex differences in otoacoustic emissions measured in rhesus monkeys (*Macaca mulatta*). *Hormones and Behavior, 50*, 274–284. https://doi.org/10.1016/j.yhbeh.2006.03.012

Phoenix, C. H., Goy, R. W., Gerall, A. A., & Young, W. C. (1959). Organizing action of prenatally administered testosterone propionate on the tissues mediating mating behavior in the female guinea pig. *Endocrinology, 65*, 369–382. https://doi.org/10.1210/endo-65-3-369

Raper, J., Stephens, S. B. Z., Sanchez, M., Bachevalier, J., & Wallen, K. (2014). Neonatal amygdala lesions after mother-infant interactions in rhesus monkeys living in a species-typical social environment. *Developmental Psychobiology, 56*, 1711–1722.

Stephens, S. B. Z., & Wallen, K. (2013). Environmental and social influences on neuroendocrine puberty and behavior in macaques and other nonhuman primates. *Hormones and Behavior, 64*, 226–239. https://doi.org/10.1016/j.yhbeh.2013.05.003

Stephens, S. B., Raper, J., Bachevalier, J., & Wallen, K. (2015). Neonatal amygdala lesions advance pubertal timing in female rhesus macaques. *Psychoneuroendocrinology, 51*, 307–317.

Tomaszycki, M., Davis, J., Gouzoules, H., & Wallen, K. (2001). Sex differences in infant rhesus macaque separation-rejection vocalizations and effects of prenatal androgens. *Hormones and Behavior, 39*, 267–276. https://doi.org/10.1006/hbeh.2001.1659

Tomaszycki, M., Gouzoules, H., & Wallen, K. (2005). Sex differences in juvenile rhesus macaque (*Macaca mulatta*) agonistic screams: Life history differences and effects of prenatal androgens. *Developmental Psychobiology, 47*, 318–327. https://doi.org/10.1002/dev.20102

Wallen, K. (1982). Influence of female hormonal state on rhesus sexual behavior varies with space for social interaction. *Science, 217*, 375–377. https://doi.org/10.1126/science.7201164

Wallen, K. (1990). Desire and ability: Hormones and the regulation of female sexual behavior. *Neuroscience and Biobehavioral Reviews, 14*, 233–241. https://doi.org/10.1016/S0149-7634(05)80223-4

Wallen, K. (1996). Nature needs nurture: The interaction of hormonal and social influences on the development of behavioral sex differences in rhesus monkeys. *Hormones and Behavior, 30*, 364–378. https://doi.org/10.1006/hbeh.1996.0042

Wallen, K. (2005). Hormonal influences on sexually differentiated behavior in nonhuman primates. *Frontiers in Neuroendocrinology, 26*, 7–26. https://doi.org/10.1016/j.yfrne.2005.02.001

Wallen, K., & Goy, R. W. (1977). Effects of estradiol benzoate, estrone, and propionates of testosterone or dihydrotestosterone on sexual and related behaviors of ovariectomized rhesus monkeys. *Hormones and Behavior, 9*, 228–248. https://doi.org/10.1016/0018-506X(77)90059-9

Wallen, K., Goldfoot, D. A., & Goy, R. W. (1981). Peer and maternal influences on the expression of foot-clasp mounting by juvenile male rhesus monkeys. *Developmental Psychobiology, 14*, 299–309. https://doi.org/10.1002/dev.420140402

Wallen, K., Winston, L. A., Gaventa, S., et al. (1984). Periovulatory changes in female sexual behavior and patterns of ovarian steroid secretion in group-living rhesus monkeys. *Hormones and Behavior, 18*, 431–450. https://doi.org/10.1016/0018-506X(84)90028-X

Wallen, K., Mann, D. R., Davis-DaSilva, M., et al. (1986). Chronic gonadotropin-releasing hormone agonist treatment suppresses ovulation and sexual behavior in group-living female rhesus monkeys. *Physiology & Behavior, 36*, 369–375. https://doi.org/10.1016/0031-9384(86)90031-4

Zehr, J. L., Maestripieri, D., & Wallen, K. (1998). Estradiol increases female sexual initiation independent of male responsiveness in rhesus monkeys. *Hormones and Behavior, 33*, 95–103. https://doi.org/10.1006/hbeh.1998.1440

Zehr, J. L., Van Meter, P. E., & Wallen, K. (2005). Factors regulating the timing of puberty onset in female rhesus monkeys (*Macaca mulatta*): Role of prenatal androgens, social rank, and adolescent body weight. *Biology of Reproduction, 72*, 1087–1094. https://doi.org/10.1095/biolreprod.104.027755

John Charteris Wingfield

38

L. Michael Romero

Abstract

John Wingfield is a British-born scientist who spent his career in the United States. He was one of the first behavioral endocrinologists to study endocrine mechanisms in wild freely behaving animals, focusing primarily on birds. His work is best known for three major contributions. First, he proposed a new paradigm, the Challenge Hypothesis, for understanding the role of testosterone in aggression. Three decades after publication, the Challenge Hypothesis continues to spur exciting research. Second, he made seminal contributions in understanding how hormonal systems regulate transitions between different stages of an animal's life, such as migration and breeding. Third, he pioneered new techniques for studying stress in free-living animals and applied those techniques to propose paradigm-shifting hypotheses about the role of stress under natural conditions. His combined work helped establish the new discipline of field endocrinology, or the study of hormones under natural conditions.

Keywords

Challenge Hypothesis · Emergency life-history · Stress, allostasis · Behavioral endocrinology

---

L. M. Romero (✉)
Department of Biology, Tufts University, Medford, MA, USA
e-mail: michael.romero@tufts.edu

R. J. Nelson, Z. M. Weil (eds.), *Biographical History of Behavioral Neuroendocrinology*, https://doi.org/10.1007/978-3-031-12970-4_38

John Wingfield's eminence started from humble beginnings. He was born in 1948 in Ashbourne, Derbyshire, a small town in the middle of England. Coming from a family of modest means, higher education was not an obvious career path. However, Wingfield was accepted to the University of Sheffield where he received a Bachelor of Science with special honors in Zoology in 1970. This led to his doctoral work (1970–1973) with Andrew S. Grimm at the University College of North Wales where he focused on identifying the steroid hormones of plaice (Wingfield & Grimm, 1976), an economically important flatfish. Many of Wingfield's future students were regaled with his stories of collecting source tissues from freshly caught fish on the decks of fishing vessels in the roiling waters of the North Atlantic. Unbeknownst to Wingfield at the time, this introduction to collecting endocrine samples from animals in their natural habitats would foreshadow a career of studying hormones from wild animals.

Academic positions were scarce in the United Kingdom in the early 1970s, so a freshly hooded Wingfield crossed the Atlantic to join Donald Farner's group as a postdoctoral fellow at the University of Washington in Seattle. It was in Seattle that Wingfield pioneered the techniques and approaches that would characterize his career. Comparative endocrinology, or the study of hormones in species not typically used in research (i.e., not laboratory rodents), at that time was generally performed by either catching these animals and bringing them back to the laboratory or taking lethal samples at the capture site. It made little sense to most researchers to study the endocrine regulation of behavior in freely behaving animals; surely the simple act of collecting samples would disrupt any natural behaviors. Wingfield decided to challenge this dogma, despite doubts from most of his peers, including Farner, that the approach would be worthwhile. The resultant study of the endocrine profiles of the breeding cycle of a freely behaving population of white-crowned sparrows proved that such work could be done (Wingfield & Farner, 1978). Wingfield captured, marked, collected blood samples, and released the same individuals multiple times over the course of the breeding season, which apparently had negligible impact on the seasonal behavior of these birds. This paper initiated a paradigm shift in the field of behavioral neuroendocrinology. For the first time, researchers could contemplate studying the endocrine regulation of behavior without the confounding variables inherent in captive animals. Endocrinologists could finally study the regulation of what animals naturally want to do.

Armed with the tools of both the field naturalist and the laboratory endocrinologist, Wingfield moved to Rockefeller University in 1981 where he continued his experiments in free-living birds and was the Associate Director of Rockefeller's Field Research Station in Millbrook, NY. Accompanying him was his wife, Marilyn Ramenofsky, who he met as a graduate student in Seattle. Ramenofsky was also a talented scientist and the two collaborated throughout their careers. It was at Millbrook that Wingfield and members of his laboratory first formulated the ideas of the "Challenge Hypothesis" (see below). The time at Rockefeller was short-lived, however, as Farner's retirement left an opening at the University of Washington that Wingfield would fill.

Back in Seattle, Wingfield continued to rise through the academic ranks until becoming Professor in 1988 and served a term as Department Chair in 1999. He also served as Editor-in-Chief for *General and Comparative Endocrinology* from 1987 to 1989 and President of the Society for Integrative and Comparative Biology (SICB) from 2003 to 2005, which was the home society for comparative endocrinology in North America. During his time in Seattle, Wingfield began a remarkable period of work on hormonal regulation of breeding in the high Arctic. For 25 years, starting in 1989, Wingfield and his students traveled to the Alaskan tundra to study how the endocrine system helps migratory birds arrive, settle, breed, raise chicks, molt, and survive in this unforgiving environment. It is here that Wingfield and people in his lab honed ideas of the hormonal regulation of breeding cycles and how the stress response allows birds to cope and survive (see below). In 2007, Wingfield retired from the University of Washington.

Retirement, however, allowed Wingfield to start a second career in 2007 at the University of California, Davis as the Professor and Endowed Chair in Physiology. In addition to starting another vibrant laboratory in California, Wingfield served a 10-month term as the Director of the Integrative Organismal Systems division of the US National Science Foundation. In this role, he helped oversee the majority of funding for organismal research in the United States, including funding for behavioral endocrinology. At the end of his term, Wingfield was promoted to Assistant Director of the National Science Foundation and the Head of the Biology Directorate. From 2011 to 2014, Wingfield oversaw and helped guide the funding for the majority of US nonmedical biological research. He also talked to the US Congress to advocate for more funding for fundamental research. In this position, he had a tremendous and long-lasting impact on nonmedical biological research in the entire United States. After his 4-year period at the National Science Foundation, Wingfield returned to University of California, Davis, until he retired a second time in 2016, but as of this writing, he was still maintaining an active lab.

Wingfield's ideas spawned many symposia at various international meetings, including a symposium in honor of his 65th birthday at the Society for Integrative and Comparative Biology in 2014. It would be impossible to do justice in this short article to Wingfield's scientific work that is documented in over 600 (and counting) primary research papers, reviews, and book chapters. Instead, I will focus on three major areas below where he has had unique impact and influence.

The Challenge Hypothesis

For many years, people knew that testosterone was not only important in sperm production, but there was also a connection between testosterone and aggression. However, the accumulating data up until the late 1980s could be confusing. Testosterone titers did not always correlate with aggression (Wingfield & Ramenofsky, 1985). Furthermore, Wingfield's early studies showed that seasonal patterns of testosterone secretion were often much more variable in free-living birds than similar studies on captive birds (Wingfield, 1983). In fact, several studies were hinting that testosterone titers were better correlated with periods of social instability (Wingfield & Ramenofsky, 1985). These data led Wingfield and several of his colleagues to posit that patterns of testosterone secretion were driven by social change, with a special emphasis on male-male interactions.

These ideas culminated in the seminal 1990 publication that introduced the Challenge Hypothesis (Wingfield et al., 1990). The main idea was that a certain baseline of testosterone titers was necessary for sperm production, but that any increase in testosterone above that baseline was a result of the frequency and intensity of male-to-male behaviors such as territorial defense and mate-guarding. I always found one of Wingfield's early papers describing territorial aggression in song sparrows (Wingfield, 1985) to be an especially good illustration of the concept. In this study, a group of male song sparrows had established a mosaic of territories in a neighborhood. The territorial boundaries were already established, but nevertheless, the males would patrol their territorial boundaries and use song to defend them against both neighbors and from "floater" males that had been unable to establish territories. Despite all this activity, testosterone titers in territorial males were quite low. Apparently, once a male established his territory and could adequately defend it against intrusion, there was little need for high testosterone titers. Wingfield then disrupted this stable territorial mosaic by capturing and removing one of the males. Within 12–24 h, significant aggression took place between the neighbors of the removed male as they competed to enlarge their territories. Further aggression was directed at floaters who would come in to attempt to claim the "abandoned" territory. Critically, testosterone in both the replacement males and the former neighbors skyrocketed. Stable maintenance of a territorial hierarchy did not require elevated titers of testosterone, but the aggression needed to challenge competing males did require more testosterone. Thus was the Challenge Hypothesis born.

The Challenge Hypothesis has had a profound impact on the study of testosterone and behavior. It stimulated hundreds of studies in many diverse taxa, and although the hypothesis originated in the study of male birds, it has subsequently been used to organize research on all the other vertebrate taxa (Moore et al., 2020), been applied to female behavior (Rosvall et al., 2020), and even been extended to insect behavior (Tibbetts et al., 2020). Perhaps most remarkably given its genesis in wild birds, it has been successfully applied to human behavior as well (Archer, 2006; Wingfield, 2017; Gray et al., 2020). In fact, the concept of the Challenge Hypothesis had such an important influence that it was the subject of an entire issue in the journal *Hormones and Behavior* in 2020 (Volume 123) in celebration of the 30-year anniversary of the publication of the original article (Maney, 2020).

Regulation of Seasonal Breeding

Wingfield's early work on seasonal breeding was performed in the forests of Washington during his postdoctoral studies and in New York while he was at Rockefeller. Much of what drove this earlier work was an attempt to understand how birds regulated transitions between different life-history stages. Life-history theory was a concept in ecology whereby energetically costly events (called stages) during an animal's life (e.g., reproduction, migration, etc.) could not occur simultaneously and therefore needed to be distributed across the annual cycle. The summation and sequence of those stages forms the species' life-history. But what regulates these life-history transitions? One obvious possibility was hormones. Wingfield's original work focused on the reproductive steroids, but later work extended to other reproductive hormones such as luteinizing hormone (LH) and follicle-stimulating hormone (FSH) from the pituitary, and gonadotropin-releasing hormone (GnRH) and gonadotropin-inhibiting hormone (GnIH) from the hypothalamus. This work culminated in the integration of seasonal hormonal patterns in free-living animals that helps explain the hormonal regulation of life-history transitions (Wingfield, 2008a, b).

Then, in 1989, Wingfield received his first grant to work in the Alaskan Arctic. The goal was to understand the endocrine mechanisms that allowed the migratory birds to breed in high-latitude habitats. The short Arctic summer, paired with constant sunlight, posed some unique challenges to regulating life-history transitions. Perhaps the extremely short life-history stage transitions necessitated by the short Arctic breeding season would help illuminate underlying endocrine mechanisms. Thus began 25 years of students, postdocs, and Wingfield himself, migrating to the Arctic every spring, thereby mimicking the migrating birds they were going to study. Upon arrival, the birds, and Wingfield, had to cope with snow on the breeding sites, cold temperatures, ephemeral abundance of food to feed young, and unpredictable storms. And, because autumn would arrive far too soon, breeding needed to commence quickly. The periods of pair formation, nest building, and egg laying often took place in days, whereas they would take weeks in lower latitudes. Work focused on the neuroendocrinology of a number of these key transition points, including migration (Ramenofsky & Wingfield, 2007), arrival (Wingfield et al., 2004), and the substages of breeding (nest building, egg laying, and chick rearing) (Wingfield & Hunt, 2002). Wingfield's research showed that, whereas the same transitions and regulatory mechanisms were present in both Arctic, temperate, and tropical birds, the hormone titers could change much more rapidly in the Arctic birds. For example, some Arctic birds arrive and pair so quickly that testosterone titers might only be elevated for a few days at most (Hunt et al., 1995). Testosterone then immediately falls to help lower aggression and allow parental behaviors to commence. In the end, the original hypothesis proved correct – the rapid life-history stage transitions helped elucidate the endocrine mechanisms regulating those transitions.

Near the end of his career, Wingfield compared the Arctic studies to sister taxa at temperate, equatorial, and Southern high-latitude field sites. Once again, his data showed that the patterns of hormone secretion are not as comparable as we would have predicted. For example, equatorial and Southern latitude birds showed similar changes in testosterone but different levels of aggression (Moore et al., 2002; Addis et al., 2013). It was studies like these that helped establish the evolutionary conservation of the endocrine mechanisms regulating behavior during different life-history stages, but at the same time the remarkable variation in the expression of those mechanisms across different species.

Coping with a Capricious Environment: The Emergency Life-History Stage

From the very beginning of his career, Wingfield recognized that free-living animals were subjected to changes in the environment that could disrupt the hormonal rhythms he was studying. For example, a major storm at one of Wingfield's Washington field sites disrupted work when the birds a graduate student (Michael Moore) was studying abandoned their nests (Wingfield et al., 1983). He was also intrigued by how the birds in the Alaskan Arctic coped with the frequent unpredictable storms. Wingfield consequently became interested in understanding the role of stress in modulating the behavior of free-living animals – an area of study he continued for his entire career. He focused much of his early work on the role of corticosterone, the primary glucocorticoid in birds.

One potentially confounding variable that he recognized early, however, is that capturing a wild animal is itself a stressor. That meant that capture, an act necessary to collect a blood sample, would inevitably alter corticosterone concentrations from those blood samples. Wingfield was one of the originators of an elegant solution to this problem. He recognized that biomedical studies indicated that it normally takes several minutes after the initiation of a stressor for corticosterone increases to be measurable in a blood sample. That meant that if a sample could be taken very quickly after an animal was captured, then the initial sample would reflect the pre-capture, or baseline, corticosterone titers. So, Wingfield did something that few other ornithologists were doing – he sat very close to his traps so that he could remove the bird and take a blood sample within 2–3 min of the animal becoming trapped. With this baseline sample in hand, Wingfield then held the bird and took subsequent samples at various intervals after capture to monitor the bird's response to being captured, held, and bled. This innovative technique mimicked a major predatory attack (being captured) by a simulated predator (Wingfield himself). The ensuing response curve could then be compared across seasons, between individuals that varied in habitat conditions and/or physiological condition at the time of capture, between species, etc. (Wingfield & Romero, 2001). Wingfield called this technique the "Capture Stress" protocol, and since its first introduction in 1982 (Wingfield et al., 1982), it has been used in thousands of studies using hundreds of different species. It was instrumental in generating data that led to the realization that adrenocortical responses to stress could be modulated in response to an animal's physiological and environmental conditions (Wingfield, 1994). The Capture Stress protocol helped revolutionize the study of stress in free-living animals.

Over time, Wingfield and members of his lab expanded the stress studies to other hormones important to the stress response, thereby providing a more-complete physiological and behavioral picture of the mechanisms and consequences of stress in wild animals. After many such studies, Wingfield created a new concept to better integrate stress responses into ecological theory. In 1998, he and members of his lab at the time proposed that stress could be conceived as an animal entering what they termed an "emergency life-history stage" (Wingfield et al., 1998). Wingfield built upon existing life-history theory to suggest that there was a further life-history stage that occurred during an emergency. This emergency stage had different physiology and behaviors than the stage just prior to the emergency, and the entrance to and exit from the emergency life-history stage is regulated by the hormones of the stress response. Prior to this paper, scientists largely viewed stress through a biomedical lens, whereby stress was inherently damaging and created problems for the animal. After this paper, scientists increasingly recognized that stress was part of the natural history of the animal and not simply a disease.

Soon after the publication of the Emergency Life-History paper (Wingfield et al., 1998), Wingfield had a fortuitous encounter with Bruce McEwen at a workshop. McEwen, Professor at Rockefeller University and a giant in the field of neuroscience and neuroendocrinology (see Chap. 19, in this volume), had been working for several years on using the concepts of allostasis as a substitute for stress. Allostasis, or the maintenance of physiological systems through change, came from biomedical work, and McEwen was interested in applying it to laboratory rodent models and human health. At the workshop, Wingfield and McEwen realized that many of the ideas underlying allostasis and the emergency life-history stage were similar. Both theories had energy regulation and allocation as their cores, so McEwen and Wingfield decided to try to meld the two ideas into a unified whole. The resultant paper, published in 2003 (Mcewen & Wingfield, 2003), adopted allostasis terminology for the combined ideas and was a major advance. Although the marriage of allostasis and emergency life-history theory was not flawless (and has generated various critiques both at the time of publication and subsequently), the use of energy as a universal metric of both healthy stress responses and disease-inducing stress responses proved revolutionary. Prior to this paper, the behavioral endocrinology of stress was almost always a post hoc explanation of behavior. If an animal did something unexpected, "stress" was often invoked as a reason for the behavior. Rarely could stress serve as a predictor of that unexpected behavior. Incidentally, this lack of an ability to a priori predict changes in physiology and behavior led to substantial disaffection of stress studies among many researchers. McEwen and Wingfield's conception of allostasis, however, provided a way to predict future behavior when an animal was exposed to adverse conditions. If an animal had sufficient resources (i.e., energy), then the animal was predicted to remain in its normal life-history stage and have few, if any, hormonal responses. If there were insufficient resources, however, then the animal would enter an emergency life-history stage (i.e., a stress response), and glucocorticoids would rise and subsequently regulate the physiological and behavioral responses necessary to cope with the adverse conditions. Allostasis provided, for the first time, a predictive theory of stress. This paper has spawned and influenced hundreds of studies in the nearly two decades since its publication.

Finally, Wingfield decided to collect and describe all the work on stress in wild animals that had been performed by him and others. He collaborated with Michael Romero (the author of this article) to consolidate and make sense of the collected work. Wingfield and Romero did not appreciate how much work had been done in this area, nor how much was actually known. The original conception of a small work ballooned into a major synthetic work that took over a decade to write. The resultant book, *Tempests, Poxes, Predators, and People: Stress in Wild Animals and How They Cope* (Romero & Wingfield, 2016), summarized and integrated information on stress theory, hormonal mechanisms of the stress response, and techniques used to study stress in free-living animals. The text also described natural stress physiology that regulates normal life-history stage transitions, such as molt, migration, and development, in nontraditional species. Finally, the text explored what was known about how free-living animals cope with adverse conditions such as famines, storms, social instability, infections, predator attacks, and human encroachment. The book was a comprehensive compendium of knowledge-to-date and was a fitting capstone to a career.

Conclusion

Wingfield would be the first to credit the many students and colleagues he worked with throughout the years. The success and power of Wingfield's approach of studying endocrine-behavior relationships in free-living wild animals captured the imagination of many young ambitious students and postdocs from many different fields. The approach appealed to behavioral ecologists in search of mechanisms, and to neuroendocrinologists in search of behaviors relevant to the animal. His work attracted over 30 graduate students and over 35 postdoctoral fellows. Astoundingly, more than 50 of these trainees went on to obtain faculty and/or research positions at colleges, universities, and nongovernmental organizations. These former trainees continued and extended Wingfield's general approach, thereby creating a formidable cohort of scientists that have made enormous progress in understanding hormone-behavior relationships. This legacy is the strongest endorsement of the lasting impact of Wingfield's career (Fig. 38.1).

Fig. 38.1 John Wingfield in the field, circa 1997. (Photo by the author)

References

Addis, E. A., Clark, A. D., Vasquez, R. A., & Wingfield, J. C. (2013). Seasonal modulation of testosterone during breeding of the rufous-collared sparrow (Zonotrichia capensis australis) in Southern Patagonia. *Physiological and Biochemical Zoology, 86*, 782–790.

Archer, J. (2006). Testosterone and human aggression: An evaluation of the Challenge Hypothesis. *Neuroscience and Biobehavioral Reviews, 30*, 319–345.

Gray, P. B., Straftis, A. A., Bird, B. M., Mchale, T. S., & Zilioli, S. (2020). Human reproductive behavior, life history, and the Challenge Hypothesis: A 30-year review, retrospective and future directions. *Hormones and Behavior, 123*, 104530.

Hunt, K., Wingfield, J. C., Astheimer, L. B., Buttemer, W. A., & Hahn, T. P. (1995). Temporal patterns of territorial behavior and circulating testosterone in the Lapland longspur and other Arctic passerines. *American Zoologist, 35*, 274–284.

Maney, D. L. (2020). The challenge hypothesis: Triumphs and caveats. *Hormones and Behavior, 123*, 104663.

Mcewen, B. S., & Wingfield, J. C. (2003). The concept of allostasis in biology and biomedicine. *Hormones and Behavior, 43*, 2–15.

Moore, I. T., Perfito, N., Wada, H., Sperry, T. S., & Wingfield, J. C. (2002). Latitudinal variation in plasma testosterone levels in birds of the genus Zonotrichia. *General and Comparative Endocrinology, 129*, 13–19.

Moore, I. T., Hernandez, J., & Goymann, W. (2020). Who rises to the challenge? Testing the Challenge Hypothesis in fish, amphibians, reptiles, and mammals. *Hormones and Behavior, 123*, 104537.

Ramenofsky, M., & Wingfield, J. C. (2007). Regulation of migration. *Bioscience, 57*, 135–143.

Romero, L. M., & Wingfield, J. C. (2016). *Tempests, poxes, predators, and people: Stress in wild animals and how they cope*. Oxford University Press.

Rosvall, K. A., Bentz, A. B., & George, E. M. (2020). How research on female vertebrates contributes to an expanded challenge hypothesis. *Hormones and Behavior, 123*, 104565.

Tibbetts, E. A., Laub, E. C., Mathiron, A. G. E., & Goubault, M. (2020). The challenge hypothesis in insects. *Hormones and Behavior, 123*, 104533.

Wingfield, J. C. (1983). Environmental and endocrine control of reproduction: An ecological approach. In S.-I. Mikami & M. Wada (Eds.), *Avian endocrinology: Environmental and ecological aspects*. Springer-Verlag.

Wingfield, J. C. (1985). Short-term changes in plasma levels of hormones during establishment and defense of a breeding territory in male song sparrows, *Melospiza melodia*. *Hormones and Behavior, 19*, 174–187.

Wingfield, J. C. (1994). Modulation of the adrenocortical response to stress in birds. In K. G. Davey, R. E. Peter, & S. S. Tobe (Eds.), *Perspectives in comparative endocrinology*. National Research Council of Canada.

Wingfield, J. C. (2008a). Comparative endocrinology, environment and global change. *General and Comparative Endocrinology, 157*, 207–216.

Wingfield, J. C. (2008b). Organization of vertebrate annual cycles: Implications for control mechanisms. *Philosophical Transactions of the Royal Society B-Biological Sciences, 363*, 425–441.

Wingfield, J. C. (2017). The Challenge Hypothesis: Where it began and relevance to humans. *Hormones and Behavior, 92*, 9–12.

Wingfield, J. C., & Farner, D. S. (1978). The endocrinology of a naturally breeding population of the white-crowned sparrow (Zonotrichia leucophrys pugetensis). *Physiological Zoology, 51*, 188–205.

Wingfield, J. C., & Grimm, A. S. (1976). Preliminary identification of plasma steroids in the plaice, Pleuronectes platessa L. *General and Comparative Endocrinology, 29*, 78–83.

Wingfield, J. C., & Hunt, K. E. (2002). Arctic spring: Hormone-behavior interactions in a severe environment. *Comparative Biochemistry and Physiology - B Biochemistry and Molecular Biology, 132*, 275–286.

Wingfield, J. C., & Ramenofsky, M. (1985). Hormonal and environmental control of aggression in birds. In R. Gilles & J. Balthazart (Eds.), *Neurobiology*. Springer-Verlag.
Wingfield, J. C., & Romero, L. M. (2001). Adrenocortical responses to stress and their modulation in free-living vertebrates. In B. S. Mcewen & H. M. Goodman (Eds.), *Handbook of physiology; Section 7: The endocrine system; Volume IV: Coping with the environment: Neural and endocrine mechanisms*. Oxford University Press.
Wingfield, J. C., Smith, J. P., & Farner, D. S. (1982). Endocrine responses of white-crowned sparrows to environmental stress. *Condor, 84*, 399–409.
Wingfield, J. C., Moore, M. C., & Farner, D. S. (1983). Endocrine responses to inclement weather in naturally breeding populations of white-crowned sparrows (*Zonotrichia leucophrys pugetensis*). *Auk, 100*, 56–62.
Wingfield, J. C., Hegner, R. E., Dufty, A. M., Jr., & Ball, G. F. (1990). The "Challenge Hypothesis": Theoretical implications for patterns of testosterone secretion, mating systems, and breeding strategies. *American Naturalist, 136*, 829–846.
Wingfield, J. C., Maney, D. L., Breuner, C. W., Jacobs, J. D., Lynn, S., Ramenofsky, M., & Richardson, R. D. (1998). Ecological bases of hormone-behavior interactions: The "emergency life history stage". *Integrative and Comparative Biology, 38*, 191–206.
Wingfield, J. C., Owen-Ashley, N. T., Benowitz-Fredericks, Z. M., Lynn, S. E., Hahn, T. P., Wada, H., Breuner, C. M., Meddle, S. L., & Romero, L. M. (2004). Arctic spring: The arrival biology of migrant birds. *Acta Zoologica Sinica, 50*, 948–960.

Jacques Balthazart

39

Charlotte A. Cornil and Gregory F. Ball

Abstract

Jacques Balthazart is a Belgian behavioral neuroendocrinologist whose work focused on the neuroendocrine regulation of reproductive behaviors of birds, primarily based on studies in Japanese quail and songbirds. He is best known for his studies of the organizational and activational actions of sex steroids on the brain and behavior, the first comprehensive identification of the distribution of aromatase-immunoreactive neurons in the brain of any vertebrate, the functional role played by these aromatase-positive cells in the neural circuits controlling male sexual behavior, the characterization of the genomic and non-genomic controls of brain aromatase activity and his work on the influence of sex steroid hormones on the development and seasonal plasticity of neural circuits underlying the production of song in songbirds. Overall, his career is a model as to how an investigator can systematically attack fundamental questions employing a wide range of methods over a long period of time.

Keywords

Neuroendocrinology · Behavior · Aromatase · Quail · Sex behavior

Jacques Balthazart was born in Liège, Belgium, in 1949. He developed an interest in birds during his very early childhood (3–4 years old) when he was trapping birds

C. A. Cornil (✉)
University of Liege, GIGA Neurosciences, Liège, Belgium
e-mail: charlotte.cornil@uliege.be

G. F. Ball
Department of Psychology, University of Maryland, College Park, MD, USA
e-mail: gball@umd.edu

R. J. Nelson, Z. M. Weil (eds.), *Biographical History of Behavioral Neuroendocrinology*, https://doi.org/10.1007/978-3-031-12970-4_39

with his father during their fall migration, an activity that was at the time very widespread in Belgium, but became illegal (for conservation reasons) in the 1970s. He had then initiated studies in zoology and his interest in birds transformed into a passion for bird watching and photography (Fig. 39.1).

Accordingly, his master's thesis was initially dedicated to the study by multivariate statistical analyses of courtship sequences in black grouse (*Lyrurus tetrix*), a species still living in relatively high numbers in the largest wildlife refuge of Belgium, the "Hautes Fagnes" plateau. The 1970–1971 winter was unusually cold and the study sites were buried under about 1 meter of snow, making it impossible for the birds to display in their leks. This thesis thus had to be reoriented and the same statistical techniques were applied to behavioral sequences of the cichlid fish, *Tilapia macrochir* (now *Oreochromis macrochir*) (Balthazart, 1972). Balthazart wanted to expand these statistical analyses during his PhD thesis, but was considering that the behavioral units recorded in these fishes were too variable and thus not recognized reliably by different observers. He therefore turned to duck displays that were considered the prototypic example of stereotypic behaviors in ethology, the fixed action patterns as described by Lorenz (Lorenz, 1950). Balthazart started to record behavioral sequences in several duck species, in particular the green teal, *Anas crecca*. However, due to disagreements with his supervisor, after a few months, he moved to another lab in the medical school, directed at the time by Ernest Schoffeniels, a comparative biochemist who worked primarily on the acetylcholine receptor of electric eels. Schoffeniels had very broad interests in zoology, and he

Fig. 39.1 Jacques Balthazart and his passion for bird watching and photography

suggested that Balthazart continue working on ducks, but on a completely different project attempting to determine the role of olfactory communication in the control of social behavior in ducks. This led to a few exciting experiments that provided tentative evidence suggesting the existence of pheromonal communication in ducks (Balthazart & Schoffeniels, 1979; Jacob et al., 1979), an idea that was only recognized more broadly in birds several decades later (e.g., (Bonadonna & Nevitt, 2004; Hagelin, 2007; Balthazart & Taziaux, 2009; Caro & Balthazart, 2010)).

During the first year of his PhD studies with Schoffeniels, Balthazart however faced an unanticipated problem: ducks are seasonal breeders and testing the contribution of pheromones to the control of sexual behavior was only feasible for a few months each year, during the breeding season. When the first reproductive season ended, he therefore decided to inject testosterone in male ducks and test whether this would activate sexual activity, which would allow for the analysis of the effects of female odors on male behaviors. This experiment turned out to be very successful (Balthazart, 1974; Balthazart & Stevens, 1975). This finding led Balthazart to realize that a compelling question in the study of the control of behavior is to understand how steroid hormones can have such profound effects on behavior. Thus, over the years, the endocrine controls of behavior occupied an increasing, almost exclusive, part of his scientific activity. This switch in focus occurred even though this discipline was not well-represented in Belgium and only very scarcely in Europe.

In his early studies, he was among the group that pioneered the use of radioimmunoassay methods to measure gonadotropins and steroid hormones in the plasma of birds. He helped develop and utilized a heterologous assay for avian follicle-stimulating hormone (FSH) (Croix et al., 1974) that has provided what to this day is some of the best data on the environmental regulation of FSH over the breeding cycle in birds (Balthazart & Hendrick, 1976). In 1978, he went to the Institute of Animal Behavior at Rutgers University in the USA for 1 year. There he worked with Mei Cheng on ring doves studying the effects of nest building on gonadotropin release (Cheng & Balthazart, 1982), and he collaborated with Jeffrey Blaustein and Harvey Feder to study changes in progesterone receptor binding in male doves as they transition from courtship to incubation behavior without a change in progesterone concentration in the blood (Balthazart et al., 1980). This work represents one of the first examples of a change in brain sensitivity to steroids in the absence of a change in hormonal concentration in the blood.

When he returned to Liège as Chargé de Recherches du F.N.R.S (a postdoctoral fellowship supported by the Belgian Science Foundation), Balthazart established a long-term research program on the neuroendocrine control of male-typical reproductive behavior in Japanese quail (e.g., (Schumacher & Balthazart, 1983; Balthazart & Schumacher, 1984)). This program is a model as to how an investigator can systematically attack a fundamental question using a wide range of methods over a long period of time (Balthazart et al., 2009b; Ball & Balthazart, 2010). Balthazart asked how testosterone can effectively activate male-typical behavior in one sex but not the other. Over the years, he has studied this problem from a developmental perspective, as well as in adulthood employing behavioral, neuroanatomical, physiological, as well as cellular/molecular methods (see Balthazart (2017) for a review).

He first addressed this question focusing on the metabolism of testosterone, a question he had started to work on before his postdoc at Rutgers in collaboration with Renato Massa, then in the laboratory of Luciano Martini in Milan (Balthazart et al., 1979). With one of his first PhD students, Michael Schumacher, he studied the actions of the androgenic and estrogenic metabolites of testosterone on aspects of male sexual behavior and in particular established the key role of estrogens derived from the aromatization of testosterone in the brain in the activation of this behavior (Balthazart, 1991). In parallel, he further characterized the endocrine mechanisms involved in the sexual differentiation of the brain, which had originally been discovered by Elizabeth Adkins-Regan (Balthazart et al., 2009a). In collaboration with GianCarlo Panzica and Carla Viglietti-Panzica (University of Torino, Italy), Balthazart learned the anatomy the preoptic area and the hypothalamus and discovered the medial preoptic nucleus (POM), a sexually dimorphic nucleus in the quail preoptic area. Together, they described the ultrastructure and neurochemical markers of this region and how they are influenced by the embryonic exposure to estrogens and adult actions of testosterone and its metabolites (Panzica et al., 1987, 1996). Using stereotaxic methods, he demonstrated that the POM constitutes a major site of local aromatase activity that plays a pivotal role in the activation of consummatory (and later appetitive as well) aspects of male-typical sexual behavior by testosterone (Balthazart & Surlemont, 1990; Balthazart et al., 1998). In collaboration with Nobuhiro Harada (Fujita Health University, Nagoya, Japan), who had developed an antibody against human placental aromatase (and who later also developed an antibody specifically directed against recombinant quail aromatase), he mapped the distribution of aromatase-containing cells in quail nervous system, their relationship with a variety of neurochemical markers (most prominently tyrosine hydroxylase, vasotocin, and nitric oxide synthase), and their connectivity to other brain regions involved in the control of social behavior (Foidart et al., 1995; Evrard et al., 2000; Carere et al., 2007; Absil et al., 2001). This set of studies on aromatase represents the first comprehensive description of the distribution of aromatase-immunoreactive cells in a well-defined neural circuit with a clear link to behavioral function.

This work inspired a series of studies with Gregory Ball at Johns Hopkins where the endocrine and neural mechanisms regulating both appetitive and consummatory dimensions of male-typical sexual behavior were characterized in quail. One important idea that came out of this work is that there is a functional topography of the POM related to the differential regulation of these two aspects of sexual behavior (Balthazart et al., 1998; Taziaux et al., 2005) in quail, but also presumably other vertebrates (Balthazart & Ball, 2007). In parallel, he conducted a comprehensive analysis of the regulation of aromatase by testosterone metabolites ranging from the quantification and visualization of the messenger RNA coding for the enzyme to the site-specific analysis of its enzymatic activity (Balthazart et al., 2011; Balthazart, 2017). This multilevel approach to the study of aromatase was and remains difficult to achieve in mammals to this day.

While optimizing the tritiated water assay to measure estrogen synthesis ex vivo, Michelle Baillien, then working with Balthazart, serendipitously discovered that brain aromatase activity is modulated within minutes by calcium-dependent phosphorylation (Balthazart et al., 2001, 2003, 2006a). This finding led to novel ideas about the possible roles estradiol may play in the brain and to the hypothesis that it acts more like a neurotransmitter than a hormone. These ideas opened a new avenue of research as many labs were accumulating results supporting the notion that steroids, estrogens in particular, are able to act on cell activity much faster (milliseconds-minutes) than anticipated based on their action on gene transcription. This work resulted in two influential papers in *Trends in the Neurosciences* published with Ball (Balthazart & Ball, 1998, 2006). With Charlotte Cornil, who had recently joined the lab, he started asking the question of the functional significance of these rapid changes in brain estrogen synthesis. Together, they demonstrated that brain aromatase activity was rapidly modulated in a region-specific manner following sexual encounters and that manipulation of local estrogen synthesis affected specifically appetitive male sexual behavior (Balthazart et al., 2006b). Along with the characterization of the genomic mode of action of testosterone on quail sexual behavior by different lab members including Thierry D. Charlier (Charlier et al., 2005), these results led to the formulation of the hypothesis of estrogens having dual actions which poses that neuroestrogens act in different timescales to control different aspects of a same behavior possibly acting on different receptors and in different subregions of the brain (Cornil et al., 2015; Fig. 39.2).

In addition to his work on quail, Jacques also conducted studies on the neuroendocrine control of reproductive behaviors in other avian species, most notably songbirds (Ball & Balthazart, 2017). This work has involved a long-term collaboration with US colleague, Gregory F. Ball (Johns Hopkins University, Baltimore, and subsequently University of Maryland, College Park), who worked with him on aspects of the quail project as well. In the 1960s and 1970s, it became clear that most vertebrate species had a similar distribution of androgen and estrogen receptors in the hypothalamus, preoptic area, amygdala, septum, midbrain, and infundibular regions (Morrell et al., 1975; Pfaff, 1976). Species in the songbird order (nearly 50% of living bird species) had, in addition to steroid receptor expression in these areas, a unique set of receptors in the forebrain areas that controlled birdsong (Arnold et al., 1976). This song system is linked specifically to song behavior. Key nuclei in this circuit, the HVC, acronym serves as the nucleus name, robust nucleus of the arcopallium (RA), and lateral magnocellular nucleus of the anterior nidopallium (LMAN), all express high densities of androgen receptors and in the case of HVC, estrogen receptors. In collaboration with Ball, he was the first investigator to use immunohistochemistry to localize androgen receptors in songbirds ((Balthazart et al., 1992); a paper cited over 260 times). Androgens activate song behavior and induce marked neuroplasticity in these forebrain song control nuclei as first measured by changes in nucleus volume by Fernando Nottebohm (Nottebohm, 1981). It was also clear that androgens can act on cells that project to the song control nuclei and his group characterized with tract-tracing studies how brainstem catecholamine nuclei project to these forebrain song nuclei and represent another site where testosterone can act to modulate song behavior (Appeltants et al., 2000, 2002). He and Ball then became interested in the role played by adult neurogenesis in contributing

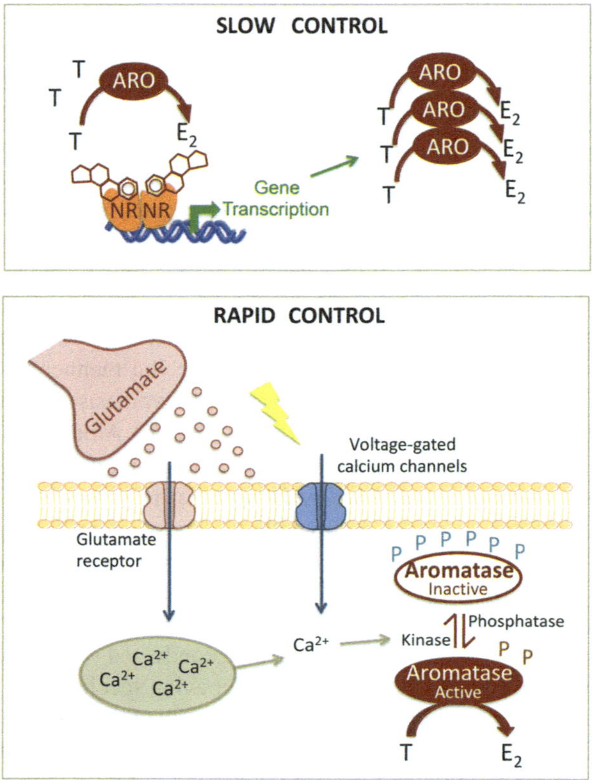

Fig. 39.2 Schematic illustration of the dual control of aromatase activity. This enzymatic activity can be regulated in the long term by changes in transcription/transduction of the enzymatic protein or by more rapid changes in phosphorylation of the enzymatic protein by calcium-regulated kinases

to seasonal changes in the song system and developed the use of doublecortin, a marker of adult neurogenesis, to characterize the role played by new neurons in these seasonal changes in nucleus volume of the song control system (Balthazart & Ball, 2014). Most recently, Balthazart has identified the possible role of perineuronal nets (PNN) in regulating sensitive periods for song learning. Correlation studies indicate that there are differences in PNN expression between open-ended and age-limited song learners, but it has still not been established causally whether there is a link between PNN and brain plasticity (Cornez et al., 2021) Altogether he and Ball published over 50 papers together on songbirds and over 150 total joint papers (Fig. 39.3).

One of us (CAC) had the privilege to witness firsthand this successful collaboration. It provides a rare example of how scientific collaboration can result in a synergy that is almost symbiotic. Both are without a doubt efficient independently, but their creativity and productivity when working together are absolutely remarkable. One example of this is provided by the regular visits Ball still pays to the lab, usually in the spring, during which they either draft full grant applications or write

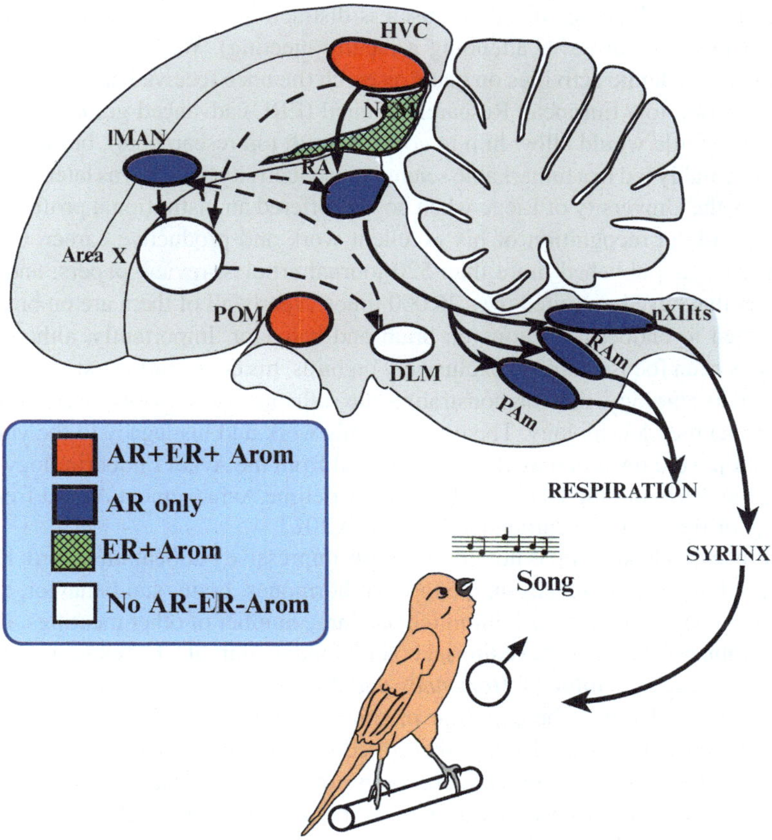

Fig. 39.3 The song control system of songbirds expresses high concentrations of androgen receptors (AR) and estrogen receptors (ER) alpha. Aromatase (Arom) expression is also present in the POM that controls singing motivation and in the NCM adjacent to HVC. This nucleus sends aromatase-positive fibers into HVC. Abbreviations: DLM dorsolateral thalamic nucleus; HVC initially hyperstriatum ventrale pars caudalis, now acronym used as proper name; LMAN lateral magnocellular nucleus of the anterior nidopallium; NCM caudal medial nidopallium; nXIIts tracheosyringeal part of the XIIth cranial nerve; PAm nucleus parambigualis; POM medial preoptic nucleus; RA nucleus robustus arcopallialis, RAm nucleus retroambigualis

several papers sometimes from scratch within a single week, all while sharing stories about the good old times, making lab walls shake with their distinctive laughter, and visiting churches and breweries all over Europe.

Balthazart followed an unusual academic track. Although he is an excellent teacher, he was never much interested in this activity. He never cared about academic titles and was not interested in ascending the academic ladder. For a long time, he thus avoided moving to the professor track but rather stayed in the assistant professor track being initially hired as Assistant (1980–1982), and successively promoted Premier Assistant (1982–1986) and Chef de travaux (1986–2007). Actually, Balthazart actively avoided teaching and all activities associated with traditional academic life, which he considered useless distractions from his research activity (he still brags about never attending a faculty meeting). We can argue about the benefits of academic activities on one's research (he once received nasty reviews for an application to a European Research Council (ERC) advanced grant questioning whether his title would allow him to compete with top researchers), but this career path undeniably led to a remarkable scientific productivity, which was later acknowledged by the University of Liège when he was offered an institutional professorship (2007–2014) in recognition of his excellent work and productive career. Overall, Balthazart has published more than 520 journal articles, review papers, and book chapters that have been cited over 26,000 times. Nearly all of them are on birds and are related to endocrine actions on brain and behavior. Importantly, although his work has been focused almost exclusively on birds, his contribution extends beyond avian endocrinology, thus demonstrating the influence of his work over the entire field of neuroendocrinology. The quality of his work and his legacy to the younger generation were rewarded by the Farner medal from the Avian Endocrinology community in 2016 and the Daniel S. Lehrman Lifetime Achievement Award from the Society of Behavioral Neuroendocrinology in 2017.

Balthazart's leadership is not limited to his impressive publication record. He has organized many conferences on the topic of hormones, brain, and behavior, and he has served on the organizing committee of a large number of other meetings. He has and continues to serve on the editorial board of many journals. For example, at present he is associate editor of *Hormones and Behavior*, where he has championed work on birds. He was the co-editor in chief of *Frontiers in Neuroendocrinology* (Elsevier) for many years. He has also edited several books notably for the Oxford series in *Behavioral Neuroendocrinology* (e.g., (Balthazart & Ball, 2013)). Beginning in 2010, he also started writing books for the lay public. Notably, he wrote an entire book on the biological basis of homosexuality in humans that placed this phenomenon firmly in a neuroendocrine perspective (Balthazart, 2011). He was president of the Society for Behavioral Neuroendocrinology in 2003–2005. His acclaimed mentorship participated in launching the career of many students or postdocs trained in his laboratory. The list of neuroendocrinologists trained in his laboratory includes Pierre Deviche (Arizona State University, Tempe, AZ, USA), Michael Schumacher (University Paris-Sud, France), Yvon Delville (University of Texas, Austin, TX, USA), Charlotte A. Cornil (University of Liège, Belgium), Lauren Riters (University of Wisconsin-Madison, WI, USA), Julie Bakker (University of Liège, Belgium), Samuel Caro (CEFE-CNRS, Montpellier, France), Henry Evrard (Tuebingen, Germany), and Thierry Charlier (University of Rennes, France). Finally, he has developed an impressive number of collaborations with scientists from all over the world and on a vast number of research topics and species highlighting again his sheer enthusiasm for science and research.

In June 2014, Jacques officially retired and organized for the occasion a 3-day scientific meeting. He had initially intended it as a light schedule meeting to have fun with his closest collaborators and friends. However, not only all invited speakers accepted the invitation within 1 or 2 days, but many more scientists, including former trainees, collaborators, and longtime friends in the community of behavioral neuroendocrinology, participated to what turned out to be a very busy scientific meeting and a memorable party. The attendance illustrates the impact Balthazart had on the field and this meeting did not mark the end of his scientific activity. He remains fully active to this day, thanks notably to continued US National Institutes of Health (NIH) funding received in collaboration with Ball. His career is undoubtedly a model of passion and dedication to science and a remarkable example of efficiency, hard work, and availability to his trainees balanced with a pronounced taste for travel and spending good time with friends/colleagues.

References

Absil, P., Baillien, M., Ball, G. F., Panzica, G., & Balthazart, J. (2001). The control of preoptic aromatase activity by afferent inputs in Japanese quail. *Brain Research Reviews, 37*, 38–58.

Appeltants, D., Absil, P., Balthazart, J., & Ball, G. F. (2000). Identification of the origin of catecholaminergic inputs to HVc in canaries by retrograde tract tracing combined with tyrosine hydroxylase immunocytochemistry. *Journal of Chemical Neuroanatomy, 18*, 117–133.

Appeltants, D., Ball, G. F., & Balthazart, J. (2002). The origin of catecholaminergic inputs to the song control nucleus RA in canaries. *Neuroreport, 13*, 649–653.

Arnold, A. P., Nottebohm, F., & Pfaff, D. W. (1976). Hormone concentrating cells in vocal control areas of the brain of the zebra finch (Poephila guttata). *The Journal of Comparative Neurology, 165*, 487–512.

Ball, G. F., & Balthazart, J. (2010). Japanese quail as a model system for studying the neuroendocrine control of reproductive and social behaviors. *ILAR Journal, 51*, 310–325.

Ball, G. F., & Balthazart, J. (2017). Neuroendocrine regulation of reproductive behavior in birds. In D. W. Pfaff & M. Joels (Eds.), *Hormones, brain and behavior* (3rd ed.). Academic Press.

Balthazart, J. (1972). Analyse factorielle du comportement agonistique chez Tilapia macrochir (Boulenger 1912). *Behaviour, 46*, 32–72.

Balthazart, J. (1974). Short-term effects of testosterone propionate on the behaviour of young intact male domestic ducks (Anas platyrhynchos). *Psychologica Belgica, 14*, 1–10.

Balthazart, J. (1991). Testosterone metabolism in the avian hypothalamus. *The Journal of Steroid Biochemistry and Molecular Biology, 40*, 557–570.

Balthazart, J. (2011). *The biology of homosexuality*. Oxford University Press.

Balthazart, J. (2017). Steroid metabolism in the brain: From bird watching to molecular biology, a personal journey. *Hormones and Behavior, 93*, 137–150.

Balthazart, J., & Ball, G. F. (1998). New insights into the regulation and function of brain estrogen synthase (aromatase). *Trends in Neurosciences, 21*, 243–249.

Balthazart, J., & Ball, G. F. (2006). Is brain estradiol a hormone or a neurotransmitter? *Trends in Neurosciences, 29*, 241–249.

Balthazart, J., & Ball, G. F. (2007). Topography in the preoptic region: Differential regulation of appetitive and consummatory male sexual behaviors. *Frontiers in Neuroendocrinology, 28*, 161–178.

Balthazart, J., & Ball, G. F. (2013). *Brain aromatase, estrogens and behavior*. Oxford University Press.

Balthazart, J., & Ball, G. F. (2014). Endogenous versus exogenous markers of adult neurogenesis in canaries and other birds: Advantages and disadvantages. *The Journal of Comparative Neurology, 522*, 4100–4120.

Balthazart, J., & Hendrick, J. C. (1976). Annual variation in reproductive behavior, testosterone, and plasma FSH levels in the Rouen duck, Anas platyrhynchos. *General and Comparative Endocrinology, 28*, 171–183.

Balthazart, J., & Schoffeniels, E. (1979). Pheromones are involved in the control of sexual behaviour in birds. *Naturwissenschaften, 66*, 55–56.

Balthazart, J., & Schumacher, M. (1984). Organization and activation of behavior in quail: Role of testosterone metabolism. *The Journal of Experimental Zoology, 232*, 595–604.

Balthazart, J., & Stevens, M. (1975). Effects of testosterone propionate on the social behaviour of groups of male domestic ducklings Anas platyrhynchos L. *Animal Behaviour, 23*, 926–931.

Balthazart, J., & Surlemont, C. (1990). Copulatory behavior is controlled by the sexually dimorphic nucleus of the quail POA. *Brain Research Bulletin, 25*, 7–14.

Balthazart, J., & Taziaux, M. (2009). The underestimated role of olfaction in avian reproduction? *Behavioural Brain Research, 200*, 248–259.

Balthazart, J., Massa, R., & Negri-Cesi, P. (1979). Photoperiodic control of testosterone metabolism, plasma gonadotrophins, cloacal gland growth, and reproductive behavior in the Japanese quail. *General and Comparative Endocrinology, 39*, 222–235.

Balthazart, J., Blaustein, J. D., Cheng, M. F., & Feder, H. H. (1980). Hormones modulate the concentration of cytoplasmic progestin receptors in the brain of male doves (Streptopelia risoria). *The Journal of Endocrinology, 86*, 251–261.

Balthazart, J., Foidart, A., Wilson, E. M., & Ball, G. F. (1992). Immunocytochemical localization of androgen receptors in the male songbird and quail brain. *Journal of Comparative Neurology, 317*, 407–420.

Balthazart, J., Absil, P., Gérard, M., Appeltants, D., & Ball, G. F. (1998). Appetitive and consummatory male sexual behavior in Japanese quail are differentially regulated by subregions of the preoptic medial nucleus. *Journal of Neuroscience, 18*, 6512–6527.

Balthazart, J., Baillien, M., & Ball, G. F. (2001). Rapid and reversible inhibition of brain aromatase activity. *Journal of Neuroendocrinology, 13*, 61–71.

Balthazart, J., Baillien, M., Charlier, T. D., & Ball, G. F. (2003). Calcium-dependent phosphorylation processes control brain aromatase in quail. *European Journal of Neuroscience, 17*, 1591–1606.

Balthazart, J., Baillien, M., & Ball, G. F. (2006a). Rapid control of brain aromatase activity by glutamatergic inputs. *Endocrinology, 147*, 359–366.

Balthazart, J., Cornil, C. A., Taziaux, M., Charlier, T. D., Baillien, M., & Ball, G. F. (2006b). Rapid changes in production and behavioral action of estrogens. *Neuroscience, 138*, 783–791.

Balthazart, J., Arnold, A. P., & Adkins-Regan, E. (2009a). Sexual differentiation of brain and behavior in birds. In D. W. Pfaff, A. P. Arnold, A. M. Etgen, S. E. Fahrbach, & R. T. Rubin (Eds.), *Hormones, brain and behavior* (2nd ed.). Academic Press.

Balthazart, J., Cornil, C. A., Charlier, T. D., Taziaux, M., & Ball, G. F. (2009b). Estradiol, a key endocrine signal in the sexual differentiation and activation of reproductive behavior in quail. *Journal of Experimental Zoology. Part A, Ecological Genetics and Physiology, 311*, 323–345.

Balthazart, J., Charlier, T. D., Cornil, C. A., Dickens, M. J., Harada, N., Konkle, A. T., Voigt, C., & Ball, G. F. (2011). Sex differences in brain aromatase activity: Genomic and non-genomic controls. *Frontiers in Endocrinology (Lausanne), 2*, 34.

Bonadonna, F., & Nevitt, G. A. (2004). Partner-specific odor recognition in an Antarctic seabird. *Science, 306*, 835.

Carere, C., Ball, G. F., & Balthazart, J. (2007). Sex differences in projections from preoptic area aromatase cells to the periaqueductal gray in Japanese quail. *The Journal of Comparative Neurology, 500*, 894–907.

Caro, S. P., & Balthazart, J. (2010). Pheromones in birds: Myth or reality? *Journal of Comparative Physiology A: Neuroethology, Sensory, Neural, and Behavioral Physiology, 196*, 751–766.

Charlier, T. D., Ball, G. F., & Balthazart, J. (2005). Inhibition of steroid receptor coactivator-1 blocks estrogen and androgen action on male sex behavior and associated brain plasticity. *Journal of Neuroscience, 25*, 906–913.

Cheng, M. F., & Balthazart, J. (1982). The role of nest-building activity in gonadotropin secretions and the reproductive success of ring doves (Streptopelia risoria). *Journal of Comparative and Physiological Psychology, 96*, 307–324.

Cornez, G., Valle, S., Dos Santos, E. B., Chiver, I., Muller, W., Ball, G. F., Cornil, C. A., & Balthazart, J. (2021). Perineuronal nets in HVC and plasticity in male canary song. *PLoS One, 16*, e0252560.

Cornil, C. A., Ball, G. F., & Balthazart, J. (2015). The dual action of estrogen hypothesis. *Trends in Neurosciences, 38*, 408–416.

Croix, D., Hendrick, J. C., Balthazart, J., & Franchimont, P. (1974). Radioimmunologic assay of follicle stimulating hormone of duck hypophysis by a mammalian system. *Comptes Rendus des Seances de la Societe de Biologie et de Ses Filiales, 168*, 136–140.

Evrard, H., Baillien, M., Foidart, A., Absil, P., Harada, N., & Balthazart, J. (2000). Localization and controls of aromatase in the quail spinal cord. *Journal of Comparative Neurology, 423*, 552–564.

Foidart, A., Reid, J., Absil, P., Yoshimura, N., Harada, N., & Balthazart, J. (1995). Critical re-examination of the distribution of aromatase-immunoreactive cells in the quail forebrain using antibodies raised against human placental aromatase and against the recombinant quail, mouse or human enzyme. *Journal of Chemical Neuroanatomy, 8*, 267–282.

Hagelin, J. C. (2007). Odors and chemical signaling. In B. G. M. Jamieson (Ed.), *Reproductive biology and phylogeny of birds, Part B*. Science Publishers.

Jacob, J., Balthazart, J., & Schoffeniels, E. (1979). Sex differences in the chemical composition of uropygial gland waxes in domestic ducks. *Biochemical Systematics and Ecology, 7*, 149–153.

Lorenz, K. (1950). The comparative method in studying innate behavior patterns. *Symposia of the Society for Experimental Biology, 4*, 221–268.

Morrell, J. I., Kelley, D. B., & Pfaff, D. W. (1975). Sex steroid binding in the brain of vertebrates. In K. M. Knigge, D. E. Scott, H. Kobayashi, S. Miura, & S. Ishii (Eds.), *Brain-endocrine interactions II* (0th ed.). Karger.

Nottebohm, F. (1981). A brain for all seasons: Cyclical anatomical changes in song-control nuclei of the canary brain. *Science, 214*, 1368–1370.

Panzica, G. C., Viglietti-Panzica, C., Calagni, M., Anselmetti, G. C., Schumacher, M., & Balthazart, J. (1987). Sexual differentiation and hormonal control of the sexually dimorphic medial preoptic nucleus in the quail. *Brain Research, 416*, 59–68.

Panzica, G. C., Viglietti-Panzica, C., & Balthazart, J. (1996). The sexually dimorphic medial preoptic nucleus of quail: A key brain area mediating steroid action on the male sexual behavior. *Frontiers in Neuroendocrinology, 17*, 51–125.

Pfaff, D. W. (1976). The neuroanatomy of sex hormone receptors in the vertebrate brain. In T. C. Anand Kumar (Ed.), *Neuroendocrine regulation of fertility* (0th ed.). Karger.

Schumacher, M., & Balthazart, J. (1983). The effects of testosterone and its metabolites on sexual behavior and morphology in male and female Japanese quail. *Physiology & Behavior, 30*, 335–339.

Taziaux, M., Cornil, C. A., Dejace, C., & Balthazart, J. (2005). Differences in neural activation following expression of appetitive and consummatory male sexual behavior in the quail brain. *Hormones and Behavior, 48*, 130.

Andrew Howard Bass

40

Luke Remage-Healey

Abstract

The path of Andrew Bass, a first-generation scientist, toward fundamental discoveries in behavioral neuroendocrinology was serpentine and serendipitous, and marked by the advice and care from established mentors, as well as his career-long ability to combine discovery with mentoring those around him to thrive in science. His work integrated nearly all major aspects of the biology of a vocalizing fish species, the plainfin midshipman, and provided a monumental set of advances in behavioral neuroendocrinology and comparative neurobiology. His research program was driven by an unending curiosity and spirit of collaboration with experts from a variety of disciplines. His lasting contributions toward educating, mentoring, and training a host of scientists in the field were accompanied by his deep, integrative perspective that championed the evolutionary, ecological, anatomical, comparative, neurobiological, physiological, and hormonal explanations of variation in natural behavior.

Keywords

Brain evolution · Comparative neuroanatomy · Mormyrid fishes · Neuroethology · Plainfin midshipman · Testosterone

L. Remage-Healey (✉)
Center for Neuroendocrine Studies, Neuroscience and Behavior Program, University of Massachusetts, Amherst, MA, USA
e-mail: lremageh@umass.edu

R. J. Nelson, Z. M. Weil (eds.), *Biographical History of Behavioral Neuroendocrinology*, https://doi.org/10.1007/978-3-031-12970-4_40

Andrew H. Bass was born to Leon and Freda Bass in Staten Island, NY, in 1951. His parents, neither of whom attended college, were married for nearly 60 years at the time of his father's death and highly supportive of all manner of educational pursuits for Bass and his older brother Geoffrey. Leon graduated from high school a year early to work with his father in the garment industry. After military service in World War II, Leon opened a tailoring and dry-cleaning business, in which Bass and his brother worked part-time during the summer. Freda worked for Bass' dad a few years and then full-time as a bookkeeper for other businesses. Bass and his brother were "latchkey kids." From a young age, Bass witnessed firsthand the long-lasting friendships they maintained from when they were teenagers in Brooklyn. Growing up in Staten Island offered the best of NYC – easy access to woods and beaches, the shoreline in Brooklyn and Long Island, and the cultural richness of Manhattan.

At an early age, Bass developed a love of swimming. Because Staten Island was surrounded by water, his familiarity and interest in the ocean also started early. He grew to love the ferryboat ride to Manhattan, especially in the summertime, reveling in the view of "the City" especially on ferry rides back home at sunset.

In grade school, Bass had inklings about his interest in science. An elementary school science fair project involved the careful dissection of amphibian heart tissue and testing the physiological effects of digitalis (aka "foxglove") on cardiac function. The exact source of the highly toxic digitalis is lost to the sands of time, but Bass' fascination with biology, comparative physiology, and extrinsic modulation were born.

In high school, he earned a Bausch and Lomb Science award at graduation, and his full slate of AP courses readied him for college-level curricula. After graduating in 1969, he attended the historic Woodstock music festival that summer in upstate NY. His personal highlight was Richie Havens' electrifying performance, and sharp-eyed visitors to his faculty office at Cornell could catch a glimpse of the framed ticket on the wall.

During the fall of 1969, Bass enrolled as a Biology major at Case Western Reserve University (CWRU) in Cleveland, a recent merger of the Case Institute of Technology (founded in 1880) and Western Reserve University (founded in 1826). Like many first-generation college students, Bass' family had high expectations for him to study medicine. Bass dutifully pursued a premed curriculum, but this interest was adjusted by a stint volunteering at Rainbow Babies and Children's Hospital affiliated with CWRU's medical school in Cleveland. His sophomore course in introductory biology was an eye-opener, and he soon came across a small pamphlet about undergraduate research opportunities on campus. The eminent neuroanatomist Glenn Northcutt was in the Department of Anatomy at the medical school, and Bass became involved in research working in Northcutt's lab in his sophomore year. His research mentor in the lab was an MD/PhD student named Michael Pritz who was working on thalamo-telencephalic pathways in *Caiman crocodilus* and remained a lifelong friend and colleague. With this, Bass' career-long love of comparative neuroanatomy had begun.

When Bass joined the lab, Northcutt handed him a medical school human neuroanatomy textbook to pour over while also exposing him to the ideas of William

Hodos (Maryland), Harvey Karten (UCSD), and others in comparative neuroanatomy. Working closely with Pritz, Bass conducted research on the brain of the African side-necked turtle (*Podocnemis unifilis*), focusing on their relatively expanded dorsal cortex. At the time, Gerald Schneider (MIT) had been advancing the idea of two visual systems in vertebrates, one "concerned with the locating of objects" and the "other concerned with the specific identification of objects" (Schneider, 1969). Bass' research topic was derived from Northcutt's keen interest in reptilian brain evolution, and they tested the existence of two visual systems in turtles, resulting in several co-authored publications (Bass et al., 1973; Pritz et al., 1973).

As the summer after his sophomore year approached, Bass' Intro Bio instructor at CWRU, Georgia Lesh Laurie, encouraged him to apply for an NSF-sponsored summer fellowship. This award provided an opportunity to dive headlong into laboratory research with Northcutt. This left a lasting impression on Bass, as he saw firsthand the pivotal role of a faculty member who takes a personal interest in a student and encourages them to pursue further research opportunities.

The summer after his junior year, Bass went back home to work at the Addiction Services Agency of the City of New York. This experience was to have a profound influence on him, working with recovering substance abusers, and mentoring young children at youth centers in Brooklyn. This interest continued back at CWRU as Bass volunteered as a math tutor for Cleveland schoolkids. Bass' deft and care with students, faculty, and his successes in administrative work later in his career were clearly forged by these early-life experiences.

During his senior year of 1973, Northcutt had taken a position at the University of Michigan while maintaining his lab at Case for some unfinished projects. These included the ongoing work of his students Mike Pritz and Andy Bass. Bass was by then functioning as a de facto graduate student, performing lesions of the turtle brain and then observing their performance in a visual discrimination task.

As graduation approached, Bass initially considered a path to graduate school to work on animal orientation. He also took the MCAT exam and started filling out medical school applications. During this process, he was nagged by the thought, "How am I going to feel about getting up in the morning to go to med school?" and realized that he would rather pursue basic research in the laboratory. So, without studying, he took the GRE and ended up applying to graduate school. He first went to Princeton but left after about 6 months to spend more time thinking about his research path. He decided to move to Ann Arbor, Michigan, and got a job working at Sears in the bike repair shop to "pay the bills." On nights and weekends, he volunteered in Northcutt's lab, "just doing comparative neuroanatomy," which had become his passion.

Bass formally entered graduate school at the University of Michigan to work with Northcutt in 1975. It was in this environment where he became fascinated by the heady combination of neuroethology and neuroanatomy. Two foundational books helped steer his initial studies: George C. Williams' *Adaptation and Natural Selection* (purchased in April 1976 for $3.76, according to the original receipt/bookmark) and Nikolaas Tinbergen's *The Study of Instinct*. It was in his first year as a graduate student that he met Margaret "Midge" Marchaterre in Northcutt's class on "Comparative Neuroanatomy." Her spirit of adventure and love of science was infectious, and no more apparent than when she agreed to his invitation in August 1975: "Want to hitchhike to Oregon with me?" That fateful trip and the resulting lifelong "amazing, supportive partnership in life and work" began. The two were married in 1979 by a rabbi and a priest under an oak tree and a Saint Francis statue in the backyard of Midge's childhood home in Florham Park, NJ.

Following up on his undergraduate projects, Bass began his graduate research by learning multiunit electrophysiology recordings in the turtle visual pallium (Bass et al., 1983). This provided an important methodological companion to his neuroanatomical and behavioral expertise. For his doctoral thesis, Bass investigated olfactory pathways and telencephalic organization in catfishes. To source animals for his research, he relied on the kindness of a local trout farmer with a catfish pond. The farmer kept the bigger catfish, but the smaller ones were for science. Over the course of this work, Bass discovered several efferent pathways from the telencephalon to the olfactory bulb and optic tectum, providing evidence for a "top-down" influence of the forebrain on chemo-sensation in fishes (Bass, 1979, 1981).

A pivotal moment in graduate school came when Bass encountered some new discoveries on sex differences in the brain in three 1976 papers from Rockefeller University on song production circuits and steroid-concentrating neurons in canaries (Nottebohm et al., 1976; Nottebohm & Arnold, 1976; Arnold et al., 1976) that together fomented a paradigm shift in the field. As a grad student, Bass was inspired by this work and began to consider how the notions of sex differences and steroid regulation might apply to vocalizing fishes. He dug into this literature with an eye toward "the singers of the sea" and hoped to work on similar topics in toadfishes. However, this provocative direction "was too impractical" at the time in land-locked Ann Arbor. Willing to wait, Bass filed away this promising idea for later in his scientific career and learned all he could about the telencephalon of fishes from Northcutt and postdoctoral associate Mark Braford (Oberlin).

While completing his PhD in 1979, Bass came across an article in *American Scientist* about African electric fish (mormyrids) by Carl Hopkins (University of Minnesota) and was attracted to the questions and the experience of observing animals in their natural environment. Bass applied for and obtained a 3-year NIH National Research Service Award for postdoctoral support for field and laboratory experiments with Hopkins on mormyrids. Their collaboration showed that waveform time-domain cues were important for species recognition (Hopkins & Bass, 1981) and revealed the reversible effects of steroid hormones on sex differences in the electric organ discharge (Bass & Hopkins, 1983). This work provided some of the earliest indications that membrane ion channels of electrically signaling cells were dynamically regulated by steroid hormones (Bass & Volman, 1987).

In 1982, Hopkins received a job offer in the Section (now Department) of Neurobiology and Behavior (NB&B) at Cornell, a renowned center for neuroethology. As part of the move, Hopkins offered to support Bass on an NIH grant as a research associate. Bass agreed, figuring he would tag along, and "spend a year or so there" while thinking about applying for faculty positions. In January 1984, Bass took a position in NB&B as a Visiting Assistant Professor, teaching comparative neuroanatomy. When a tenure-track faculty positioned opened later that year, Northcutt encouraged him to apply for the job, and Bass was offered the position to start as an Assistant Professor in the fall of 1984. Needless to say, he immediately embarked on a project on vocal fishes.

For the first three summers of his faculty position at Cornell, Bass made the drive to Woods Hole, MA, to teach in the Neural Systems and Behavior course at the Marine Biological Laboratory (MBL). He was recruited into the course by Darcy Kelley (Columbia). Kelley was studying androgenic actions on the vocal system of amphibians and taught him how to make steroid hormone pellets for his work in mormyrid fishes, and she became a mentor and colleague. Bass also met Neil Segil (USC), a technician at the time in Darcy's lab who went on to study hair cell regeneration and became a lifelong colleague and great friend (Bass et al., 1986). In the summer of 1986 at Woods Hole, Bass was introduced to Bob Baker (NYU) by Harvey Karten (UCSD). Baker became a valued career-long mentor to Bass, and he learned several neurophysiology approaches under Baker's tutelage. Bass and Marchaterre often stayed with Harriet (Weill Cornell) and Bob Baker in their Woods Hole summer home during many visits thereafter. Baker taught Bass the surgery and neurophysiology preparation for a fictive vocal preparation at first with sea robins (*Prionotus carolinus*), a group of sonic fishes that were particularly tricky in a surgical setting, presenting a delicate combination of cranial thickness and dense brain vasculature (Bass & Baker, 1991). Once mastered, this approach became a mainstay in Bass' lab.

In addition to developing projects on vocal fishes, Bass also began studying reproductive plasticity in sex-reversing fish soon after joining the Cornell faculty. Diving trips to St. Croix and Puerto Rico provided exposure to the natural habitat of wrasses. Following his interests in steroid hormone regulation of sex differences, Bass saw the opportunity to further test ideas about brain, behavior, and hormonal action by examining different reproductive morphs in bluehead wrasses (Grober & Bass, 1991; Grober et al., 1991). He was therefore already in hot pursuit of the neurobiology and neuroendocrinology of between- and within-sex plasticity when he had a fateful encounter with a different fish species on the California coast.

Bass had become especially familiar with toadfishes and their sound production through the literature and using them as a study species in the MBL course. There were descriptions dating back to the 1920s of one species in particular, the "California singing fish" or plainfin midshipman (*Porichthys notatus*; named for the arrangement of photophores extending down the body, resembling buttons on a Navy midshipman's coat). Midshipman were already infamous among the residents of a houseboat community in Sausalito, CA, who were often kept awake by their loud, droning "hums" in the wee hours of summer mornings. Midshipman had even inspired the Sausalito Humming Toadfish Festival in the 1980s. Bass first started collecting midshipman from commercial dealers, when his Cornell colleague Ed Brothers, a fish hearing expert, introduced him to John McCosker, then director of the Steinhart Aquarium in San Francisco, following a McCosker seminar at Cornell. McCosker invited Bass out to the Steinhart in August 1986 to visit their midshipman exhibit. Bass, Midge, and graduate student Richard Brantley came along and visited a midshipman breeding site near Tomales, CA. Bass and other Bass lab members were to return to that same beach every summer up to and including the time of this writing.

Bass knew from earlier work on midshipman that muscles surrounding the swim bladder were used to "drum" the loud humming sounds heard by the Sausalito houseboaters. While examining specimens at the Steinhart, he found that relatively large-bodied males, but not females, had well-developed sonic muscles (Bass & Marchaterre, 1989a). Unexpectedly, there were also a set of small fish that he assumed were females, externally, since they were half the size of the males. Dissecting the swim bladder also conveniently revealed the gonads. He was "blown away" to see a pair of large testes contained in these otherwise female-like males, but he immediately knew the implications of this discovery from his background with wrasses. They had hit the scientific jackpot. Here was a single species that was a champion of both sound production and sexual plasticity! Bass credited his discovery to serendipity, preparation, and engagement with fellow scientists, fishermen, and locals about the lore and lure of this peculiar fish species.

Bass' subsequent decades of studies of the midshipman species became expansive. He and his lab investigated all major aspects of midshipman biology in a species he referred to as "forever interesting." The "ground truth" of Bass' explorations of midshipman biology was their natural behavior. Field collecting and observation trips were paramount to understand the ecological basis for their vocal communication system as well as their sexual plasticity (Bass, 1996). Bass' team developed evidence that larger "type I" males were the "hummers," essentially acoustic beacons attracting females from offshore sites to nests in the unforgiving intertidal zone where they spawned (McKibben & Bass, 1998). Type I males guarded and cared for the eggs as they developed in the nest. By contrast, they found that the smaller "type II" males became sexually mature at earlier ages than type Is and did not excavate nests or acoustically court females. Instead, they invested in hypertrophied testes compared to type I's and adopted an alternative reproductive tactic (ART), mimicking female size and coloration as they "sneak" into the type I's nest, competing with them for fertilizations (Brantley & Bass, 1994).

Bass' work dove into the central question of whether alternative morphs differed not only in reproductive physiology, muscle adaptations, and behavior but also in the structure and function of the nervous system (Bass & Marchaterre, 1989b). Guided by a study showing transneuronal transport of biocytin in the retina, Bass did a post-tenure sabbatical in Baker's lab at NYU where he applied crystalline biocytin to transected axons of the motoneurons innervating the midshipman's sonic muscles. They hoped to uncover the full distribution of hindbrain neurons that they recently showed innervate the motoneurons and generate pacemaker-like action potentials (Bass & Baker, 1990). The biocytin fills outperformed their wildest expectations, revealing an expansive brainstem vocal circuit including pools of midbrain tegmental and medullary pacemaker and pre-pacemaker neurons (Bass et al., 1994). Subsequent studies confirmed involvement of these neurons in patterning the duration, frequency, and amplitude modulation of vocalizations (Chagnaud et al.,

2021). Moreover, the three adult reproductive morphs differed in multiple circuit aspects. The involvement of steroids in midshipman ARTs became clear from Richard Brantley's doctoral work with Bass that included a collaboration with John Wingfield (UW-Seattle). Type II males had higher plasma testosterone concentrations than type Is and females, whereas type I males had much higher plasma concentrations of the primary fish androgen, 11-ketotestosterone (11-KT), and neither type II males nor females had detectable 11-KT levels (Brantley et al., 1993a, b; Knapp et al., 1999). Later studies using steroid implants showed that when type II males were treated with 11-KT, their sonic muscles enlarged but they did not engage in courtship behaviors as was seen in type I males (Brantley et al., 1993a; Lee & Bass, 2005). The way steroids and other hormones regulated the expression of morph-specific characters in midshipman became an intriguing puzzle.

Midshipman breed from late May to August along the Pacific coast of the United States, while spending the rest of the year in deep ocean. The neuroendocrine mechanisms that accompanied this transition became a primary topic of interest in Bass' lab. Early studies included changes in size of GnRH somata in the preoptic area during juvenile to adult transitions in the three adult morphs (Grober et al., 1994) and retinal influences on GnRH expression in the thalamus (Foran et al., 1997). In collaboration with Barney Schlinger (UCLA), Bass discovered that the activity of the estrogen-synthesis enzyme aromatase diverged in the vocal hindbrain region of type I vs type II male midshipman (Schlinger et al., 1999), establishing notions of neuro-steroidal control of midshipman neural circuits and behaviors.

Bass subsequently led a team with postdoc Joe Sisneros (UW-Seattle) and grad student Paul Forlano (Brooklyn College) collecting fish at multiple times throughout the year and found in collaboration with another former postdoc, Rosemary Knapp (U Oklahoma), that the primary steroid hormones 11-KT, testosterone, and estradiol varied according to the seasonal migration and breeding cycle, with steroid concentration peaking during the summer breeding months (Sisneros et al., 2004b). Forlano's doctoral work with Bass revealed pronounced expression of aromatase in glial cells throughout the brain, including a dense concentration enshrouding the sonic motoneurons (Forlano et al., 2001), and further showed seasonal regulation of aromatase and estrogen signaling in both the motor nucleus and preoptic area (Forlano & Bass, 2005a, b; Forlano et al., 2005). This remarkable capacity for neural estrogen signaling epitomized evidence for the "brain as an endocrine organ" and the power of steroids in shaping the development and activity of a vocal communication system. Seasonal plasticity also extended to the auditory system of midshipman, as Bass and Sisneros discovered that estrogens expanded the hearing sensitivity of female midshipman during the breeding season to enable them to detect the upper harmonics of type I male vocalizations (Sisneros et al., 2004a). Therefore, the motor and sensory facets of the vocal communication system were each sensitive to seasonal rhythms via sex steroid-dependent mechanisms that also varied according to reproductive morphotype.

Bass' exploration of the neuroendocrine control of behavior was simultaneously propelled by experiments using the fictive vocal preparation. A foundational study (Bass & Baker, 1990) recorded vocal motor volleys from occipital nerve roots that innervated the sonic muscles and showed that they determined the muscle contraction rate and, in turn, the pulse repetition rate/fundamental frequency of vocalization in all adult morphotypes. Bass and Baker also used sharp electrode intracellular recordings to show that synchronous motoneuron firing could account for the vocal motor volley's large compound potentials; intracellular HRP fills revealed morph differences in vocal motor and pacemaker neuron size (Bass & Chagnaud, 2012; Chagnaud et al., 2011, 2012). The vocal pattern generator neurons were up to 300% larger in type I males, as compared to type IIs and females, consistent with enlargement of the sonic muscles they drove and the actions of hormones in establishing the neural and behavioral differences (Bass, 1996).

In the 1980s and 1990s, the actions of nonapeptides like arginine vasopressin and oxytocin had been linked to sex differences in many courtship and aggressive behaviors in vertebrate species (Crews & Moore, 1986; Goodson, 1998; Insel et al., 1998; Insel & Young, 2000). Bass therefore began working with James Goodson (UCSD, Indiana) when he joined Bass' lab as a postdoctoral associate to investigate possible neuropeptide influences on the midshipman vocal pattern generator circuit, and how it might differ both within and between sexes. They showed that the teleost fish orthologs, arginine vasotocin (AVT) and isotocin (IT), could rapidly control the rhythmic vocal motor activity in midshipman via actions directly in the anterior hypothalamus-preoptic area (Goodson & Bass, 2000). Type I male fictive vocalizations were rapidly sensitive to AVT, and not regulated by IT. Remarkably, this modulation pattern was completely reversed in *both* type II males and females, in that their fictive calls were rapidly sensitive to IT, but not AVT. This meant that gonadal sex (testis vs. ovary) was dissociable from the neuropeptide actions in the hypothalamus in this species "allowing an evolutionary labile patterning of sexual behaviors" (Goodson & Bass, 2000).

The role of steroids in shaping vocal motor and auditory systems in midshipman, coupled with the observation of rapid regulation of the vocal pattern generator by hormones, was then combined in a series of laboratory and field studies. As a graduate student, the author worked with Bass on how steroids shape vocal motor output in midshipman and the closely related Gulf toadfish (*Opsanus beta*) on a rapid timescale. The divergent levels of circulating sex steroids among the three midshipman morphotypes provided a set of predictions to guide these studies. During bouts of vocal advertisement calling, male toadfish and type I male midshipman had elevated levels of plasma 11-KT, as compared to non-calling contexts (Knapp et al., 2001; Remage-Healey & Bass, 2005). Moreover, male toadfish "challenged" with an artificial playback of territorial male calls showed parallel rapid increases in plasma 11-KT and vocal signaling (Remage-Healey & Bass, 2005). In the lab, electrophysiology recordings showed that 11-KT caused similarly rapid (5–10 min) increases in the duration of fictive vocalizations in both male toadfish and type I male midshipman, but did not change responses in females of either species or in type II male midshipman (Remage-Healey & Bass, 2004, 2006, 2007). Instead, as with the decoupling of neuropeptide actions from gonadal sex, type II male and female midshipman were rapidly responsive to testosterone, whereas both sexes in toadfish and type I male midshipman were all unresponsive to testosterone. Estrogens were also shown to rapidly regulate fictive vocalizations in all three midshipman morphs (Remage-Healey & Bass, 2007). Therefore, the evolution and diversification of neuroendocrine regulation of vocal communication behaviors had multiple axes of differentiation and plasticity. Finally, back in the field, when 11-KT was noninvasively administered to male toadfish in their natural environment, calling behavior was rapidly altered, providing a field-based test of the neurophysiological results (Remage-Healey & Bass, 2006). During these field experiments, the author and Bass routinely marveled at the power of the sounds emitted by Gulf toadfish near the Florida State University Marine Lab, with Bass pointing out with his famous "whoa!" that the earth was shaking: the vocalizations emitted at low tide could be detected through ground-borne substrate vibrations.

The nocturnal calling pattern of midshipman became a source of curiosity for Bass and his lab. Working with fresh-caught individuals, Bass and grad student Jon Lee established seminatural, captive tank environments at a UW field station in Seabeck, WA. When type I males were group-housed in these enclosures, they established nests and began humming and engaging in agonistic contests (Lee & Bass, 2004). This provided a means to observe many natural behaviors and reinforced the heightened activity level for midshipman vocal communication at nighttime. Bass and grad student Tine Rubow later discovered that seasonal and circadian housing conditions affected the durations and types of fictive calls elicited in the vocal motor activity preparation (Rubow & Bass, 2009). They found that midbrain-evoked fictive calls in type I males housed under long days and tested in the scotophase of their circadian rhythm were twice as long as calls from those of type I males tested during the photophase, or as compared to calls from other type I males housed under short-day photoperiods. Bass and grad student Ni Feng later provided a mechanism for this when they discovered that daily treatments targeting the melatonin receptor could account for the effects of photoperiod on fictive and natural calling duration from type I males (Feng & Bass, 2014, 2016; Feng et al., 2019). Therefore, Bass' team showed that the vocal motor network could be modulated by a broad range of neuromodulators to include neuropeptides, neurotransmitters, steroids, and monoaminergic hormones. The most recent work on the neuroendocrine control of vocalization in midshipman by former graduate student Joel Tripp (UT Austin) showed that this list now included the preoptic area neuropeptide galanin, expressed in the vocal motor and auditory systems and like other hormonal characters diverged in its actions between type I males and type II males and was associated with mating behaviors in type I males only (Tripp et al., 2020).

Along the way, Bass' mentoring and support of his current and former trainees had a huge impact on behavioral neuroendocrinology. Bass and Marchaterre fostered a collegial and lively atmosphere of discovery and support within and outside the lab, celebrating major milestones by hosting events for the lab at their house in Ithaca, NY, and joining in the discoveries in the field in coastal sites in California and Florida. Bass' kindness, humor, and voracious appetite for the literature, writing, and editing and his essential humanity inspired success in his trainees with a highly personal touch.

The scientific career of Andrew Bass has spanned nearly four decades, still active and engaged at the time of this writing in late 2021. The scope of Bass' contribution to comparative neuroendocrinology, neuroethology, and neuroanatomy and his integrated understanding of midshipman in particular are staggering to consider in toto. This chapter highlights his work in neuroendocrinology while leaving off his substantial contributions in sensory biology, comparative anatomy, and evolutionary/developmental neurobiology. Bass' love of teaching, mentoring students and faculty, along with his unending curiosity, sense of wonder, collegiality and congeniality, and collaborative and integrative nature are all hallmarks of an outstanding scientific journey.

Acknowledgments The author is thankful for dedicated mentorship, scientific training, and friendship of Andrew Bass and Margaret Marchaterre, as well as their gentle nudges about the accuracy of this biography. This chapter is dedicated to the memory of Rosemary Knapp and James L. Goodson.

References

Arnold, A. P., Nottebohm, F., & Pfaff, D. W. (1976). Hormone concentrating cells in vocal control and other areas of brain of zebra finch (Poephila-Guttata). *The Journal of Comparative Neurology, 165*, 487–511.

Bass, A. H. (1979). Telencephalic efferents in the channel catfish, ictalurus punctatus. *The Anatomical Record, 193*, 478–478.

Bass, A. H. (1981). Telencephalic efferents in the channel catfish, ictalurus- punctatus - Projections to the olfactory-bulb and optic tectum. *Brain, Behavior and Evolution, 19*, 1–16.

Bass, A. H. (1996). Shaping brain sexuality. *American Scientist, 84*, 352–363.

Bass, A. H., & Baker, R. (1990). Sexual dimorphisms in the vocal control-system of a teleost fish – Morphology of physiologically identified neurons. *Journal of Neurobiology, 21*, 1155–1168.

Bass, A., & Baker, R. (1991). Evolution of homologous vocal control traits. *Brain, Behavior and Evolution, 38*, 240–254.

Bass, A. H., & Chagnaud, B. P. (2012). Shared developmental and evolutionary origins for neural basis of vocal-acoustic and pectoral-gestural signaling. *Proceedings of the National Academy of Sciences, 109*(Suppl 1), 10677–10684.

Bass, A. H., & Hopkins, C. D. (1983). Hormonal-control of sexual-differentiation – Changes in electric organ discharge waveform. *Science, 220*, 971–974.

Bass, A. H., & Marchaterre, M. A. (1989a). Sound-generating (sonic) motor system in a teleost fish (Porichthys-Notatus) – Sexual polymorphism in the ultrastructure of myofibrils. *The Journal of Comparative Neurology, 286*, 141–153.

Bass, A. H., & Marchaterre, M. A. (1989b). Sound-generating (sonic) motor system in a teleost fish (Porichthys-Notatus) – Sexual polymorphisms and general synaptology of sonic motor nucleus. *The Journal of Comparative Neurology, 286*, 154–169.

Bass, A. H., & Volman, S. F. (1987). From behavior to membranes – Testosterone-induced changes in action-potential duration in electric organs. *Proceedings of the National Academy of Sciences of the United States of America, 84*, 9295–9298.

Bass, A. H., Pritz, M. B., & Northcut, R. G. (1973). Effects of telencephalic and tectal ablations on visual behavior in side-necked turtle, podocnemis-unifilis. *Brain Research, 55*, 455–460.

Bass, A. H., Andry, M. L., & Northcutt, R. G. (1983). Visual activity in the telencephalon of the painted turtle, chrysemys-picta. *Brain Research, 263*, 201–210.

Bass, A. H., Segil, N., & Kelley, D. B. (1986). Androgen binding in the brain and electric organ of a mormyrid fish. *Journal of Comparative Physiology A, 159*, 535–544.

Bass, A. H., Marchaterre, M. A., & Baker, R. (1994). Vocal-acoustic pathways in a teleost fish. *The Journal of Neuroscience, 14*, 4025–4039.

Brantley, R. K., & Bass, A. H. (1994). Alternative male spawning tactics and acoustic-signals in the plainfin midshipman fish Porichthys-Notatus Girard (Teleostei, Batrachoididae). *Ethology, 96*, 213–232.

Brantley, R. K., Marchaterre, M. A., & Bass, A. H. (1993a). Androgen effects on vocal muscle structure in a teleost fish with inter-sexual and intra-sexual dimorphism. *Journal of Morphology, 216*, 305–318.

Brantley, R. K., Wingfield, J. C., & Bass, A. H. (1993b). Sex steroid-levels in porichthys-notatus, a fish with alternative reproductive tactics, and a review of the hormonal bases for male dimorphism among teleost fishes. *Hormones and Behavior, 27*, 332–347.

Chagnaud, B. P., Baker, R., & Bass, A. H. (2011). Vocalization frequency and duration are coded in separate hindbrain nuclei. *Nature Communications, 2*, 346.

Chagnaud, B. P., Zee, M. C., Baker, R., & Bass, A. H. (2012). Innovations in motoneuron synchrony drive rapid temporal modulations in vertebrate acoustic signaling. *Journal of Neurophysiology, 107*, 3528–3542.

Chagnaud, B. P., Perelmuter, J. T., Forlano, P. M., & Bass, A. H. (2021). Gap junction-mediated glycinergic inhibition ensures precise temporal patterning in vocal behavior. *eLife, 10*, e59390.

Crews, D., & Moore, M. C. (1986). Evolution of mechanisms controlling mating-behavior. *Science, 231*, 121–125.

Feng, N. Y., & Bass, A. H. (2014). Melatonin action in a midbrain vocal-acoustic network. *The Journal of Experimental Biology, 217*, 1046–1057.

Feng, N. Y., & Bass, A. H. (2016). "Singing" fish rely on circadian rhythm and melatonin for the timing of nocturnal courtship vocalization. *Current Biology, 26*, 2681–2689.

Feng, N. Y., Marchaterre, M. A., & Bass, A. H. (2019). Melatonin receptor expression in vocal, auditory, and neuroendocrine centers of a highly vocal fish, the plainfin midshipman (Porichthys notatus). *The Journal of Comparative Neurology, 527*, 1362–1377.

Foran, C. M., Myers, D. A., & Bass, A. H. (1997). Modification of gonadotropin releasing hormone (GnRH) mRNA expression in the retinal-recipient thalamus. *General and Comparative Endocrinology, 106*, 251–264.

Forlano, P. M., & Bass, A. H. (2005a). Seasonal plasticity of brain aromatase mRNA expression in glia: Divergence across sex and vocal phenotypes. *Journal of Neurobiology, 65*, 37–49.

Forlano, P. M., & Bass, A. H. (2005b). Steroid regulation of brain aromatase expression in glia: Female preoptic and vocal motor nuclei. *Journal of Neurobiology, 65*, 50–58.

Forlano, P. M., Deitcher, D. L., Myers, D. A., & Bass, A. H. (2001). Anatomical distribution and cellular basis for high levels of aromatase activity in the brain of teleost fish: Aromatase enzyme and mRNA expression identify glia as source. *The Journal of Neuroscience, 21*, 8943–8955.

Forlano, P. M., Deitcher, D. L., & Bass, A. R. H. (2005). Distribution of estrogen receptor alpha mRNA in the brain and inner ear of a vocal fish with comparisons to sites of aromatase expression. *The Journal of Comparative Neurology, 483*, 91–113.

Goodson, J. L. (1998). Territorial aggression and dawn song are modulated by septal vasotocin and vasoactive intestinal polypeptide in male field sparrows (Spizella pusilla). *Hormones and Behavior, 34*, 67–77.

Goodson, J. L., & Bass, A. H. (2000). Forebrain peptides modulate sexually polymorphic vocal circuitry. *Nature, 403*, 769–772.

Grober, M. S., & Bass, A. H. (1991). Neuronal correlates of sex-role change in labrid fishes – Lhrh-like immunoreactivity. *Brain, Behavior and Evolution, 38*, 302–312.

Grober, M. S., Jackson, I. M. D., & Bass, A. H. (1991). Gonadal-steroids affect Lhrh preoptic cell number in a sex-role changing fish. *Journal of Neurobiology, 22*, 734–741.

Grober, M. S., Fox, S. H., Laughlin, C., & Bass, A. H. (1994). Gnrh cell-size and number in a teleost fish with 2 male reproductive morphs – Sexual-maturation, final sexual status and body-size allometry. *Brain, Behavior and Evolution, 43*, 61–78.

Hopkins, C. D., & Bass, A. H. (1981). Temporal coding of species recognition signals in an electric fish. *Science, 212*, 85–87.

Insel, T. R., & Young, L. J. (2000). Neuropeptides and the evolution of social behavior. *Current Opinion in Neurobiology, 10*, 784–789.

Insel, T. R., Winslow, J. T., Wang, Z., & Young, L. J. (1998). Oxytocin, vasopressin, and the neuroendocrine basis of pair bond formation. *Advances in Experimental Medicine and Biology, 449*, 215–224.

Knapp, R., Wingfield, J. C., & Bass, A. H. (1999). Steroid hormones and paternal care in the plainfin midshipman fish (Porichthys notatus). *Hormones and Behavior, 35*, 81–89.

Knapp, R., Marchaterre, M. A., & Bass, A. H. (2001). Relationship between courtship behavior and steroid hormone levels in parental male plainfin midshipman fish. *Hormones and Behavior, 39*, 335.

Lee, J. S. F., & Bass, A. H. (2004). Does exaggerated morphology preclude plasticity to cuckoldry in the midshipman fish (Porichthys notatus)? *Naturwissenschaften, 91*, 338–341.

Lee, J. S. F., & Bass, A. H. (2005). Differential effects of 11-ketotestosterone on dimorphic traits in a teleost with alternative male reproductive morphs. *Hormones and Behavior, 47*, 523–531.

McKibben, J. R., & Bass, A. H. (1998). Behavioral assessment of acoustic parameters relevant to signal recognition and preference in a vocal fish. *The Journal of the Acoustical Society of America, 104*, 3520–3533.

Nottebohm, F., & Arnold, A. P. (1976). Sexual dimorphism in vocal control areas of the songbird brain. *Science, 194*, 211–213.

Nottebohm, F., Stokes, T. M., & Leonard, C. M. (1976). Central control of song in canary, serinus-canarius. *The Journal of Comparative Neurology, 165*, 457–486.

Pritz, M. B., Bass, A. H., & Northcut, R. G. (1973). Simple apparatus and training techniques for teaching turtles to perform a visual discrimination task. *Copeia, 1973*(1), 181–183.

Remage-Healey, L., & Bass, A. H. (2004). Rapid, hierarchical modulation of vocal patterning by steroid hormones. *The Journal of Neuroscience, 24*, 5892–5900.

Remage-Healey, L., & Bass, A. H. (2005). Rapid elevations in both steroid hormones and vocal signaling during playback challenge: A field experiment in Gulf toadfish. *Hormones and Behavior, 47*, 297–305.

Remage-Healey, L., & Bass, A. H. (2006). From social behavior to neural circuitry: Steroid hormones rapidly modulate advertisement calling via a vocal pattern generator. *Hormones and Behavior, 50*, 432–441.

Remage-Healey, L., & Bass, A. H. (2007). Plasticity in brain sexuality is revealed by the rapid actions of steroid hormones. *The Journal of Neuroscience, 27*, 1114–1122.

Rubow, T. K., & Bass, A. H. (2009). Reproductive and diurnal rhythms regulate vocal motor plasticity in a teleost fish. *The Journal of Experimental Biology, 212*, 3252–3262.

Schlinger, B. A., Creco, C., & Bass, A. H. (1999). Aromatase activity in the hindbrain vocal control region of a teleost fish: Divergence among males with alternative reproductive tactics. *Proceedings of the Royal Society of London. Series B: Biological Sciences, 266*, 131–136.

Schneider, G. E. (1969). Two visual systems. *Science, 163*, 895–902.

Sisneros, J. A., Forlano, P. M., Deitcher, D. L., & Bass, A. H. (2004a). Steroid-dependent auditory plasticity leads to adaptive coupling of sender and receiver. *Science, 305*, 404–407.

Sisneros, J. A., Forlano, P. M., Knapp, R., & Bass, A. H. (2004b). Seasonal variation of steroid hormone levels in an intertidal-nesting fish, the vocal plainfin midshipman. *General and Comparative Endocrinology, 136*, 101–116.

Tripp, J. A., Salas-Allende, I., Makowski, A., & Bass, A. H. (2020). Mating behavioral function of preoptic galanin neurons is shared between fish with alternative male reproductive tactics and tetrapods. *Journal of Neuroscience: The Official Journal of the Society for Neuroscience, 40*, 1549–1559.

Judith M. Stern

41

Joseph S. Lonstein

Abstract

Judith M. Stern was born in 1944 in Brooklyn, NY. As a student at Brooklyn College, she was exposed to the field of physiological psychology and developed an interest in how the brain controls behavior. Her scientific path involved graduate training at the Rutgers Newark Institute of Animal Behavior, postdoctoral work at Stanford University, and then an 35-year career as a faculty member at Rutgers University in New Brunswick. Much of Stern's scientific research focused on the behavioral neuroendocrinology of parenting and it involved a broad array of animals (birds, rodents, humans). Stern's research provided critical early knowledge regarding where steroid hormones bind in the brain to regulate parenting, how motherhood alters the hypothalamic-pituitary-adrenal axis and vice versa, how suckling and other offspring cues affect maternal prolactin levels, the indispensable role of somatosensation for maternal caregiving, and how motherhood elicits neuroplasticity in the female cerebral cortex. Threads of Stern's work can still be seen in a number of current research subfields studying the endocrinology and neurobiology of non-human and human motherhood.

Keywords

Adrenal glands · Estrogen · Lactation · Maternal behavior · Nursing · Rats · Ring doves · Somatosensation

J. S. Lonstein (✉)
Department of Psychology & Neuroscience Program, Interdisciplinary Science & Technology Building (ISTB), Michigan State University, East Lansing, MI, USA
e-mail: lonstein@msu.edu

R. J. Nelson, Z. M. Weil (eds.), *Biographical History of Behavioral Neuroendocrinology*, https://doi.org/10.1007/978-3-031-12970-4_41

Early Life and Education

Judith Marlena Stern was born in Brooklyn, New York on 4 April 1944 to a New York City-born mother of Russian Jewish descent and an Egyptian-born father who as a child, immigrated to the United States with his Ashkenazi Jewish family, which eventually settled in New York City. Stern's father died prematurely from a heart attack when Judy was just six months old, after which she and her two older brothers, then aged 8 and 14 years old, were raised by their poorly educated and financially struggling mother. It was against this background that Stern found respite in her education. Despite little formal exposure to science in elementary school, and living in a densely urban environment, Stern's interest in the natural world was sparked by neighborhood trees, city parks, a pet dog, and her fifth-grade teacher bringing a radio into the classroom for the students to listen to a science program for children regularly broadcast on WNYC radio. After performing strongly on standardized tests of intellectual functioning often used at the time to determine 6th-grade students' upcoming junior high school placement, Stern was enrolled in accelerated science and other courses that would further develop her young, curious mind.

During Stern's early education in the late 1940s–1950s, girls' interest in science was often, if anything, expected to lead to a career as a high school teacher. Her beliefs about this likely trajectory began to change while attending the renowned James Madison High School in Brooklyn that had educated thousands of first- and second-generation Americans and boasts as graduates a U.S. Supreme Court Justice (Ruth Bader Ginsburg), five Nobel Laureates, numerous US senators (e.g., Chuck Schumer, Berine Sanders), and many prominent authors and academics. Stern thrived in this environment with its advanced curriculum and many honors courses. After graduating high school at age 16 in 1960, she enrolled at Brooklyn College because of its free tuition and strong pressure to continue living at home with her widowed mother. As a double major in biology and psychology, Stern was particularly inspired by a course she took in Physiological and Comparative Psychology, which led to the search for research opportunities studying the biological basis of behavior.

As a rising senior, Stern responded to a flyer in the biology building on campus advertising National Science Foundation undergraduate research assistantships. She was accepted into this competitive 9-month program at New York's Museum of Natural History Department of Animal Behavior, directed by Lester Aronson who studied reproduction in fish and cats. Stern was offered a choice between two research opportunities, one studying sexual behavior in male cats guided by Aronson, and the other with a researcher examining the role of the cerebellum in avoidance learning in *Tilapia macrocephala*. Stern chose the latter based on her interest in the brain control of behavior and worked on the retention of avoidance conditioning (light/shock two-way shuttle) following cerebellum ablation. Although avoidance of shock was recorded automatically, Stern's notes revealed an otherwise undetected partial memory retention, such that the ablated fish waited at the hole between compartments when the light came on, quickly going through to the other side when the shock began. This was her first appreciation of the critical importance of careful observation for understanding animal behavior.

Also during her senior year, Stern attended a lecture at Brooklyn College given by Danny Lehrman, the charismatic founder and director of the Institute for Animal Behavior (IAB) at Rutgers University (Newark campus) and its first National Academy of Sciences member. Lehrman had conducted his doctoral research at the New York Museum of Natural History Department of Animal Behavior and earned early fame with his critique of Konrad Lorenz' theory of instinctive behavior (Lehrman, 1953). He continued to be well known for his theoretical and empirical research on animal behavior, including the behavioral endocrinology of reproduction in ring doves. It was after his lecture at Brooklyn College, which included his unforgettable demonstration illustrating the male ring dove's courting bow-coo, that Stern met Lehrman to discuss her desire to conduct graduate work with him, which she began at the IAB in 1964.

Rutgers' Institute of Animal Behavior: Hormonal Control of Ring Dove Reproduction

At the IAB, Stern was surrounded by an unparalleled intellectual community of faculty (including Lehrman, Jay Rosenblatt, Colin Beer, Barry Komisaruk, and Harvey Feder), postdoctoral fellows (including Ron Barfield, Ben Sachs, and Jack Hailman), and fellow graduate students (including Joe Terkel, Celia Moore, George Michel, Rae Silver, Alison Fleming, and David Crews). Stern was lastingly impressed by Lehrman's critical role in the influx of women scientists to the IAB and the field of behavioral endocrinology more broadly. This intellectual community was further enriched by monthly seminars by preeminent researchers from all over the world, with opportunities for graduate students to interact with them, and where Stern was always keen to hear Lehrman's invariably incisive questions and comments. One year Lehrman also arranged a special course on Neuropsychology with a memorable lineup of guest lecturers that included Lester Aronson, Philip Teitelbaum, Hans-Lukas Teuber, and Philip Zeigler.

Stern's first project in the Lehrman lab involved working with a senior graduate student on nestbuilding in ring doves, but Lehrman encouraged Stern to conceptualize and work on an independent project. Stern had hypothesized, and then demonstrated, a synergistic role for testosterone in progesterone-induced incubation behavior in male ring doves, which was in contrast to progesterone's inhibition of the male's testosterone-dependent bow-cooing. This resulted in Stern's first scientific publication, and her only one with Lehrman before his premature death a few years after Stern completed her Ph.D. under his mentorship (Stern & Lehrman, 1969). When developing her doctoral research, Stern sought further insights into how hormones act on the brain. The mid-to-late 1960s was an exciting time to be working in the nascent fields of neuroscience and neuroendocrinology. Eisenfeld and Axelrod's groundbreaking work at the National Institutes of Health had recently documented the selective distribution of radioactive estradiol binding in the central nervous system (e.g., Eisenfeld & Axelrod, 1965, 1966), which made it possible to much better investigate specific brain sites where hormones acted to influence behavior. With Lehrman's enthusiastic support, Stern visited Eisenfeld in his new laboratory at the Yale University School of Medicine to learn how to perform steroid receptor binding in male laboratory rats. This led to a fruitful collaboration, including work they conducted together in the summer of 1967 at Yale, showing inhibitory effects of progesterone on ^{3}H-testosterone distribution and metabolism in the periphery and brain of laboratory rats (Stern & Eisenfeld, 1969, 1971).

When back at the IAB, Stern applied this new technique for the first time in male ring doves to examine relative testosterone binding across brain regions and demonstrated that binding was inhibited by pre-treatment with progesterone but not corticosterone (Stern, 1972). Then, at the first Eastern Regional Conference on Reproductive Behavior hosted in East Lansing, Michigan (where Stern was one of very few women attendees), a meeting with Bruce McEwen and Richard Zigmond of Rockefeller University led to a collaboration showing that the radioactive testosterone in the male ring dove hypothalamus was localized to cell nuclei, but not other cellular compartments (Zigmond et al., 1972). The foundations established by this and other similar research are still evident in studies continuing to ask where and how steroid hormone receptors and their associated proteins drive cellular and molecular outcomes in the brain (see Balthazart et al., 2018; Diotel et al., 2018; Levin & Hammes, 2016).

Postdoc at Stanford University: Postpartum Stress and Developmental Psychobiology

For her postdoctoral research (1970–1973), Stern joined the laboratory of Seymour (Gig) Levine in the Department of Psychiatry at Stanford University, whom she had met after a seminar he delivered at the IAB in 1969. Joining Levine's laboratory offered Stern numerous opportunities – to expand her knowledge in behavioral neuroendocrinology, to leave the New York metropolitan area, to interact with other notable Stanford faculty (e.g., Julian Davidson, David Hamburg, Karl Pribram), to start considering the clinical applications of her basic research results, and to begin participating in the newly formed International Society of Developmental Psychobiology (ISDP), of which she would years later be elected President (1992–93).

Stern found Gig's enthusiasm for research and keen "nose" for interesting projects to be infectious. The support and flexibility she was afforded in the Levine lab allowed Stern to meld her existing interests in parenting behavior with the lab's broad interests in stress, resulting in a line of research on how motherhood affected HPA axis function in female rats and, conversely, how adrenal hormones regulated their maternal caregiving behaviors. This work was some of the very first to establish that early-to-mid lactation in laboratory rats involves a blunting of the HPA axis response to many stressors and that suckling by a sufficient number of pups was required for this phenomenon (Stern & Levine, 1972, 1974; Stern et al., 1973b).

These studies further demonstrated that neither prolactin nor progesterone were responsible for this change in postpartum HPA function (Stern & Levine, 1974), as well as showed that the blunted HPA axis response during lactation was not intimately tied to dams' postpartum "emotional" behaviors (Stern et al., 1973a). Understanding how female reproduction alters basal and stress-induced HPA axis activity, and how glucocorticoids regulate a mother's interaction with their young, has since become a large scientific field involving both basic and clinical researchers, often with relevance to women's peripartum mental illness (e.g., Brunton, 2016; Dickens & Pawluski, 2018; Hillerer et al., 2012). When Levine passed away, Stern helmed a special edition of *Developmental Psychobiology* in tribute to his long and distinguished career, edited along with his former graduate student, Joanne Weinberg, and his former postdoctoral fellow, Michael Hennessey (Stern et al., 2010).

Independent Research Career and Professional Struggles

When Stern was seeking a tenure-track faculty position, she heard about one at the New Brunswick campus of Rutgers University through an old-boys network – specifically, information that was casually passed from the renowned developmental psychobiologist Victor Denenberg to her postdoc advisor Gig Levine. In the early 1970s, the all-men's Rutgers College was undergoing gender integration of its faculty and student body, and since many of its departments had few or no women faculty members, federal and state affirmative action lines were created to remedy this. In 1973, Stern became the second woman ever appointed to the Rutgers College Psychology Department as Assistant Professor. But in stark contrast with the nurturing environments provided by her mentors Lehrman at the IAB and Levine at Stanford, Stern almost immediately found herself in a largely unfriendly, misogynistic environment. Sadly consistent with the experiences of many new faculty from underrepresented groups still today, Stern was excluded from opportunities to socialize with her male colleagues, her involvement in the nascent academic women's groups on campus was dismissed if not denigrated by her male peers, she suffered with repeated ogling and sexually suggestive comments made by a male colleague, and soon realized she was being paid less than newly hired men on campus. Nonetheless, Stern was successful in establishing her lab, in obtaining grant support, and in publishing her early independent research on steroid hormones and maternal motivation in laboratory rats (Stern & Mackinnon, 1976; Mackinnon & Stern, 1977). Other experiments driven by student thesis projects during this time included studying the effects of maternal stress during pregnancy on offspring sexual behavior and emotionality (Chapman & Stern, 1978, 1979) and determining the consequences of in utero alcohol on offspring reactivity and learning in rats (Anandam et al., 1980; Anandam & Stern, 1980). Furthermore, Stern was appointed as a full member of the National Institute of Mental Health's Neuropsychology Research Review Committee ("study section") from 1974 to 1978, which was a tremendous honor for a junior faculty member and provided her with invaluable learning and networking experiences.

Given Stern's academic record and accomplishments since joining the Psychology Department at Rutgers College, it came as a great shock in 1977 when her department did not recommend her for promotion and tenure, overriding her concurrent positive recommendation by the university-level Appointment and Promotion Committee. Thus began a several-years long appeals process, which took a horrendous toll on Stern personally and professionally. It was not at all unusual in that era for women faculty to experience maltreatment by their academic institutions, but Stern's case was thought by her and others at Rutgers and elsewhere to be particularly egregious. She learned that there were inaccuracies about her academic performance presented by colleagues at her departmental tenure review, but the most sordid aspect was the solicitation of an evaluation letter from a man who had been her supportive professor in college but was later sexually spurned by Stern after propositioning her at an academic conference. This personally tainted letter had undue influence on her department colleagues, but after a university-directed review in 1980 by a national ad hoc committee consisting of prominent individuals in Stern's field (including Eliot Stellar at the University of Pennsylvania who was a "founder" of the field of behavioral neuroscience), she was granted promotion and tenure retroactive to 1978.

A silver lining of Stern's tenure woes was the unexpected research that came next during her first sabbatical. With little prior planning after her promotion, Stern sought respite at Boston Children's Hospital, where her graduate-school friend George Michel was working with the infant developmentalist Peter Woolf. Because Stern was unable to find an appropriate 1-year project there, she contacted Melvin Konner at Harvard's Anthropology Department, having met him in the spring at Rutgers after reading his accounts of nursing and birth-spacing among !Kung hunter-gatherers (Konner & Worthma, 1980). Konner suggested contacting Seymour Reichlin, a prominent neuroendocrinologist at Tufts-New England Medical Center. There, Stern was able to translate her previous studies on suckling, prolactin, and nursing behavior in laboratory rats to a unique study of nursing behavior during prolonged lactation in American women, study its relationships to circulating prolactin levels and the duration of postpartum amenorrhea, to compare some of these features with a group of pre-industrialized (i.e., !Kung) mothers (Stern et al., 1986) and to examine circadian rhythm effects (Stern & Reichlin, 1990).

Sensory Regulation of Maternal Behavior

After returning to Rutgers, Stern began to rebuild resources and momentum. Her research exploring various facets regulating the onset of maternal behavior in laboratory rats eventually led to the insight that the mothers' perioral somatosensory inputs – but not olfaction, vision, and hearing – were singularly essential for the motivation and ability to retrieve pups (reviewed in Stern (1989, 1996)). Perioral inputs were also essential for the characteristically high aggression that postpartum dams exhibit toward intruders (Kolunie & Stern, 1995; Stern & Kolunie, 1991, 1993). Indeed, prior work by Kenyon et al. (1981, 1983) had reported that perioral anesthesia with lidocaine or infraorbital denervation abolished pup retrieval by postpartum rats. Analogously, Jacquin and Zeigler (1983) had demonstrated the critical role of trigeminal orosensation (but not the sense of taste) in rat feeding motivation and behavior. Stern further made a key observation that although periorally anaptic dams were disinterested in pups, if the hungry pups were mobile enough to gather under their mother, they could successfully attach to nipples, begin to suckle, and elicit their dams' upright crouched nursing (Stern & Johnson, 1989). This ventroflexion nursing posture, reminiscent of the *lordosis* posture of female sexual receptivity involving the opposite dorsiflexion of the spinal column, was subsequently termed *kyphosis* (Stern, 1996). Another series of experiments by Stern manipulating the pups' ability to provide tactile stimulation to their dam, or the dam's ability to detect those pup inputs directed to her ventrum, definitively demonstrated that effective ventral somatosensation elicits kyphotic nursing (Stern & Johnson, 1990; Stern et al., 1992). Also akin to the active and inactive components of female sexual behavior in rats, Stern's research during this time found that dopamine receptor inhibition interferes with maternal motivation, retrieval, and licking (Stern & Keer, 1999), but enhances immobile nursing behavior (Stern and Taylor, 1991), by acting on the nucleus accumbens (Keer & Stern, 1999).

Establishing the somatosensory regulation of retrieval and nursing behaviors in rats further enabled Stern to revise the much earlier theory of Frank Beach that inaccurately emphasized the equipotentiality of the senses in regulating reproductive behaviors (Beach & Jaynes, 1956; Stern, 1990). It also led Stern and her students to refute a more recent theory that nursing bouts in rats were terminated by maternal hyperthermia (e.g., Leon et al., 1978), a contention based on automatic monitoring of nest bout duration without direct observations of maternal behavior or measures of litter weight gains that would have revealed some of its flaws (Stern & Johnson, 1990; Stern & Lonstein, 1996; Stern & Azzara, 2002; Stern, 1996).

While on a colloquia tour in California in 1989, Stern visited Michael Merzenich at the University of California San Francisco due to her respect and interest in his seminal work on somatosensation and cortical neuroplasticity (e.g., Merzenich et al., 1988). This meeting led to a collaboration on cortical plasticity and nursing behavior, which Merzenich considered to be "a natural use experiment," which was in contrast to his and others' previous work in this area involving extensive training of particular body parts in monkeys (e.g., Jenkins et al., 1990). In subsequent research trips, the team found that the somatosensory cortex representation of the ventral skin surrounding the nipples was almost twice as large in lactating rats compared to either virgin females or postpartum rats deprived of their pups, while receptive field sizes shrunk (Xerri et al., 1994). This was the very first study of a function-related neuroplastic change within the maternal brain, and this groundbreaking work has since led to an "explosion" of research on various types of neuroplasticity in maternal laboratory rodents and women (for recent reviews, see Duarte-Guterman et al. (2019), Martínez-García et al. (2021), and Pawluski et al. (2022)).

Soon thereafter, the advent of visualizing activity of the immediate-early gene c-*fos* as an indirect marker of acute neural stimulation provided neuroscientists with a way to identify or confirm brain sites involved in behaviors of interest. Many forebrain sites were known to express Fos protein in response to the display of active maternal behaviors in rats (Fleming et al., 1994; Numan & Numan, 1994; Lonstein et al., 1998); Stern's thoughts about the similarities between lordosis and kyphosis – both bilaterally symmetrical postures, involved immobility and rigid leg support, and naturally elicited by dorsal or ventral trunk tactile stimulation resulting in dorsiflexion or ventroflexion – led us to use visualization of *c-fos* followed by site-specfic lesions in the brain identify a particular subregion of the midbrain periaqueductal gray [PAG; known for its role in lordosis (e.g., Pfaff et al., 1994)] as the first brain site revealed to be selectively involved in the postural control of kyphotic nursing (Lonstein & Stern, 1997a, b).

Conclusion

Judith M. Stern's curiosity, resilience, ambition, and hard work were nurtured by an era of broad societal support for scientific inquiry, by women's increasing participation in that endeavor, and especially by numerous teachers, mentors and collaborators who provided support, encouragement, and intellectual structure. Consequently, Stern spent decades at the forefront of the field studying the behavioral neuroendocrinology of parental caregiving. Of course, her role as a faculty member at Rutgers University also involved her teaching of graduate and undergraduate courses, including a course in Hormones and Behavior (with admirable help of Randy Nelson's textbook, *An Introduction to Behavioral Endocrinology*) as well as a very popular undergraduate course on the Psychology of Sex and Gender. Stern's research contributed to some of the earliest knowledge about where and how steroid hormones could influence the brain to drive behavior and presumably vice versa; understanding the reciprocal relationships among motherhood, hormones, and sensory cues from infants; and lastly the sensory and neurobiological control of retrieval, nursing, and other postpartum caregiving behaviors. Threads emerging from her contributions can easily be traced to current-day research continuing to study very similar questions regarding how and why parents interact with their infants, often using molecular and genetic tools that Stern and her associates in graduate school could only have only dreamed about. It is an understatement to say that considerable progress had been made over the past 60 years by many scientists, including Stern and her many collaborators, on unraveling why and how parents display their highly complex caregiving behaviors.

Acknowledgments The author would like to thank Judith M. Stern for the numerous interviews and editorial contributions that contributed to this chapter.

References

Anandam, N., & Stern, J. M. (1980). Alcohol *in utero*: Effects on preweaning appetitive learning. *Neurobehavioral Toxicology, 2*, 199–205.

Anandam, N., Felegi, W., & Stern, J. M. (1980). In utero alcohol heightens juvenile reactivity. *Pharmacology, Biochemistry, and Behavior, 13*, 531–535.

Balthazart, J., Choleris, E., & Remage-Healey, L. (2018). Steroids and the brain: 50 years of research, conceptual shifts and the ascent of non-classical and membrane-initiated actions. *Hormones and Behavior, 99*, 1–8.

Beach, F. A., & Jaynes, J. (1956). Studies of maternal retrieving in rats. III. Sensory cues involved in the lactating female's response to her young. *Behaviour, 10*, 104–125.

Brunton, P. J. (2016). Neuroactive steroids and stress axis regulation: Pregnancy and beyond. *The Journal of Steroid Biochemistry and Molecular Biology, 160*, 160–168.

Chapman, R. H., & Stern, J. M. (1978). Maternal stress and pituitary-adrenal manipulations during pregnancy in rats: Effects on morphology and sexual behavior of male offspring. *Journal of Comparative and Physiological Psychology, 92*, 1074–1083.

Chapman, R. H., & Stern, J. M. (1979). Failure of severe maternal stress or ACTH during pregnancy to affect emotionality of male rat offspring: Implications of litter effects for prenatal studies. *Developmental Psychobiology, 12*, 255–267.

Dickens, M. J., & Pawluski, J. L. (2018). The HPA axis during the perinatal period: Implications for perinatal depression. *Endocrinology, 159*, 3737–3746.

Diotel, N., Charlier, T. D., Lefebvre d'Hellencourt, C., Couret, D., Trudeau, V. L., Nicolau, J. C., Meilhac, O., Kah, O., & Pellegrini, E. (2018). Steroid transport, local synthesis, and signaling within the brain: Roles in neurogenesis, neuroprotection, and sexual behaviors. *Frontiers in Neuroscience, 12*, 84.

Duarte-Guterman, P., Leuner, B., & Galea, L. A. (2019). The long and short term effects of motherhood on the brain. *Frontiers in Neuroendocrinology, 53*, 100740.

Eisenfeld, A. J., & Axelrod, J. (1965). Selectivity of estrogen distribution in tissues. *Journal of Pharmacology and Experimental Therapeutics, 150*, 469–475.

Eisenfeld, A. J., & Axelrod, J. (1966). Effect of steroid hormones, ovariectomy, estrogen pretreatment, sex and immaturity on the distribution of 3H-estradiol. *Endocrinology, 79*, 38–42.

Fleming, A. S., Suh, E. J., Korsmit, M., & Rusak, B. (1994). Activation of Fos-like immunoreactivity in the medial preoptic area and limbic structures of maternal and social interactions in rats. *Behavioral Neuroscience, 108*, 724.

Hillerer, K. M., Neumann, I. D., & Slattery, D. A. (2012). From stress to postpartum mood and anxiety disorders: How chronic peripartum stress can impair maternal adaptations. *Neuroendocrinology, 95*, 22–38.

Jacquin, M. F., & Zeigler, H. P. (1983). Trigeminal orosensation and ingestive behavior in the rat. *Behavioral Neuroscience, 97*, 62–97.

Jenkins, W. M., Merzenich, M. M., Ochs, M. T., Allard, T., & Guic-Robles, E. (1990). Functional reorganization of primary somatosensory of in adult owl monkeys after behaviorally controlled tactile stimulation. *Journal of Neurophysiology, 6*, 82–104.

Keer, S. E., & Stern, J. M. (1999). Dopamine receptor blockade in the nucleus accumbens inhibits maternal retrieval and licking, but enhances nursing behavior in lactating rats. *Physiology & Behavior, 67*, 659–669.

Kenyon, P., Cronin, P., & Keeble, S. (1981). Disruption of maternal retrieving by perioral anesthesia. *Physiology & Behavior, 27*, 313–321.

Kenyon, P., Cronin, P., & Keeble, S. (1983). Role of the infraorbital nerve in retrieving behavior in lactating rats. *Behavioral Neuroscience, 97*, 255–269.

Kolunie, J. M., & Stern, J. M. (1995). Maternal aggression in rats: Effects of olfactory bulbectomy, $ZnSO_4$-induced anosmia, and vomeronasal organ removal. *Hormones and Behavior, 29*, 492–518.

Konner, M., & Worthma, C. (1980). Nursing frequency, gonadal function, and birth spacing among !Kung hunter-gatherers. *Science, 207*, 788–791.

Lehrman, D. S. (1953). A critique of Konrad Lorenz's theory of instinctive behavior. *The Quarterly Review of Biology, 28*, 337–363.

Leon, M., Croskerry, P. G., & Smith, G. K. (1978). Thermal control of mother-young contact inrats. *Physiology & Behavior, 21*, 793–811.

Levin, E. R., & Hammes, S. R. (2016). Nuclear receptors outside the nucleus: Extranuclear signalling by steroid receptors. *Nature Reviews Molecular Cell Biology, 17*, 783–797.
Lonstein, J. S., & Stern, J. M. (1997a). Role of the midbrain periaqueductal gray in maternal nurturance and aggression: c-fos and electrolytic lesion studies in lactating rats. *The Journal of Neuroscience, 17*, 3364–3378.
Lonstein, J. S., & Stern, J. M. (1997b). Somatosensory contributions to c-fos activation within the caudal periaqueductal gray of lactating rats: Effects of perioral, rooting, and suckling stimuli from pups. *Hormones and Behavior, 32*, 155–166.
Lonstein, J. S., Simmons, D. A., Swann, J. M., & Stern, J. M. (1998). Forebrain expression of *c-fos* due to active maternal behaviour in lactating rats. *Neuroscience, 82*, 267–281.
Mackinnon, D. A., & Stern, J. M. (1977). Pregnancy duration and fetal number: Effects on maternal behavior in rats. *Physiology & Behavior, 18*, 793–797.
Martínez-García, M., Paternina-Die, M., Desco, M., Vilarroya, O., & Carmona, S. (2021). Characterizing the brain structural adaptations across the motherhood transition. *Frontiers in Global Women's Health, 2*, 742775.
Merzenich, M. M., Recanzone, G., Jenkins, W. M., Allard, T., & Nudo, R. J. (1988). Cortical representational plasticity. In P. Rakic & W. Singer (Eds.), *Neurobiology of neocortex* (pp. 42–67). Wiley.
Numan, M., & Numan, M. J. (1994). Expression of Fos-like immunoreactivity in the preoptic area of maternally behaving virgin and postpartum rats. *Behavioral Neuroscience, 108*, 379.
Pawluski, J. L., Hoekzema, E., Leuner, B., & Lonstein, J. S. (2022). Less can be more: Fine tuning the maternal brain. *Neuroscience and Biobehavioral Reviews, 133*, 104475.
Pfaff, D. W., Schwartz-Giblin, S., Mccarthy, M. M., & Kow, L. M. (1994). Cellular and molecular mechanisms of female reproductive behaviors. In E. Knobil & J. D. Neil (Eds.), *The physiology of reproduction* (pp. 107–220). Raven Press.
Stern, J. M. (1972). Androgen accumulation in hypothalamus and anterior pituitary of male ring doves; influence of steroid hormones. *General and Comparative Endocrinology, 18*, 439–449.
Stern, J. M. (1989). Maternal behavior: Sensory, hormonal, and neural determinants. In F. R. Brush & S. Levine (Eds.), *Psychoendocrinology* (pp. 105–226). Academic Press.
Stern, J. M. (1990). Multisensory regulation of maternal behavior and masculine sexual behavior: A revised view. *Neuroscience and Biobehavioral Reviews, 14*, 183–200.
Stern JM (1996). Somatosensation and maternal care in Norway rats. In *Advances in the study of behavior* (Vol. 25, pp. 243–294). Academic Press.
Stern, J. M., & Azzara, A. V. (2002). Thermal control of mother–young contact revisited: Hyperthermic rats nurse normally. *Physiology & Behavior, 77*, 11–18.
Stern, J. M., & Eisenfeld, A. J. (1969). Androgen accumulation and binding to macromolecules in seminal vesicles: Inhibition by cyproterone. *Science, 166*(3902), 233–235.
Stern, J. M., & Eisenfeld, A. J. (1971). Distribution and metabolism of 3H-testosterone in castrated male rats; effects of cyproterone, progesterone and unlabeled testosterone. *Endocrinology, 88*, 1117–1125.
Stern, J. M., & Johnson, S. K. (1989). Perioral somatosensory determinants of nursing behavior in Norway rats (*Rattus norvegicus*). *Journal of Comparative Psychology, 103*, 269–280.
Stern, J. M., & Johnson, S. K. (1990). Ventral somatosensory determinants of nursing behavior in Norway rats. I. Effects of variations in the quality and quantity of pup stimuli. *Physiology & Behavior, 47*, 993–1011.
Stern, J. M., & Kolunie, J. M. (1991). Trigeminal lesions and maternal behavior in Norway rats: I. Effects of cutaneous rostral snout denervation on maintenance of nurturance and maternal aggression. *Behavioral Neuroscience, 105*, 984.
Stern, J. M., & Kolunie, J. M. (1993). Maternal aggression of rats is impaired by cutaneous anesthesia of the ventral trunk, but not by nipple removal. *Physiology & Behavior, 54*, 861–868.
Stern, J. M., & Lehrman, D. S. (1969). Role of testosterone in progesterone-induced incubation behaviour in male ring doves (*Streptopelia risoria*). *The Journal of Endocrinology, 44*, 13–22.
Stern, J. M., & Levine, S. (1972). Pituitary-adrenal activity in the postpartum rat in the absence of suckling stimulation. *Hormones and Behavior, 3*, 237–246.
Stern JM, Levine S (1974). Psychobiological aspects of lactation in rats. In *Progress in brain research* (Vol. 41, pp. 433–444). Elsevier.

Stern, J. M., & Lonstein, J. S. (1996). Nursing behavior in rats is impaired in a small nestbox and with hyperthermic pups. *Developmental Psychobiology, 29*, 101–122.
Stern, J. M., & Mackinnon, D. A. (1976). Postpartum, hormonal, and nonhormonal induction of maternal behavior in rats: Effects on T-maze retrieval of pups. *Hormones and Behavior, 7*, 305–316.
Stern, J. M., & Reichlin, S. (1990). Prolactin circadian rhythm persists throughout lactation in women. *Neuroendocrinology, 51*, 31–37.
Stern, J. M., & Taylor, L. A. (1991). Haloperidol inhibits maternal retrieval and licking, but enhances nursing behavior and litter weight gains in lactating rats. *Journal of Neuroendocrinology, 3*, 591–596.
Stern, J. M., & Keer, S. E. (1999). Maternal motivation of lactating rats is disrupted by low dosages of haloperidol. *Behavioural Brain Research, 99*, 231–239.
Stern, J. M., Erskine, M. S., & Levine, S. (1973a). Dissociation of open-field behavior and pituitary-adrenal function. *Hormones and Behavior, 4*, 149–162.
Stern, J. M., Goldman, L., & Levine, S. (1973b). Pituitary-adrenal responsiveness during lactation in rats. *Neuroendocrinology, 12*, 179–191.
Stern, J. M., Konner, M., Herman, T. N., & Reichlin, S. (1986). Nursing behaviour, prolactin and postpartum amenorrhoea during prolonged lactation in American and !Kung Mothers. *Clinical Endocrinology, 25*, 247–258.
Stern, J. M., Dix, L., Bellomo, C., & Thramann, C. (1992). Ventral trunk somatosensory determinants of nursing behavior in Norway rats: 2. Role of nipple and surrounding sensations. *Psychobiology, 20*, 71–80.
Stern, J. M., Weinberg, J., & Hennessy, M. B. (2010). From Freud to a modern understanding of behavioral, physiological, and braindevelopment. *Developmental Psychobiology, 52*, 609–615. *Developmental Psychobiology 50*, 423–426.
Xerri, C., Stern, J. M., & Merzenich, M. M. (1994). Alterations of the cortical representation of the rat ventrum induced by nursing behavior. *The Journal of Neuroscience, 14*, 1710–1721.
Zigmond, R. E., Stern, J. M., & McEwen, B. S. (1972). Retention of radioactivity in cell nuclei in the hypothalamus of the ring dove after injection of 3H-testosterone. *General and Comparative Endocrinology, 18*, 450–453.

Hidden Gems in Neuroendocrinology and Behavior

42

Carmel Martin-Fairey and Jennifer M. Swann

Abstract

The preceding chapters have been dominated, in terms of numbers, by white men from European backgrounds with a smaller number of women. However, minority researchers have also been a part of behavioral neuroendocrinology. African-American men and women can be found at the genesis of the field, drawing from endocrinology, neuroscience, and behavior to craft current research in behavioral neuroendocrinology. Their contributions are precious as they represent the resilience and tenacity of talented investigators in a segregated society that ignored and repressed their work. This chapter celebrates their contributions and highlights the struggles they faced as minorities in a white world.

Keywords

Diversity · Percy Lavon Julian · Lilian Burwell Lewis · Eleanor Lutia Ison-Franklin · Ruth Smith Lloyd · Robert W Harris III · Antonio Alberto Nuñez · Gregory Florant

Introduction

The proceeding chapters tell one story about the development and history of behavioral neuroendocrinology as a scientific discipline. In terms of total numbers, this story is dominated by white men of European extraction with smaller numbers of

C. Martin-Fairey (✉)
Department of Biology, Harris-Stowe State University, St. Louis, MO, USA
e-mail: MartinCa@hssu.edu

J. M. Swann
Department of Biological Sciences, Lehigh University, Bethlehem, PA, USA
e-mail: Jms5@lehigh.edu

R. J. Nelson, Z. M. Weil (eds.), *Biographical History of Behavioral Neuroendocrinology*, https://doi.org/10.1007/978-3-031-12970-4_42

women. The reasons for this are of course numerous and speak as much to the history of science, race, and power as of the individuals profiled. However, it also does not tell a complete story. Minority researchers have always been a part of behavioral neuroendocrinology. African-American men and women can be found at the genesis of the field, drawing from medicine, endocrinology, neuroscience, and behavior to craft current research in behavioral neuroendocrinology. Their contributions are precious as they represent the resilience and tenacity of talented investigators in a segregated society that ignored and repressed their work. This chapter celebrates their contributions and highlights the struggles they faced as minorities in a white world.

Here we highlight individuals who in various ways contributed to the field we now consider to be behavioral neuroendocrinology. These individuals were not afforded opportunities to delve into the many places where minorities now exist in behavioral, endocrinology, or neuroscience. These unfair times during our history restricted access to the wide-open spaces we hopefully see in the field today. Given that these pioneers, in the face of all that, persevered and left an indelible mark on many of these areas ought to be celebrated today. We posit here that not only are there contributions worth noting, but have truly paved the way for students, research scientists, and faculty of color across the globe today.

A Solid Beginning Early researchers focused on hormones and their role in development.

Percy Lavon Julian (Fig. 42.1)

Summary Percy Lavon Julian was an enterprising steroid chemist and an entrepreneur, who participated actively in the synthesis and large-scale production of steroids from plant compounds. Heading the call of his time to synthesize and create sex and cortical hormones in the lab, Percy Julian cleverly figured out how to synthesize important medicinal compounds such as steroid hormones, progesterone, cortisone, and hydrocortisone from abundant plant sources, making them more affordable to produce. His work forever changed the possibilities of research and analysis of these hormones for research and physician scientists alike around the globe.

What Are Steroids and How Do They Affect Us? Steroids are powerful hormones that regulate both physiology and behavior. We now know that they work by slipping through cell membranes to bind to DNA. We owe much of this knowledge to Percy Julian.

As an organic chemist, Julian was able to mimic nature by creating these steroids in the lab and later by modifying them to make them safer and more effective as drugs. Drug discovery for many hormones we use today in understanding has

Fig. 42.1 Percy Lavon Julian

benefited from Dr. Julian's work. Julian was also an avid activist for scientists of color at every level. He was an outspoken advocate for people of color and a change agent for many chemical societies within and beyond the academy.

Education and Early Career

The grandson of enslaved people, Julian was born in Montgomery, Alabama, in 1899. After high school, he was accepted at DePauw University in Greencastle, Indiana. Majoring in chemistry, he graduated Phi Beta Kappa and as valedictorian of his class in 1920. Despite his outstanding college performance, Fisk University was the only option for postgraduate employment, where he taught chemistry for 2 years before winning an Austin Fellowship to Harvard where he completed a master's degree in organic chemistry. After completing his master's degree at Harvard, he returned to teaching, this time at West Virginia State College and Howard University. During his short tenure at Howard, he was awarded a fellowship that would give him leave to study natural product chemistry in Vienna, Austria, and which would finally afford him the opportunity to earn his doctorate.

Free of teaching and administrative duties, Julian was able to devote himself to research in a manner never afforded him previously. His thesis research involved isolating the active ingredient from and identifying the chemical structure of the component of *Corydalis cava* responsible for its ability to treat pain and heart palpitation. His PhD advisor described Julian as an extraordinary student the likes of which he had never before encountered in all of his career. He would return to Howard with his PhD determined to publish his thesis work, and he would become the first black chemist to publish a first author paper (Julian, 1933). The research he conducted for his doctorate became the foundation of work that fueled his career (Cobb, 1971).

Julian's next research challenge was the Calabar bean brought from Nigeria. An alkaloid had been isolated from the bean, called physostigmine, which was used to treat glaucoma. Using skills he honed in Vienna, he would embark on the characterization and synthesis as an effort that might be his hope for the future (Julian & Pikl, 1933, 1935a, b). He took this on with everything he had, even battling with Robert Robertson, and the papers were immediately recognized as total synthesis in ACS. His work in these experiments was described as elegant. He had now made a worldwide discovery that was acknowledged throughout the globe.

Unfortunately, he remained unsuccessful finding faculty positions in academia. It was now that he would pivot to industry to support his family. Finally, he was offered a job as director of research at the Glidden paint company where he would focus on the soybean. His first assignment, as director of research in the Soya products division, was to isolate the protein from the soybean, which had never been done on an industrial scale. Julian and his team of chemists would indeed isolate the "Alpha protein" protein (the first vegetable protein) and were the first to produce it in bulk in America. Glidden made millions from the protein as an industrial paper coating and later as an ingredient in latex paints. The foundations of impacts on endocrinology started here. Intellectually, he was tired of the applications of his work to textiles and longed to get back to the natural world. Perhaps due to the experiences his family had had with miscarriages, he was enamored with progesterone. At that time, one out of six pregnancies ended in miscarriage or preterm birth. Julian saw this as an opportunity to engage his science to help at-risk mothers who could use progesterone to carry their babies to term. Julian realized that plants had steroids and that these could be converted for use in animals. He stumbled across stigmasterol while at DePauw. The new Glidden Soya department environment was rich with this substance after being discovered in a leak from a soybean tank. He discovered that water would precipitate stigmasterol from soybean oil. Julian was the first person to scale up the process of converting stigmasterol to progesterone. A German group was the first to characterize the steps needed to transfer stigmasterol to progesterone. In 1940, Julian sent the first shipment of an artificial sex hormone produced in America. At that time, the shipment was valued at around $70,000; Julian shipped, under armed guard, a package of progesterone to the Upjohn pharmaceutical company. Testosterone and other steroid hormones such as hydrocortisone and cortisone would follow, making millions for Glidden and thrusting the field of scientific investigation of steroid hormones into high gear (Cullen, 2006; Kyle & Shampo, 1996).

Later Life

While he struggled for consistency in many ways during his formative years, Julian would reap many benefits from his stalwart work effort and dedication to his research in the later years of his life. Being in industry would set him up to benefit from the techniques he pioneered. Julian died in 1975 of cancer. He was ultimately unable to secure a faculty position and thus remained in industry. His private lab was noted to have been the primary employer of black chemists of the day. Prior to his death, he founded Julian Laboratories of Franklin Park, Illinois, and Mexico City (which he eventually sold to Smith, Kline & French) shortly after he left Glidden in 1954. He is reported to have left Glidden with 109 patents. He remained active in organizations dedicated to the advancement of underrepresented minorities through donations and service (Weissmann, 2005).

Impact on the Field

The advances that followed from Julian's work opened an entirely new world of opportunity for scientist interested in the study of steroid hormones. The influences of the industrialization of the isolation and synthesis of these hormones have rippled through all hormonal studies. The commercial availability of steroid hormones led to medical treatments and in some cases almost real-time interventions for serious condition in the reproductive and inflammatory sectors of both human and animal research. The work of Julian would unshackle the hands of many scientists interested in understanding not only how active ingredients in plant affected the physiology of mammals but also a tool to manipulate the byproducts to ascertain the role of chemical structure on function within the behavioral, neural reproductive, and endocrine system. His monumental efforts certainly yielded processes and strategies still in use today. Despite the efforts to give credit to his contributions, many textbooks do not mention figures such as Percy Julian in their pages, so generations of students continue to come forth with no knowledge of the tremendous contributions of his work on methods we use today in behavioral neuroendocrinology.

Lilian Burwell Lewis (Fig. 42.2)

Summary Lillian Burwell Lewis was among the first to study the development of the gonads (Lewis, 1946). Her work with Lincoln Domm (a force in the hormonal regulation of gonadal development and reproductive behavior) focused on the role of hormones in the differentiation of the gonads in ducks (Lewis & Domm, 1948). She broke ground as the first black woman to earn a doctorate at the University of Chicago.

Lilian was born in Meridian, Mississippi, on 13 August 1904 and began her academic studies as an undergraduate with Ernest Everett Just, a seminal biologist and African American at Howard University. After graduating with her bachelor's degree, Burwell held an associate professorship at South Carolina State A&M (1926–1929) and taught at Morgan State while obtaining her degree at the University of Chicago (1929–1931) whereupon she became an associate professor at Tillotson

Fig. 42.2 Lillian Burwell Lewis

College until 1947. Her doctorate earned her a position as a full professor at Winston-Salem State University until she retired in 1971.

Eleanor Lutia Ison-Franklin (Fig. 42.3)

Summary Eleanor Lutia Ison-Franklin was an endocrinologist and medical physiologist. Ison-Franklin's research focused on three areas: cardiac performance in conscious animals, experimental hypertension, and left ventricular hypertrophy. Dr. Ison-Franklin's list of awards and funded grants includes grants from NASA, the National Institutes of Health, and the Washington Heart Association. She would earn the distinction of being the first woman, either black or white, as an associate dean for the administration of the Howard University College of Medicine in 1970.

What's Stress Got to Do with It? We all experience stress in our everyday lives. For some stress takes its toll on their cardiovascular system leading to stroke and heart attack. The basis of our understanding includes the contributions of Eleanor Lutia Ison-Franklin.

Fig. 42.3 Eleanor Franklin

Education and Early Career

Born in Dublin, Georgia Eleanor attended segregated schools graduating as valedictorian from Carver High School in Monroe, GA. She went on to attend the all-female Spelman College in Atlanta where she studied under Eugenia V. Dunn and received a BS in biology *magna* cum *laude* in 1948. After a year as a biology instructor at Spelman, she entered a master's program in zoology at the University of Wisconsin in 1949. She was awarded the MS in 1951. Franklin would again return to Spelman as a biology instructor where she remained until 1953. Obtaining a general education board fellowship from the Rockefeller Foundation for graduate study would permit her to enter the University of Wisconsin at Madison graduate program in endocrinology and medical physiology. Immediately after completing the doctorate in 1957, she became an assistant professor in the department of Physiology and Pharmacology at the Tuskegee Institute School of Veterinary Medicine.

In 1963, she accepted a position as assistant professor of Physiology at Howard University College of Medicine. Here she would make history. She steadily rose through the ranks, being promoted to professor in 1971. At the same time, Ison-Franklin's talent at administration was noted, such that in 1970 she was named Associate Dean for Administration for the College of Medicine at Howard, the first woman to hold a deanship in the 103-year history of the university.

Later Life

As a professor, Franklin received research grants from the National Aeronautics and Space Administration Ames Research Center, the National Institutes of Health, and the Washington Heart Association. Franklin served on many panels for the National

Science Foundation, the National Academies, and the MARC program of the National Institutes of Health. Dr. Franklin was the first woman, black or white, to serve as the head of a university medical department in the USA. In the last decade of her life, Franklin published a symposium of her findings, "The structure and function of the myocardial cell as it relates to myocardial mass myocardial hypertrophy," and instructional videos and other articles (1984).

She retired from these positions in July 1997. For her many accomplishments and great service to Howard University, Franklin was designated a "Magnificent Professor" in May 1998. Franklin's research focused on three areas: cardiac performance in conscious animals, experimental hypertension, and left ventricular hypertrophy. Dr. Franklin died a year after her retirement of a heart attack. A rare honor, the *Physiologist* posted an extensive obituary of her teaching and research accomplishments (1998). She was held in such high esteem by the American Physiological Society that there is a competitive fellowship in her honor. The *Eleanor Ison Franklin Fellow* is awarded for having the highest ranked renewal application, receiving a second year of funding through the Porter Physiology Fellowship Continuation Award. One of her greatest rewards was the mentoring and teaching relationship she had with students.

Impacts on the Field

Franklin worked on hypertension (Vaishnav et al., 1990), the basis for our current studies on stress. Franklin had an active career as a research scientist and educator. During her career as endocrinologist and medical physiologist, Franklin became involved in cardiovascular research after developing an interest in studying the relationship between hypertension and the autonomic nervous system. She began a series of animal model investigations into the behavior of blood circulation in response to hypertension, with a particular regard to various mechanisms associated with modulation of the left vertical mass. Additionally, Franklin was active in efforts to ease the entry of black women into scientific and medical professions transitioning later in her career from research to policy making to facilitate that goal.

Ruth Smith Lloyd (Fig. 42.4)

Summary Ruth S. Lloyd was a native Washingtonian, anatomist, reproductive endocrinologist, and geneticist. Her areas of research were endocrinology, sex-related hormones, and medical genetics. She studied the fertility of macaque monkeys, becoming the first African-American woman to gain a PhD in anatomy with her dissertation, "Adolescence of macaques *(Macacus rhesus)*" (1941). Lloyd's work on fertility and sex-related hormones in the mammalian ova was one of few and the first published by an African American in the field at the time.

Fig. 42.4 Ruth Smith Lloyd

What is Sex Really? The genetic and hormonal basis of sex behavior and phenotype continues to be a topic of discussion. We owe no small part of our current knowledge to the pioneering work of Ruth Smith Lloyd.

Education and Early Career

Ruth Smith Lloyd was born on 17 January 1917. Smith graduated from Dunbar High School in the District of Columbia. Smith then attended Mount Holyoke College in Massachusetts where she majored in zoology, graduating cum laude with a Bachelor of Arts degree in 1937. From 1937 to 1938, Lloyd studied for a master's degree in zoology at Howard University supported by a fellowship, under Ernest Everett Just. With funding from the Rosenwald Fund, Lloyd undertook doctoral studies under Boris Rubenstein at Western Reserve University in Cleveland, Ohio. She studied the fertility of macaque monkeys. Lloyd in 1941 became the first African-American woman to gain a PhD in anatomy with her dissertation, "Adolescence of Macaques (*Macacus rhesus*)," in 1941.

After earning her doctorate, Lloyd taught at Hampton Institute in Virginia from 1941 to 1942. She would join the medical faculty of Howard University in 1942. Here she would stay until retiring in 1977. Her areas of research were sex-related hormones, genetics, and endocrinology. She published papers investigating the reproductive system centered on ovulation and published review papers of diagnostics commonly used in female reproductive health such as the vaginal smear prior to her retirement.

Later Life

As she approached retirement, Lloyd began to more involved in university administrative, service positions, and medical collaborations. She earned the nickname "Mama Lloyd," from her time working in student engagement. She was committed to sustaining and retaining students at Howard. Lloyd was also engaged in university service through her position as chair of the university's Committee on Student Guidance and as director of the Academic Reinforcement Program. Posthumously she was honored as a pioneer by the Association for Women in Science (AWIS) and Black in Anatomy (BIN). Dr. Lloyd died in her Washington home of cancer on 5 February 1995 at the age of 78 (Obituary, 1995).

Impacts on the Field

Lloyd began to specialize her studies in the fertility of female macaques. Her work on ovulation in juvenile macaques coupled with her human medicine work was fertile ground for concepts we study today in reproductive endocrinology and physiology. Some of her findings we now know inform processes that at the time of her studies may have been ambiguous due to the current advances of the day.

Robert W Harris III (Fig. 42.5)

Summary: The Powerful Role of the Stress Hormones Glucocorticoids play a major role in the body and mind's response to stress. Robert Harris is a clinical researcher whose work was seminal to our understanding of hormonal regulation of protein production, a critical aspect of the stress response.

The Powerful Role of Stress Hormones

Glucocorticoids play a major role in the body and mind's response to stress. The basis of our current knowledge of the nature of that role and the mechanisms it utilizes was formed by the work of Robert W Harris III.

Education and Early Career

Professor Harrison was born in Mississippi and graduated with a BS from Tougaloo College in 1961 and an MD from Northwestern University Medical School in 1966.

Fig. 42.5 Robert Harris

He completed postgraduate training in internal medicine and subspecialty training in endocrinology and metabolism at Vanderbilt University.

Later in Life

During almost three decades in academic medicine, Dr. Harrison ran a clinical practice and led a laboratory that generated over 50 peer-reviewed papers, review articles, and book chapters while being supported by grants from the NIH and Department of Defense. Professor Harris was an investigator of the Howard Hughes Medical Institute from 1977 to 1982, which is a prestigious honor. Of the 297 investigators, only three have been African American. He was also appointed to Rochester's police accountability board by Rochester's city council. He has been a member of the FDA Advisory Committee on Endocrine and Metabolic Drugs and chaired the FDA Science Advisory Board to the Air Force Health Study.

Impact on the Field: The Powerful Role of the Stress Hormones

Harris showed that glucocorticoids influence protein production through their receptors which bind to DNA. In 1984 he published his work characterizing a monoclonal antibody to the receptor and used it to visualize its cellular and subcellular locations (Gametchu & Harrison, 1984). Harrison then used the antibody to determine its cellular distribution. By examining the receptor in animals with and without the steroid, he found that the receptor translocated to the nucleus when the hormone was present and that a separate and distinct receptor was located on the cell membrane. Harrison also showed that the receptor was present and active in the paraventricular nucleus indicating that corticosterone acted on neural tissues as well as those in the periphery.

As with other minority researchers, Harris advocated for diversity in his profession. He served as a consultant to Project I.M.P.A.C.T., a program designed to increase minority participation in clinical trials. His goal was to increase their participation so that African Americans could fully benefit from pharmaceutical science. He has also advised the medical community on the social determinants of health particularly the role of society in disproportionately increasing diabetes among minority populations in the USA (Jovanovic & Harrison 3rd, 2004).

Antonio Alberto Nuñez (Fig. 42.6)

Summary Antonio Nuñez is a dedicated functional neuroanatomist, who was a part of the Stephan lab that discovered what we now know as the master clock of the brain: the suprachiasmatic nucleus (SCN). Some of his early experiments would weave the fabric of our current understanding of how behaviors and hormones are controlled by the SCN. These early experiments set the stage for the field we now know as biological and circadian rhythms. He is also a champion of ongoing diversity, equity, and inclusion efforts within and around the field of neuroscience as a whole.

Fig. 42.6 Antonio Nuñez

How Do We Measure Time? We live in a world that is regulated by the movement of the earth. We set our behaviors by its rotation on its axis and its movement around the sun. How do we do this? With our circadian system. And we have Dr. Antonio Nuñez to thank for much of what we know about the neurobiology of the system and its regulation of our behavior.

Early Career

Antonio Nuñez was born in Cuba and has lived much of his life in the USA. He completed his bachelor's degree in psychology at Florida Atlantic University where he graduated *summa* cum *laude* in 1970 and obtained his master's in psychology in 1973 at the same institution. He then matriculated to the doctoral program in psychology at Florida State University receiving his degree in just 4 years! He conducted postdoctoral studies with George Wade, 1978–1980, at the University of Massachusetts, Amherst, MA, Department of Psychology, as a NIH NRSA postdoctoral fellow.

Nuñez's doctoral work concerned the anatomy of the suprachiasmatic nucleus (SCN), the powerful master oscillator in the circadian system. He utilized delicate knife cuts to delineate the role of the projections of the SCN on rhythms and their development. His generous, open nature and scientific prowess generated a work with George Whitney's where he was introduced to the power of hormones. This led to his work with George Wade at Amherst investigating the role of hormones on food intake and weight gain. His work did not go unnoticed, and he landed a position at Michigan State immediately after his postdoc where he remained until his retirement in 2019.

Later in Life/Impacts on the Field

Nuñez's research returned again and again to his roots in circadian and photoperiodic time measurement. His work redefined the anatomy of the SCN using more sophisticated techniques. He also explored the role of sex differences in ultrasonic communication, a technique he acquired while working with the Whitney lab (Nunez & Tan, 1984). His collaborations with Laura Smale (Smale et al., 2003) brought insights into the neurobiology of diurnal rodents; his work with Lyn Clemens defined the neuroendocrinology of female sex behavior, partner preference, and maternal care (Henley et al., 2011). His work was supported by numerous grants from NIH, NSF, and NIEH. He received awards for his teaching, research, and leadership and served as reviewers on their panels.

Nuñez remained committed to diversifying and supporting the profession. He moved from the Associate Chair and Coordinator of Graduate Programs in the Psychology Department to Vice Chair of the University Graduate Council to Associate Dean for Academic Affairs in the Graduate School where he remained while serving as Director of the MSU Postdoctoral Office until he retired. Through these offices, he obtained millions of dollars in training grants from NSF, NSF-AGEP to support the training of graduate students, postdocs, and faculty of color in STEM fields. He is internationally known for his work and was recruited to discuss conflict and careers at the Society for Research on Biological Rhythms and the National Institutes of Health.

Gregory Florant

Summary Gregory Florant is a research giant who has made significant contributions to our understanding of the diets, hormones, and neuroscience underlying the behavior of hibernators. His work spans four decades, has been cited over 3000 times, and includes studies on birds, marmots, squirrels, and woodchucks (Florant et al., 1993).

What to Do When It Gets Cold: A Seasonal Solution

The movement of the earth in its rotation around the sun creates wide swings in temperature during the seasons, a characteristic of temperate climates. One solution is to hibernate, setting metabolism to impossibly low levels to conserve energy normally committed to maintaining body temperature to preserve physiologic functions. But how is body temperature regulated seasonally? Science has the answer, thanks to an African-American Professor – Gregory Florant. Florant is a research giant who has made significant contributions to our understanding of the diets, hormones, and neuroscience underlying the behavior of hibernators.

His research on the fat composition of wild ruminants, a food source for primitive humans, has implications for our understanding of current obesity issues; his recent work links endocannabinoids to hibernation in the regulation of fat composition, food intake, and bone formation.

Education and Early Career

Florant began his research at a young age. Growing up in Palo Alto, he became fascinated by the birds of prey he saw during his family's frequent trips to Yosemite National Park. He bought and trained falcons and joined the North American Falconry Association. His hobby earned him a job at the Palo Alto Junior Museum and Zoo, a research project with Dr. Risebrough at the University of California at Berkeley, and his first publication (Risebrough et al., 1970) all before graduating from high school.

As with many scientists, Florant's career was aided by numerous mentors. While working on the project with Risebrough, he encountered a visiting professor from Cornell University who encouraged Florant to apply to his home institution. Florant followed the suggestion and was accepted at Cornell University where he began his graduate work studying the golden eagle at the Rocky Mountain Biological Laboratory in Gothic, Colorado. There he met Paul Ehrlich, a world-renowned biologist, who proposed that Florant apply to Stanford cementing his pursuit of research rather than a medical career. There he was awarded a Ford Foundation fellowship which provided the freedom to continue his studies at the institution of his choice. He was accepted at Stanford where he continued his research on birds. When his subjects escaped, Florant moved to a new mentor – Professor Craig Heller – whose focus on hibernation set the stage for his career.

Later in Life

Florant went on to become NIH postdoctoral fellow at the Montefiore Hospital where he studied the mechanisms that underlie sleep cycles and the insulin's regulation of metabolism. He then secured a tenure track position at Swarthmore where he rose to associate professor. While at Swarthmore, Dr. Florant continued his studies collaborating with professors as a Fulbright scholar in France, a National Research Council fellow in Alaska, DAAD Fellow in Germany lead to his appointment as a fellow of the American Association for the Advancement of Science in 1989. Florant enjoyed his time at Swarthmore and wanted to provide training for future researchers, so he joined the graduate faculty at Temple as a full professor in 1990 and received a second Fulbright award in 2000. Florant served as a Professor of Biology in the College of Natural Sciences at Colorado State University, beginning in 1995 where he has been awarded two additional Fulbright scholarships. In addition, Florant's research has been supported by numerous grants from federal and private research institutions.

Impact on the Field

Florant's work spans four decades, and includes over 140 papers, abstracts, books, book chapters, and presentations, and has been cited over 3000 times; his work includes studies on birds, marmots, squirrels, and woodchucks. His research on the fat composition of wild ruminants, a food source for primitive humans, has implications for our understanding of current obesity issues; his recent work links endocannabinoids to hibernation in the regulation of fat composition, food intake, and bone formation (Mulawa et al., 2018). His prolific profile is all the more compelling because his studies required dedication. Hibernators are studied under natural conditions and as the process follows an annual cycle, studies can span months or years. Florant showed his commitment at the beginning of his graduate career when he trapped ten marmots in a single night. Before he joined that lab, his mentor had only trapped a few a year (Kessler et al., 1996)! As one of the few minorities to earn degrees at predominantly white institutions, Florant has contributed significantly to increasing diversity in STEM fields. He drafted a letter to the administration at Swarthmore supporting the students' requests to increasing the number of diverse faculty and staff. While at Colorado State, he served as the Director of the Graduate School Program for Diversity and Access. He also served as the liaison for the Ford Foundation Fellowship program recruiting minority faculty to institutions in the Rocky Mountain States, and as a mediator for Ford Fellows in the Rocky Mountain Area (Colorado, Wyoming), and a mentor for minority faculty.

As the senior member of the NIH-NIDDK minority network committee, he met once a year to discuss scientific research, and minority issues in science, and mentor young assistant professors. He was an active member of the black student service office and Colorado State University's minority caucus which established a Native American Women in Science Scholarship Program. His work has earned him numerous awards for mentoring. And he does outreach every Ground Hog Day bringing his marmot "Baby" to meet with students in the local schools.

Barrier Breakers

The pioneers and trailblazers mentioned did not follow a single, linear road to success. These scientists did indeed carve the lay of the land for minority scientists and scholars. Most were first-generation students and many if not all started at historically black universities and colleges. During this time in history, HBCUs represented the safest place to learn and train for blacks and other minorities. These university settings presented a variety of challenges including those stemming from racism and racial violence from the outside world. While respected and championed by their mentors, the policies and practices adopted by the USA had little room for those with darker skin. Thus, another commonality these individuals share is the lack of opportunities afforded them in spaces that were not created for them. Many returned to HBCUs to work and flourish. They all also managed to work in Diversity, Equity, and Inclusion work simultaneously with their scientific endeavors. One must wonder how much more prolific their scholarly and research activities would have been if they were permitted to maintain a singular focus like their majority counterparts. And we mourn the lack of contributions we will never reap from those who did not survive the structural racism of their era (for more see McNeill (2020)).

Historical Roles for Black Colleges and Universities

HBCUs are the bedrock of past and current opportunity on which all students, scholars, and scientists from underrepresented backgrounds rest in this nation. Without a training ground like Hampton University, Howard University, Tuskegee University, and Spelman University, the pioneers we amplify here may have not succeeded in finding either employment or advancement beyond earning their degrees granted by their home institutions. For those mentioned in this chapter that were not directly affiliated with an HBCU, it is likely that they benefited from the road they trod being worn down by an HBCU alumni. HBCUs provide the majority of minority scientists in the nation. Unfortunately, in society today, this fact has not changed with the evolution of racism and bias prevalent in America. HBCUs are federally underfunded and often overlooked as scientifically competitive institutions of higher learning but still hold the key to success for many students, scientists, and faculty of color.

Societal Restraints on Jobs and Research Funding

We also recognize that the constraints on scientists of color have not disappeared. The latest data have shown that still scientists of color are publishing and are cited less, receiving less fellowship and grants awards at every level. To that end, we are still celebrating firsts at each level at majority serving institutions exposing underlying policies and cultures, not in line with DEI. Much work is yet to be done; this chapter is a beginning.

Stony the Road We Trod: To a Brighter Future

The future looks bright for minorities in behavioral neuroendocrinology. As we reflect on the journey for the marginalized and underrepresented scientist we have evidence that when a society, organization or institution exhibits clear and precise intentional efforts to supporting brown scientist or scientists in training, the field of study benefits generation after generation. The concept of intentionality extends to funding institutions that provide the life blood to all scientist. The grant funding institutions who have been intentional and aggresive about providing supports and barriers against the systemic racism that is embedded into all academic and scientific settings are making strides as evidenced in the latest generations of Behavioral Neuroendocrinologists. For example, Juan Dominguez at the University of Texas is an expert in the neurobiology that underlies hormonal regulation of male sex behavior. Erica Glasper has carved out her status as the go-to for the neuroendocrine factors that mediate parental influence in stress and resilience. Dr. Kelli Duncan is defining role of steroid hormones on repair of the nervous system following traumatic brain injury (TBI). Farrah Madison is ferreting out hormonal regulation of brain plasticity. Finally, Johnathan Borland is one to watch as the 2021 winner of SBN's WC Young recent graduate award. Although these individuals put us on the road, more are necessary to achieve the true promise of a diverse society.

References

1998. Obituary: Eleanor Lutia Ison-Franklin (1929-1998). *The Physiologist*, 41.

Cobb, W. M. (1971). Percy Lavon Julian, Ph.D., Sc.D., LL.D., L.H.D., 1899- ? *Journal of the National Medical Association, 63*, 143–150.

Cullen, K. E. (2006). Chapter 8: Percy Julian (1899–1975): Synthesis of glaucoma drug and sterols from natural plant products. In *Chemistry: The people behind the science (pioneers in science)*. Infobase Publishing.

Florant, G., Hester, L., Ameenuddin, S., & Rintoul, D. A. (1993). The effect of a low essential fatty acid diet on hibernation in marmots. *American Journal of Physiology-Regulatory, Integrative and Comparative Physiology, 264*, R747–R753.

Gametchu, B., & Harrison, R. W. (1984). Characterization of a monoclonal antibody to the rat liver glucocorticoid receptor. *Endocrinology, 114*, 274–279.

Henley, C., Nunez, A., & Clemens, L. (2011). Hormones of choice: The neuroendocrinology of partner preference in animals. *Frontiers in Neuroendocrinology, 32*, 146–154.

Jovanovic, L., & Harrison, R. W., 3rd. (2004). Advances in diabetes for the millennium: Diabetes in minorities. *MedGenMed, 6*, 2.

Julian, P. L. (1933). On the progenitors of certain plant alkaloids and the mechanism of their formation in the plant structure. *Proceedings of the Indiana Academy of Science*, 122–124.

Julian, P. L., & Pikl, J. (1933). Studies in the indole series. I. The synthesis of Alpha-Benzylindoles. *Journal of the American Chemical Society, 55*, 2105–2110.

Julian, P. L., & Pikl, J. (1935a). Studies in the indole series VII. The course of the Fischer reaction with ketones of the type R CH2 CO CH3. Alpha-propyl and alpha-homoveratryl indole. *Proceedings of the Indiana Academy of Science*, 145–150.

Julian, P. L., & Pikl, J. (1935b). Studies in the indole series. V. The complete synthesis of physostigmine (eserine). *Journal of the American Chemical Society, 57*, 755–757.

Kessler, J. H., Morin, K. A., Kidd, J., & Kidd, R. A. (1996). *Distinguished African American scientists of the 20th century*. Greenwood Publishing Group.

Kyle, R. A., & Shampo, M. A. (1996). Stamp vignette on medical science. Percy Lavon Julian--industrial chemist. *Mayo Clinic Proceedings, 71*, 1170.

Lewis, L. B. (1946). A study of some effects of sex hormones upon the embryonic reproductive system of the white pekin duck. *Physiological Zoology, 19*, 282–329.

Lewis, L. B., & Domm, L. V. (1948). A sexual transformation of the osseus bulla in duck embryos following administration of estrogen. *Physiological Zoology, 21*, 65–69.

Lloyd, R. S. (1941). *Adolescence of macaques (macacus Rhesus)*. Western Reserve University.

Mulawa, E., Packer, R., Kirkwood, J., Wolfe, L., Wojda, S., Prenni, J., Florant, G., & Donahue, S. (2018). Bone properties and the endocannabinoid system observed with neurectomy and hibernation in marmots (Marmota flaviventris). *Journal of Bone and Mineral Research, 33*, 236–236.

Nunez, A. A., & Tan, D. T. (1984). Courtship ultrasonic vocalizations in male Swiss-Webster mice: Effects of hormones and sexual experience. *Physiology & Behavior, 32*, 717–721.

Obituary. 1995. RUTH SMITH LLOYD Howard U. Professor. *Washington Post*, February, 9th 1995.

Risebrough, R., Florant, G., & Berger, D. (1970). Organochlorine pollutants in Peregrines and Merlins migrating through Wisconsin. *Can Field Natur.*

Smale, L., Lee, T., & Nunez, A. A. (2003). Mammalian diurnality: Some facts and gaps. *Journal of Biological Rhythms, 18*, 356–366.

The structure and function of the myocardial cell as it relates to myocardial mass. In E. I. Ison-Franklin (Ed.), *Myocardial hypertrophy*. Symposium held in honor of E. W. Hawthorne at the heart house auditorium, American College of Cardiology, 1984. Washington DC. Howard University Press.

Vaishnav, R. N., Vossoughi, J., Patel, D. J., Cothran, L. N., Coleman, B. R., & Ison-Franklin, E. L. (1990). Effect of hypertension on elasticity and geometry of aortic tissue from dogs. *Journal of Biomechanical Engineering, 112*, 70–74.

Weissmann, G. (2005). Cortisone and the burning cross. The story of Percy Julian. *The Pharos of Alpha Omega Alpha-Honor Medical Society. Alpha Omega Alpha, 68*, 13–16.

Index

R. J. Nelson, Z. M. Weil (eds.), *Biographical History of Behavioral Neuroendocrinology*, https://doi.org/10.1007/978-3-031-12970-4

The manufacturer's authorised representative in the EU is Springer Nature Customer Service Centre GmbH, Europaplatz 3, 69115 Heidelberg, Germany. If you have any concerns regarding our products, please contact ProductSafety@springernature.com

Printed and bound by CPI Group (UK) Ltd, Croydon, CR0 4YY
28/09/2026
02224056-0012